Heat Exchangers
Volume III

Heat Exchangers Volume III: Operation, Performance, and Maintenance covers heat exchanger installation, commissioning and operation, and maintenance and performance monitoring in service.

Focusing on in-service issues like flow-induced vibration, corrosion, and corrosion control, and fouling and fouling control, the book explores performance deterioration in service, maintenance issues, defects, tube failures, and how to detect these issues with NDT methods. It discusses various cleaning processes and repair methods. The book also considers boilers, utility boilers, coal-based thermal power plants, boiler corrosion, and boiler degradation mechanisms. It discusses different types of cooling systems, feedwater treatment, deaerators, feedwater heaters, economizers, condensers, cooling towers, and cooling-water management.

The book serves as a useful reference for researchers, graduate students, power plant engineers, and engineers in the field of heat exchanger design, including pressure vessel manufacturers.

Heat Exchangers
Volume III

Operation, Performance, and Maintenance

Third Edition

Kuppan Thulukkanam

CRC Press
Taylor & Francis Group
Boca Raton London New York

CRC Press is an imprint of the
Taylor & Francis Group, an **informa** business

Designed cover image: Shutterstock

Third edition published 2024
by CRC Press
2385 Executive Center Drive, Suite 320, Boca Raton, FL 33431

and by CRC Press
4 Park Square, Milton Park, Abingdon, Oxon, OX14 4RN

CRC Press is an imprint of Taylor & Francis Group, LLC

© 2024 Kuppan Thulukkanam

First edition published by Marcel-Dekker 2000
Second edition published by CRC Press 2013

ISBN: 978-1-032-39936-2 (hbk)
ISBN: 978-1-032-39937-9 (pbk)
ISBN: 978-1-003-35206-8 (ebk)

DOI: 10.1201/9781003352068

Typeset in Times
by Newgen Publishing UK

Dedicated to

my parents, S. Thulukkanam and T. Senthamarai,
my wife, Tamizselvi Kuppan,
and my mentor, Dr. Ramesh K. Shah

Contents

Preface

INTRODUCTION

This volume discusses recent advances in the heat exchanger and pressure vessels operation, and their maintenance, flow-induced vibration (FIV) mechanisms of shell and tube heat exchangers and their prevention methods, fouling of heat exchangers, types of fouling, fouling model, mechanisms of fouling, gas-side fouling, fouling prevention and control, foulant cleaning methods and fouling in cooling-water system; basics of corrosion, polarization, forms of corrosion, corrosion prevention and control, inhibitors, use of corrosion-resistant alloys, protective coatings, cathodic and anodic protection, corrosion monitoring, cooling-water system corrosion, etc.; boiler, steam power plant, corrosion in boilers, water chemistry, boiler pressure parts' degradation mechanisms, function, construction, performance of deaerator, feedwater heater, steam surface condenser, power plant cooling system types, open recirculating systems with cooling tower are discussed in detail. Heat exchanger installation, operation, performance monitoring and maintenance, equipment damage mechanisms, periodical inspection of pressure vessels and heat exchangers, condition monitoring, pressure vessel failure, tube bundle repair, NDT methods to inspect and assess the condition of heat exchanger and pressure vessel components, etc., are discussed in detail.

This book is illustrated with approximately 225 figures, images, and tables.

The book will be a centerpiece of information for thermal power plant engineers, plant maintenance engineers, practicing engineers, process plant engineers, research engineers, academicians, designers, and manufacturers, etc.

CHAPTER SUMMARIES

Chapter 1 Flow-Induced Vibration of Shell and Tube Heat Exchangers

Flow-induced vibration (FIV) of shell and tube heat exchangers cause heat exchanger tubes to vibrate under the influence of crossflow velocities, and if the amplitude of vibration becomes large enough, the tubes can be damaged and tube failures may result in plant shutdown and hence repairs.

This chapter discusses principles of FIV and their mechanisms, most probable regions of tube failure, tube failure mechanisms, tube response curve due to crossflow velocity and dynamical behavior of tube arrays, flow mediums, approaches to FIV analysis, vortex shedding, Strouhal number, vortex shedding against a tube and tube bundle, criteria to avoid vortex shedding, turbulence-induced excitation mechanism, Owen's expression for turbulent buffeting frequency, fluid elastic instability (FEI), Connors' FEI analysis, design recommendations, various criteria to avoid FEI, stability diagrams, shellside acoustic vibration, acoustic resonance, acoustic resonance excitation mechanisms, expressions for acoustic resonance frequency, solutions to vibration problems, vibration problems as addressed in heat exchanger standards, vibration evaluation procedure, design guidelines for vibration prevention, methods to increase tube natural frequency and reduce crossflow velocity, acoustic vibration mitigation methods, baffle damage and collision damage, impact and fretting wear, tube wear prediction, determination of hydrodynamic mass and natural frequency, antivibration tools, antivibration tube stakes, etc.

Chapter 2 Fouling

Fouling is defined as the formation of undesired deposits on heat transfer surfaces, which impede the heat transfer and increase the resistance to fluid flow; the growth of the deposits causes the thermohydraulic performance of heat exchanger to degrade with time. This chapter discusses the effect of fouling on the thermohydraulic performance of heat exchangers, costs of fouling, the need for oversizing, additional energy costs, treatment cost to lessen corrosion and fouling, lost production due to maintenance schedules, etc., fouling curves/modes of fouling, fouling model, mechanisms of fouling, such as particulate, reaction, corrosion, precipitation, biological, and solidification fouling are discussed; parameters that influence incidence of fouling, alternative heat exchanger design types to reduce fouling, such as gasketed and spiral plate exchangers, nonsegmental baffle types such as Philips RODbaffle, Twisted Tube®, HELIXCHANGER®, and EMbaffle® heat exchangers are discussed; it also discusses gas-side fouling of heat exchangers, fouling data, fouling monitoring, fouling analysis, tube fouling monitors, expert system, fouling prevention and control, selection of cleaning methods and cleaning procedure, various offline mechanical cleaning methods, thermal cleaning, chemical cleaning, offline chemical cleaning methods, various online mechanical cleaning methods including, sponge rubber balls cleaning system, brush and cage system, insert technology, grit cleaning and self-cleaning fluidized-bed exchangers, foulant control by chemical additives, fouling in cooling-water systems, macrofouling, methods of cooling-water fouling control, scaling tendencies, scaling indices and scaling control, cleaning of scales, iron oxide removal, monitoring of fouling, etc.

CHAPTER 3 CORROSION

Most common metals and their alloys are attacked by environments, such as the atmosphere, soil, water, or aqueous solutions. This destruction of metals and alloys is known as corrosion. It is generally agreed that metals are corroded by an electrochemical mechanism. This chapter discusses basics of corrosion, forms of electrochemical corrosion, bimetallic cell, corrosion potential and current, corrosion kinetics, polarization, passivation, factors affecting corrosion of a material in an environment, high-temperature corrosion, forms of corrosion—uniform or general, galvanic, pitting, crevice, intergranular, dealloying or selective leaching, erosion–corrosion, stress corrosion cracking, fretting, corrosion fatigue and microbiologically influenced corrosion; corrosion of weldments, hydrogen damage, hydrogen-induced cracking, stress-oriented hydrogen-induced cracking, hydrogen embrittlement, hydrogen-assisted cracking, hydrogen blistering, pressure vessel steels for sour environments, detecting hydrogen damage, material selection for hydrogen sulfide environment high-temperature, hydrogen attack, Nelson curves, etc.

Other types of corrosion like touch point corrosion, corrosion under insulation, pipeline corrosion, soil and concrete corrosion; corrosion prevention and control, inhibitors, corrosion-resistant alloys, bimetal concept, cladding, duplex tubing, protective coatings, cathodic and anodic protection, etc., corrosion measurement, corrosion monitoring, corrosion management system, cooling-water system corrosion, corrosion of individual metal in cooling-water systems, forms of corrosion in cooling water, etc.

CHAPTER 4   BOILER, THERMAL POWER PLANT, AND HEAT EXCHANGERS OF COOLING AND FEEDWATER SYSTEMS

Boiler is a pressure vessel into which water is fed, and by the application of heat, it is evaporated into steam. Boilers are in use in numerous process industries, workshops, and coal-based power plants. Major parts and assemblies of a boiler, classification like water tube and fire tube, industrial, utility and waste heat recovery boiler, etc., boiler accessories, boiler code, utility boilers like drum and once-through boilers are discussed. Basics of coal-based thermal power plant, steam generators, supercritical technology, air preheaters, economizers, environmental considerations, emissions control, performance indicators, etc., are discussed.

Corrosion in boilers, types of corrosion, corrosion control and prevention, industrial boiler water treatment methods, corrosion and deposition control, carryover, corrosion in utility boilers, condensate polisher systems, various boiler pressure parts degradation mechanisms, boiler tube failure mechanisms, boiler tube leaks, NDT techniques for crack detection, etc.

Deaerator function, construction, types and working, closed feedwater heater, construction, working, performance issues, failures, etc., steam surface condenser, types, parts, performance, operation and maintenance, condenser tube leaks, leak detection methods, plugging condenser tube leaks, etc.

Power plant cooling system types, open recirculating systems with cooling tower, cooling tower function, types, construction details, psychometric analysis, dry cooling system, "Heller system," etc., components of a cooling tower, water system concerns, cycles of concentration, relation between make-up, cycles of concentration and blowdown, corrosion and scaling tendencies, common cooling tower problems, drift, water loss, environmental protection measures, etc., are discussed.

CHAPTER 5 HEAT EXCHANGER INSTALLATION, OPERATION, AND MAINTENANCE

The major problems faced with heat exchangers in service are inadequate heat transfer and excessive pressure drop; improper maintenance procedures and lack of preventative maintenance schedules, cause failures. To ensure the designed performance level and to prevent leaks and failures, the heat exchanger installation, operation, maintenance and periodical inspection/schedule maintenance shall be as per manufacturers' guidelines/maintenance manual. This chapter discusses storage, installation, operation of heat exchangers, equipment damage mechanisms within process equipment as per API RP 571 fatigue and fatigue life, S-N curve, creep, brittle fracture, basic theory of fracture mechanics, fracture toughness, dew point corrosion, fuel ash corrosion, hydrogen damage, water hammer, etc. It also discusses periodical inspection of pressure vessels and heat exchangers, condition monitoring, pressure vessel failure and their causes, asset integrity management, pressure vessel inspection, stress analysis, inspection of piping systems, etc. It also discusses removal of tube bundle, transportation, cleaning tube bundles, repair, upgrade, tube plugging, primary tube failure mechanisms of ACHE and STHE and their inspection methods, NDT methods to inspect and assess the condition of heat exchanger and pressure vessel components, residual life assessment of heat exchangers by NDT techniques on-site metallurgy services, professional service providers for heat exchangers are discussed.

DISCLAIMER

The text of this book is based upon open literature resources like Standards, Codes, authentic books on heat exchangers and pressure vessels, technical literature from leading heat exchanger and pressure vessel manufacturers, technical articles, and technical information from many websites, etc. No Indian Railways related technical information is adopted in this book.

Acknowledgments

A large number of my colleagues from Indian Railways, well-wishers, and family members have contributed immensely toward the preparation of the book. I mention a few of them here, as follows: Jothimani Gunasekaran, V. R. Ventakaraman, Amitab Chakraborty (ADG), O. P. Agarwal (ED) and M. Vijayakumar (Director), RDSO Lucknow; Member Mechanical and senior Officials of Railway Board, New Delhi, V Baskaran, and T. Adikesavan of Southern Railway; K Narayanan for CAD drawings, Satheeh kumar S, Sundar Raj A, V. Baskaran, and Er. K. Praveen for their assistance. I have immensely benefited from the contributions of scholars such as Dr. Ramesh K. Shah, Dr. K. P. Singh, and Dr. J. P. Gupta; I also acknowledge Ministry of Railways, and the library facilities of IIT-M, IIT-K, IIT-D, and RDSO, Lucknow. A large number of heat exchanger manufacturers and research organizations have spared photos and figures, and their names are acknowledged in the respective figure captions.

About the Author

Kuppan Thulukkanam, Indian Railway Service of Mechanical Engineers (IRSME), Ministry of Railways, retired as Principal Executive Director, CAMTECH, Gwalior, RDSO, has authored an article in the *ASME Journal of Pressure Vessel Technology*. His various roles have included being an experienced administrator, staff recruitment board chairman for a zonal railway, and joint director, Engine Development Directorate of RDSO, Lucknow (Min. of Railways). He was also involved in design and performance evaluation of various types of heat exchangers used in diesel electric locomotives and has served as chief workshop engineer for the production of rolling stocks like coaches, diesel and electric multiple units, wagons, electric locomotive, etc., and as Director, Public Grievances (DPG) to the Minister of State for Railways, Railway Board, Government of India. Kuppan received his BE (Hons) in 1980 from the PSG College of Technology, Coimbatore, Madras University, and his MTech in production engineering in 1982 from the Indian Institute of Technology, Madras, India.

1 Flow-Induced Vibration of Shell and Tube Heat Exchangers

1.1 PRINCIPLES OF FLOW-INDUCED VIBRATION

A shell and tube heat exchanger (STHE) consists of a shell with an internal tube bundle that is typically supported by tubesheets at the ends and intermittent tube support plates known as baffles. Two fluids, of different inlet temperatures, flow through the heat exchanger. One flows through the tubes (the tubeside) and the other flows outside the tubes but inside the shell (the shellside). The tube bundle may be composed of several types of tubes supported by baffles. The baffle configuration shall be segmental or nonsegmental. Based on the baffles' configuration, the types of STHEs are classified as given below:

1. With segmental baffles:
   i. Shell and tube heat exchanger.
   ii. Disk and doughnut heat exchanger.
2. With nonsegmental baffles:
   i. Phillips RODbaffle heat exchanger.
   ii. EMbaffle® heat exchanger.
   iii. HELIXCHANGER® heat exchanger.
   iv. Twisted Tube® heat exchanger (no baffle is used).

Flow-induced vibration of shell and tube heat exchangers (STHEs) has been known for a long time. Heat exchanger tubes tend to vibrate under the influence of crossflow velocities, and if the amplitude of vibration becomes large enough, the tubes can be damaged by one or more of several mechanisms: (1) thinning due to repeated midspan collision; (2) impact and fretting wear at baffle plate and tube interface; and (3) fatigue or corrosion fatigue due to high wear rate. Tube failures are costly because they result in plant shutdown to effect expensive repairs. These problems can be very serious in nuclear heat exchangers. FIV can cause severe damage to the tubes and other structural components of the heat exchanger. In surface condensers, tube damages because of vibration occurs when high-velocity steam travels through the tube bundle. Whether by design, a change of tube materials, or the result of a power uprate, the tubes are susceptible to all types of flow-induced vibration mechanisms. Therefore, it is important to ensure that the modern STHEs are free from FIV problems at all operating conditions.

In the past, heat exchangers were designed conservatively. With the success of state-of-the-art computer programs, the trend is to design an efficient and compact heat exchanger. Higher thermal performance and the desirability of low fouling generally require higher flow velocities, while fewer baffle plates are desirable to minimize pressure drop. Higher flow velocities and reduced structural supports can lead to severe FIV problems. In addition to these, the incorporation of new materials and processes without adequate considerations of the effects on the structural dynamics contributed

DOI: 10.1201/9781003352068-1

to more FIV problems, many of which led to tube failures [1]. It is essential to avoid such costly tube failures by a detailed FIV analysis, preferably at the design stage after thermal design is over. The subject continues to receive increasing attention because of its significance in heat exchanger applications, since as much as 60% of the heat exchangers in process industries are shell and tube type. References [1–18] bring about a better understanding of the FIV phenomenon in STHEs. In this chapter, the mechanisms that cause FIV and their evaluation, acceptance criteria, and design guidelines for vibration prevention are presented. Design guidelines included in TEMA Standards [19] and ASME Code Section III [20] are presented at appropriate places.

1.1.1 PRINCIPLES OF FLOW-INDUCED VIBRATION

Tube vibration is manifested by the periodic movement of the tube from its equilibrium position. With increasing crossflow velocity, the tube movement has the following three manifestations [1]:

1. At low crossflow velocities, the tubes vibrate with low-amplitude random motion.
2. As the flow velocity is increased, rattling of tubes within the baffle holes takes place.
3. As the flow velocity exceeds a threshold value, high-amplitude motion takes place.

When the natural frequencies of the tubes are closer to the exciting frequency, resonance takes place. The relative motion between the tubes and the rigid structures like baffle supports and shell boundary can cause impact and fretting wear of tubes.

1.1.2 FLOW-INDUCED VIBRATION MECHANISMS

The excitation mechanisms generally regarded as responsible for FIV are as follows:

1. Vortex shedding or flow periodicity.
2. Turbulent buffeting.
3. Fluid elastic instability (FEI).
4. Acoustic resonance.

Tube vibration mechanisms in order of their increasing severity are turbulent buffeting, vortex shedding, and fluid elastic instability. Vortex shedding, turbulent buffeting, and acoustic excitation are due to resonance phenomena. Resonance occurs when the excitation frequency synchronizes with the natural frequency of the tubes. FEI sets in for tubes in a crossflow at a critical flow velocity or threshold velocity, resulting in the amplitude of tube response large enough to collide with the adjacent tubes and cause failure. Below the critical velocity, FEI will not take place. Instability attains when the energy input to the tube mass-damping system exceeds the energy dissipated by the system.

(a) Vortex shedding frequency: this is a flow-induced frequency, which can excite vibrations when it matches with natural frequency of the tubes.
(b) Turbulent buffeting: a random turbulence can excite tubes into vibration at their natural frequency by selectively extracting energy from a highly turbulent flow of gas across the bundle.
(c) Fluid elastic instability: there is a critical velocity above which the fluid elastic instability vibrations become a problem.

Excessive tube vibrations caused by turbulent buffeting or vortex shedding resonances can result in thinning and eventual failure of tubes at the baffles or leaks at the tube-to-tubesheet joints.

Fluid elastic instability tends to have a run-away effect; once crossflow velocities approach a critical velocity the tubes may vibrate uncontrollably and can fail very rapidly. The shellside may be subjected to acoustic resonances termed "acoustic vibrations." Acoustic vibrations can vibrate the shell and the exchanger's foundations, but are more often associated with high levels of radiated sound [10].

1.1.3 Most Probable Regions of Tube Failure

Although the tubes can fail anywhere in the exchanger, the regions more susceptible for FIV are the high-velocity regions such as the following:

1. Largest unsupported midspan between two baffles.
2. Tubes located in the baffle window region at periphery of the tube bundle.
3. U-bend regions of U-tube bundle.
4. Tubes located beneath the inlet nozzle.
5. Tubes located in the tube bundle bypass area, next to pass partition lanes.
6. Regions/interfaces where there is a relative movement between the tube and the heat exchanger structural components. Such regions include tube and baffle support interfaces and tube and tubesheet interfaces.

Figure 1.1 shows few probable regions of tube failure in a shell and tube heat exchanger.

1.1.4 Tube Failure Mechanisms

The primary failure mechanisms that cause tube failure are as follows [1]:

1. Impact wear (tube-to-tube and tube-to-baffle).
2. Fretting wear at the tube–baffle interfaces as a result of impact and/or sliding motion at the baffle support.
3. Combination of impact and fretting wear.

1.1.5 Possible Damaging Effects of FIV on Heat Exchangers

Most commonly, mechanical failures occur as a result of midspan collision damage, circumferential fracture at the tubesheet or baffle plates (baffle damage). Tube vibration, if left unresolved, may result in catastrophic tube failures and the subsequent unplanned shutdown of the unit [11,12]. These failures are discussed next.

Midspan collision: if the amplitude of response at the midspan is great enough, collision with adjacent tubes takes place. The resulting wear causes failure of the tube wall under pressure. Tubes located in the bundle periphery may impact the vessel wall. The tube wall is worn thin and eventually splits open.

TEMA Baffle damage: baffle holes are drilled such that a diametral clearance of 1/64" (in tight fit designs) to 1/32" between tube and baffle hole exists.

Wear damage at the tube interface with the baffle support: heat exchangers are generally designed with a clearance between the tube and the baffle plates. This clearance is required for ease of manufacture and design considerations. Thus, the tubes are free to laterally displace inside the baffle holes. Tubes that suffer lower amplitude vibration close to baffle plates may fail by impact and fretting wear or fatigue.

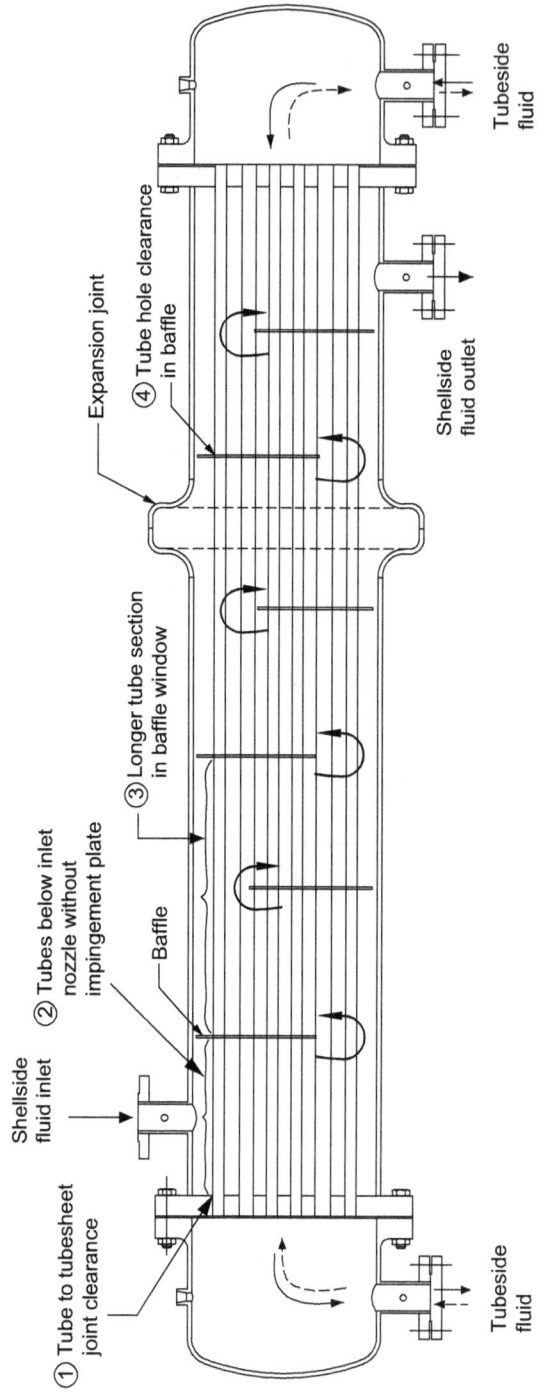

FIGURE 1.1   Probable regions of tube failure due to FIV in a shell and tube heat exchanger.

Fatigue failures: if the contact stress due to impact or collision is greater than the allowable fatigue stress, fretting wear takes place.

Excessive operating noise level: when the shellside medium is a gas, steam, or air, acoustic vibration will be induced within the tube bank containment. The acoustic vibration is characterized by pure-tone, low-frequency intense noise.

Severe pressure drop: since the vibration of tube requires the energy from the shellside fluid, the shellside pressure drop increases. If the vibration is severe, destructive pressure fluctuations take place.

Intensified stress corrosion: due to repeated impact with the baffle supports, intensive tensile stresses are induced on the tube surface. Susceptible tube material can fail due to the accelerated stress corrosion cracking in the shellside medium. However, corrosion due to FIV is second to the failure of tube material due to the corrosive nature of shellside fluid and/or tubeside fluid.

1.1.6 TUBE RESPONSE CURVE

Figure 1.2 shows the tube response due to FIV of tubes in tube bundles as a result of the three excitation mechanisms, namely, vortex shedding, turbulent buffeting, and FEI [3]. Each of them manifests itself only over a given range of flow parameters. However, it is believed that turbulent buffeting is operative in the entire range of flow parameters.

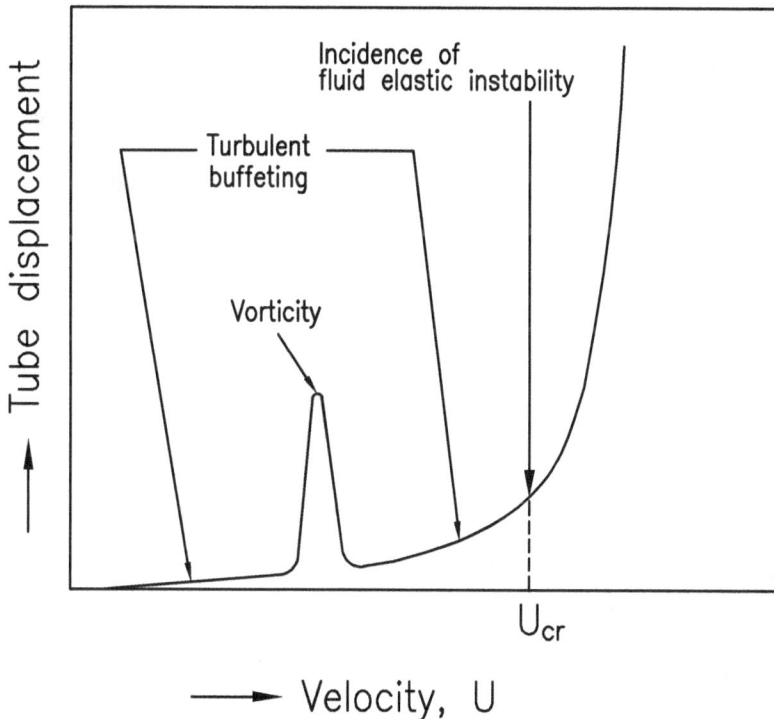

FIGURE 1.2 Tube response due to FIV mechanism—ideal diagram.

Source: Adapted from Paidoussis, M.P., *J. Sound Vib.* 76, 329–359, 1981.

1.1.7 Dynamical Behavior of Tube Arrays in Crossflow

Flow of fluid over an array of elastic tubes results in (1) hydrodynamic effects or fluid oscillation (acoustic vibration) and (2) fluid–structure coupling. These effects cause hydrodynamic forces and fluid–structure coupling forces. The dynamical behavior of an array of cylinders in increasing cross-flow velocity (U) is considered to have three distinct manifestations as follows [3]:

1. At low flow velocities, the cylinder responds principally to turbulent buffeting; with increasing flow velocity, the amplitude of tube vibration goes up roughly as U^2.
2. At higher flow velocities, various kinds of resonance conditions may arise, such as vortex shedding, turbulent buffeting, and acoustical oscillation of gas column.
3. At sufficient high-flow velocities, FEI will generally develop, and the amplitude of vibration increases rapidly with the flow velocity without a limit.

1.1.8 Hydrodynamic Forces

The hydrodynamic forces that contribute to FIV mainly fall into three groups:

1. Forces arising due to turbulent fluctuations of the pressure field.
2. Forces resulting from periodic vortex shedding from tubes and formation of von Kármán's streets in their wakes.
3. Motion-dependent fluid forces arising when the tubes are shifted elastically from their equilibrium within the bundle due to interaction with the flow.

1.1.9 FIV Mechanisms Versus Flow Mediums

Of the different excitation mechanisms of FIV, only FEI is a primary concern in all flow mediums. Vortex shedding is possible in liquid and gas medium but unlikely in two-phase medium. Other mechanisms have less importance in certain flow media. Turbulent buffeting is possible in liquid and gas medium, which may result in fretting wear.

Hence, design restrictions are imposed to limit acoustic resonance and FEI [9].

1.1.10 Approaches to FIV Analysis

Two approaches are normally followed to predict FIV effects of STHEs:

1. Finite-element modeling technique. This model simulates the time-dependent motion of a multispan heat exchanger tube in the presence of tube and baffle plate clearance, and the resulting wear is determined. This approach is normally followed for heat exchangers and steam generators used in very critical services, such as nuclear energy generation.
2. Limiting amplitude of vibration. This approach linearizes the structural model by assuming tubes as classical beams with baffle offering simple support at the intermediate points and clamped at the tubesheet ends. The designer then predicts the amplitude of vibration or instability thresholds and selects an acceptance criterion that will conservatively limit the vibration. This approach is used in this chapter. This procedure can be used to provide conservative designs or troubleshoot an existing heat exchanger and is widely accepted in the field.

1.1.11 Empirical Nature of Flow-Induced Vibration Analysis

Before discussing FIV excitation mechanisms in detail, it is important to note that FIV of STHEs is a physical phenomenon that cannot be explained by simple empirical correlations [9,13]. It is most difficult to analyze [1,7] due to reasons like the following:

1. Tube bank dynamics is a multibody problem. The tubes are supported by multiple baffles with holes slightly larger than the tube diameter.
2. The interaction between the tube and the support plates is characterized by impacting as well as sliding motion. This makes the system nonlinear in nature.
3. The tubes and the surrounding fluid form a fluid–structure coupling that results in motion-dependent fluid forces that give rise to added mass, coupled modes, and damping.
4. Generally, the flow field is quite complex, nonuniform, and quite unsteady, and the incidence of flow on the tubes is at variable angles to the longitudinal axis [21].
5. Structural complexity arises due to time-variant flow-dependent boundary conditions. Mechanical tolerances, initial straightness, fit-up, and tube buckling due to the manufacturing process add complexities in defining boundary conditions [22].
6. Effects of tube bundle parameters, such as transverse and longitudinal pitches, tube layout pattern, pass partition lanes, shell-to-tube bundle clearance, number of tube rows, etc., on the occurrence of FIV cannot be correctly evaluated. The effects of some parameters have been studied by Gorman [23].

For these reasons, most of the methods in the analysis of tube bank dynamics are semiempirical in nature. To render the problem amenable for most analytical studies and experimental investigations, the flow conditions are idealized as follows:

1. The flow is uniform and steady.
2. The incident of the flow is either axial or normal to the tubes.
3. The tube motion is linearized, and it is assumed that the frequencies are well defined.
4. The baffle supports provide a simply supported condition.

1.2 DISCUSSION OF FLOW-INDUCED VIBRATION MECHANISMS

1.2.1 VORTEX SHEDDING

1.2.1.1 Vortex Shedding against a Single Tube

Consider a bluff body such as a circular cylinder in crossflow with the tube axis perpendicular to the flow. As the fluid flows past the tube, the wake behind the tube is no longer regular, but contains distinct vortices of the pattern shown in Figure 1.3a. The periodic shedding of vortices alternately from each side of the body in a regular manner gives rise to alternating lift and drag forces. This causes periodic movement of the tube. The familiar example is the von Kármán vortex street behind a circular cylinder in crossflow.

1.2.1.2 Strouhal Number

The vortex shedding phenomenon can be characterized by a nondimensional parameter known as the Strouhal number S_u, and it is related to vortex shedding frequency f_v by

$$S_u = \frac{f_v D}{U_\infty} \quad or \quad f_v = \frac{U_\infty S_u}{D} \tag{1.1}$$

where
D is the tube outer diameter,
U_∞ the upstream velocity.

When the vortex shedding frequency f_v is sufficiently close to the natural frequency of the tubes f_n, the following will occur [4]:

(a)

(b)

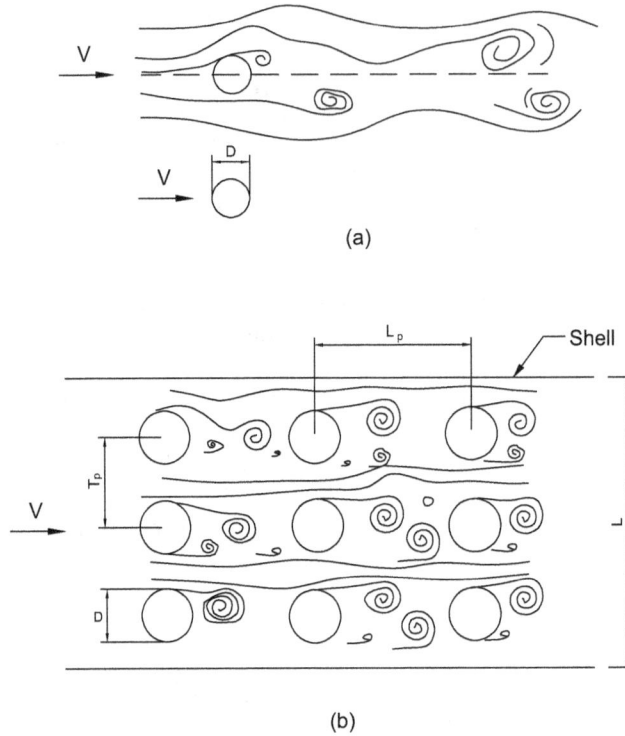

FIGURE 1.3 Principle of vortex shedding. (a) Vortex shedding past a single cylinder and (b) vortex shedding in a tube array.

1. The vortex shedding frequency shifts to the structural natural frequency, developing the condition called "lock-in" or "synchronization." The lock-in phenomenon leads to high-amplitude vibration with substantial energy input to the tube.
2. The lift force becomes a function of structural amplitude.
3. The drag force on the structure increases.
4. The strength of the shed vortices increases.

When the vortex shedding frequency coincides with the tube natural frequency or close to the natural frequency, resonance takes place. This mechanism has been variously referred to as vortex shedding, periodic wake shedding, Strouhal periodicity, or Strouhal excitation.

Since the vortex shedding drag force in the streamwise direction (drag) occurs at twice the vortex shedding frequency and the magnitude of drag force is smaller than the oscillating lift force, normally the analysis is carried out for lift forces only.

The vortex shedding phenomenon for a single cylinder with a peak response is well defined and has been dealt with by various researchers. Information on the lift and drag force coefficients and the Strouhal number over the complete range of Reynolds numbers of interest has been reviewed and presented by Chen and Weber [24]. For crossflow over a single tube, the Strouhal number is an almost constant value of 0.2 for Reynolds numbers starting from 300 to the lower critical Reynolds number of 2×10^5. After this point, the Strouhal number seems to increase due to the narrowing of the wake. But as the Reynolds number exceeds the value of 3.5×10^5 and when the supercritical range is reached, the wakes become completely turbulent. No regular vortex shedding exists anymore. This exceptional case lasts only to a Reynolds number of 3.5×10^6. After exceeding this value, a von Kármán vortex can be formed again. In the transcritical range, the Strouhal number is about 0.27 [25].

1.2.1.3 Vortex Shedding against a Tube Bundle

Vortex shedding past a tube bank is shown in Figure 1.3b. Owen [26] disputed the existence of vortices deep within a tube bank. Deep within the tube bank, the dominant spectral frequency for both lift and drag forces of vortex shedding and turbulent buffeting coincides. According to Blevins [8] and Zukauskas [13], within a closely spaced tube arrays with pitch ratio less than 2.0, the vortex shedding degenerates into broadband turbulent eddies rather than a single distinct frequency. Such a mechanism is referred to as turbulent buffeting and is described in the next section. In the light of this discussion, it may be concluded that vortex shedding is a potential design problem in the front tube rows of a tube bank in liquid flows or may be a source of acoustic noise in gas flows [6]. Hence, the possibility of the first few tube rows being excited by vortex shedding must be determined. Within the array, vortex shedding can be regarded as a special case of turbulent buffeting and is analyzed by the method of random vibration, as explained in the next section.

The expression for the Strouhal number for a tube bank is same as Equation 1.1, but the velocity term U_∞ should be replaced by the crossflow velocity, U. The crossflow velocity is calculated by Tinker's method [27], Bell's method [28], the stream analysis method [29], proprietary programs such as HTRI [30] and HTFS [31, 32], or by any other standard programs.

1.2.1.4 Avoiding Resonance

The design criterion for the possibility of vortex shedding as an excitation source involves the parameter of reduced frequency ($f_n D/U$) and/or the determination of Strouhal number, S_u. Determination of the Strouhal number is discussed next.

1.2.1.5 Calculation of the Strouhal Number for Tube Arrays

The Strouhal number may be determined from Chen's Strouhal maps [25] and Fitz-Hugh [33]. These maps are plotted with various pitch ratios. Alternatively, it can be determined from correlations of Zukauskas [13] or Zukauskas and Katinas [34] and Weaver et al. [35]. The correlation of Weaver et al. is given next. Blevins [8] presents Fitz-Hugh's map.

Correlations of Weaver et al. [35]: the expressions for the Strouhal number S_u for various tube layout patterns (Figure 1.4) are given by

$$S_u = \frac{1}{1.73\, X_p} \text{ for } 30° \text{ layout} \tag{1.2a}$$

$$= \frac{1}{1.16\, X_p} \text{ for } 60° \text{ layout} \tag{1.2b}$$

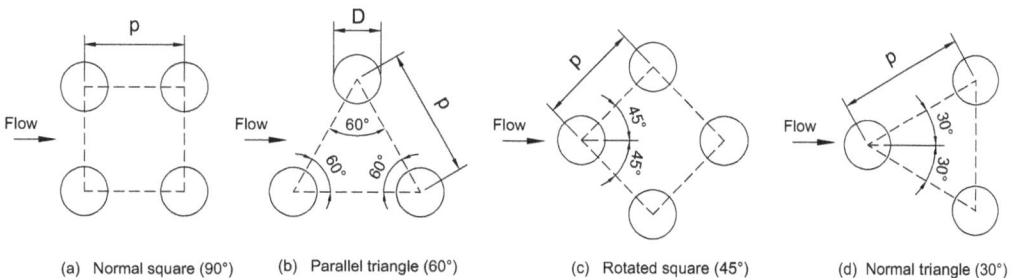

(a) Normal square (90°)   (b) Parallel triangle (60°)   (c) Rotated square (45°)   (d) Normal triangle (30°)

FIGURE 1.4 Tube layout patterns.

$$= \frac{1}{2X_p} \text{ for } 90° \text{ and } 45° \text{ layout} \tag{1.2c}$$

where X_p is the pitch ratio, p/D. The original expressions were in terms of upstream velocity. They were corrected for gap velocity by multiplying the expressions by $(p - D)/p$.

1.2.1.6   Criteria to Avoid Vortex Shedding

Criterion of Pettigrew and Gorman: the criterion to avoid resonance due to vortex shedding is expressed in terms of reduced frequency, $(f_n D/U)$, by Pettigrew and Gorman [36] as

$$\frac{f_n D}{U} > 2S_u \tag{1.3}$$

Criterion of Au-Yang [4]: to avoid resonance, the Strouhal number S_u must be less than 20% of the reduced frequency [7]:

$$\frac{f_n D}{U} < 0.2 S_u \tag{1.4}$$

Criteria of Au-Yang et al. [4]: the criteria for avoiding lock-in due to vortex shedding in the first two to three rows in a tube bank are given by Au-Yang et al. [4]. They are the following:

1. If the reduced velocity for the fundamental vibration mode ($n = 1$) is satisfied by the relation

$$\frac{U}{f_n D} < 1 \quad \text{for } n = 1 \tag{1.5}$$

   both lift and drag direction lock-in are avoided.
2. For a given vibration mode, if the reduced damping C_n is large enough,

$$C_n > 64$$

   then lock-in will be suppressed in that vibration mode.
3. If for a given vibration mode

$$\frac{U}{f_n D} < 3.3 \tag{1.6}$$

and $C_n > 1.2$, then lift direction lock-in is avoided and drag direction lock-in is suppressed. The reduced damping C_n is calculated by the equation

$$C_n = \frac{4\pi\xi_n M_n}{\rho_s D^2 \int_0^{L_e} \phi_n^2(x)\,dx} \tag{1.7}$$

where
M_n is the modal mass,
L_e is the tube length subjected to vortex shedding,
$\phi_n(x)$ is the mode shape.

The expression for M_n is given by

$$M_n = \int_0^{L_e} m(x)\phi_n^2(x)\,dx \tag{1.8}$$

where $m(x)$ is the tube mass per unit length. Substituting the expression for the modal mass and normalizing the mode shape, the expression for reduced damping C_n is given by

$$C_n = \frac{4\pi\xi_n m}{\rho_s D^2} \text{ when } M_n = m(x) = m \tag{1.9}$$

These guidelines are included in ASME Code Section III.

4. Effective tube mass per unit length. In the preceding equations, f_n is the tube natural frequency, m is the effective tube mass per unit length, and ξ_n is the critical damping ratio. The effective tube mass is the sum of structural mass, fluid-added mass due to the contribution of shellside fluid displaced by the vibrating tube, and the mass of the contained fluid per unit length. In simple terms,

$$m = \text{added mass} + \text{contained fluid mass} + \text{structural mass}$$
$$= m_a + m_i + m_t \tag{1.10}$$

$$= \frac{\pi D^2 C_m \rho_s}{4} + \frac{\pi D_i^2 \rho_i}{4} + \frac{\pi(D^2 - D_i^2)\rho}{4} \tag{1.11}$$

where
m_a is the added mass per unit length,
m_i the contained fluid mass per unit length,
m_t the structural mass per unit length.

The terms ρ_i, ρ_s, and ρ are density of tubeside fluid, shellside fluid, and tube metal, respectively. The added mass involves a term called added mass coefficient, C_m. The determination of f_n, C_m, and critical damping ratio ξ_n are discussed at the end of this section.

1.2.1.7 Response due to Vortex Shedding Vibration Prediction by Dynamic Analysis

If resonance occurs, the maximum tube response can be obtained by a forced response analysis as described in Refs. [4,36]. Sandifer [9] describes this from first principles. The generalized equation for the tube response $y(x)$ for any mode number j is given by [36]

$$y(x) = \frac{C_L \rho_s D}{16\pi^2 \xi_{n,j} f_{n,j}^2 M_j} \phi_j(x) \int_0^{L_j} U^2(x)\phi_j(x)\,dx \tag{1.12}$$

The mode shape in Equation 1.12 is normalized by

$$\int_0^{L_j} \phi^2(x)\,dx = 1 \tag{1.13}$$

and evaluation of this integral gives

$$\phi(x) = \sqrt{\frac{2}{L_i}} \sin\left(\frac{\pi x}{L_i}\right)$$

$$= \sqrt{\frac{2}{L_i}} \text{ (maximum value)}$$

(1.14)

After normalizing the modal mass, the maximum response y_{max} is given by

$$y_{max} = \frac{C_L \rho_s D U^2}{4\pi^3 \xi_n f_n^2 M} \text{ where } M = m$$

(1.15)

The peak lift coefficients C_L (peak) for various tube layout patterns are tabulated in Refs. [9,37]. A conservative design can be obtained with C_L(peak) = 0.091. According to these, as long as the peak amplitude of tube response is less than 2% of the tube diameter D, it is unlikely that the tube motion would be sufficient to control and correlate wake shedding along the tube. Accordingly, the acceptance criterion is given by

$$y_{max} < 0.02\, D$$

(1.16)

1.3   TURBULENCE-INDUCED EXCITATION MECHANISM

1.3.1   TURBULENCE

In general, higher flow rates promote and maintain high turbulence in the fluid, which is desirable for enhanced heat transfer, but the high turbulence is a source of structural excitation. Heat exchanger tubes respond in a random manner to turbulence in the flow field. In addition to structural excitation, turbulence in the flow can affect the existence and strength of other excitation mechanism, namely, vortex shedding.

1.3.2   TURBULENT BUFFETING

Turbulent buffeting in a tube bank, sometimes called subcritical vibration, refers to the low-amplitude response before the critical velocity is reached and away from the vortex lock-in velocity region due to unsteady forces developed on a body exposed to a high turbulence in the flow field.

When the dominant central frequency in the flow field coincides with the lowest natural frequency of the tube, a considerable amount of energy transfer takes place, leading to resonance and high-amplitude tube vibration. Even in the absence of resonance, turbulent buffeting can cause fretting wear and fatigue failure. With a design objective of 40 years' codal life for nuclear power plant steam generators and heat exchangers, even relatively small tube wear rates cannot be acceptable [6,38].

1.3.3   OWEN'S EXPRESSION FOR TURBULENT BUFFETING FREQUENCY

Based on the experimental study of gas flow normal to a tube bank, Owen [26] correlated an expression for the central dominant frequency of the turbulent buffeting, f_{tb}, as

$$f_{tb} = \frac{U}{DX_l X_t} \left[3.05 \left(1 - \frac{1}{X_t} \right)^2 + 0.28 \right] \tag{1.17}$$

where
  X_l = longitudinal pitch ratio = L_p/D (L_p is longitudinal pitch),
  X_t = transverse pitch ratio = T_p/D (T_p is transverse pitch).

Weaver and Grover [39] reviewed various works and observed that Owen's approach is most reliable for predicting the peak frequency in the turbulence, provided the minimum gap velocity is used in the expression. The preceding correlation is applicable for a tube bank with a transverse pitch ratio of more than 1.25. Since this correlation has not been tested for liquids, it should be restricted to gases only. TEMA has included this expression.

1.3.4 TURBULENT BUFFETING EXCITATION AS A RANDOM PHENOMENON

By assuming that the tube vibrations represent steady-state random process, expression for rms amplitude of tube response has been developed by Au-Yang et al. [4], Au-Yang [7], and Pettigrew and Gorman [36]. Sandifer [9] describes the equation for tube response from first principles. The mean square resonant response of a lightly damped structure is given by

$$\overline{y}^2(x) = \frac{\rho_s^2 D^2 L_i}{256 \pi^3} \sum_j \frac{\left[C_R\left(f_j \right) U_j^2 \phi_j(x) \right]^2}{\xi_{n,j} f_{n,j}^3 M_j^2} \tag{1.18}$$

After normalizing the mode shape over the span length and for the first mode of vibration, the maximum mean square response is given by

$$\overline{y}^2_{max} = \frac{L_i \left[C_R(f) \rho_s U^2 D \sqrt{\frac{2}{L_i}} \right]^2}{256 \pi^3 \xi_n f_n^3 M^2} \quad \text{where } M = m \tag{1.19a}$$

$$\overline{y}^2_{max} = \frac{\left[C_R(f) \rho_s U^2 D \right]^2}{128 \pi^3 \xi_n f_n^3 M^2} \quad \text{where } M = m \tag{1.19b}$$

$$y_{rms} = \sqrt{\overline{y}_{max}^2} \tag{1.19c}$$

A recommended acceptance criterion is

$$y_{rms} \leq 0.010 \text{ in.or } 0.254 \text{ mm} \tag{1.20}$$

The parameter $C_R(f)$ can be determined from Ref. [36] or TEMA Table V-11.3 [19].

1.4 FLUID ELASTIC INSTABILITY

A group of circular cylinders submerged in crossflow can be subjected to dynamic instability, typically referred to as FEI. Fluid elastic vibration sets in at a critical flow velocity and can become of large amplitude if the flow is increased further. The familiar examples of FEI vibration are aircraft wing flutter, transmission line galloping, and vibration of tube arrays in heat exchangers.

A sudden change in vibration pattern within the tube array indicates instability and is attained when the energy input to the tube mass-damping system exceeds the energy dissipated by the system. FEI has been recognized as a mechanism that will almost lead to tube failure in a relatively short period of time, and this is to be avoided at any cost by limiting the crossflow velocity [40]. However, some tube responses due to turbulent buffeting or vortex shedding cannot be avoided, and this may lead to long-term fretting failure [41]. If vortex shedding resonances are predicted at velocities above the fluid elastic critical velocity, then vortex shedding is not a concern, and it is not necessary to predict the associated amplitudes of vibration.

1.4.1 FLUID ELASTIC FORCES

Flow of fluid over an array of elastic tubes results in the following fluid elastic forces or fluid damping forces with or without fluid–structure coupling. The forces induced on the tube fall into the following three major groups [4,5]:

1. Forces that vary approximately linearly with displacement of a tube from its equilibrium position. The resulting instability mechanism is known as displacement mechanism.
2. Velocity-dependent fluid forces such as fluid inertia, fluid damping, and fluid stiffness forces. The resulting instability mechanism is known as velocity mechanism.
3. The combination of the foregoing forces.

Instability may result from any or all of these fluid forces, which are functions of the tube motion.

1.4.2 GENERAL CHARACTERISTICS OF INSTABILITY

Before understanding the FEI mechanism, it is pertinent to know the features of instability. A sudden change in vibration pattern within the tube array is indicative of instability. The general characteristics of tube vibrations during instability include the following:

1. Large vibration amplitude.
2. Synchronization between adjacent tubes.
3. Fluid–structure coupling.

1.4.3 CONNORS' FLUID ELASTIC INSTABILITY ANALYSIS

The pioneering work on FEI was initiated by Connors [42]. According to Connors, the FEI results when the amount of energy input to tubes in crossflow exceeds the amount of energy that can be dissipated by the system damping. As a result of energy imbalance, the tube vibration will intensify to the point that clashing with adjacent tubes takes place. From the model, Conners measured quasi-steady force coefficients and developed a semiempirical stability criterion for predicting the onset of FEI of tube arrays. The stability criterion relates the critical flow velocity U_{cr} to the properties of the fluid and structures of the form

$$\frac{U_{cr}}{f_n D} = K\left(\frac{m\delta}{\rho_s D^2}\right)^a \tag{1.21}$$

where

U_{cr} is the critical velocity,

δ is equal to $2\pi\xi_n$.

For this single-row experimental model with $p/D = 1.41$, the value of K is 9.9 and $a = 0.5$. Accordingly, the expression for instability is given by

$$\frac{U_{cr}}{f_n D} = 9.9 \left(\frac{m\delta}{\rho_s D^2} \right)^{0.5}$$

(1.22)

In this expression, the two main parameters are $U_{cr}/f_n D$, the reduced velocity, and $m\delta/\rho_s D^2$, the mass-damping parameter. Connors' vibration mechanism was later referred to as a displacement mechanism, and the model is known as a quasi-static model.

For tube bundles, the parameter $K = 9.9$ does not hold good, and hence numerous investigators conducted extensive experiments to form a more appropriate instability parameter K. Several new models—the analytical model [43–45] and the unsteady model [15,16,46,47]—have been proposed. A few researchers [48–50] refined the quasi-static model. A mechanism called the velocity mechanism was suggested by Blevins [8]. Many reviews [3,6,14] and the state of the art [6] have been presented. However, most of the researchers proposed a stability criterion similar to Connors' stability equation form. In the following sections, various instability models are discussed briefly. Subsequently, the design guidelines and acceptance criteria are presented.

1.4.4 ANALYTICAL MODEL

The analytical model was proposed by Lever and Weaver [43–45]. From a series of flow visualization experiments conducted in airflow to investigate the underlying mechanism responsible for the FEI of tube arrays in crossflow, the authors assumed that a single elastic tube surrendered by rigid tubes becomes independently unstable; the effect of neighboring tubes is primarily to define the flow field unique to each array pattern. The model is known as the tube-in-channel model.

1.4.5 UNSTEADY MODEL

Prior to 1980, the mechanism thought to cause FEI was a displacement mechanism, and the dominant force causing instability was referred to as fluid elastic stiffness force. From the unsteady fluid dynamic forces of Tanaka and Takahara [46,47] and Refs. [15,16,51–53], it is known that the instability is caused not only by a displacement mechanism but also by an additional mechanism called the velocity mechanism. This instability model is called the unsteady model. The existence of velocity mechanism was demonstrated analytically and verified experimentally by Chen et al. [54,55]. In this model, the FEI is caused by the following mechanisms:

1. Position-dependent FEI, which occurs at relatively high-flow velocities. The instability mechanism is called the displacement mechanism.
2. Fluid-damping-type instability occurring at low-flow velocities. The instability mechanism is called the velocity mechanism.

In most cases, either the velocity mechanism or the displacement mechanism or a combination of both will be dominant. The displacement mechanism and velocity mechanism are discussed in detail in Refs. [5,16,55]. They are briefly explained next.

1.4.5.1   Displacement Mechanism

According to this mechanism, the instability is due to fluid elastic force, which is proportional to the displacement of tubes. This mechanism is dominant for high values of reduced flow velocity corresponding to gas flows. The instability criterion is given by [5]

$$\frac{U_{cr}}{f_n D} = \alpha_d \left(\frac{m\delta}{\rho_s D^2} \right)^{0.5} \tag{1.23a}$$

where α_d is a function of fluid elastic stiffness coefficients.

1.4.5.2   Velocity Mechanism

According to this mechanism, the dominant fluid force is a flow-velocity-dependent damping force that is proportional to the velocity of the tubes. At low reduced velocity, the fluid damping force may act as an energy dissipation mechanism, whereas, at high reduced velocities, it acts as an excitation mechanism. When it acts as an excitation mechanism, the system damping is reduced. The tube loses stability once the modal damping of a mode becomes negative. This type of instability is called fluid-damping-controlled instability, and it is dominant for low values of reduced velocities that correspond to liquid flow. The instability criterion is given by [5]

$$\frac{U_{cr}}{f_n D} = \alpha_v \left(\frac{m\delta}{\rho_s D^2} \right)^{0.5} \tag{1.23b}$$

where α_v is a function of fluid elastic stiffness coefficients.

1.4.5.3   Unsteady Model

Based on the displacement mechanism and velocity mechanism, the dynamics of the tube array in simple form is written as [16]

$$\left[M_s + M_f \right] \{\ddot{q}\} + \left[C_s + C_f + C_v \right] \{\dot{q}\} + \left[K_s + K_f \right] \{q\} = \left[F \right] \tag{1.24}$$

where
  $[M_s]$ is the structural mass,
  $[M_f]$ is the added mass of the fluid,
  $[C_s]$ is the structural damping of the system,
  $[C_f]$ is the viscous damping of the still fluid,
  $[C_v]$ is the velocity-dependent damping of the fluid,
  $[K_s]$ is the structural stiffness,
  $[K_f]$ is the fluid elastic stiffness,
  $[F]$ is the fluid forces independent of the tube motion,
  $\{q\}$ is the structural displacement,
  $\{\dot{q}\}$ is the structural velocity,
  $\{\ddot{q}\}$ is the structural acceleration.

In the preceding expression, the terms in square brackets are matrices and those in braces are vectors. The constituents of fluid dynamic forces are as follows:

1. $[M_f]\{\ddot{q}\}$, fluid inertia force.
2. $[C_f + C_v]\{\dot{q}\}$, fluid damping force.
3. $[K_f]\{q\}$, fluid elastic force.

The major obstacle in the analysis of the instability mechanism by the unsteady model is to determine the flow-velocity-dependent damping $[C_v]$ and fluid stiffness $[K_f]$ for every tube layout pattern and for the entire range of reduced flow velocity of interest, either analytically or experimentally. These forces were measured experimentally by Tanaka and Takahara [47].

1.4.6 DESIGN RECOMMENDATIONS

1.4.6.1 Chen's Criterion

From the data of many investigators, Chen [5] recommended a criterion for the lower bound of critical velocity. Later, Chen [56] revised the criteria and the same is included in TEMA [19]. For 45° layout and $0.1 \leq \chi \leq 300$, the equation is given by

$$\frac{U_{cr}}{f_n D} = 4.13 \left\{ \left(\frac{p}{D} \right) - 0.5 \right\} \chi^{0.5} \tag{1.25a}$$

where the expression for χ (same as the damping parameter included in Equation 1.21) is given by

$$\chi = \frac{m\delta}{\rho_s D^2} \tag{1.25b}$$

The notable feature of Chen's criterion is the presence of pitch ratio (p/D) in the equation for 30° and 45° layouts.

1.4.6.2 Au-Yang et al. Criteria

From the instability database of Ref. [5], Au-Yang et al. [4] developed the following criteria:

$$\frac{U_{cr}}{f_n D} = K \left(\frac{m\delta}{\rho_s D^2} \right)^a \tag{1.26}$$

1. For the displacement mechanism where $m\delta/\rho_s D^2 > 0.7$, the stability equation is similar to Connors' equation with $a = 0.5$. The values for K are given in Table 1.1.
2. For the velocity mechanism, where $m\delta/\rho D^2 < 0.7$, the criteria of displacement mechanism may be used. However, this will give a conservative value.

TABLE 1.1
Au-Yang et al. [4] Criteria for FEI

Pitch Angle	30°	60°	45°	90°	All Arrays
K	4.5	4.0	5.8	3.4	4.0

Conservative guideline: if the crossflow velocity is less than the critical velocity calculated using $K = 2.1$ and $a = 0.5$ with $\xi_n = 0.5\%$ in gas flow and 1.5% in liquid flow, instability is almost certainly not a problem.

1.4.6.3 Guidelines of Pettigrew and Taylor

From a parametric study of nearly 300 data points on FEI of the flexible tube bundles subjected to single-phase crossflow, a design criterion was evolved by Pettigrew and Taylor [14]. The criterion is

$$\frac{U_{cr}}{f_n D} = 3.0 \left(\frac{m\delta}{\rho_s D^2} \right)^{0.5} \tag{1.27}$$

The reasons for variation among the various acceptance criteria are discussed in Ref. [14].

1.4.7 ACCEPTANCE CRITERIA

To avoid FEI, the acceptance criteria are thus as follows:

1. Normal criterion

$$U_{cr} > U \tag{1.28a}$$

2. Conservative criterion [9]

$$\frac{U}{U_{cr}} = < 0.5 \tag{1.28b}$$

1.4.8 STABILITY DIAGRAMS

The presentation of the stability criteria in a graphical form (log-log plot) with mass-damping parameter on the *x*-axis and reduced velocity on the *y*-axis is referred to as a stability diagram. It shows relatively distinct "stable" and "unstable" regions [48]. A typical stability diagram is shown in Figure 1.5.

1.5 SHELLSIDE ACOUSTIC VIBRATION

The natural frequencies of the shellside gas volume are dictated by the speed of sound and the size of the exchanger. The most offending mode is transverse to the longitudinal axes of the tubes and perpendicular to the crossflow. The wavelength of this mode is twice the shell diameter. As the diameter increases, the resonant frequency of the transverse mode decreases, and opposite sides of the shell approach parallel planes. Both of these effects increase the likelihood of a standing wave.

1.5.1 ACOUSTIC RESONANCE

Acoustic resonance is excited by the crossflow of air, gas, or steam. It is due to the vibration of the standing waves surrounding the tubes. The existence of standing waves is described as follows. The gas flow across the tube bank has, in addition to its mean velocity in the flow direction, a fluctuating velocity transverse to the mean flow direction. This fluctuating velocity is associated with the

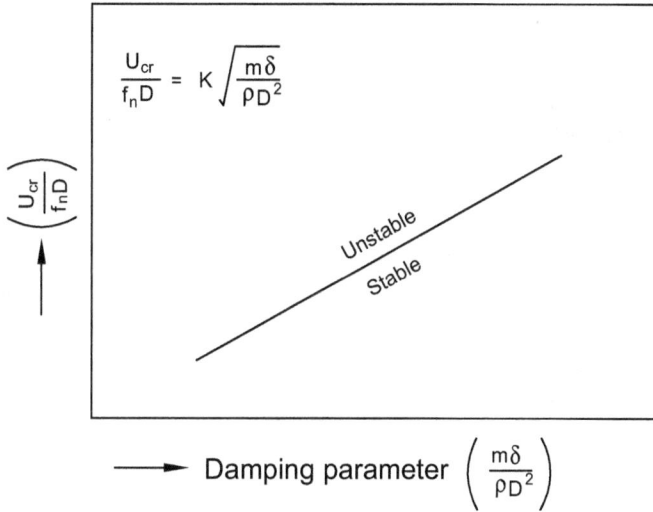

$$\frac{U_{cr}}{f_n D} = K \sqrt{\frac{m\delta}{\rho D^2}}$$

Unstable

Stable

$\left(\dfrac{U_{cr}}{f_n D}\right) \uparrow$

Damping parameter $\left(\dfrac{m\delta}{\rho D^2}\right)$

FIGURE 1.5 FIV stability diagram.

standing waves (gas column). Standing waves (Figure 1.6) occur transverse to both the tube axis and the flow direction. The resonant vibration of the standing waves surrounding the tubes is commonly called acoustic resonance or acoustic vibration. It is usually characterized with intense, low-frequency, pure-tone noise. When the excitation frequency is closer to the standing wave frequency or both coincide in a heat exchanger, there is a possibility of acoustic vibration. Also, if the standing wave frequency coincides with natural frequencies of structural components, such as casing, tubes, etc., it may be structurally harmful [57], and it can affect the performance characteristics of the heat exchanger by increasing the pressure drop. Acoustic resonance occurred in-line and staggered tube banks, single rows of tubes, helically coiled tubular heat exchangers, air heater economizers, superheaters, ducts of rectangular and circular shells, and others [58–61].

1.5.2 PRINCIPLE OF STANDING WAVES

Standing waves will develop if the distance L_a between the bounding walls is $\lambda/2$ or a multiple of $\lambda/2$, i.e., $L_a = n\lambda/2$ where $n = 1, 2, 3, \ldots$ and λ is the wavelength of the standing wave. The standing wave frequencies are given by

$$f_a = \left(\frac{nC}{2L_a}\right) \text{ for } n = 1, 2, 3, \ldots \tag{1.29}$$

where
  C is the velocity of sound in the shellside medium,
  L_a is the characteristic dimension (normally the enclosing walls of the flow passages),
  n is the mode number.

The velocity of sound C in the shellside medium is given by

$$C = \sqrt{\frac{g_c Z_\gamma R_c T}{M_g}} \tag{1.30}$$

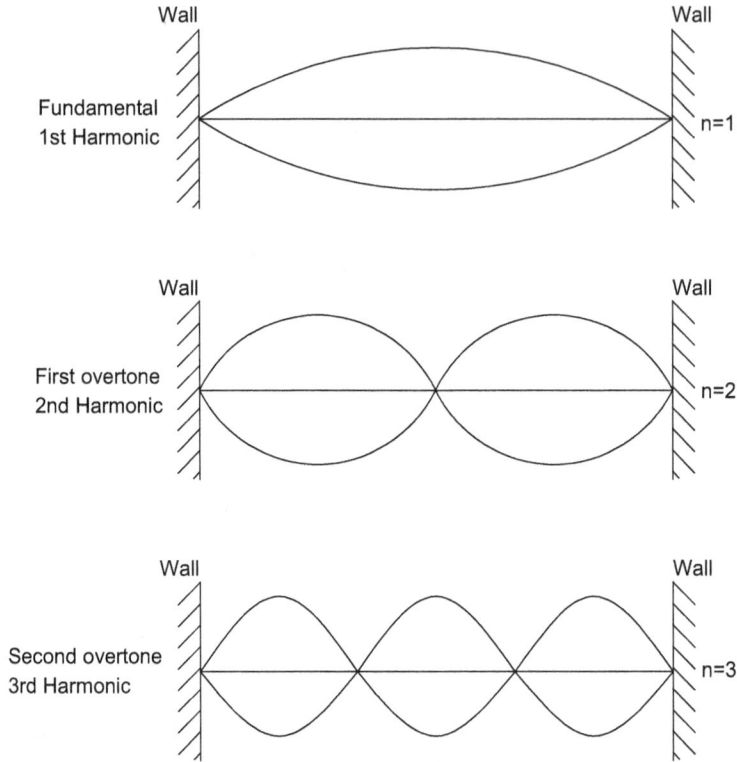

FIGURE 1.6 Variation of the fluctuating transverse velocity associated with standing waves with one, two, and three half-waves.

where
    g is the acceleration due to gravity (32.174 ft/s^2 or 9.81 m/s^2),
    Z is the gas compressibility factor,
    R_c is the universal gas constant, 1545.32 lbf · ft/lb mole or 847.6 kgf · m/kg mole K,
    T is the absolute temperature of the gas, K = °F + 459.69°F or °C + 273.16,
    M_g is the molecular weight of the gas, lbmol or kg mol,
    γ is the ratio of specific heat of gas at constant pressure to constant volume.

The value of M_g (molecular weight of the gas) for air or flue gas is 28.97 lbmol and for steam 18.02 lbmol, and γ, the ratio of the specific heat of the gas at constant pressure to that at constant volume, is 1.4 for air or flue gas and 1.328 for steam [62].

Typical standing waves with fundamental and second modes are shown in Figure 1.6. Normally, the standing wave will form in open lanes of 45° or 90° layout angle geometries, since the least exciting energy is required to form in these tube layout patterns. According to Barrington [63], acoustic vibration occurs most frequently with a rotated square (45°) tube layout compared to other tube layouts. Although the rotated square geometry exhibited the greatest resistance to FEI, this was marred by the presence of intense acoustic standing waves. Hence, they may not be suitable for shellside gaseous medium [40].

1.5.3 EFFECT OF TUBE SOLIDITY ON SOUND VELOCITY

For STHEs, Parker [64] and Burton [65] have shown that the actual speed of sound C in the shellside fluid is reduced due to the presence of the tubes. The rate of decrease in the sound speed

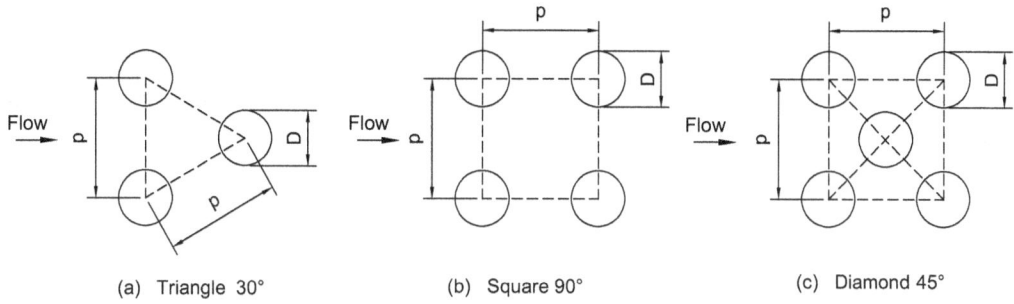

(a) Triangle 30°   (b) Square 90°   (c) Diamond 45°

FIGURE 1.7 Tube layout parameters to calculate solidity factor.

Source: From Blevins, R.D. and Bressler, M.M., *Trans. ASME J. Pressure Vessel Technol.*, 109, 275, 1987.

is mainly dependent on the solidity ratio σ of the tube layout and a weak function of added mass coefficient, C_m. Accordingly, formulas for effective speed of sound C_{eff} through the tube bundle are given by

$$C_{eff} = \frac{C}{\sqrt{(1+\sigma)}} \quad \text{Parker[64]} \tag{1.31}$$

$$C_{eff} = \frac{C}{\sqrt{(1+C_m\sigma)}} \quad \text{Burton[65]} \tag{1.32}$$

For widely spaced arrays, $C_m = 1$ (approximately).

The solidity ratio σ is defined as the ratio of free flow area to the frontal area of the tube layout. The expression for σ for the various tube layout patterns shown in Figure 1.7 is given by Blevins [8]:

$$\rho = 0.9069 \left(\frac{D}{P}\right)^2 \quad \text{for 30 or 60° layout}$$

$$= 0.7853 \left(\frac{D}{P}\right)^2 \quad \text{for 90° layout}$$

$$= 1.5707 \left(\frac{D}{P}\right)^2 \quad \text{for 45° layout} \tag{1.33}$$

In heat exchangers of normal size, either the fundamental mode or the second mode is most likely to occur. However, in large exchangers with shell diameter of the order of 20–30 m, the acoustic vibration can be excited up to fifth or sixth mode.

Principles of acoustic vibration, their evaluation, and prediction methods are discussed by Grotz and Arnold [66], in Refs. [24,25,57–62], by Barrington [63,67], and by Fitzpatrick [68]. The better understanding of acoustic resonance is mostly due to Dr. Blevins. In this section, the excitation mechanisms are discussed and acceptance criteria are defined.

1.5.4   Expressions for Acoustic Resonance Frequency

1.5.4.1   Blevins Expression

Circular shell: for a circular shell of radius R, the expression for standing wave frequency is given by [8,62]

$$f_a = \frac{C_{eff} \lambda_i}{2\pi R} \tag{1.34}$$

where
  $\lambda_1 = 1.84,$
  $\lambda_2 = 3.054.$

Rectangular shell: consider a closed rectangular volume of dimension L_x, L_y, and L_z with tubes as shown in Figure 1.8. The indices i, j, and k give the number of acoustic waves in the flow direction (x), transverse direction (y), and tube axial (z) direction, respectively. The acoustic modes typically excited by crossflow are the transverse modes (modes j = 1, 2, 3, …) in the y direction, which are perpendicular to the flow direction and tube axis. Longitudinal modes i in the flow direction x and transverse modes k in the direction of the tube axis z are rarely excited by crossflow. Assuming that both the tubes and the duct are rigid structures and the dimension of the wavelength of the standing waves is of comparable size to the duct width and much higher than the tube diameters, the expression for acoustical frequency of the duct is given by [8]

$$f_a = \frac{C_{eff}}{2}\left[\left(\frac{i}{L_x}\right)^2 + \left(\frac{j}{L_y}\right)^2 + \left(\frac{k}{L_z}\right)^2\right]^{0.5} \tag{1.35}$$

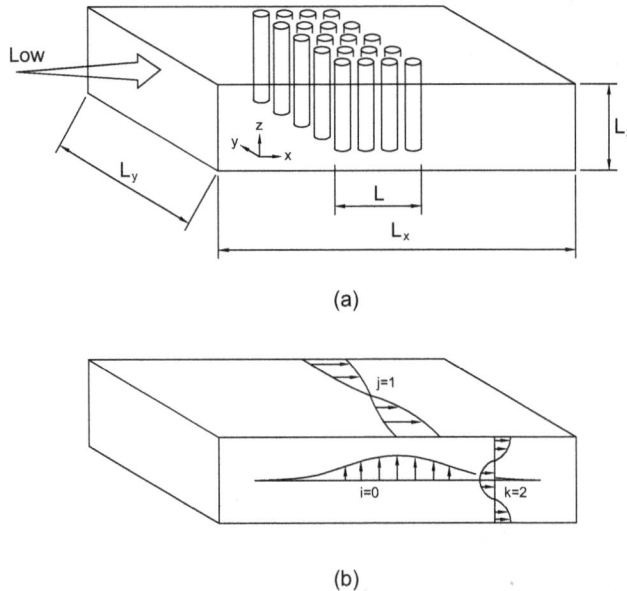

(a)

(b)

FIGURE 1.8 Acoustic modes in a closed rectangular chamber with tubes. (a) Tube bundle in duct and (b) typical mode shape of pressure in cutoff mode.

Source: From Blevins, R.D. and Bressler, M.M., *Trans. ASME J. Pressure Vessel Technol.*, 109, 275, 1987.

In the absence of any standing waves in x and z directions ($i = k = 0$), the equation simplifies to the following [62]:

$$f_a = \frac{jC_{eff}}{2L_y} \tag{1.36}$$

where L_y is the chamber dimension in the y direction, which is equal to the shell width. For other expressions, refer to TEMA.

1.5.5 Acoustic Resonance Excitation Mechanisms

The occurrence of acoustic resonance in tubular heat exchangers is caused by either a vortex shedding mechanism or turbulent buffeting.

1.5.5.1 Vortex Shedding Mechanism

According to vortex shedding theory, if the frequency of vortex shedding coincides with the standing wave frequency, a strong acoustic oscillation of the gas column is possible and resonance is said to occur. Thus the resonance criterion is

$$f_a = f_v \tag{1.37}$$

The mechanism of vortex shedding is shown schematically in Figure 1.9, which shows an in-line heat exchanger tube bank with its containment; and f_{a1} and f_{a2} represent the frequencies of the first and second acoustical modes of the containment [69]. Normally it is considered that for the lock-in phenomenon to occur, the acoustic resonance should be within ±20% of vortex shedding frequency.

1.5.5.2 Turbulent Buffeting Mechanism

According to this theory, acoustic resonance takes place if the dominant frequency of the turbulent eddies coincides with the standing wave frequency of the gas column. Thus, the criterion for resonance is given by

$$f_a = f_{tb} \tag{1.38}$$

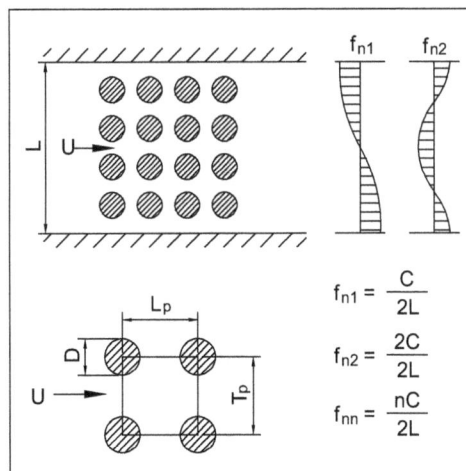

FIGURE 1.9 Acoustic resonance due to vortex shedding.

The turbulent buffeting frequency f_{tb} for fundamental mode is determined using the criteria of Owen [26] or Fitzpatrick [68]. TEMA included Owen's buffeting frequency criteria.

1.5.6 Acceptance Criteria for Occurrence of Acoustic Resonance

1.5.6.1 Vortex Shedding

Eisinger criterion [25, 57]. Eisinger expressed a criterion in terms of the Chen number, ψ, for in-line arrays (Figure 1.10a). The Chen number is a function of the Reynolds number, Strouhal number, and longitudinal and transverse pitch ratios. The expression for the Chen number is

$$\Psi = \frac{\text{Re}}{S_u} \frac{\left(2X_l - 1\right)^2}{4X_l^2 X_t} \tag{1.39}$$

where the Reynolds number is given by

$$\text{Re} = \frac{UD}{\nu} \tag{1.40}$$

Acoustic resonance criteria are as follows:

   $\psi < 2000$ no vibration,
   $\psi = 2000{-}4000$ low likelihood of vibration,
   $\psi > 4000$ high likelihood of vibration.

The TEMA condition is as follows: acoustic resonance is possible if $\psi > 2000$.
   For staggered arrays as shown in Figure 1.10b, replace L_p by $2L_p$ [60].

Blevins criterion: acoustic resonance is possible if [8]

$$\frac{\left(1-\alpha\right)S_u U}{D} < f_a < \frac{\left(1+\beta\right)S_u U}{D} \tag{1.41}$$

where the parameters α and β take their values as follows.

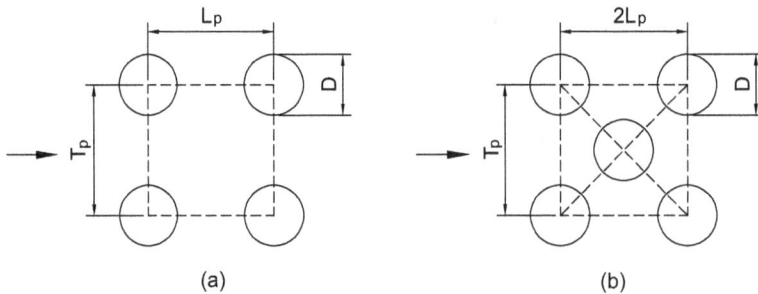

(a)   (b)

FIGURE 1.10 Tube layout parameters for calculating SPL as per Equation 1.39 (a) inline array and (b) staggered array.

Normal criterion: for normal design criteria, $\alpha = \beta = 0.2$, and the resulting expression for the Blevins criterion is

$$\frac{0.8S_u U}{D} < f_a < \frac{1.2S_u U}{D} \tag{1.42}$$

Conservative criterion: for the conservative criterion, $\alpha_{max} = 0.40$ and $\beta_{max} = 0.48$, and the resulting expression for the Blevins criterion is

$$\frac{0.6S_u U}{D} < f_a < \frac{1.48S_u U}{D} \tag{1.43}$$

TEMA prescribes $\alpha = \beta = 0.2$ in the conditionality equation.

The acoustical Strouhal number due to vortex shedding is determined either from Chen's plot [25] or from the Fitz-Hugh [33] map or from the correlations of Weaver et al. [35].

Blevins [8] *sound pressure level (SPL) at resonance*: if resonance is predicted, calculate SPL from the following equation [8]:

$$\text{SPL} = 20 \log_{10}\left(\frac{P_{rms}}{0.00002}\right) \tag{1.44}$$

or determine from the SPL maps of Blevins [8]. These maps are valid for Reynolds number up to 95.1×10^3.

Check: if SPL < 140 dB, there is no damage either to the structural components or to the surroundings.

1.5.7 TURBULENT BUFFETING

Owen's [26] *criterion for resonance*:

$$0.8f_{tb} < f_a < 1.2f_{tb} \tag{1.45}$$

Criterion of Rae and Murray [70]: resonance will not occur for

$$\frac{U}{f_a D} < 2(X_1 - 0.5) \tag{1.46}$$

This criterion is included in TEMA.

Grotz and Arnold's [66] *criterion for in-line arrays*: calculate the slenderness ratio τ from the following equation:

$$\tau = \frac{L_a}{Dn(X_1 - 1)} \tag{1.47}$$

where
L_a is the duct width,
n is the mode number.

Resonance will occur for $\tau = 62–80$. Application to staggered arrays is questionable [62].

Other methods for prediction of acoustic resonance have been suggested by Fitzpatrick et al. [71] and Zaida et al. [72].

1.6   FIV EVALUATION

Evaluation of the vibration potential requires the determination of several important parameters:

i.   tube natural frequencies,
ii. acoustic natural frequencies,
iii. turbulent buffeting frequency,
iv. vortex shedding frequency,
v.   acoustic vibration frequency, and
vi. the ratio of excitation frequencies to natural frequencies along with the vibration amplitude.

Excitation frequencies within a range of ± 20 % of the natural frequency have "lock-on" potential, and may lead to mechanical or acoustical resonance. Additionally, the critical velocity, which is the threshold for fluid elastic instability, and the crossflow velocities at both the midspan and at the baffle tips are important parameters to understand. Vulnerable tubes can include those at the bundle entrance and exit, at the baffle tips, in the baffle window region (where tubes are supported at every second baffle), and at the bundle perimeter.

1.6.1   SOLUTIONS TO VIBRATION PROBLEMS

Vulnerability to flow-induced vibration depends on flow rate, tube and baffle materials, baffle thickness, unsupported tube spans, tube field layout, shell diameter, and shellside inlet nozzle configuration, presence of impingement plate protection below the nozzle, etc. There are three approaches the designer can use if they anticipate vibration problems:

i.   Reduce the crossflow velocity.
ii. Increase the tube natural frequency.
iii. Increase damping.

Tube natural frequency can be increased by adding extra baffles and reducing the span length; but care must be exercised since extra baffles may increase crossflow velocities. Velocities can be reduced by increasing the shellside diameter, increasing tube pitch, creating bypass lanes in the direction of the flow, enlarging inlet/outlet nozzle diameters, or changing the tube field layout angle. Damping can be increased by reducing the tube-to-baffle clearance, at the possible expense of increased fretting.

1.6.1.1   Increase the Tube Natural Frequency

The natural frequency of a tube is functions of:

i.   The modulus of elasticity of the tube material.
ii. The moment of inertia of the tubular section.

iii. The number and length of spans.
iv. The effective tube mass, which includes the mass of the tube, the tubeside fluid, and a portion of the shellside fluid.

The lowest natural frequency is termed the "fundamental frequency."

1.6.1.2 Fluid Damping Effect

In a heat exchanger, vibrational energy dissipation or damping is desirable, but not affected much by conventional design choices. Dampening depends on the fluid density, viscosity, and surface tension, the vapor volume fraction, the tube diameter, number and thickness of tube supports, the tube mass, and the fundamental natural frequency of the tube.

1.6.2 VIBRATION PROBLEMS AS ADDRESSED IN HEAT EXCHANGER STANDARDS

The problem of vibration is discussed in both the HEI Standards for Steam Surface Condensers and the Standards of the Tubular Exchanger Manufacturers Association [19]. These standards give criteria on certain aspects of the design that can affect vibration, such as baffle spacing, baffle thickness, inlet velocity, etc.

1.6.3 VIBRATION EVALUATION PROCEDURE

Basically, the vibration evaluation procedure involves an estimation of certain parameters for various flow-induced excitation mechanisms. The estimated parameters are compared with their respective limiting values to check whether or not vibrations from such excitation mechanisms can cause potential damage to the tubes and the shell. There is a need to individually examine various zones of interest, namely, the nozzle inlet zone, U-bend region, and baffle window region, since there is a likelihood of high turbulence and high crossflow velocity in these regions and the variations of span lengths between baffle supports compared to the central baffle region.

1.6.3.1 Steps of Vibration Evaluation Procedure

1. Calculate the effective mass per unit length.
2. Identify the zones of interest (inlet, baffle window, central baffle zones, U-bend, etc.) to calculate the natural frequency.
3. Calculate the natural frequency for spans in various regions of interest.
4. Calculate the damping parameter.
5. Calculate crossflow velocity for the TEMA shell type under consideration.

For Liquid Flow
Vortex shedding:
  a. Calculate vortex shedding frequency. Check for the acceptance criteria.
  b. If resonance takes place, calculate tube response and check whether it is not exceeding the limiting value.

Turbulent buffeting: if resonance is predicted, calculate tube response due to random excitation and check for the acceptance criterion.

Fluid elastic instability: calculate the critical velocity and compare with crossflow velocity. Keep the maximum crossflow velocity below the critical velocity.

For Gas Flow

In addition to the criteria given for liquid flow, the following check for acoustic resonance due to standing waves may be carried out.

  a. Calculate acoustic resonance frequency.
  b. Calculate vortex shedding frequency. Check for various vortex shedding criteria.
  c. Calculate turbulent buffeting frequency. Check for various turbulent buffeting criteria.

Fill these values into the FIV specification sheet for acoustic resonance given in Table 1.2.

TABLE 1.2
Flow-Induced Vibration Evaluation for Gases—Acoustic Resonance

---

1. Effective sound velocity
   a. Sound velocity

   $$c = \sqrt{g_c Z \gamma R T / M_g}$$

   b. Solidity ratio

   $$\sigma = 0.9069 \left(\frac{D}{p}\right)^2 \text{ for } 30° \text{ or } 60° \text{ layout}$$

   $$= 0.7853 \left(\frac{D}{p}\right)^2 \text{ for } 90° \text{ layout}$$

   $$= 1.5707 \left(\frac{D}{p}\right)^2 \text{ for } 45° \text{ layout}$$

   c. Effective sound velocity

   $$C_{eff} = \frac{C}{\sqrt{(1+\sigma)}}$$

   or

   $$C_{eff} = \frac{C}{\sqrt{(1+C_m \sigma)}}$$

2. Acoustic resonance frequency Blevins
   i. For circular shell

   $$f_a = \frac{C_{eff} \lambda_i}{2\pi R} \left(\lambda_1 = 1.84, \lambda_2 = 3.054\right)$$

   ii. For rectangular shell

   $$f_a = \frac{j C_{eff}}{2 L_y} \quad j = 1, 2, 3 \ldots$$

3. Check for vortex shedding lock-in
   a. Normal criteria
      SPL
      If SPL < 140, safe

   $$0.8 \frac{S_u U}{D} < f_a < 1.2 \frac{S_u U}{D}$$

   b. Conservative criteria

   $$0.6 \frac{S_u U}{D} < f_a < 1.48 \frac{S_u U}{D}$$

4. Check for turbulent excitation by Own criteria

   $$f_{tb} = \frac{U}{D X_1 X_1} \left[3.05 \left(1 - \frac{1}{X_t}\right)^2 + 0.28 \right]$$

   for resonance:
   $$8.8 f_{tb} < f_a < 1.2 f_{tb}$$

5. Chen No. Ψ and Eisinger criteria

   $$Re = \frac{U D \rho_s}{\mu}, \Psi = \frac{Re}{S_u} \frac{(2X_1 - 1)^2}{4 X_1^2 X_t}$$

   $\Psi < 2000$ no vibration

   $\Psi < 2000\text{–}4000$ low likelihood of vibration
   $\Psi > 4000$ high likelihood of vibration

6. Bryce criteria

   $$\frac{U}{f_a D} < 2(X_1 - 0.5)$$

   No resonance
   6.1 Blevins criteria for $U/f_a D > 2$ resonance will occur
   6.2 Arnold's slenderness ratio

   $$\tau = \frac{L_a}{Dn(X_1 - 1)}$$

   If, $\tau < 62\text{–}80$, resonance will occur

1.6.4 CAUTION IN APPLYING EXPERIMENTALLY DERIVED VALUES FOR VIBRATION EVALUATION

The vibration prediction methods are mostly based on linearized models that require input data reflecting a particular flow pattern and a structural configuration. The guidelines are derived from simplified test conditions, and they cannot be generalized for actual heat exchangers. The parameters, such as damping, Strouhal number, FEI constant, etc., are mostly known for idealized test conditions and may be different in the real environment of the heat exchanger due to interaction between various mechanisms of excitation [57]. If the outcome of the experimentally derived design criteria is applied to specific heat exchangers, it may lead to conservative or marginal designs.

Accurately predicting the critical velocity requires testing a single-tube mock-up with end conditions and supports as nearly identical as possible to the actual heat exchanger unit, or a part scale model [73] or a full scale model if it is economically feasible. Hence, it can be concluded that the judgment of the designer, precautions taken by the heat exchanger fabricator, and model test conducted by the researchers are the means available to provide reasonable assurance that a design will not fail due to FIV [73].

1.7 DESIGN GUIDELINES FOR VIBRATION PREVENTION

FIV problems are better prevented rather than corrected. After the thermal design stage, a check for FIV is carried out, and if the tubes are susceptible to FIV, excitation sources are identified and corrective action is taken. Tube vibration can be reduced to within limits by either increasing the tube natural frequency or reducing the crossflow velocity or suppressing the source of excitation as discussed next.

1.7.1 TEMA GUIDELINES FOR TUBE BUNDLE VIBRATION PREVENTION

Shellside flow may produce excitation forces, which result in destructive tube vibrations. The vulnerability of an exchanger to flow-induced vibration depends on the flow rate, tube and baffle materials, unsupported tube spans, tube field layout, shell diameter, and inlet/outlet configuration. Section 6 of this Standard addresses FIV problems.

1.7.2 TEMA GUIDELINES FOR SHELLSIDE IMPINGEMENT PROTECTION REQUIREMENTS

An impingement plate, or other means to protect the tube bundle against impinging fluids, shall be provided for all shellside inlet nozzle(s), unless the product of ρV^2 in the inlet nozzle does not exceed the following limits:

   i. 1500 (2232) for nonabrasive, single phase fluids (liquids, gases, or vapors);
   ii. 500 (744) for all other liquids, including a liquid at its boiling point;

where V is defined as the linear velocity of the fluid in ft/s (m/s) and ρ is its density in lb/m^3 (kg/m^3).

For all other gases and vapors, including steam and all nominally saturated vapors, and for liquid vapor mixtures, impingement protection is required. A properly designed diffuser may be used to reduce line velocities at the shell entrance.

Shell or bundle entrance and exit areas. For shell or bundle entrance and exit areas, in no case can ρV^2 be over 4000 (5953). This requirement is independent of impingement protection.

Shell entrance or exit area with impingement plate. When an impingement plate is provided, the flow area shall be considered as the unrestricted area between the inside diameter of the shell at the nozzle and the face of the impingement plate. Fitment of impingement plate below the shellside nozzle inlet is shown in Figure 1.11.

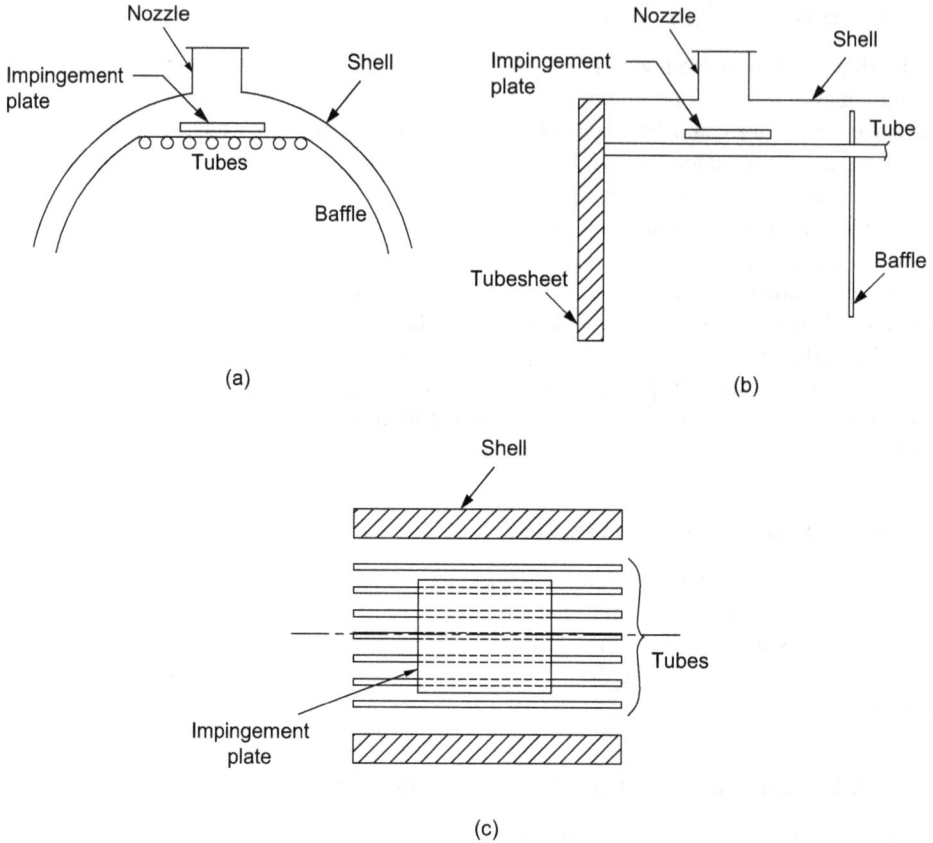

FIGURE 1.11   Impingement plate below the shellside inlet nozzle.

Consideration shall be given to the need for special devices to prevent erosion of the tube ends under any of the following conditions:

(1)   Use of an axial inlet nozzle.
(2)   Liquid ρV^2 in the inlet nozzle is in excess of 6000 (8928).
(3)   For two-phase flow.

Tie rods and spacers. Tie rods and spacers, or other equivalent means of tying the baffle system together, shall be provided to retain all transverse baffles and tube support plates securely in position.

Number and size of tie rods. Refer to TEMA Table R-4.7.1 for suggested tie rod count and diameter for various sizes of heat exchangers. Other combinations of tie rod number and diameter with equivalent metal are permissible; however, no fewer than four tie rods, and no diameter less than 9.5 mm (3/8") shall be used. Any baffle segment requires a minimum of three points of support.

1.7.3   Methods to Increase Tube Natural Frequency

Tube natural frequency is represented by

$$f_n \propto \frac{1}{L_i^2} \left(\frac{EIg_c}{m} \right)^{0.5} \tag{1.48}$$

A careful review of this natural frequency equation indicates that the tube natural frequency can be increased by reducing the tube span L_i, increasing the ratio I/m, or increasing Young's modulus E. Since the moment of inertia and tube mass are limited by the standard tube dimensions and E is limited by the material chosen, in most of the cases, the only option available to increase the natural frequency is to reduce the tube span L_i [11]. For single-segmental baffles, reducing the tube span will increase the natural frequency proportionally to $1/L_i^2$ and increase the crossflow velocity proportionally to $1/L_i$. Hence, the tube span may be reduced to an optimum level [17]. The influence of E and thin tubing on FIV is explained next with reference to a retubed unit.

1.7.4 FIV of Titanium Alloy Tubes Unit Retubed

The importance of tube support spacing and Young's modulus has often come to light when a condenser that was initially fitted with copper alloy tubes is retubed with thin titanium tubes, which have a relatively low modulus of elasticity. Titanium's hardness and corrosion fatigue resistance act to minimize vibration damage, but its lower modulus of elasticity (than steel or copper-nickel alloys) must be considered in design to keep deflection within acceptable limits. Proper baffle design and spacing should be incorporated into the designs of both new and retrofit titanium tube bundles to avoid flow-induced vibration. Since the thin-wall titanium tubing is not as stiff as the copper tubing it replaces, so that it usually requires supplementary support to avoid FIV problems.

1.7.5 Other Guidelines

1. If the tube vibration problem is anticipated in the window zone due to longer tube span between baffles $(2L_{bc})$ and higher velocity than the tubes at central zone, eliminate the tubes in the window zone, resulting in a no-tubes-in window design (NTIW). To offset any loss in heat transfer, introduce support plates to divert the fluid across the bundle.
2. An alternative to the NTIW design is to employ strip baffles in the window zone. This additional support given to the tubes in this zone increases the stiffness of the tubes and hence increases the natural frequency.
3. Increase the structural damping of the tube bundle by means of thicker baffle plates and reduce the fretting wear by employing softer baffle material.
4. Lace the tube bundle using wire laces to increase the tube bundle stiffness.
5. U-bend support: the natural frequency of the U-bend is considerably increased by providing a support known as antivibration bars (AVB) to the U-bend. Principal types of U-bend support configurations are (a) scalloped bars; (b) wiggle bars; (c) lattice bars; and (d) flat bars, and they are as shown in Figure 1.12 [41].
6. Tubes are likely to be under some axial loading as a result of fabrication and operation. Tensile stresses increase the natural frequency, whereas compressive stress decreases the natural frequency. Since axial compressive stress in the tube decreases the tube natural frequency, rolling-in of the tubes into the tubesheet should be done carefully. Care should be also taken to allow for the free thermal expansion of the tubes. When significant compressive axial loads are induced on the tubes in a multipass exchanger, these tubes shall be located in the baffle overlap regions to minimize the unsupported span length [11].
7. When only few tubes are exposed to high localized velocities, sealing strips may be applied to provide flow resistance and reduce flow in critical areas [74].
8. *Plugging of tubes*: the affected tubes may be omitted or plugged if proper precautions are taken to ensure that collision will not endanger the adjacent tubes [74].
9. *Detuning of tubes*: detuning is defined as the method of producing differences among the natural frequencies of the neighboring tubes in a tube array. The tubes are detuned by installation of additional supports, such as clips, wedges, flat bars, helical spacers, strip baffles, or

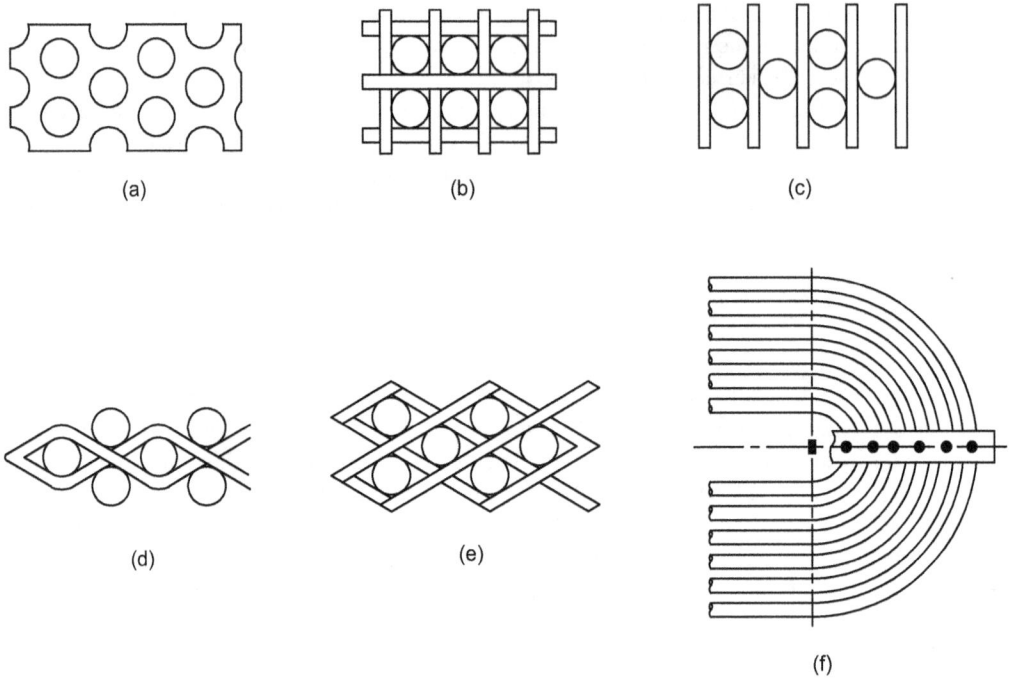

FIGURE 1.12 U-bend supports.

Source: Based on Weaver, D.S. and Schneider, W. *Trans. ASME J. Eng. Power*, 105, October, 775, 1983.

any other means of increasing the tube fixity. Detuned tube arrays lose stability at a higher flow velocity; the incidence of instability is more gradual in a tube bank with differences in tube-to-tube frequency than in a bank with identical tube frequency [8].

10. *Detuning effect*: for the displacement mechanism associated with the coupled motion of adjacent tubes, detuning will have a significant effect, whereas, for the velocity mechanism, the effect is little since the individual tube motion is independent of neighboring tubes and the instability is controlled by fluid damping forces [55]. Lever and Weaver [43] found that for airflow, detuning of neighboring tubes produced a maximum increase in critical velocity about 40% at a frequency difference of 3%. For frequency differences greater than about 10%, there was no significant improvement in threshold instability. Tanaka et al. [46] report that detuning has increased the critical velocities for airflow by about 60%.

11. Avoid longer unsupported tube spans. Follow maximum unsupported tube length as specified in TEMA, RCB-4.52 [19].

1.7.6 METHODS TO REDUCE CROSSFLOW VELOCITY

1.7.6.1 Tube Vibration Mitigation

1. At the design stage, excessive tube vibrations can be dampened by shortening the span lengths that increase the natural frequencies of the tubes. Intermediate supports in vulnerable areas like the inlet or the U-bend region can be very effective and have minimal thermal hydraulic effect.

2. The selection of the baffle type, e.g., NTIW (no tube in the window) or rod supports, and adequate baffle spacing, can keep tube vibrations within acceptable limits (for details refer to Chapter 4, Heat Exchangers: Classification, Selection, and Thermal Design).

3. For operating units, tube vibration problems can be addressed by stiffening the tubes with stakes, clips, or inserts of various types.

4. Removing vulnerable tubes, reducing shellside flow rate, and replacing the tube bundle are additional mitigation techniques.

5. Reduce the crossflow velocity either by decreasing the flow rate or by increasing the shell diameter. While evaluating the critical velocity, if such operational transients are anticipated, it may be desirable to reduce the allowable flow rate significantly below the instability threshold [74].

6. Increase the tube pitch such that the crossflow area is increased. However, this will be at the cost of lower shellside heat transfer.

7. If vibration damage to the tubes located beneath the nozzle is predicted, increase the nozzle openings to decrease the mass velocity or install the impingement plate.

8. If the vibration potential is predicted in the middle zone of an E shell, change the shell design into either an X shell or J shell. Theoretically speaking, for the same mass flow rate, the crossflow velocity on the shellside of a J shell will be half that of the E shell [75].

9. Use double or multiple segmental baffles to divide the flow into two or more streams. In an exchanger with single segmental baffles, the total flow, except for leakages and bypass streams, passes through the tube bank between baffles, whereas in an exchanger with double segmental baffles barring the leakages, the flow divides into two streams on either side of the shell and into three streams in triple segmental baffles. Under ideal conditions, the crossflow velocity in a double-segmental baffle unit is half that of a comparable single-segmental baffle unit and one-third with triple-segmental baffles. Hence, it is expected that double-segmental baffles will double the critical velocity compared to single-segmental baffles. However, experiments on an industrial-size heat exchanger at Argonne National Laboratory [76] show that the threshold velocity increased by 24%–67%. The full theoretical potential of approximately doubled threshold velocities was not realized. This was attributed to the effects of localized end zone conditions, which were probably not eliminated.

10. Limit the high localized velocities in the bundle bypass flow region or in the pass partition lanes by carefully selecting the sealing strips and pass portion lane widths, respectively, without introducing a penalty on heat transfer on the shellside.

11. Avoid too small or too large baffle cuts, since both yield poor velocity distribution and high localized velocities.

12. Maintain small tube-to-baffle hole clearances; follow TEMA Standard RCB-4.2 [19].

13. Ensure uniform baffle spacings.

14. Alternative tube bundle design. Since the crossflow velocity is the major cause of FIV, one of the design alternatives is to substitute the existing segmental baffles design either during design stage or replacement of bundle, by Philips RODbaffle heat exchangers [77], EMbaffle® heat exchanger, HELIXCHANGER® heat exchanger or Koch Twisted Tube® heat exchanger, etc. HELIXCHANGER® heat exchanger (which uses helical baffles), and Twisted Tube® heat exchanger without baffle (baffle-free tube support) eliminates flow induced vibration. Like RODbaffle supports, other tube support arrangements or heat exchangers that produces axial flow over the tubes include NESTS™ [78], EMbaffle® design (which uses expanded metal baffles made of plate material that has been slit and expanded),

1.7.6.2 Philips RODbaffle Heat Exchangers

In a Philips RODbaffle heat exchanger the longitudinal flow pattern of the shellside fluid and the multipoint support structure of the bundle, the potential for flow-induced damaging vibration is

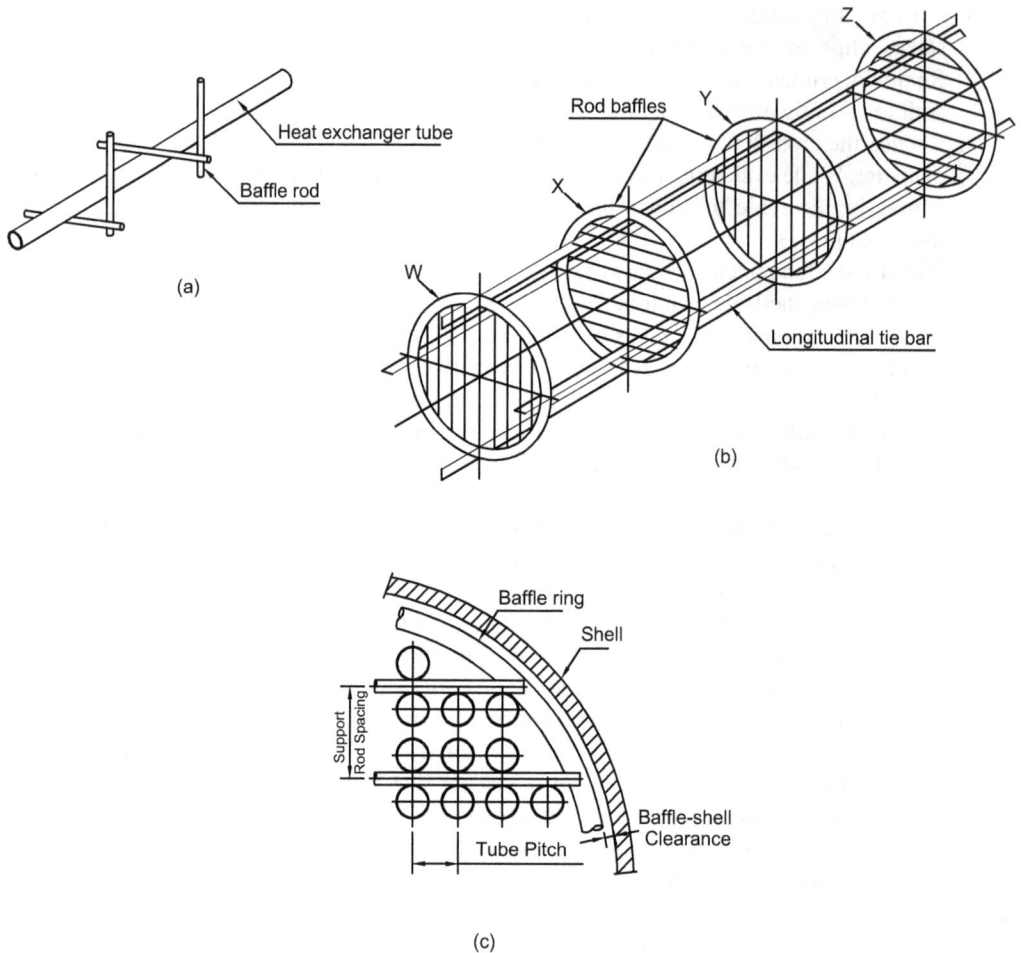

FIGURE 1.13 Concept of Philips RODbaffle heat exchanger.

mitigated. Concept of Philips RODbaffle heat exchanger is shown in Figure 1.13. The RODbaffle bundle eliminates harmful flow-induced vibration by using these major design innovations:

1. Have longitudinal flow fields.
2. Each tube is supported in all four directions.
3. Positive four-point confinement of the tubes.
4. Minimum convex point contact between rods and tubes.

1.7.6.3 EMbaffle® Heat Exchanger[1]

Use expanded metal baffles as shown in Figure 1.14 as against segmental baffles. The patented EMbaffle design uses expanded metal baffles made of plate material that has been slit and expanded.

Figure 1.15 shows a section of EMbaffle. With this new EMbaffle technology, the shellside fluid flows axially along the tubes, but in the vicinity of the baffles, the flow area is reduced. As a direct consequence of the longitudinal flow on the shellside, tube vibration in an EMbaffle heat exchanger is effectively eliminated, significantly reducing the risk of mechanical damage.

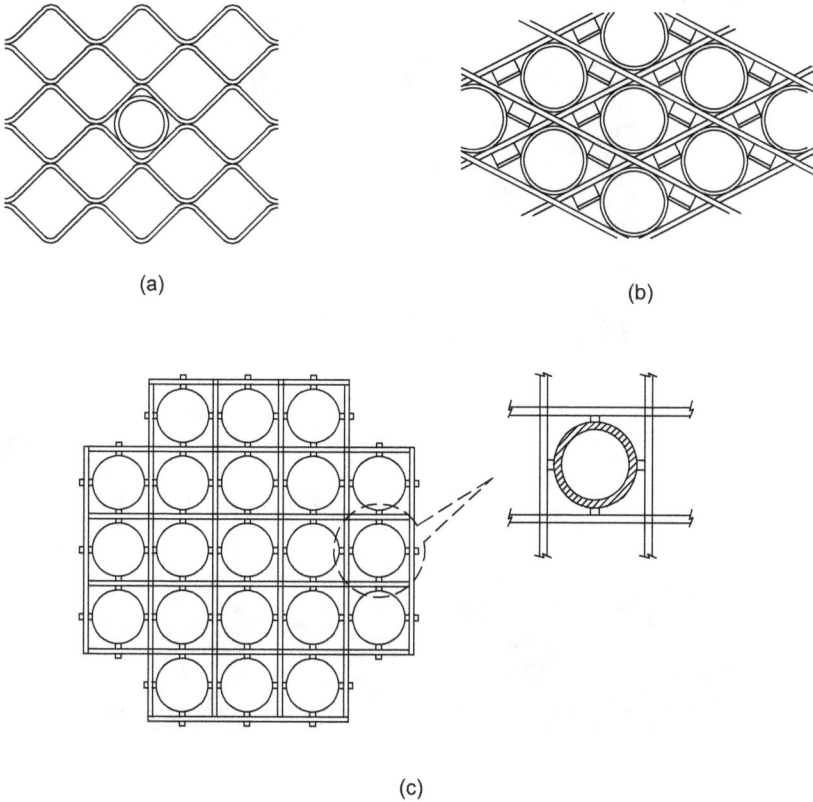

(a)

(b)

(c)

FIGURE 1.14 Expanded metal baffles.

1.7.6.4 HELIXCHANGER® Heat Exchanger

The HELIXCHANGER heat exchanger is a high-efficiency heat exchanger and a proprietary product of Lummus Technology Heat Transfer, a division of Lummus Technology. In a HELIXCHANGER heat exchanger, the conventional segmental baffle plates are replaced by quadrant-shaped baffle plates positioned at an angle to the tube axis in a sequential arrangement to create a helical flow pattern creating a uniform velocity through the tube bundle. Figure 1.16 shows helical flow pattern on the shellside and orientation of helical baffles of a tube bundle. Shellside helical flow offers higher thermal efficiency as well as lower fouling rates as compared to the conventional segmental baffled heat exchangers. Effective protection against flow-induced vibrations is achieved by both the single- and the double-helix baffle arrangement. In a double-helix arrangement, two strings of helical baffles are intertwined to reduce the unsupported tube spans offering greater integrity against vibrations without compromising the thermohydraulic performance.

1.7.6.5 Koch Twisted Tube® Heat Exchanger[2]

Koch Heat Transfer Company's innovative Twisted Tube design avoids the need for baffles. The unique helix-shaped tubes are arranged in a triangular pattern. Each tube is firmly and frequently supported by adjacent tubes as shown in Figure 1.17, yet fluid swirls freely along its length. Due to the generally longitudinal flow pattern of the shellside fluid and the multipoint support structure of the bundle, the potential for flow-induced damaging vibration is mitigated.

FIGURE 1.15   A section of EMbaffle® heat exchanger bundle.

1.8   ACOUSTIC VIBRATION MITIGATION—SUPPRESSION OF STANDING WAVE VIBRATION

The acoustic standing waves can be suppressed by any of the following devices or methods:

1. Antivibration baffles: solid or porous baffle.
2. Resonators.
3. Fin barriers.
4. Helical spacers.
5. Detuning or structural modification.

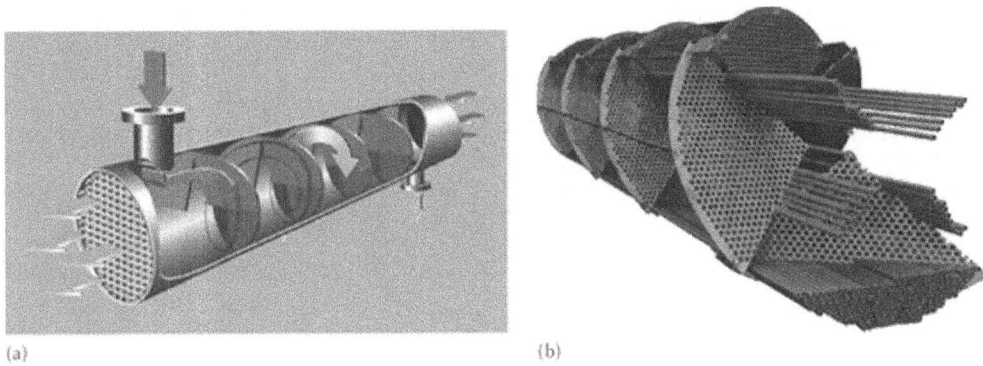

FIGURE 1.16 Helical baffle heat exchanger (a) Helical flow pattern on the shellside and (b) orientation of helical baffles of a tube bundle.

6. Removal of tubes.
7. Surface modification.
8. Irregular lateral spacing of tubes.
9. Changing the mass flow rate.

1.8.1 ANTIVIBRATION BAFFLES

The occurrence of acoustic resonance in a crossflow heat exchanger is related to the development of standing waves. If the standing waves are inhibited from developing, then the associated acoustic resonance will not take place. This is achieved by incorporation of either a solid baffle or a porous baffle inside the heat exchanger, in a direction parallel to the flow and perpendicular to the direction of standing waves. The baffles are also called detuning baffles. The placement of solid baffles is shown schematically in Figure 1.18.

Number of solid baffles: as per theory, if solid baffles are to be installed in a tubular heat exchanger in which the maximum number of half-waves that can occur between the walls is N_w, then N_w solid baffles should be installed to prevent the acoustic vibration. The major drawback associated with the solid baffles is that the number of baffles required is very high for a heat exchanger of spatial size of 20–30 m used in the fossil fuel utilities, like an economizer, reheater, etc. However, the number of baffles' requirement can be minimized by following Chen's method [25]. The method consists of placing solid baffles in preferred locations as shown in Figure 1.19. The distances b_1, b_2, b_3, b_4, b_5, and b_6 are all to be kept different from each other and their ratios b_2/b_1, b_4/b_3, and b_6/b_5 should deviate considerably from the integers corresponding to the modes of the vibrations to be expected. While placing the baffles, the length of the baffle should be large in comparison with the wavelength of the standing waves to be expected in the heat exchanger. The shortcomings associated with the solid baffles are discussed by Eisinger [57]. They include the following:

1. Expensive for retrofitting.
2. Inhibition of periodical inspection and maintenance operations.
3. Limited life if the medium of flow is corrosive in nature.
4. Tendency to vibrate and susceptible to thermal distortion in service.

Porous baffle: the porous baffle concept consists of a single flow-resistive element inserted between the tubes so that it is parallel to the flow direction [79]. Byrne [79] has shown that, in general, if the maximum number of half-waves between the walls is N_w, then a single porous baffle should be located at a distance from one wall equal to the wall-to-wall span divided by $2N_w$. The

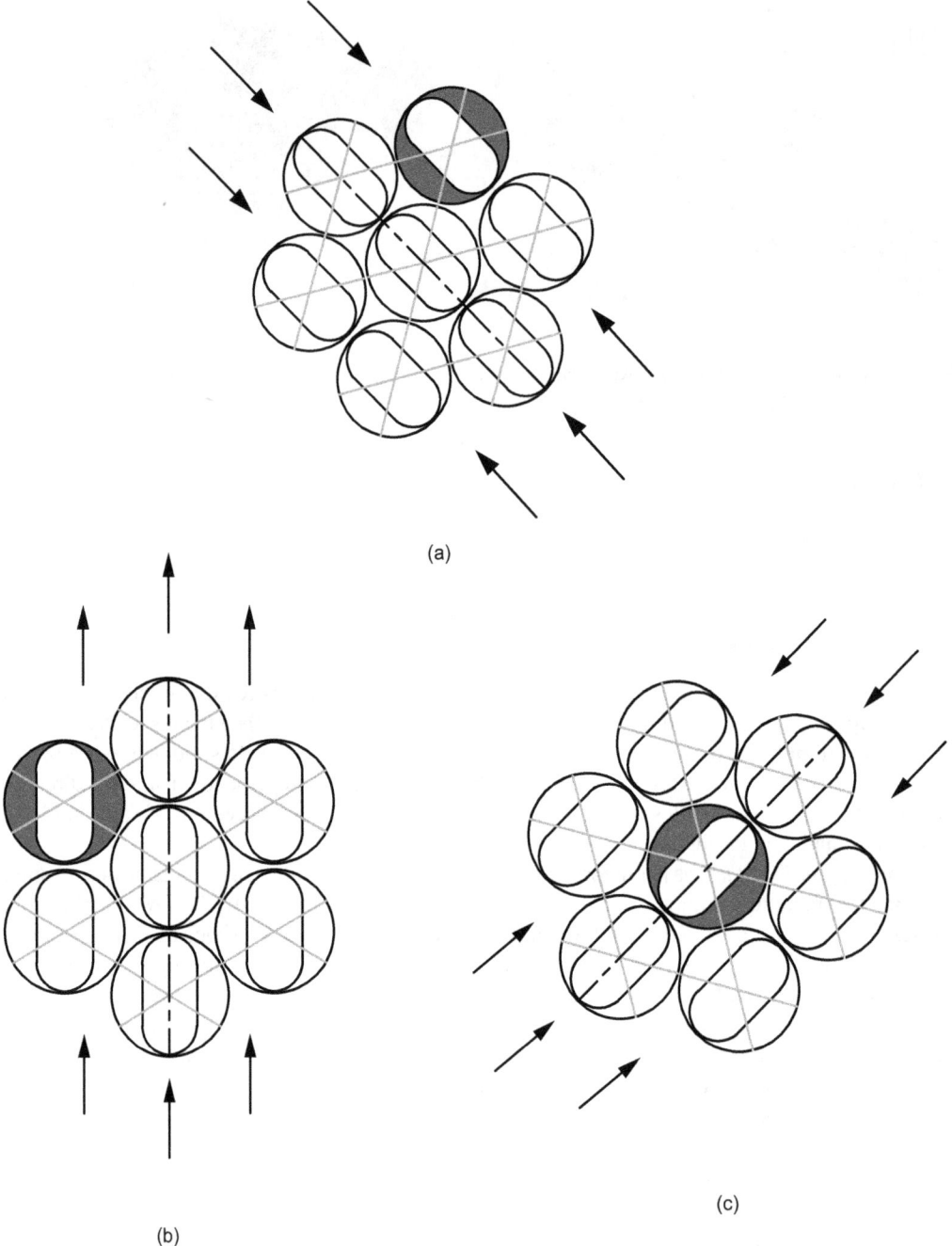

FIGURE 1.17   Adjacent tubes contact in a Twisted Tube layout heat exchanger (image depicts an approximate contact pattern and not the actual design).

Note: no baffle is used in twisted tube heat exchanger.

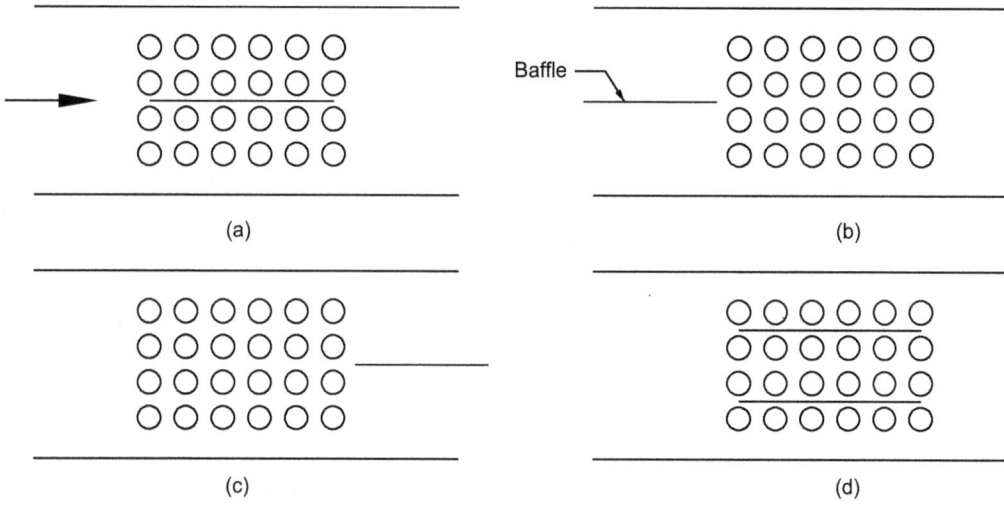

FIGURE 1.18 Detuning solid baffle arrangement. (a) Central baffle; (b) upstream baffle; (c) downstream baffle; and (d) two baffles at 1/3 and 2/3 transverse locations.

Source: From Blevins, R.D. and Bressler, M.M., *Trans. ASME J. Pressure Vessel Technol.*, 109, 282, 1987.

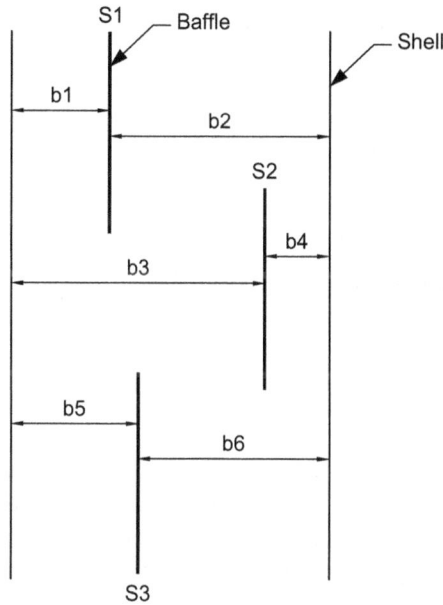

FIGURE 1.19 Chen's method of solid baffle arrangement to prevent acoustic vibration. Insertion of the slightly overlapped detuning baffles into the heat exchanger to make the natural frequencies of the transverse gas columns in the spaces, b_1, b_2, b_3, b_4, b_5, and b_6 as different as possible from one another.

Source: From Chen, Y.N., *Trans. ASME J. Eng. Industry*, February, 134, 1968.

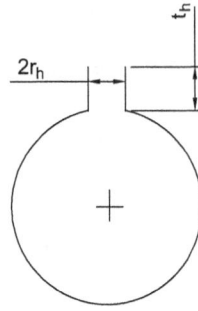

FIGURE 1.20   Helmholtz resonator.

advantage of a porous baffle over that of solid baffles is that a single porous baffle can be used to inhibit the development of a number of standing waves, whereas the number of solid baffles required is equal to the maximum number of half-waves that can develop between the walls.

1.8.2   HELMHOLTZ CAVITY RESONATOR

Installation of an acoustic damper such as a tuned Helmholtz cavity resonator on the shellside will suppress the acoustic vibration. A simple Helmholtz resonator is shown in Figure 1.20. The damping is maximum when the resonator is tuned with the acoustic mode of interest [62]. When tuned, there is a large oscillating mass flow through the resonator neck that expends energy by passing through the neck and/or the screen placed in the neck. An approximate formula for the resonator natural frequency is given by [62]

$$f = \frac{C}{2\pi}\left(\frac{A}{V_h l_h}\right)^{0.5}$$

(1.49)

where
  A is the area of cross section of opening in the resonator,
  V_h the volume of the cavity resonator,
  l_h the length of the opening in the resonator.

1.8.3   CONCEPT OF FIN BARRIER

To overcome the drawbacks associated with solid baffles, a new method of solving standing wave vibration was suggested by Eisinger [57]. The method consists of using fin barriers welded to the heat exchanger tubes, forming a thin wall parallel to the flow direction and perpendicular to the direction of propagation of standing waves. The various forms of fin barriers are shown schematically in Figure 1.21.

1.8.4   CONCEPT OF HELICAL SPACERS

A novel idea for overcoming FIV in a closely spaced tube array is to introduce helical spacers inside the tube bank area [57]. This method is well suited to the initial design stage or for use in field modifications. The method is applicable both for staggered and for in-line tube layout, and spacer placement is shown schematically in Figure 1.22.

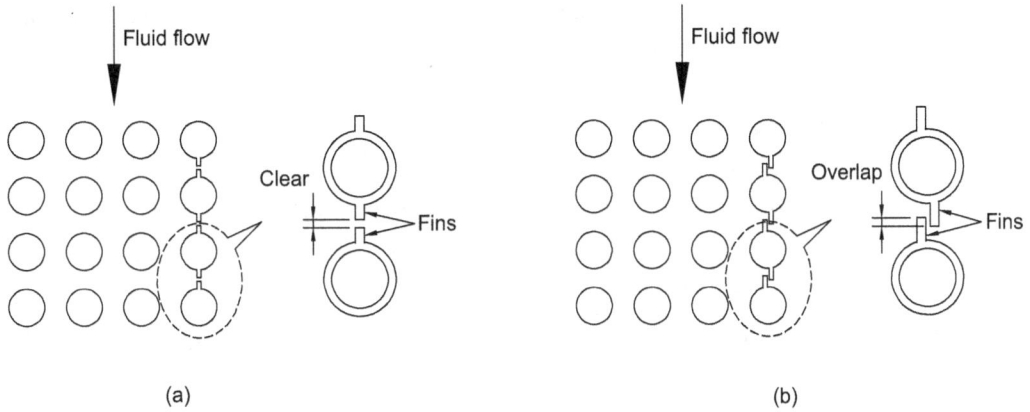

FIGURE 1.21 Arrangements of fin barriers for in-line array to solve the standing wave problem. (a) Inline layout and (b) staggered layout.

Source: From Eisinger, F.L., *Trans. ASME J. Pressure Vessel Technol.*, 102, 138, 1980.

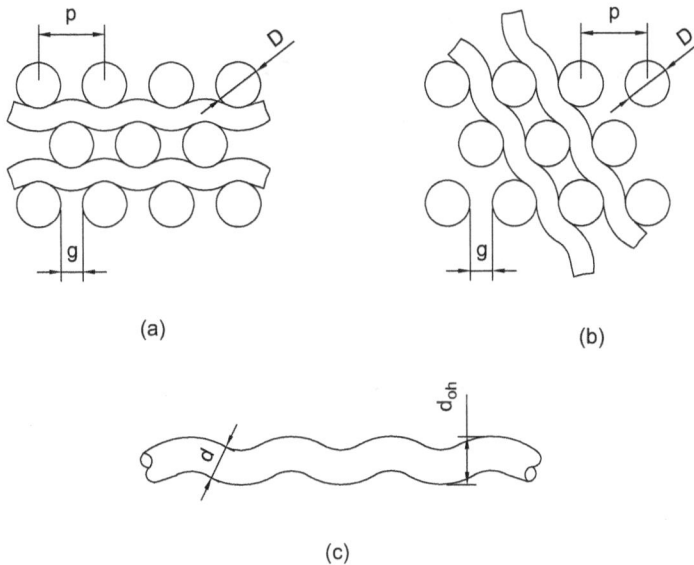

FIGURE 1.22 Helical spacers to overcome FIV of a staggered tube array.

Source: From Eisinger, F.L., *Trans. ASME J. Pressure Vessel Technol.*, 102, 138, 1980.

1.8.5 DETUNING

This has been already discussed.

1.8.6 REMOVAL OF TUBES

Selective removal of a few tubes at the displacement antinodes of the standing waves will eliminate the vortex shedding and thus avoid the coupling between vortex shedding and acoustic resonance. This also upsets the vortex shedding regularity at pressure nodes. Walker and Reising [80] and

Barrington [63,67] have observed that by removing 3%–10% of tubes near the center of the tube bank, acoustic resonance can be diminished or eliminated.

1.8.7 SURFACE MODIFICATION

Fouling of heat exchanger tubes by dirt and carbon soot has been associated with the reduction in sound level, and hence the deposition increases the acoustic damping [8,81].

1.8.8 IRREGULAR SPACING OF TUBES

Destroy the phasing of vortex streets by having irregular spacing of tubes. This will scramble the vortex shedding frequency. Zdravkovich and Nuttal [82] found the nonoccurrence of standing waves by displacing the susceptible tubes either laterally or longitudinally. However, this method is attractive only if the thermal performance is not affected.

1.8.9 CHANGE THE MASS FLOW RATE

Modify vortex shedding frequency by changing the mass flow rate. This may not be very attractive from a thermal performance point of view.

1.9 BAFFLE DAMAGE AND COLLISION DAMAGE

1.9.1 EMPIRICAL CHECKS FOR VIBRATION SEVERITY

From the prevalent pattern of tube failures occurring at the baffle supports and at the midspan, Thorngren [83] has attributed tube failures mainly to (1) baffle-type damage and (2) collision-type damage. He proposed two empirical correlations, called baffle damage number, and collision damage number to check the severity of tube damage due to FIV.

1.10 IMPACT AND FRETTING WEAR

Heat exchanger tubes are supported at regular intervals by passing through oversized holes in baffle plates (Figure 1.23). When the tube vibrates against the baffle plates, material can be fretted away from both the tube and the support plate as shown in Figure 1.24. If the fretting wear of tube is sufficiently large, the tubes will eventually rupture due to the fluid pressure [38]. When one considers that a design objective for a nuclear power plant steam generator or a heat exchanger is a 40-year codal life, even relatively small tube wear rates cannot be tolerated [6,38]. Hence in recent years, significant attention has been given to study the effect of tube-to-tube support plate clearances on tube vibration and wear, by (1) laboratory experiments; (2) theoretical simulations; and (3) analytical models. Vibration excitation mechanisms responsible for fretting wear are (1) vortex shedding; (2) turbulent buffeting; and (3) FEI. References [38,84–93] brought about a greater understanding of the tube-to-tube support plate interaction characteristics and resultant tube wear.

1.10.1 TUBE WEAR PREDICTION BY EXPERIMENTAL TECHNIQUES

Among the experimental techniques, the notable are that of Blevins [84,85] and Haslinger et al. [86]. The fundamental assumption of the Blevins [84] experimental fretting wear model is that fretting wear is the result of relative motion between the tube and the support plates. Blevins assumed that

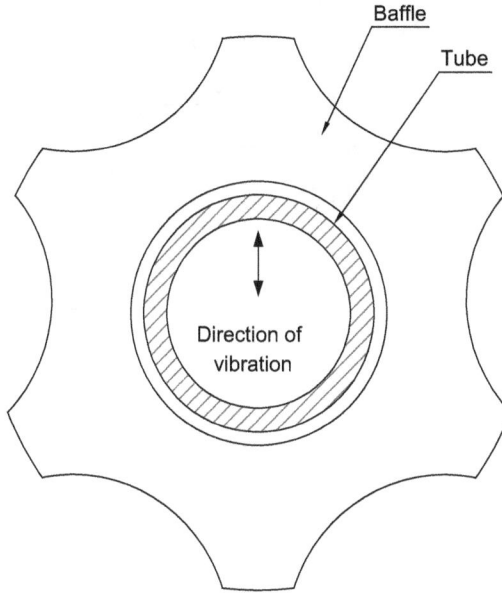

FIGURE 1.23 Vibrating tube in baffle hole.

the fretting wear is a function of parameters, such as (1) tube geometry; (2) baffle plate geometry; (3) tube and baffle plate material properties; (4) amplitude, frequency, and mode shape of tube vibration; (5) preload between the tube and support plate; and (6) nature of environment and operating temperature.

Blevins [94] conducted experiments to predict that wear induced by vibration on a heat exchanger model with alloy 800H and 2.25Cr–1Mo steel heat exchanger tubes loosely held supports in a helium environment at temperatures up to 650°C (1200°F). From the conclusions derived from his experiments, it is seen that (1) impact wear can be minimized by reducing the impact contact stresses below the fatigue allowable stress; (2) decreasing the vibration amplitude or baffle hole clearance markedly decreases wear rate and subsurface deformation at all temperatures; and (3) application of chromium coating both to the tube and the plate specimens greatly reduces the surface self-adhesion and surface deformation that would otherwise occur above 500°C (930°F). Below this temperature, a coating is probably not required.

1.10.2 THEORETICAL MODEL

Nonlinear finite-element modeling techniques to predict tube wear due to turbulence excitation, FEI, or vortex shedding are discussed in Refs. [38,89–93]. The wear parameters central for the theoretical models are (1) sliding distance; (2) contact time; (3) average contact time; (4) threshold crossings; and (5) work rate.

1.11 DETERMINATION OF HYDRODYNAMIC MASS, NATURAL FREQUENCY, AND DAMPING

FIV analysis of heat exchangers requires the knowledge of damping, hydrodynamic mass, and natural frequency of tube bundle. Their determination is discussed next.

(a)

(b)

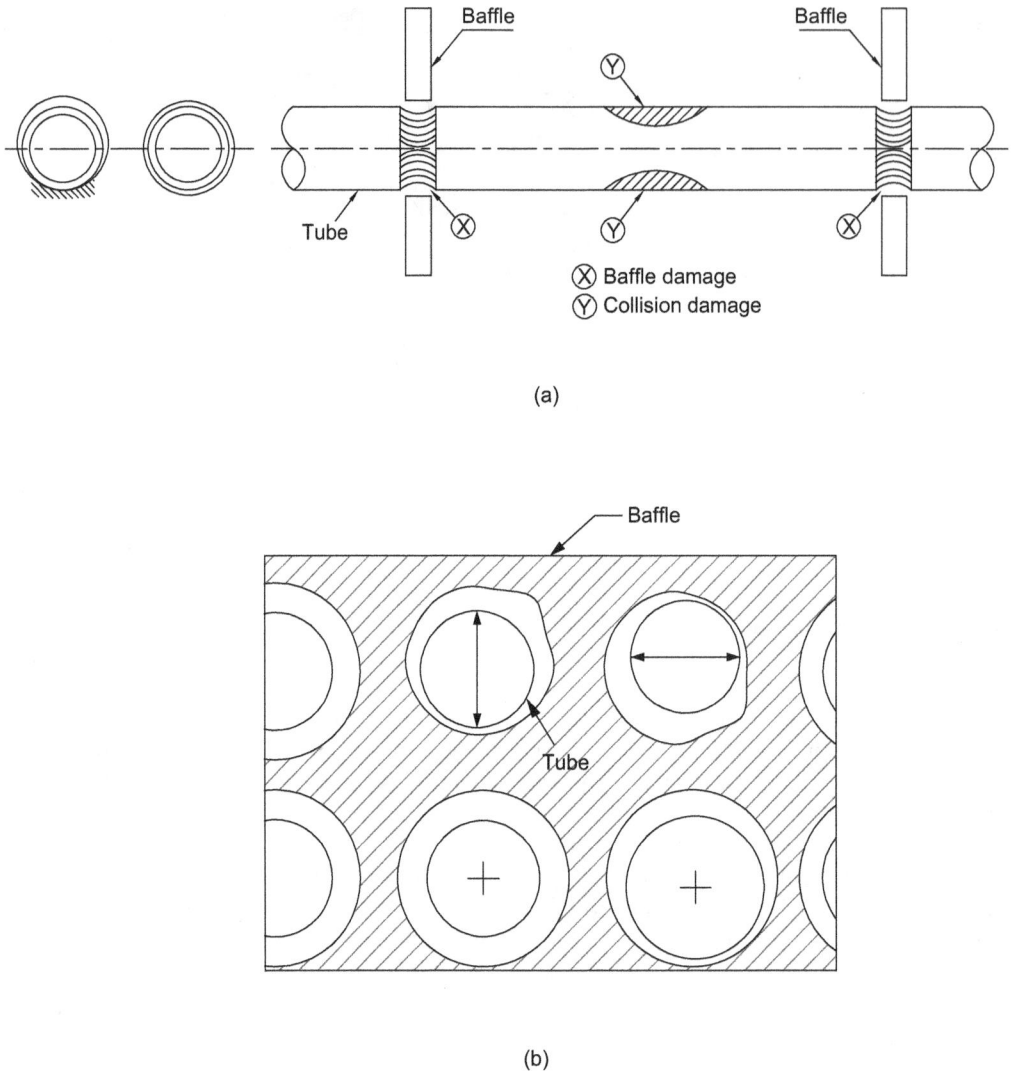

FIGURE 1.24 Fretting wear of tube.

1.11.1 ADDED MASS OR HYDRODYNAMIC MASS

During FIV, the vibrant tubes displace the shellside fluid. When the fluids involved are liquids or very dense gases, the inertia of the fluid will have a substantial effect on the natural frequency of the tubes. Hence, while calculating the natural frequency of the tube, the influence of the displaced fluid is taken care of by augmenting the mass of the vibrating tube by including hydrodynamic mass or added mass. The added mass is defined as the displaced fluid mass times an added mass coefficient C_m. Since the added mass augments the vibrating tube mass, the natural frequency of the tube will be reduced compared to when the tube is vibrating in vacuum or very low-density gas.

1.11.2 DETERMINATION OF ADDED MASS COEFFICIENT, C_M, FOR SINGLE-PHASE FLOW

The added mass coefficient is estimated either by the analytical method of Blevins [8] or from the experimental database of Moretti et al. [95].

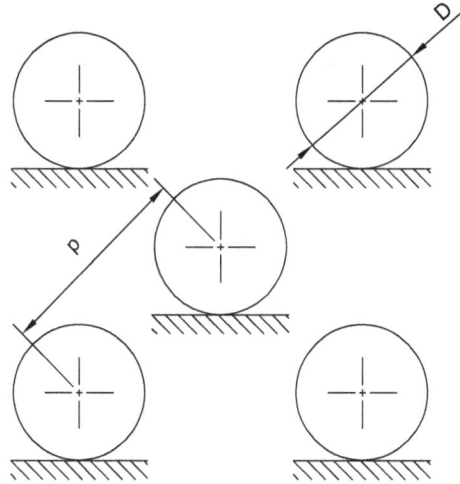

FIGURE 1.25 Tube arrangement for determining added mass coefficient.

Source: From Moretti, P.M. and Lowery, R.L., *Trans. ASME J. Pressure Vessel Technol.*, 98, 190 1976.

1.11.2.1 Blevins Correlation

Blevins [8] gave an analytical formula to determine added mass coefficient of a single flexible tube surrounded by an array of rigid tubes as shown in Figure 1.25. This is a reasonable approximation of the more complex case of all flexible cylinders. The expression for added mass coefficient C_m is given as

$$C_m = \frac{\left(D_e/D\right)^2 + 1}{\left(D_e/D\right)^2 - 1} \qquad (1.50)$$

where $D_e/D = ((1 + 0.5\ p/D)p/D)$ in which D_e is the equivalent diameter of a tube array. The equivalent diameter D_e is used to represent the confinement due to surrounding tubes.

1.11.2.2 Experimental Data of Moretti et al.

The experimental results of Moretti et al. [95] for a single flexible tube surrounded by rigid tubes in a hexangular array or in a square array with a pitch-to-diameter ratio from 1.25 to 1.50 are given in Figure 1.26. Their test results for C_m for four pitch-to-diameter ratios are given in Table 1.3. TEMA included this figure for determination of added mass coefficient [19].

1.11.3 Natural Frequencies of Tube Bundles

The tubes in a heat exchanger are slender beams and the most flexible members. They vibrate at discrete frequencies when they are excited. The natural frequency depends on their geometry, material properties like Young's modulus, density and material damping, tube-to-support interactions, etc. The lowest frequency at which the tubes vibrate is known as the natural frequency. The coincidence of the exciting frequency with the natural frequency is called resonance. The conventional STHEs have either straight tubes or U-tubes with baffle supports at intermediate points and fixed at their ends. The determination of the natural frequencies of the straight tubes and U-tubes is dealt with next.

TABLE 1.3
Added Mass Coefficient, C_m

Pitch-to-Diameter Ratio	C_m	
	Triangular Pitch	Square Pitch
1.25	1.1756	1.519
1.33	1.429	1.381
1.42	1.347	1.286
1.50	1.274	1.272

Source: Interpreted from Moretti, P.M. and Lowery, R.L., *Trans. ASME J. Pressure Vessel Technol.*, 98, 190–193, 1976.

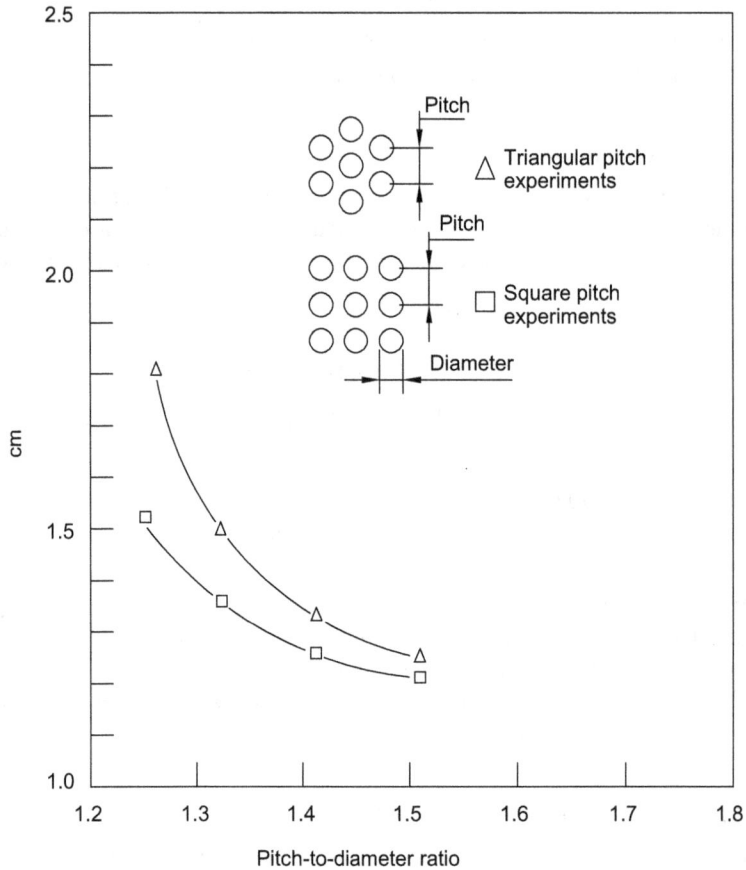

FIGURE 1.26 Determination of added mass coefficient.

Source: From Moretti, P.M. and Lowery, R.L., *Trans. ASME, J. Pressure Vessel Technol.*, 98, 190 1976.

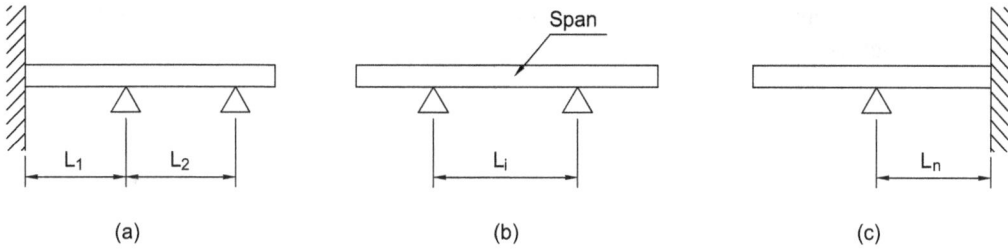

FIGURE 1.27 Heat exchanger tube span modeling—multispan uniform beam with fixed ends.

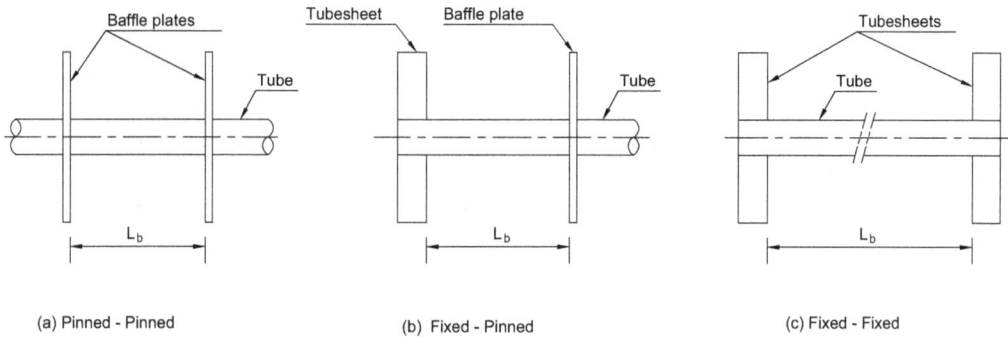

FIGURE 1.28 Tube span. (a) Pinned–pinned; (b) fixed–pinned; and (c) fixed–fixed end conditions.

1.11.3.1 Estimation of Natural Frequencies of Straight Tubes

Hand calculation method: though a rigorous analysis considering the multispan (Figure 1.27) modeling is possible, the state of the art on vibration prediction method recommends a simpler but a slightly conservative approach treating the individual spans separately [11]. For example, the simpler hand calculation methods of Timoshenko and Young [96] basically consider a tube as a continuous elastic beam on multiple supports with fixed end conditions and simply supported at the intermediate points (baffle locations). The individual spans (Figure 1.28) are each treated as a separate beam with its own end conditions (fixed–pinned, pinned–pinned, etc.). Natural frequencies of the individual spans are calculated, and the lowest frequency of the spans is taken as the representative figure for the whole tube and hence for the tube bundle.

The Timoshenko and Young method of treating the individual spans separately to find the natural frequency is recommended by Singh and Soler [11], Zukaskus [13], TEMA [19], and others. The frequency equation in a generalized form including the effect of axial stress is given by

$$f_n = \frac{1}{2\pi} \frac{\chi_\beta \lambda_n^2}{L_i^2} \left(\frac{EIg_c}{m} \right)^{0.5} \quad n = 1, 2, 3, \ldots \tag{1.51}$$

where
I = moment of inertia of the tube cross section, m⁴(ft⁴).

$$= \frac{\pi}{64} \left(D^4 - D_i^4 \right)$$

The expression for axial stress factor χ_B in terms of tube axial load F and Euler's critical buckling load F_{cr} is given by

$$\chi_B = \sqrt{1 \pm \frac{F_a}{F_{cr}}} \qquad (1.52)$$

where

$$F_a = A_t \sigma_t \qquad (1.53)$$

In Equation 1.53, F_a is tube axial load, A_t is tube metal cross-sectional area, and σ_t is tube longitudinal stress. In Equation 1.52, the plus sign is for tensile load and the negative sign is for compressive load included in the tubes. The expressions for frequency constant λ_n for the nth mode and for Euler's buckling load for various end conditions are given in Table 1.4, and the values of λ_n up to fifth mode are given in Table 1.5.

TEMA formula for straight span tube natural frequency

Fundamental natural frequency for various tube spans geometries is given by

$$f_n = 10.838 \frac{AC}{L_l^2} \left(\frac{EI}{m} \right)^{0.5} \qquad (1.54)$$

where A is tube axial stress multiplier and C is constant that depend on tube section end conditions. For values A and C refer to TEMA V-5.3.

Rigorous approaches to calculate straight-tube natural frequencies: these include the methods of Timoshenko and Young [96] and ABAQUS [97].

1.11.3.2 U-Tube Natural Frequency

The different span geometries for U-tube baffle spacing is shown in Figure 1.29 [19]. The expression for the lower bound of the natural frequency of U-tubes included in TEMA for various forms of tube symmetry and U-bend support configuration is of the form

TABLE 1.4
Frequency Constant, λ_n, and Euler Buckling Load, F_{cr}

	Pinned–Pinned	Fixed–Pinned	Fixed–Fixed
λ_n	$n\pi$	$0.25(4n+1)\pi$	$0.5(2n+1)\pi$
F_{cr}	$\pi^2 EI/L^2$	$2\pi^2 EI/L^2$	$4\pi^2 EI/L^2$

TABLE 1.5
Numerical Values for Frequency Constant λ_n up to Fifth Mode

End Conditions	First Mode	Second Mode	Third Mode	Fourth Mode	Fifth Mode
Pinned–pinned	3.1416	6.2832	9.4248	12.5664	15.7080
Fixed–pinned	3.9269	7.0686	1.2102	13.3518	16.4934
Fixed–fixed	4.7124	7.8540	1.9956	14.1372	17.2788

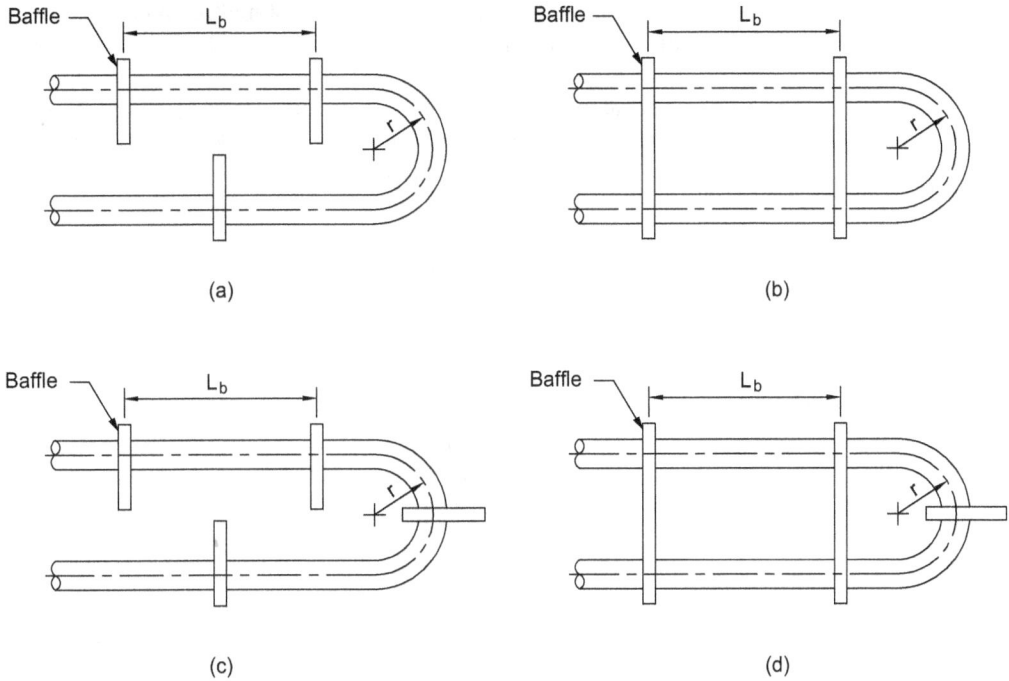

FIGURE 1.29 U-tube baffle support and U-bend support

$$f_{n,u} = 68.06 \frac{C_u}{r^2} \left(\frac{EI}{m} \right)^{0.5} \tag{1.55}$$

Detailed charts are given in TEMA [19] to determine C_u for various U-tube span geometries.

1.11.4 DAMPING

Damping limits the tube response when the tube is excited by any one of the excitation mechanisms. The higher the damping, the lower will be the tube response. Damping determines the critical flow velocity for FEI. The critical velocity increases with increasing damping. Damping in a vibrating system is due to several possible energy dissipation mechanisms. According to Pettigrew et al. [98], various energy dissipation mechanisms are as follows:

Internal or material damping.
Viscous damping.
Squeeze film damping in the baffle hole clearance.
Flow-dependent damping.
Frictional damping.
Joint damping at the tube and tubesheet interface.
Energy dissipated by tube impact on the support and by the traveling waves.

1.11.4.1 Determination of Damping

Determination of damping by experimental methods is very difficult. Empirical correlations for determining damping for gases and liquids are given next.

Gases: from the review and analysis of the available data on damping of multispan heat exchanger tubes in gases, Pettigrew et al. [98] recommended empirical correlations to calculate damping ratio ξ_n. The damping ratio is defined as the ratio of actual damping over critical damping. The logarithmic decrement of damping δ is equal to $2\pi\xi_n$. According to these authors, for gas flow, frictional damping and impacting are the dominant energy dissipation mechanisms, with baffle thickness a dominant parameter, whereas diametral clearance between the tube and baffle support is much less significant.

Liquids: according to Pettigrew et al. [99], the most important damping mechanisms in liquid flow are viscous damping (tube to fluid), squeeze film damping, and friction damping. They fitted experimental data with an empirical model considering viscous and squeeze film damping resulting in the following empirical correlation. The semiempirical expressions recommended to estimate damping for design purposes in single-phase fluids, gases, and liquids are given.

Gas flow on the shellside [98]:

$$\xi_n = \frac{0.7}{100}\frac{N-1}{2N}\frac{t_b}{12.7} \text{ for } t_b < 12.7 \tag{1.56}$$

$$\therefore \delta = 3.463\times10^{-3}\frac{N-1}{N}t_b \tag{1.57}$$

$$\xi_n = \frac{5}{100}\frac{N-1}{N}\left(\frac{t_b}{L_i}\right)^{0.5} \text{ for } t_b \geq 12.7 \tag{1.58}$$

$$\therefore \delta = 0.314\frac{N-1}{N}\left(\frac{t_b}{L_i}\right)^{0.5} \tag{1.59}$$

where
  t_b is the tube support thickness,
  L_i the characteristic span length,
  N the number of tube supports for the tube under consideration.

These expressions are valid for D = 12–15 mm, t_b = 6–25 mm, f = 20–600 Hz, and baffle hole clearances 0.4–0.8 mm.

TEMA included Equation 1.59 for all baffle thicknesses.
Liquid flow on the shellside: the determination of δ involves either dynamic (absolute) viscosity μ in the expression suggested in TEMA or kinematic viscosity ν in the expression of Pettigrew et al. [99]. The relation between them is

$$\text{Kinematic viscosity } \nu = \frac{\text{absolute viscosity } \mu}{\text{specific gravity}}$$

$$\therefore \mu = \nu \times \text{specific gravity} \tag{1.60}$$

The units are centistokes for v and centipoise for μ. *Note:* TEMA uses the symbol v instead of μ.

Pettigrew et al. [99] give the equation

$$\xi_n = \frac{\pi}{\sqrt{8}} \frac{1+(D/D_e)^3}{\left[1-D/D_e\right)^2\right]^2} \left(\frac{\rho_s D^2}{m}\right)\left(\frac{2v}{\pi f_n D^2}\right)^{0.5} + \left(\frac{N-1}{N}\right)\left(\frac{22}{f_n}\right)\left(\frac{\rho_s D^2}{m}\right)\left(\frac{t_b}{L_i}\right)^{0.6} \tag{1.61}$$

where

$$\frac{D_e}{D} = 1.7\frac{p}{D} \text{ for triangular arrays} \tag{1.62a}$$

$$= 1.9\frac{p}{D} \text{ for square arrays} \tag{1.62b}$$

Note: if ξ_n is less than 0.006, assume $\xi_n = 0.006$ [9] to account for friction damping.

TEMA included the following expressions for liquid flow. The damping parameter δ is the greater of δ_1 and δ_2:

$$\delta_1 = \frac{3.41D}{mf_n} \tag{1.63a}$$

$$\delta_2 = \frac{0.012D}{m}\left(\frac{\rho_s \mu}{f_n}\right)^{0.5} \tag{1.63b}$$

$$\therefore \delta = \max\{\delta_1, \delta_2\} \tag{1.64}$$

Note: use TEMA dimensional units, D = in., ρ_s = lbm/ft^3, μ = centipoise, and m = lbm/ft.

Fill the values of added mass coefficient, natural frequency, and damping parameter into the format given in Table 1.6.

1.12 NEW TECHNOLOGIES OF ANTIVIBRATION TOOLS

1.12.1 Nonsegmental Baffles

The nonsegmental baffle is a tube support device providing isotropic support to the tube against lateral vibration and permits axial flow of the shellside fluid.

1.12.2 Antivibration Tube Stakes

Tube vibration in the condenser can cause serious tube damage when increased steam flows are introduced or when retubing with thinner wall tubes or those with low modulus of elasticity. As its name implies, the antivibration stake (AVS) is used to protect tubes in tubular heat exchangers from vibrating vigorously inside the shellside space from fluid flow. The frequency of a tube span is inversely proportional to its length. In a manufactured heat exchanger where additional cross baffles cannot be added, the antivibration stakes (AVS) provide an effective means to reduce the

TABLE 1.6
Hydrodynamic Mass Coefficient, Tube Mass per Unit Length, Tube Natural Frequency, and Damping Parameter

1. Hydrodynamic mass coefficient, C_m
   (a) Blevins correlation, Equation 1.50

$$\frac{D_e}{D} = \left(1 + 0.5\frac{p}{D}\right)\frac{p}{D}$$

$$C_m = \frac{(D_e/D)^2 + 1}{(D_e/D)^2 - 1}$$

2. Tube mass per unit length

$$m = \frac{\pi D^2 C_m \rho_m}{4} + \frac{\pi D_i^2 \rho_i}{4} + \frac{\pi(D^2 - D_i^2)\rho}{2}$$

3. Tube natural frequency, f_n
   i. For straight tube (Equation 1.51)

$$L_i = \ ,\ A_i = ...,\ \sigma_t = ...,\ F = A_t \sigma_t,$$
$$F_{cr} = ...,\ \chi_B = ...,\ I = ...,\ \lambda_n = ...$$

$$f_n = \frac{1}{2\pi}\frac{\chi_B \lambda_n^2}{L_i^2}\left(\frac{EIg_c}{m}\right)^{0.5} \quad n = 1,2,3,...$$

   ii. U-tube as per TEMA
   $l_b/r = s/r =$
   $R =$
   $I =$
   $C_u =$

$$f_{n,u} = \frac{68.06 C_u}{R^2}\left[\frac{EI}{m}\right]^{0.5}$$

4 Damping parameter TEMA (Equation 1.63a and 1.63b)

$$\delta_1 = \frac{3.41D}{mf_n} \text{ and } \delta_2 = \frac{0.012D}{m}\left(\frac{\rho_s \mu}{f_n}\right)^{0.5}$$
$$\delta = \max\{\delta_1, \delta_2\}$$

If δ is less than 0.0377, then assume $\delta = 0.0377$.

(b) Moretti et al.
   Read from Figure 1.26 or from Table 1.3

Others
Gases

$$\delta = 3.463 \times 10^{-3}\frac{N-1}{N}t_b \text{ for } t_b < 12.7$$

$$\delta = 0.314\frac{N-1}{N}\left(\frac{t_b}{L_i}\right)^{0.5} \text{ for } t_b \geq 12.7$$

Liquid

$$\frac{D}{D_e} = (1.7p/D)^{-1} \text{ for triangular array}$$

$$= (1.9p/D)^{-1} \text{ for square array}$$

i. $\xi_n = \frac{\pi}{\sqrt{8}}\frac{1 + (D/D_e)^3}{\left[(1 - D/D_e)^2\right]^2}\left(\frac{\rho_s D^2}{m}\right)\left(\frac{2v}{\pi f_n D^2}\right)^{0.5}$

$$+ \left(\frac{N-1}{N}\right)\left(\frac{22}{f_n}\right)\left(\frac{\rho_s D^2}{m}\right)\left(\frac{t_b}{L_i}\right)^{0.6}$$

ii. $\delta = 2\pi\xi_n$

Note: TEMA dimensional units are E = psi, I = in.^{4}, m = lbm/ft, R = in.

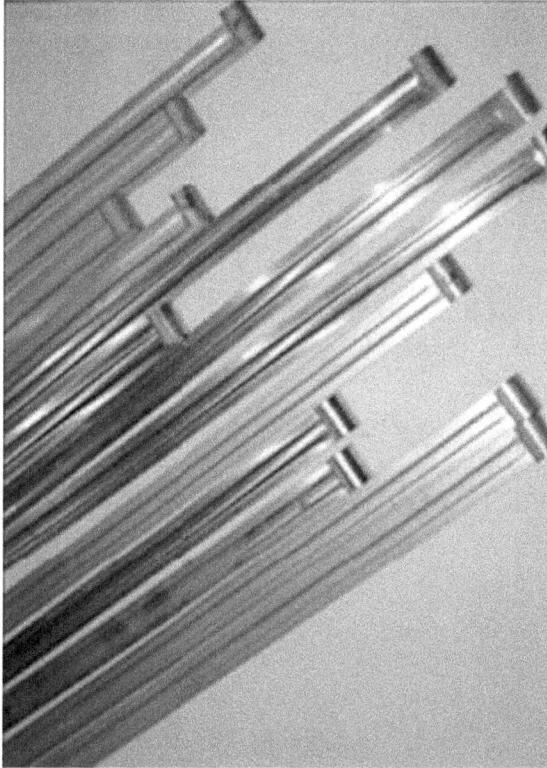

FIGURE 1.30 Antivibration tube stakes. (Courtesy of Holtec International, Marlton, NJ).

unsupported tube span and thus increase its natural frequency. An interference fit of the stake between tubes in the open lanes dampens each tube by pressing against the adjacent tubes. This dampening significantly reduces the fluid elastic instability that leads to tube failure. The clip design resists movement of the stakes during condenser operation, preventing them from backing out or working from midspan to a support plate. Addition of the AVS would add to increased shellside pressure drop, which should be considered by the heat exchanger designer. Holtec's antivibration tube stakes, as shown in Figure 1.30, prevent collisions between adjacent tubes in condenser and heat exchanger bundles that may be susceptible to FIVs. These tube stakes can be used in new heat exchanger bundles or installed in existing equipment to negate harmful vibrations occurring during a unit's operating life. The stakes are fabricated with a "Shepard's Crook" at one end, which locks onto the peripheral tubes to prevent any movement of the stakes during operation. These precision formed stakes do not require any lubrication or special tools for installation. Introduced in 1986, Holtec stakes have proved their effectiveness in numerous power plant condensers and a variety of tubular heat exchangers.

1.12.2.1 Cradle-Lock® Anti-Vibration Tube Stakes

American Efficiency Services (AES) introduced the Cradle-Lock® Anti-Vibration Tube Stake for the industry's solution for vibration induced tube failures in steam surface condensers and heat exchangers. The Cradle-Lock® design "cradles" the tube on both sides of the stake, which prevents horizontal vibration and limits vertical movement. Cradle-Lock® stakes are easier to install as they virtually eliminate the necessity of wiring, bolting, bands, or other parts [100].

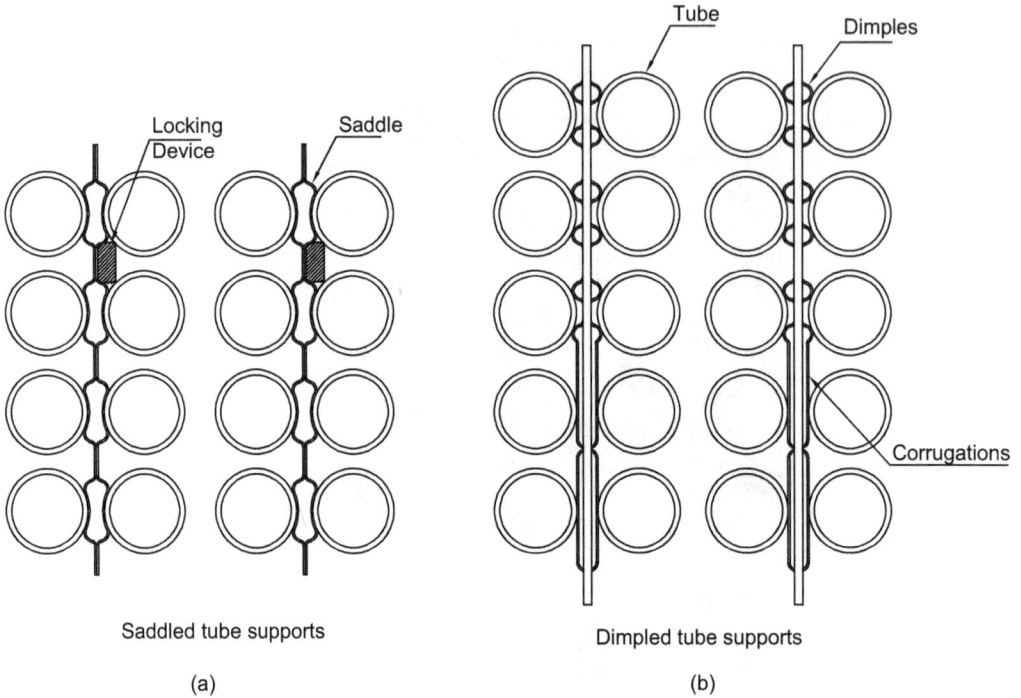

FIGURE 1.31   ExxonMobil Research and Engineering antivibration tube support-dimpled tube supports.

Source: Adapted from Amar W. and Ruzek, Z.F., Antivibration technologies for heat exchangers, ExxonMobil Research and Engineering and EPRI, August 31, 2009.

1.12.2.2   Antivibration Measures of ExxonMobil Research and Engineering

ExxonMobil Research and Engineering has developed a suite of new technologies/tools to overcome antivibration as given hereunder [101]:

  a. Dimpled tube supports (DTS™).
  b. Saddled tube supports (STS™).
  c. Slotted baffle exchangers (SBX™).

Dimpled tube support saddled tube supports are shown in Figure 1.31.

1.13   SOFTWARE PROGRAMS FOR ANALYSIS OF FIV

Proprietary programs of HTRI [30] and HTFS [32], among others, perform flow-induced vibration analysis of a single tube in a heat exchanger bundle. The program uses a rigorous structural analysis approach to calculate the tube natural frequencies for various modes and offers flexibility in the geometries it can handle. The analysis can be performed for straight or U-tubes. Results of flow-induced vibration analysis for each mode normally include (1) natural frequency of the tube; (2) maximum ratio of fluid gap velocity and critical velocity; and (3) maximum tube deflections due to vibration by vortex shedding or other mechanisms.

NOMENCLATURE

A	area of cross section of opening in the Helmholtz cavity resonator, m^2 (ft.^{2})
A_1	tube cross-sectional area, m^2 (ft.^{2})
C	speed of sound in the shellside medium, m/s (ft/s)
C_{eff}	effective speed of sound in the shellside fluid, m/s (ft/s)
C_L	lift coefficient for vortex shedding
C_m	added mass coefficient
C_n	reduced damping in n^{th} mode
$C_R(f)$	turbulence force coefficient
C_u	constant to determine U-tube natural frequency
D	tube outside diameter, m(ft)
D_B	baffle diameter, m(ft)
D_e	equivalent confinement diameter in a tube array $(1 + 0.5p/D)p$ to calculate C_m
D_e/D	1.7p/D for triangular array to calculate ξ_n
	1.9p/D for square array to calculate ξ_n
D_i	tube inside diameter, m(ft)
E	modulus of elasticity of the tube material, N/cm^2
F_a	axial load on tubes, N
F_{cr}	critical buckling load of the tube span analyzed, N
f	Helmholtz cavity resonator resonance frequency, Hz
f_a	acoustic resonance frequency, Hz
f_n	tube natural frequency, Hz
$f_{n,u}$	frequency of U-bend region, Hz
f_s	vortex shedding frequency, Hz
f_{tb}	turbulent buffeting frequency in a tube array, Hz
g	acceleration due to gravity, m/s^2 (ft/s^2)
g_c	proportionality constant,
	9.81 m/s^2
	1.0 in SI units and dimensionless
	32.174 lbmft/lbf s
I	moment of inertia of the tube cross section, m^4(ft^4)
j, n	vibration mode number; $j = n = 1$ is fundamental mode
i, j, k	acoustic modal indices
K	fluid elastic instability constant
L_a	characteristic dimension for acoustic standing wave, m(ft)
L	unsupported tube span length for FIV, m(ft)
L_b	U-tube section spacing between baffles, m(ft)
L_i	unsupported straight tube span for FIV or characteristic tube span length, m (ft.)
L_{BC}	central baffle spacing, m(ft)
L_{BI}	baffle spacing inlet region, m(ft)
L_{BO}	baffle spacing outlet region, m(ft)
L_e	characteristic tube length, m(ft)
L_p	longitudinal tube pitch, mm (ft)
L_t	total tube length, m(ft)
L_x, L_y, L_z	acoustic chamber dimensions, m(ft)
l_h	length of Helmholtz cavity resonator neck, m(ft)
M, M_n	modal mass, kg(pound)
M_g	molecular weight of shellside gas or vapor, kg mol(lbmol)
M	mass per unit length of tube $(m_a + m_c + m_s)$, kg(lb)
$m(x)$	mass per unit length of tube at a distance x along the tube axis, kg(lb)
m_a	fluid-added mass per unit length, kg(lb)
m_i	contained fluid mass per unit length, kg(lb)
m_t	structural mass per unit length, kg(lb)
N	number of tube spans for the tube being analyzed

N_w	number of solid baffles
n	mode number
p_s	shellside pressure, Pa (psi)
p_{rms}	rms value of shellside pressure, Pa (psi)
p	tube pitch, m (ft)
Q_g	volume flow rate of gas phase in two-phase flow, m³/s
Q_L	volume flow rate of liquid phase in two-phase flow,m³/s
r	U-tube bend radius, m (ft.)
R_c	universal gas constant, J/K/ mole(BTU·lbmol^{-1}°R^{-1}),
Re	Reynolds number based on the tube outside diameter, dimensionless, UD/v
r_h	Helmholtz cavity resonator neck radius, m(ft)
SPL	sound pressure level, dB
S_u	Strouhal number, dimensionless
T	absolute temperature of shellside gas, (K/°R)
T_p	transverse tube pitch, m(ft)
t	Time, s
t_b	baffle plate thickness, m(ft)
$U, U_{Bell,}$ $U_{Tinker,}$ U_{others}	reference crossflow velocity, m(ft)
U_{cr}	critical fluid velocity for fluid elastic excitation, m(ft)
$U(x)$	crossflow velocity distribution, m(ft)
U_∞	upstream velocity, m(ft)
V_h	volume of Helmholtz cavity resonator, m³(ft³)
v	kinematic viscosity of the fluid, m²/s, centistokes
X_1	longitudinal pitch ratio, L_p/D
X_p	pitch ratio = p/D
X_t	transverse pitch ratio, T_p/D
x	axial distance, m(ft)
$y(x)$	amplitude of vibration m(ft)
y_{rms}	rms amplitude of tube vibration, m(ft)
$\bar{y}^2_{rms}$	mean square amplitude of vibration, m(ft)
y_{max}	maximum tube response due to resonance due to vortex shedding, m(ft)
Z	shellside gas or vapor compressibility factor
λ_n,λ_i	frequency constant
σ	tube bank solidity, dimensionless
δ	logarithmic decrement of damping = $2\pi\xi_n$
ψ	Chen number
χ	nondimensional parameter
χ_B	factor to account for tube axial load on tube natural frequency
$\phi_n(x)$	mode constant for tube natural frequency, dimensionless
ξ_n	critical damping ratio
ρ	density of tube material, kg/m³ (lb/ft³)
ρ_i	density of tubeside fluid, kg/m³ (lb/ft³)
ρ_s	density of shellside fluid, kg/m³ (lb/ft³)
σ_T	shell longitudinal stress, +ve for tensile stress and −ve for compressive stress, Pa (psi)
τ	Arnold's slenderness ratio
γ	ratio of specific heat of gas at constant pressure to constant volume
μ	absolute viscosity, Pa.s (centipoises)

NOTES

1   EMbaffle® is a registered trademark of EMbaffle B.V., Alphen a/d Rijn, the Netherlands.
2   Twisted Tube is a registered trademark of Koch Heat Transfer Company, LP, and is registered in the United States and other countries worldwide.

REFERENCES

1. Wambsganss, M. W., Tube vibration and flow distribution in shell and tube heat exchangers, *Heat Transfer Eng.*, *8*, 62–71 (1987).

2. Paidoussis, M. P., Flow induced vibrations in nuclear reactors and heat exchangers: Practical experiences and state of knowledge, in *Practical Experiences with Flow Induced Vibration* (E. Naudascher and D. Rockwell, eds.), Springer-Verlag, Berlin, Germany, 1980, pp. 1–18.

3. Paidoussis, M. P., Fluid elastic vibration of cylinder arrays in axial and crossflow: State of art, *J. Sound Vib.*, *76*, 329–359 (1981).

4. Au-Yang, M. K., Blevins, R. D., and Mulcahy, T. M., Flow induced vibration analysis of tube bundles— A proposed Section III Appendix N nonmandatory code, *Trans. ASME J. Pressure Vessel Technol.*, *113*, 257–267 (1991).

5. Chen, S. S., Guidelines for the instability flow velocity tube arrays in crossflow, *J. Sound Vib.*, *93*(3), 439–455 (1984).

6. Weaver, D. S. and Fitzpatrick, J. A., A review of flow induced vibration in heat exchangers, in *International Conference on Flow Induced Vibration*, sponsored by BHRA, Bowness-on-Windermere, England, May 12–14, 1987, pp. 1–17.

7. Au-Yang, M. K., Flow induced vibration: Guidelines for design, diagnosis and troubleshooting of common power plant components, *Trans. ASME J. Pressure Vessel Technol.*, *107*, November, 326–334 (1985).

8. Blevins, R. D., *Flow Induced Vibration*, 2nd edn., Van Nostrand Reinhold, New York, 1990.

9. Sandifer, J. B., Guidelines for flow induced vibration prevention in heat exchangers, Welding Research Council Bulletin 372, Welding Research Council, New York, May 1992.

10. API 660, *Shell and Tube Heat Exchangers,* 9th edn., American Petroleum Institute, Washington, DC, 2015.

11. Singh, K. P. and Soler, S. I., *Mechanical Design of Heat Exchangers and Pressure Vessel Components*, Arcturus, Cherry Hill, NJ, 1984.

12. Gupta, J. P., *Fundamentals of Heat Exchanger and Pressure Vessel Technology*, Hemisphere, Washington, DC, 1984.

13. Zukauskas, A. A., Vibration of tubes in heat exchangers, in *High Performance Single Phase Heat Exchangers*, Hemisphere, Washington, DC, 1989, pp. 316–348.

14. Pettigrew, M. J. and Taylor, C. E., Fluid elastic instability of heat exchanger tube bundles: Review and design recommendations, *Trans. ASME J. Pressure Vessel Technol.*, *113*, 242–255 (1991).

15. Chen, S. S. and Jendrzejczyk, J. A., Stability of tube arrays in crossflow, *Nucl. Eng. Des.*, *75*(3), 351– 374 (1983).

16a. Chen, S. S., Some issues concerning fluid elastic instability of a group of circular cylinders in cross-flow. *Trans. ASME J. Pressure Vessel Technol.*, *111*, 507–518 (1989).

16b. Chen, S. S., *Flow Induced Vibration of Circular Cylindrical Structures*, Hemisphere, New York, 1987.

17. Shah, R. K., Flow induced vibration and noise in heat exchangers, *Proceedings of Seventh National Heat and Mass Conference*, Kharagpur, India, 1983, pp. 89–108.

18. Chenoweth, J. M. and Kistler, R. S., Tube vibration in shell and tube heat exchangers, *AIChE Symposium Series*, No. 174, *Heat Transfer: Research and Application*, vol. 74, 1978, pp. 6–14.

19. Standards of the Tubular Exchanger Manufacturers Association (TEMA), 10th edn., 2019, Tubular Exchanger Manufacturers Association, Inc., Tarrytown, NY.

20. ASME Boiler and Pressure Vessel Code*, Section VIII, Division 2, Pressure Vessels—Alternative Rules*, American Society of Mechanical Engineers, New York, 2021.

21. Paidoussis, M. P. and Besancon, P., Dynamics of arrays of cylinders with internal and external axial flow, *J. Sound Vib.*, *76*(3), 361–379 (1981).

22. Goyder, H. G. D., The structural dynamics of tube bundle vibration problem, in *International Conference on Flow Induced Vibrations*, Bowness on Windermere, England, UK, May 12–14, 1987.

23. Gorman, D. J., Experimental study of peripheral problems in heat exchangers and steam generator, Paper V.F4/g, *Proceedings of the Fourth International Conference on Structural Mechanics in Reactor Technology*, San Francisco, CA, 1977.

24. Chen, Y. N. and Weber, M., Flow induced vibrations in tube bundle heat exchangers with crossflow and parallel flow, in *Flow Induced Vibrations in Heat Exchangers, Proceedings of a Symposium Sponsored by the ASME*, New York, 1970, pp. 57–77.

25. Chen, Y. N., Flow induced vibration and noise in tube bank heat exchangers due to Von Karman streets, *Trans. ASME J. Eng. Ind.*, *90*, February, 134–146 (1968).

26. Owen, P. R., Buffeting excitation of boiler tube vibration, *J. Mech. Eng. Sci.*, *4*, 431–439 (1965).

27. Tinker, T., Shellside characteristics of shell and tube heat exchangers—A simplified rating system for commercial heat exchangers, *Trans. ASME, 80*, 36–52 (1958).

28. Bell, K. J., Delaware method for shellside design, in *Heat Transfer Equipment Design* (R. K. Shah, E. C. Subbarao, and R. A. Mashelkar, eds.), Hemisphere, Washington, DC, 1988, pp. 145–166.

29. Palen, J. W. and Taborek, J., Solution of shellside flow pressure drop and heat transfer by stream analysis method, *Chemical Engineering Progress Symposium Series*, No. 92, *Heat Transfer*, Philadelphia, PA, vol. 65, 1969, pp. 53–63.

30. Xvib, HTRI Xchanger Suite 6.0, Heat Transfer Research, Inc., Alhambra, CA.

31. Aspen Shell & Tube Mechanical, Heat Transfer and Fluid Flow Services, Oxon, UK.

32. Sing, K. P. and Soler, A. I., *HEXDES User Manual*, Arcturus, Cherry Hill, NJ, 1984. B-JAC International, Inc., Midlothian, VA.

33. Fitz-Hugh, J. S., Flow induced vibration in heat exchangers, in *International Symposium on Vibration Problems in Industry*, Keswick, UK., April 10–12, 1973.

34. Zukauskas, A. and Katinas, V., Flow induced vibration in heat exchanger tube banks, in *Proceedings of Symposium of Practical Experiences with Flow Induced Vibration* (E. Naudascher and D. Rockwel, eds.), Springer, New York, 1980, pp. 188–196.

35. Weaver, D. S., Fitzpatrick, J. A., and El Kashlan, M. L., Strouhal numbers of heat exchanger tube arrays in crossflow, in *ASME Symposium on Flow Induced Vibration* (S. S. Chen, J. C. Simonis, and Y. S. Shin, eds.), Chicago, IL, Vol. 104, 1986.

36. Pettigrew, M. J. and Gorman, D. J., Vibration of heat exchanger tube bundles in liquid and two phase crossflow, in *Flow Induced Vibration Guidelines* (P. Y. Chen, ed.), PVP Vol. 52, ASME, New York, 1981.

37. Pettigrew, M. J. and Gorman, D. J., Vibration of heat exchanger components in liquid and two phase crossflow, presented at *British Nuclear Society International Conference on Vibration in Nuclear Plant*, Keswick, UK., May 9–12, 1978.

38. Rao, M. S. M., Gupta, G. D., Eisinger, F. L., Hibbitt, H. D., and Steininger, D. A., Computer modelling of vibration and wear of multispan tubes with clearances at the supports, presented at *Flow Induced Vibration Conference*, sponsored by BHRA, Bowness-Windermere, England, May 12–14, 1987, pp. 437–448.

39. Weaver, D. S. and Grover, L. K., Crossflow induced vibration in a tube bank—Turbulent buffeting and fluid elastic instability, *J. Sound Vib.*, *59*(2), 277–294 (1978).

40. Soper, B. M. H., The effect of tube layout on the fluid elastic instability of tubes bundle in cross-flow, *Trans. ASME J. Heat Transfer, 105*, 744–750 (1983).

41. Weaver, D. S. and Schneider, W., The effect of flat bar supports on the crossflow induced response of heat exchanger U-tubes, *Trans. ASME J. Eng. Power*, *105*, 775–781 (October, 1983).

42. Connors, H. J. Jr., Fluid elastic vibration of tube arrays excitation by crossflow, in *Flow Induced Vibration in Heat Exchangers* (D. D. Reill, ed.), ASME, New York, 1970, pp. 42–56.

43. Lever, J. H. and Weaver, D. S., A theoretical model for fluid elastic instability in heat exchangers tube bundles, *Trans. ASME J. Pressure Vessel Technol.*, *104*, 147–158 (1982).

44. Lever, J. H. and Weaver, D. S., On the stability of heat exchanger tube bundles: Part 1—Modified theoretical model, *J. Sound Vib.*, *107*(3), 375–392 (1986).

45. Lever, J. H. and Weaver, D. S., On the stability of heat exchanger tube bundles: Part 2—Numerical results and comparison with experiments, *J. Sound Vib.*, *107*(3), 393–411 (1986).

46. Tanaka, H. and Takahara, S., Fluid elastic vibration of a tube array in crossflow, *J. Sound Vib.*, *77*, 19–37 (1981).

47. Tanaka, H. and Takahara, S., Unsteady fluid dynamic force on tube bundle and its dynamic effects on vibration, in *Flow Induced Vibration of Power Plant Components*, ASME Special Publication No. PVP 141 (M. K. Au-Yang, ed.), ASME, New York, 1980, pp. 77–92.

48. Price, S. J. and Paidoussis, M. P., Fluid elastic instability of a double row of circular cylinders subject to a crossflow, *Trans. ASME J. Vib. Acoust. Stress Reliab. Des.*, *105*, 59–66 (1983).

49. Price, S. J. and Paidoussis, M. P., An improved mathematical model for the stability of cylinder rows subjected to crossflow, *J. Sound Vib.*, *97*(4), 615–640 (1984).

50. Whiston, G. S. and Thomas, G. D., Whirling instabilities in heat exchanger tube arrays, *J. Sound Vib.*, *81*(1), 1–31 (1982).

51. Chen, S. S., Instability mechanisms and stability criteria of a group of circular cylinders subjected to crossflow, Part I, ASME Paper No. 81-Det 21, ASME, New York.

52. Chen, S. S., Instability mechanisms and stability criteria of group of circular cylinders subjected to crossflow, Part II: Numerical Results and Discussions, ASME Paper No. 81-Det 22, New York.

53. Chen, S. S., Instability mechanism and stability criteria of a group of cylinders subjected to crossflow, Part I: Theory; Part II: Numerical results and discussions, *J. Vibration Acoustics Stress Reliability Designs*, *105*, 253–260 (1983).

54. Chen, S. S. and Jendrzejczyk, J. A., Experiments and analysis of instability of tube rows subjected to liquid crossflow, *Trans. ASME J. Appl. Mech.*, *49*, 704–709 (1982).

55. Chen, S. S. and Jendrzejczyk, J. A., Experiments on fluid elastic instability in tube banks subjected to liquid crossflow, *J. Sound Vib.*, *78*, 355–381.

56. Chen, S. S., Flow induced vibration of circular cylindrical structures, Argonne National Laboratory, Report No. ANL-CT-85-51, Argonne, IL.

57. Eisinger, F. L., Prevention and cure of flow induced vibration problems in tubular heat exchangers, *Trans. ASME J. Pressure Vessel Technol.*, *102*, 138–145 (1980).

58. Blevins, R. D., Review of sound induced by vortex shedding from cylinders, *J. Sound Vib.*, *92*(4), 455–470 (1984).

59. Blevins, R. D., Acoustic modes of heat exchanger tube bundles, *J. Sound Vib.*, *109*(1), 19–30 (1986).

60. Blevins, R. D. and Bressler, M. M., Acoustic resonance in heat exchanger tube bundles, Part 1: Physical nature of the phenomenon, *Trans. ASME J. Pressure Vessel Technol.*, *109*, 275–281 (1987).

61. Blevins, R. D. and Bressler, M. M., Acoustic resonance in heat exchanger tube bundles—Part II: Prediction and suppression of resonance, *Trans. ASME J. Pressure Vessel Technol.*, *109*, 282–288 (1987).

62. Blevins, R. D., Acoustic resonance in heat exchanger tube bundles, Welding Research Bulletin No. 389, Welding Research Council, New York, 1994.

63. Barrington, E. A., Experience with acoustic vibration in tubular heat exchangers, *Chem. Eng. Prog.*, *69*(7), 62–68 (1973).

64. Parker, R., Acoustic resonance in passages containing bank of heat exchanger tubes, *J. Sound Vib.*, *57*, 245–260 (1978).

65. Burton, T. E., Sound speed in a heat exchanger tube bank, *J. Sound Vib.*, *71*, 157–160 (1980).

66. Grotz, B. J. and Arnold, F. R., Flow induced vibration in heat exchangers, Department of Mechanical Engineering Tech. Rep. No. 31, DTIC No. 104568, Stanford University, Stanford, CA, 1956.

67. Barrington, E. A., Cure exchanger acoustic vibration, *Hydrocarbon Process.*, 193–198 (1978).

68. Fitzpatrick, J. A., The prediction of flow induced noise in heat exchanger tube arrays, *J. Sound Vib.*, *99*, 425–435 (1985).

69. Ziada, S. and Oengoren, A., Flow induced acoustical resonances of in-line tube bundles, *Sulzer Tech. Rev.*, 45–47 (1990).

70. Rae, G. J. and Murray, B. G., Flow induced acoustic resonances in heat exchangers, *International Conference on Flow Induced Vibrations*, Bowness-on-Windermere, England, May 12–14, 1987, pp. 221–231.

71. Fitzpatrick, J. A. and Donaldson, J. A., Effects of scale on parameters associated with flow induced noise in tube arrays, in *Symposium on Flow Induced Vibrations* (M. P. Paidoussis, ed.), American Society of Mechanical Engineers, New York, pp. 243–250.

72. Ziada, S., Oengoren, A., and Buhlmann, E. T., Acoustical resonance in tube arrays, in *International Symposium on Flow Induced Vibration and Noise* (M. P. Paidoussis et al., eds.), American Society of Mechanical Engineers, New York, 1988, pp. 219–254.

73. Kerner Smith, H., Vibration in nuclear reactor heat exchangers—One manufacturers viewpoint, in *Flow Induced Vibration in Heat Exchangers, Proceedings of ASME* (D. D. Reill, ed.), 1987, pp. 1–7.

74. Halle, H., Chenoweth, J. M., and Wambsganss, M. W., Shellside water flow induced tube vibration in heat exchanger configurations with tube pitch to diameter ratio of 1.42, *Trans. ASME J. Pressure Vessel Technol.*, *111*, 441–449 (1989).

75. Saunders, E. A. D., Shellside flow induced tube vibration, in *Heat Exchangers: Design, Selection and Construction*, Addison Wesley Longman, Reading, MA, pp. 245–273, 1989.

76. Halle, H., Chenoweth, J. M., and Wambsganss, M. W., Flow induced vibration in shell and tube heat exchanger with double segmental baffles, in *Heat Transfer 1986, Proceedings of Eighth International Heat Transfer Conference*, San Francisco, CA, Vol. 6, 1986, pp. 2763–2768.

77. Gentry, C. C., Young, R. K., and Small, M. W., RODbaffle heat exchanger thermal-hydraulic predictive methods, in *Heat Transfer 1982*, Vol. 6, Hemisphere, Washington, DC, 1982, pp. 197–202.

78. Boyer, R. C. and Pase, G. K., The energy saving NESTS concept, *Heat Transfer Eng.*, *2*, 19–27 (1980).

79. Byrne, K. P., The use of porous baffles to control acoustic vibration in crossflow tubular heat exchangers, *Trans. ASME, J. Heat Transfer*, *105*, 751–757 (1983).

80. Walker, E. M. and Reising, G. F. S., Flow induced vibrations in crossflow heat exchangers, *Chem. Process Eng.*, *49*, 95–103 (1968).

81. Rogers, J. D. and Penterson, C. A., Predicting sonic vibration in crossflow heat exchangers—Experience and model testing, American Society of Mechanical Engineers, New York, Paper 77-FE-7, 1977.

82. Zdravkovich, M. and Nuttal, J. A., On elimination of aerodynamic noise in staggered tube bank, *J. Sound Vib.*, *34*, 173–177 (1974).

83. Thorngren, J. T., Predict exchanger tube damage, *Hydrocarbon Process.*, Vol. 49, April, 129–131 (1970).

84. Blevins, R. D., Fretting wear of heat exchanger tubes, Part I: Experiments, *Trans. ASME J. Eng. Power*, *101*, 625–629 (1979).

85. Blevins, R. D., Fretting wear of heat exchanger tubes, Part II: Models, *Trans. ASME J. Eng. Power*, *101*, 630–633 (1979).

86. Haslinger, K. H., Martin, M. L., and Steininger, D. A., Pressurized water reactor steam generator tube wear prediction utilizing experimental techniques, in *Proceedings of the International Conference on Flow Induced Vibration*, Bowness-on-Winderemere, England, May 12–14, 1987, pp. 437–448.

87. Ko, P. L. and Basista, H., Correlation of support impact force and fretting wear for heat exchanger tube, *Trans. ASME J. Pressure Vessel Technol.*, *106*, 69–77 (1984).

88. Cha, J. H., Wambsganss, M. W., and Jendrzejcyk, J. A., Experimental study of impact/fretting wear in heat exchanger tubes, Argonne National Laboratory Publication ANL-85-38, Argonne, IL., 1985.

89. Hofmann, P. J., Schettler, T., and Steininger, D. A., Pressurized water reactor steam generator tube fretting and fatigue wear characteristics, in *Proceedings of the ASME Pressure Vessel and Piping Conference*, ASME 86PVP-2, Chicago, IL, July 1986.

90. Fisher, N. J., Olsen, M., Rogers, R. J., and Ko, P. L., Simulation of tube-to-support dynamic interaction of heat exchange equipment, *Trans. ASME J. Pressure Vessel Technol.*, *111*, 378–384 (1989).

91. Eisinger, F. L., Rao, M. S. M., Steininger, D. A., and Haslinger, K. H., Numerical simulation of fluid elastic vibration and comparison with experimental results, in *Flow Induced Vibration and Wear Book* No. H00632, 1991, pp. 101–111. Reprinted from PVP Vol. 206.

92. Rao, M. S. M., Steininger, D. A., and Eisinger, F. L., Computer simulation of vibration and wear of steam generator and tubular heat exchanger tubes, in *International Symposium on Advanced Computers for Dynamics and Design '89*, Tsuchiura, Japan, Mechanical Division, JSME. September 6–8, 1989.

93. Rao, M. S. M., Steininger, D. A., and Eisinger, F. L., Numerical simulation of fluid elastic vibration and wear of multispan tubes with clearances at supports, in *Flow Induced Vibration in Heat Transfer Equipment* (J. M. Paidoussis, S. S. Chenoweth, J. R. Chen, Stenner, and W. J., Bryan, eds.), Vol. 5, Book No. G00445, 1988, pp. 235–251.

94. Blevins, R. D., Vibration induced wear of heat exchanger tubes, *Trans. ASME J. Eng. Mater. Technol.*, *107*, 61–67 (1985).

95. Moretti, P. M. and Lowery, R. L., Hydrodynamic inertia coefficients for a tube surrounded by rigid tube, *Trans. ASME J. Pressure Vessel Technol.*, *98*, 190–193 (1976).

96. Timoshenko, S. P. and Young, D. H., Vibration problems in engineering, in *Vibration of Elastic-Bodies*, Van Nostrand Reinhold, Toronto, Canada, 1965, Chapter 5.

97. SIMULIA 6.2 (Formerly known as ABAQUS), DassaultSystèmesSimulia Corp., 2005, *Providence, RI.*
98. Pettigrew, M. J., Goyder, H. G. D., Qiao, Z. L., and Axisa, F., Damping of multispan heat exchanger tubes, Part 1: In gases, in *Flow Induced Vibration 1986* (S. S. Chen, J. C. Simon, and Y. S. Shin, eds.), ASME PVP Vol. 104, ASME, New York, 1986, pp. 81–87.
99. Pettigrew, M. J., Rogers, R. J., and Axisa, F., Damping of multispan heat exchanger tubes, Part 2: In liquids, in *Flow Induced Vibration 1986* (S. S. Chen, J. C. Simon, and Y. S. Shin, eds.), ASME, New York, PVP Vol. 104, pp. 89–98.
100. https://americanefficiency.com/anti-vibration-tube-stakes/
101. Amar W. and Ruzek, Z. F, Antivibration technologies for heat exchangers, ExxonMobil Research and Engineering and EPRI, August 31, 2009.

SUGGESTED READINGS

Bolleter, U. and Blevins, R. D., Natural frequencies of finned heat exchanger tubes, *J. Sound Vibration, 80,* 367–371 (1982).

Chenoweth, J. M., Flow induced vibration phenomena, in *Heat Exchanger Design Handbook* (E. U. Schlunder, ed.), Vol. 4, Hemisphere, Washington, DC, 1983, Section 4.6.4.

Gorman, D. J., *Free Vibration Analysis of Beam and Shafts*, John Wiley & Sons, New York, 1975.

Lowery, R. L. and Moretti, P. M., Natural frequencies and damping of tubes on multiple supports, ed. J.C. Chen, *AIChE Symposium Series*, No. 174, *Heat Transfer: Research and Application*, vol. 74, 1978, pp. 1–5.

MacDuff, J. N. and Felgar, R. P., Vibration design charts, *Trans. ASME, 79,* 1459–1474 (1957).

Moretti, P. M., Fundamental frequencies of U-tubes in tube bundles, *Trans. ASME J. Pressure Vessel Technol., 107,* 207–209 (1985).

Naguyen, D. C., Lester, T., Good, J. K., Lowery, R. L., and Moretti, P. M., Lowest natural frequencies of multiply supported U-tubes, *Trans. ASME J. Pressure Vessel Technol., 106,* 414–416 (1984).

2 Fouling

Fouling is defined as the formation of undesired deposits on heat transfer surfaces, which impede the heat transfer and increase the resistance to fluid flow, resulting in higher pressure drop. Industrial heat exchangers rarely operate with nonfouling fluids. Low-temperature cryogenic heat exchangers are perhaps the only exception [1]. The growth of the deposits causes the thermohydraulic performance of heat exchanger to degrade with time. Fouling affects the energy consumption of industrial processes, and it can also decide the amount of extra material required to provide extra heat transfer surface employed in heat exchangers to compensate for the effects of fouling. In addition, where the heat flux is high, as in steam generators, fouling can lead to local hot spots and ultimately it may result in mechanical failure of the heat transfer surface [2]. Bott [3], Chenoweth [4], and Epstein [5] give general and specific information on fouling. Figure 2.1 shows a fouled heat exchanger.

2.1 EFFECT OF FOULING ON THE THERMOHYDRAULIC PERFORMANCE OF HEAT EXCHANGERS

Effects can include the following:

1. The fouling layers on the inside and the outside surfaces are known generally to increase with time as the heat exchanger is operated. Since the fouling layers normally have lower thermal conductivity than the fluids or the conduction wall, they increase the overall thermal resistance. For the common case of a cylindrical tube with flow and fouling at both the inside and outside surface represented in Figure 2.2, the total resistance R_T to heat transfer between the two streams is given by

$$R_T = \frac{1}{U_o A_o} = \frac{1}{h_i A_i} + \frac{R_i}{A_i} + \frac{\Delta t_w}{k_w A_m} + \frac{R_o}{A_o} + \frac{1}{h_o A_o} \tag{2.1}$$

where the symbols have the usual meaning, and R_i and R_o are fouling resistance on the inside and outside surfaces, respectively. The overall heat transfer coefficient in fouled conditions may be expressed in terms of the overall heat transfer coefficient in clean condition and an overall fouling resistance R_f involving the terms related to fouling:

$$\frac{1}{U_{o,f}} = \left[\frac{1 A_o}{h_i A_i} + \frac{\Delta t_w A_o}{k_w A_m} + \frac{1}{h_o} \right] + \left[R_i \frac{A_o}{A_i} + R_o \right] \tag{2.2a}$$

DOI: 10.1201/9781003352068-2

FIGURE 2.1(a,b) Fouled shell and tube heat exchangers. (Courtesy of Merusonline, MERUS GmbH, Sindelfingen, Germany, Also MERUS California LLC.)

$$= \frac{1}{U_{o,c}} + R_F \tag{2.2b}$$

or

$$R_f = \frac{1}{U_{o,f}} - \frac{1}{U_{o,c}} \tag{2.3}$$

2. There is an increase of the surface roughness, thus increasing frictional resistance to flow, and fouling blocks flow passages; due to these effects, the pressure drop across the heat exchanger increases.
3. Fouling may create a localized environment where corrosion is promoted.
4. Fouling will reduce the thermal effectiveness of heat exchangers, which in turn affect the subsequent processes or will increase the thermal load on the system.
5. An additional goal becomes prevention of contamination of a process fluid or product.

2.2   COSTS OF HEAT EXCHANGER FOULING

In addition routine maintenance costs, the presence of fouling on heat exchange surfaces causes additional costs due to the following reasons:

1. Increased capital expenditure due to oversizing.
2. Energy losses associated with poor performance of the equipment.
3. Treatment cost to lessen corrosion and fouling.
4. Costs towards periodical manual/chemical/mechanical cleaning and environmental considerations due to effluent discharge.
5. Lost production due to maintenance schedules.

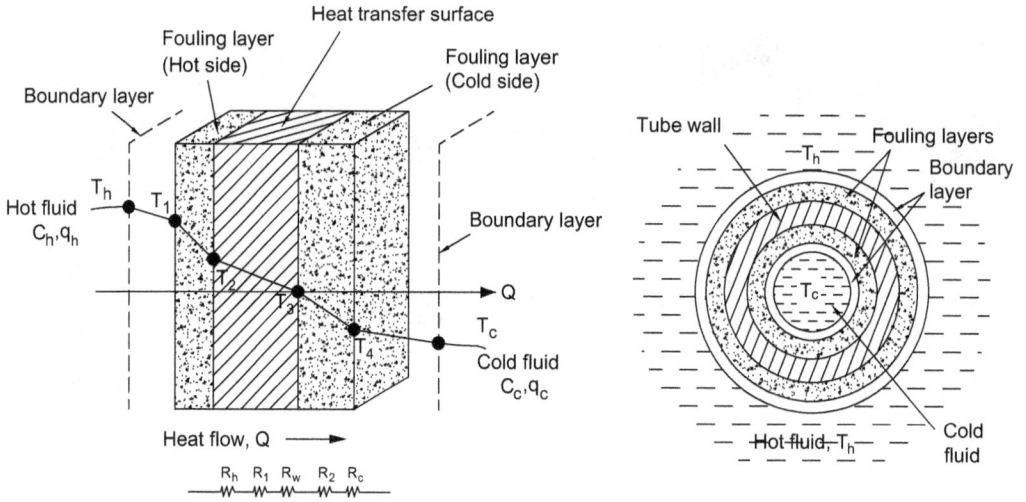

(a) Plane wall with fouling layers
    and boundary layers.

(b) Fouled tube surface.

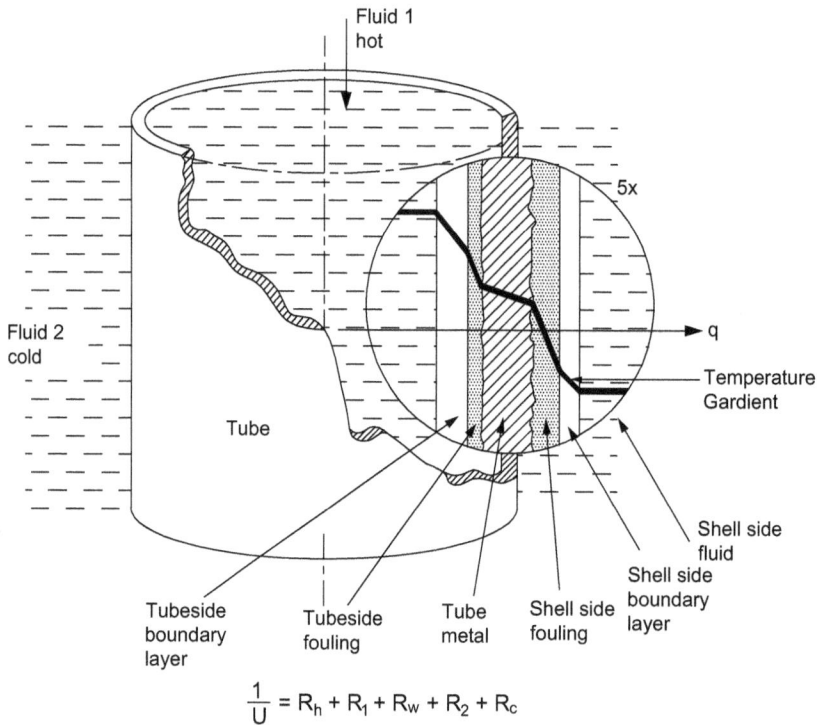

$$\frac{1}{U} = R_h + R_1 + R_w + R_2 + R_c$$

(c) Thermal resistances of fouled tube surface.

FIGURE 2.2 Illustration of fouling deposits build up on heat transfer surface(Schematic).

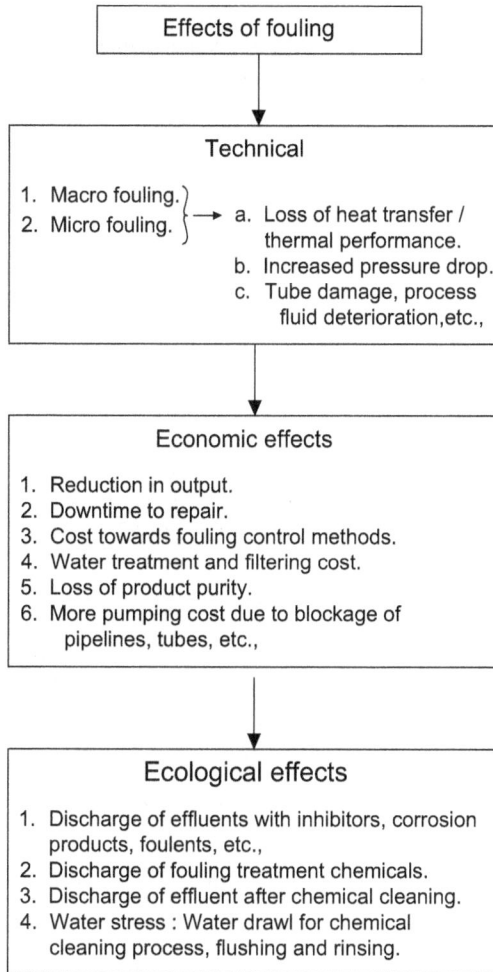

FIGURE 2.3 The effects of fouling.

Economic considerations should be among the most influential parameters for determining appropriate allowances for fouling. It is important to determine a strategy as to whether first cost, operating and maintenance cost, or total cost over a period of years is the objective.

The effects of fouling is shown in Figure 2.3

2.2.1 OVERSIZING

While sizing a heat exchanger, it is a normal practice to oversize the heat transfer surface area to account for fouling, and the oversizing is normally of the order of 20%–40%.

2.2.2 ADDITIONAL ENERGY COSTS

Since fouling reduces the heat transfer rates, additional energy is expended to increase the heat transfer rate; fouling also increases the pressure drop across the core and hence more pumping power is required to meet the heat load.

In the refining industry, the heat exchangers used for energy recovery suffer a progressive heat transfer efficiency loss due to fouling. The immediate consequences of this loss are a major consumption of energy in the furnaces and, in certain cases, the need to reduce throughput to compensate the low efficiency of the preheat train.

2.2.3 TREATMENT COST TO LESSEN CORROSION AND FOULING

The formation of fouling deposits on heat transfer surfaces necessitates periodical cleaning, which costs money for cleaning materials and process, personnel required, and recently environmental problems to discharge the effluents.

2.2.4 LOST PRODUCTION DUE TO MAINTENANCE SCHEDULES AND DOWN TIME FOR MAINTENANCE

Periodical cleaning requires plant shutdown and hence the unavailability of system for productive purposes. Critical industries that cannot afford for downtime and loss in production will maintain standby units, which again raises additional capital costs for spares.

2.3 FOULING CURVES/MODES OF FOULING

The amount of material deposited per unit area m_f is related to the fouling resistance R_f and the density of the foulant ρ_f, thermal conductivity k_f, and thickness of the deposit x_f by the following equation [1]:

$$m_f = \rho_f x_f = \rho_f k_f R_f \tag{2.4}$$

The incidence of fouling with reference to time is normally defined by the following three modes [1]:

1. Linear mode—increase of m_f (or R_f) with time t.
2. Falling rate mode—the rate of deposition decreases with increasing time.
3. Asymptotic mode—the value of m_f (or R_f) is not time variant after initial fouling.

The representation of various modes of fouling with reference to time is known as a fouling curve. Typical fouling curves are shown in Figure 2.4. This figure also shows deterioration of U due to onset of fouling and U value when automatic tube cleaning system is installed (U_{ATC})-schematic. The linear modes and falling rate modes may be the incidence of fouling in the early stages of asymptotic behavior. The asymptotic mode is of particular interest to heat exchangers, since the incidence of fouling raises the possibility of continued operation of the equipment without additional fouling. The time delay period, t_d, is time required for the formation of initial fouling substrata. Smooth and nonwetting surfaces like glass and Teflon will extend the delay period. Figure 2.4 also shows the falling rate of the overall heat transfer coefficient with incidence of time but its increase with an automatic tube cleaning system instituted in the system.

2.4 STAGES OF FOULING

For all the categories of fouling, the successive number of events that commonly occur in most situations are up to five. They are initiation of fouling, transport to surface, attachment to surfaces, removal from surfaces, and aging of deposit [5].

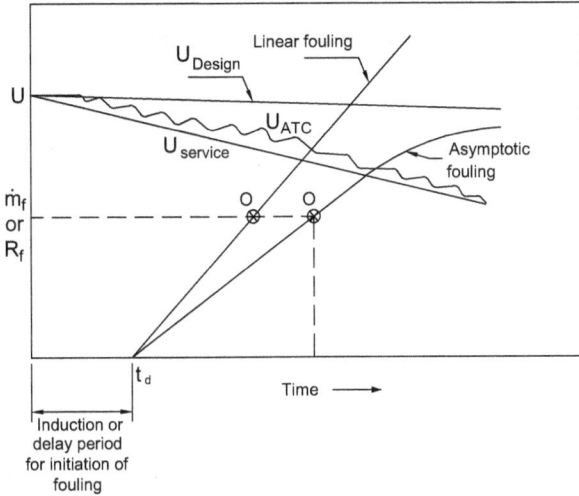

Note :   O = Begin initial cleaning.

U_{ATC} = Overall heat transfer coefficient with automatic tube cleaning.

U_{Design} = Theoretical overall heat transfer coefficient.

$U_{Service}$ = Heat transfer coefficient in service.

$\dot{m}_f$ = Fouling deposition rate.

R_f = Resistance due to fouling.

t_d = Time delay for fouling initiation.

FIGURE 2.4   Fouling curves. (Ideal).

Source: Modified from Collier, J.G., Heat exchanger fouling and corrosion, in *Heat Exchangers: Thermal-Hydraulic Fundamentals and Design*, Kakac, S., Bergles, A.E., and Mayinger, F., eds., Hemisphere, Washington, DC, pp. 999–1011, 1982.

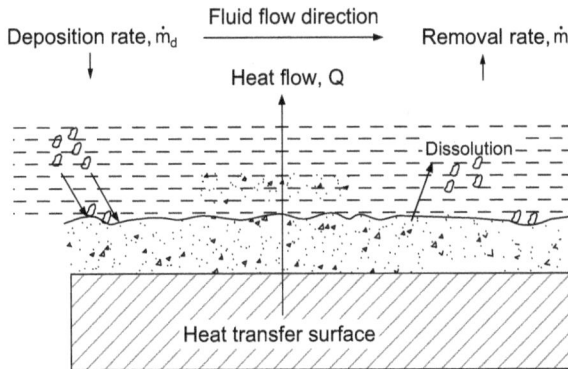

FIGURE 2.5   Basic fouling process (Schematic representation).

2.5   FOULING MODEL

Fouling is usually considered to be the net result of two simultaneous processes: a deposition process and a removal (re-entrainment process) [6]. A schematic representation of fouling process model is shown in Figure 2.5. The net rate of fouling can be expressed in terms of the rate of deposition of fouling mass ($\dot{m}_d$) and the rate of removal ($\dot{m}_r$) from the heat transfer surface. Thus,

rate of fouling = rate of deposition of fouling mass − rate of removal from the heat transfer surface

$$\frac{dm_f}{dt} = \dot{m}_d - \dot{m}_r \tag{2.5}$$

One of the earliest models of fouling was that by Kern and Seaton [7]. In this model, it was assumed that $\dot{m}_d$ remained constant with time t but that $\dot{m}_r$ was proportional to m_f and therefore increased with time to approach $\dot{m}_d$ asymptotically. Thus $\dot{m}_d = \beta m_f$; then integration of Equation 2.5 from the initial condition $m_f = 0$ at $t = 0$ gives

$$m_f = m_f^* \left(1 - e^{-\beta t}\right) \tag{2.6}$$

where

m_f^* is the asymptotic value of m_f
$\beta = 1/t_c$

The time constant t_c represents the average residence time for an element of fouling material at the heating surface. Equation 2.7 can be expressed in terms of fouling resistance R_f at time t in terms of the asymptotic value R_f^* by

$$R_f = R_f^* \left(1 - e^{-\beta t}\right) \tag{2.7}$$

2.6 MECHANISMS OF FOULING

It is of great importance to understand the fouling mechanisms in principle, as they will indicate the causes and conditions of fouling and hence give clues how fouling can be minimized. Based on the mechanism of fouling, Epstein [5] classifies fouling into six types:

1. Particulate fouling.
2. Reaction fouling.
3. Corrosion fouling.
4. Precipitation fouling.
5. Biological fouling.
6. Solidification fouling.

A possible fouling process model is shown in Figure 2.6

2.6.1 PARTICULATE FOULING

Particulate fouling may be defined as the accumulation of particles suspended in the process streams onto the heat transfer surfaces. This type of fouling includes sedimentation of settling under gravitation as well as deposition of colloidal particles by other deposition mechanisms onto the heat transfer surfaces or gas-side fouling of waste gas heat recovery heat exchangers, etc. Various forms of particulate fouling are as follows:

1. Fouling that occurs in once-through cooling-water systems using sea, river, and lake water containing mud, silt, and sediments. They are capable of depositing in low-flow areas.
2. Gas-side fouling: gas-side fouling is mostly by particulate fouling by dirty gas streams, airborne contaminants, etc. In some cases, gas-side fouling may also be accompanied by corrosion, particularly when condensation of corrosive acids take place from combustion gases. Gas-side fouling is discussed later.

2.6.2 CHEMICAL REACTION FOULING (POLYMERIZATION)

Deposits formed by chemical reactions at the heat transfer surface in which the surface material itself is not a reactant are known as chemical reaction fouling. Polymerization, cracking, and coking

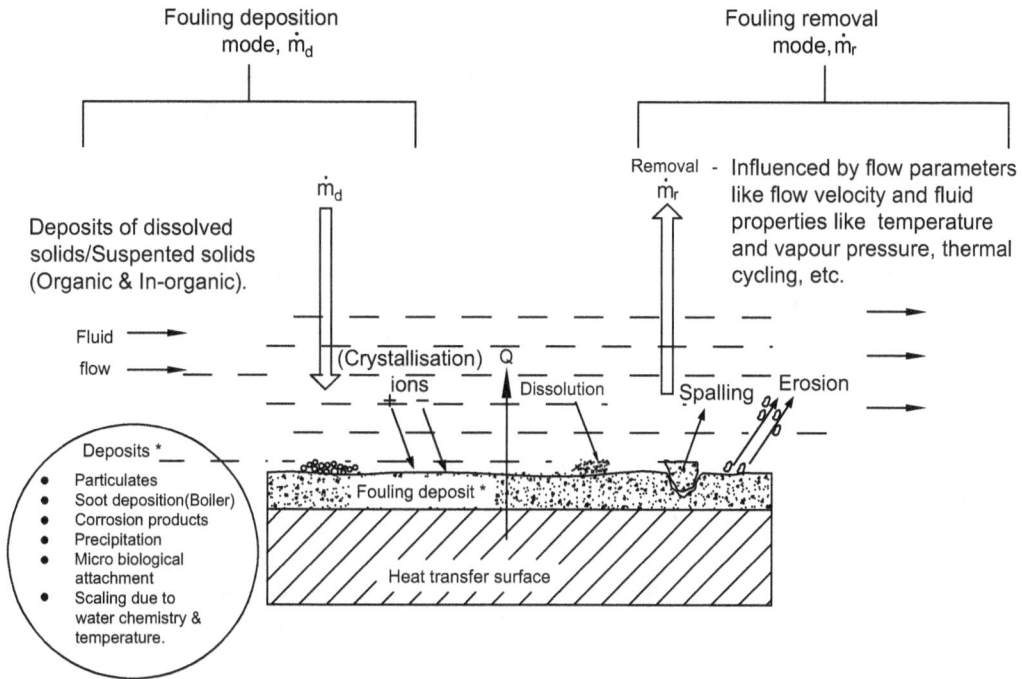

FIGURE 2.6 Schematic representation of detailed fouling process.

Source: Modified from O'Callaghan, M., Fouling of heat transfer equipment: Summary review, in *Heat Exchangers: Thermal-Hydraulic Fundamentals and Design*, S. Kakac, A. E. Bergles, and F. Mayinger, eds., Hemisphere, Washington, DC, pp. 1037–1047, 1982.

of hydrocarbons are prime examples of reaction fouling. The factors likely to affect reaction fouling include the following [3,5]:

Temperature is the most sensitive variable. It is usual that below a certain surface temperature polymerization does not initiate, but increases rapidly above that. The reaction rate is related to the temperature by the following Arrhenius law:

$$K_r = A_{cr} e^{E/RT_s} \tag{2.8}$$

where
  E is the activation energy,
  R is the gas constant,
  A_{cr} is the rate constant,
  T_s is the heat transfer surface temperature.

The presence of most sulfur compounds, nitrogen compounds, and the presence of trace elements (metallic impurities) such as Mo and Va in hydrocarbon streams significantly increases the fouling rates.

Composition of the process stream, including contaminants and, especially, oxygen ingress will affect reaction fouling.

Prevention measures for chemical reaction fouling should include the following [3,5]:

1. Avoidance of feed contact with air or oxygen by nitrogen blanketing.
2. Elimination or reduction of unsaturates, which are particularly high in cracked stocks.

3. The use of antioxidation additives that inhibit the polymerization reaction, along with steps taken to minimize oxygen ingress.
4. Use of additives known as metal coordinators, which react with the trace elements and prevent them from functioning as fouling catalysts. Other additives recommended are corrosion inhibitors and dispersion agents [3].

2.6.3 Corrosion Fouling

Corrosion fouling is due to the deposition of corrosion products on heat transfer surfaces. In this category of fouling process, the heat transfer surface material itself reacts to produce corrosion products, which foul the heat transfer surface. The most common forms of this type of fouling are material loss due to general thinning, iron oxide on carbon steel tubes in cooling-water systems, and fouling of soldered radiator tube ends on the water side by solder bloom corrosion. Measures such as the use of inhibitors, cathodic protection, and surface treatment such as passivation of stainless steel will minimize corrosion and hence corrosion fouling.

2.6.4 Crystallization or Precipitation Fouling

This type of fouling mostly takes place in cooling-water systems, when water-soluble salts, predominantly calcium carbonates, become supersaturated and crystallize on the tube wall to form scaling. Such scaling occurs because many of the dissolved salts in water exhibit inverse solubility effects, a condition that reverses the normal solubility (increasing with temperature) into one that decreases with temperature. Thus an inverse solubility solution will crystallize when heated (e.g., cooling water), while normal solubility salts will crystallize when cooled. Chemical additives can be helpful to reduce fouling problems due to crystallization and freezing in a number of ways. Broadly there are four groups of chemicals to control crystallization [3]: distortion agents, dispersants, sequestering agents, and threshold chemicals.

2.6.4.1 Modeling for Scaling

According to Hasson [8], scaling is due to diffusion of calcium and carbonate ions from the bulk of the fluid, followed by crystallization of $CaCO_3$ on the hot wall surfaces. Their model for predicting $CaCO_3$ scaling rates is given by

$$m_s = K_R \left[\left(Ca^{2+} \right)_i \left(CO_3^{2-} \right)_i - K_{sp} \right] \tag{2.9}$$

where
m_s is the scale deposition rate (kg/m$^2 \cdot$ s),
K_R the constant for crystallization rate,
K_{sp} the solubility product of $CaCO_3$ (mol/m^3)^2.

The principle of fouling and the factors promoting scaling are discussed in the section on cooling-water corrosion at the end of this chapter.

Scaling is discussed further in Section 2.32.

2.6.5 Biological Fouling

The attachment of microorganisms (bacteria, algae, and fungi) and macro-organisms (barnacles, sponges, fishes, seaweed, etc.) on heat transfer surfaces where the cooling water is used in as drawn condition from river, lake, sea and coastal water, etc., is commonly referred to as biological fouling.

Concentration of microorganisms in cooling-water systems may be relatively low before problems of biofouling are initiated. For open recirculating systems, bacteria concentrations of the order of 1 × 10^5 cells/mL and fungi of 1 × 10^3 cells/mL may be regarded as limiting values [3]. Corrosion due to biological attachment to heat transfer surfaces is known as microbiologically influenced corrosion (MIC). MIC is discussed in detail in Chapter 3. The techniques that can be effective in controlling biological fouling include the following:

1. Select materials that possess good biocidal properties.
2. Mechanical cleaning techniques like upstream filtration, air bumping, backflushing, passing brushes, sponge rubber balls, grit coated rubber balls, and scrapers [4].
3. Chemical cleaning techniques that employ biocides, such as chlorine, chlorine dioxide, bromine, ozone, surfactants, pH changes, and/or salt additions.
4. Thermal shock treatment by application of heat, or deslugging with steam or hot water.
5. Techniques like ultraviolet radiation.

2.6.6 SOLIDIFICATION FOULING OR FREEZING FOULING

The freezing of a liquid or of higher-melting constituents of a multicomponent solution on a subcooled heat transfer surface is known as solidification fouling. Notable examples include frosting of moisture in the air, freezing of cooling water in low-temperature processes, and paraffin wax deposition during cooling of hydrocarbon streams. There are various remedies for dealing with duties where solidification occurs on the product side [3]:

1. Do not include very large fouling resistances in the design. This will result in an oversized unit, which presents problems in the clean condition.
2. Use concurrent flow instead of counterflow.

2.6.7 FREEZING FOULING

Freezing fouling may occur where the temperature in the region of the heat transfer surface is reduced to below the freezing point of the fluid being processed. The deposition of wax from waxy hydrocarbons by cooling is often considered to represent freezing fouling, but it is probably better defined as crystallization fouling.

2.6.8 MIXED MECHANISMS OF FOULING

Although six mechanisms of fouling have been briefly described, it is rare for practical heat exchanger fouling to be the result of a single mechanism. In most process streams where fouling occurs, two or probably more mechanisms are involved. In cooling-water systems, it is likely that, in addition to microorganisms, the circulating water will contain dissolved solids, suspended particulate matter and also aggressive chemicals. The accumulated deposit on the equipment surfaces may therefore contain microorganisms, particles deposition, scale and products of corrosion [3]. In fouling associated with combustion gas, the fouling on heat exchanger external surfaces may be due to particle deposition, chemical reactions, corrosion and dew-point corrosion.

2.7 PARAMETERS THAT INFLUENCE INCIDENCE OF FOULING

Many operational and design variables have been identified as having most pronounced and well-defined effects on fouling. These variables are reviewed here in principle to clarify the fouling problems and because the designer has an influence on their modification. These parameters include the following:

1. Properties of fluids and their propensity for fouling.
2. Surface temperature.
3. Velocity and hydrodynamic effects.
4. Tube material.
5. Fluid purity and freedom from contamination.
6. Surface roughness.
7. Suspended solids.
8. Placing the more fouling fluid on the tubeside.
9. Shellside flow.
10. Type of heat exchanger.
11. Heat exchanger geometry and orientation.
12. Equipment design.
13. Seasonal temperate changes.
14. Heat transfer processes like sensible heating, cooling, condensation, vaporization, etc.
15. Shell and tube heat exchanger with improved shellside performance.

2.7.1 PROPERTIES OF FLUIDS AND USUAL PROPENSITY FOR FOULING

The most important consideration is the fluid and the conditions conducive for fouling. At times a process modification can result in conditions that are less likely to cause fouling.

2.7.2 TEMPERATURE

A good practical rule to follow is to expect more fouling as the temperature rises. This is due to a "baking on" effect, scaling tendencies, increased corrosion rate, faster reactions, crystal formation and polymerization, and loss in activity by some antifoulants [9]. Lower temperatures produce slower fouling buildup, and usually deposits that are easily removable [4]. However, for some process fluids, low surface temperature promotes crystallization and solidification fouling. For those applications, it is better to use an optimum surface temperature to overcome these problems. For cooling water with a potential to scaling, the desired maximum surface temperature is about 140°F (60°C). Biological fouling is a strong function of temperature. At higher temperatures, chemical and enzyme reactions proceed at a higher rate with a consequent increase in cell growth rate [10]. According to Mukherjee [10], for any biological organism, there is a temperature below which reproduction and growth rate are arrested and a temperature above which the organism becomes damaged or killed. If, however, the temperature rises to an even higher level, some heat-sensitive cells may die.

2.7.2.1 Effect of Temperature on Fouling

Low temperatures will

1. reduce the effects of chemical reaction and corrosion rate since rates of reaction are generally temperature sensitive; high temperatures favor accelerated reactions;
2. reduce the activity of micro- and macro-organisms;
3. in cooling-water system, reduce the opportunity for supersaturation conditions to occur for inverse solubility salts and hence propensity for scaling will be in control.

High temperatures will

1. reduce the incidence of biofouling where the temperature is above that for optimum growth of microorganisms;
2. avoid conditions that could lead to freezing fouling.

2.7.3 Velocity and Hydrodynamic Effects

Hydrodynamic effects such as flow velocity and shear stress at the surface influence fouling. Within the pressure drop considerations, the higher the velocity, the higher the thermal performance of the exchanger and lesser the fouling. Uniform and constant flow of process fluids past the heat exchanger favors less fouling. Foulants suspended in the process fluids will deposit in low-velocity regions, particularly where the velocity changes quickly, as in heat exchanger water boxes and on the shellside [9]. Higher shear stress promotes dislodging of deposits from surfaces. Maintain relatively uniform velocities across the heat exchanger to reduce the incidence of sedimentation and accumulation of deposits.

2.7.4 Tube Material

The selection of tube material is significant to deal with corrosion fouling.

Carbon steel is corrosive but least expensive.

Copper exhibits biocidal effects in water. Environmental protection limits the use of copper in river, lake, and ocean waters, since copper is poisonous to aquatic life.

Noncorrosive materials such as titanium and nickel will prevent corrosion, but they are expensive and have no biocidal effects.

Glass, graphite, and Teflon tubes often resist fouling and/or improve cleaning.

Although the construction material is more important to resist fouling, surface treatment by plastics, vitreous enamel, glass, and some polymers will minimize the accumulation of deposits [3].

2.7.5 Impurities

Seldom are fluids pure. Intrusion of minute amounts of impurities can initiate or substantially increase fouling. They can either deposit as a fouling layer or act as catalysts to the fouling processes [4]. For example, chemical reaction fouling or polymerization of refinery hydrocarbon streams is due to oxygen ingress and/or trace elements such as Va and Mo. In crystallization fouling, the presence of small particles of impurities may initiate the deposition process by seeding. The properties of the impurities form the basis of many antifoulant chemicals. Sometimes impurities such as sand or other suspended particles in a cooling water may have a scouring action, which will reduce or remove deposits [3].

2.7.6 Surface Roughness

The surface roughness is supposed to have the following effects [3]:

1. The provision of "nucleation sites" that encourage the laying down of the initial deposits.
2. The creation of turbulence effects within the flowing fluid and, probably, instabilities in the viscous sublayer.

Better surface finish has been shown to influence the delay of fouling and ease cleaning. Similarly, nonwetting surfaces delay fouling. Rough surfaces encourage particulate deposition. After the initiation of fouling, the persistence of the roughness effects will be more a function of the deposit itself [3]. Even smooth tubes may become rough in due course due to scale formation, formation of corrosion products, or erosion.

2.7.7 Suspended Solids

Suspended solids promote particulate fouling by sedimentation or settling under gravitation onto the heat transfer surfaces. Since particulate fouling is velocity dependent, prevention is achieved if stagnant areas are avoided. For water, high velocities (above 1 m/s) help prevent particulate fouling. Often it is economical to install an upstream filtration.

2.7.8 Placing More Fouling Fluid on the Tubeside of STHE

As a general guideline, the fouling fluid is preferably placed on the tubeside for ease of cleaning. Also, there is less probability for low-velocity or stagnant regions on the tubeside.

2.7.9 Shellside Flow

Velocities are generally lower on the shellside than on the tubeside, less uniform throughout the bundle, and limited by flow-induced vibration. Zero- or low-velocity regions on the shellside serve as ideal locations for the accumulation of foulants. If fouling is expected on the shellside, then attention should be paid to the selection of baffle design. Segmental baffles have the tendency for poor flow distribution if spacing or baffle cut ratio is not in correct proportions. Too low or too high a ratio results in an unfavorable flow regime that favors fouling and possible fouling effect is shown in Figure 2.7.

2.7.10 Low-Finned Tube Heat Exchanger

There is a general apprehension that low Reynolds number flow heat exchangers with low-finned tubes will be more susceptible to fouling than plain tubes. Fouling is of little concern for finned surfaces operating with moderately clean gases. According to Silvestrini [11], fin type does not affect the fouling rate, but the fouling pattern is affected for waste heat recovery exchangers.

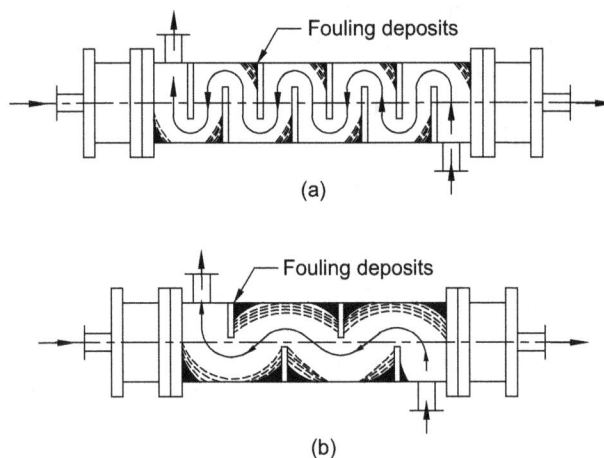

FIGURE 2.7 Effect of baffle spacing and baffle cut on fouling: (a) moderate baffle spacing and baffle cut and (b) wide baffle spacing and large baffle cut. Note: dark areas represent stagnant areas with heavy fouling.

2.7.11 Heat Transfer Augmentation Devices

Fouling is an important consideration in whether enhanced surface tubes should be used on heat exchangers to enhance the performance. Experimental evidence favors most of the heat transfer augmentation devices for improved heat transfer without penalty due to fouling. Fouling aspects of heat transfer augmentation devices are covered in the chapter on heat transfer augmentation (Chapter 8, Heat Exchangers: Classification, Selection, and Thermal Design).

2.7.12 Gasketed Plate Heat Exchangers (PHE)

High turbulence, absence of stagnant areas, uniform fluid flow, and the smooth plate surface reduce fouling, and the need for frequent cleaning. Hence the fouling factors required in plate heat exchangers are normally 10%–25% of those used in shell and tube heat exchangers. Fluid flow pattern between two adjacent plates of a PHE is shown Figure 2.8.

2.7.13 Spiral Plate Exchangers

High turbulence and scrubbing action minimize fouling on the spiral plate exchanger. This permits the use of low fouling factors. Figure 2.9 shows fluid flow pattern between two adjacent plates of a SPHE.

2.7.14 Seasonal Temperature Changes

When cooling-tower water is used as coolant, considerations are to be given for winter conditions where the ambient temperature may be near zero or below zero on the Celsius scale. The increased temperature driving force during the cold season contributes to more substantial overdesign and hence overperformance problems, unless a control mechanism has been instituted to vary the water/air flow rate as per the ambient temperature.

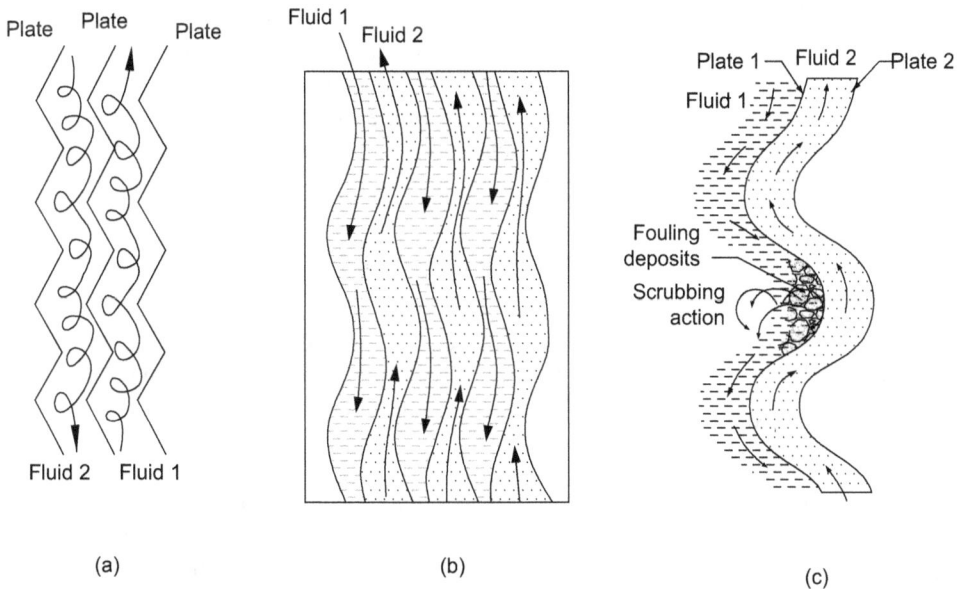

FIGURE 2.8 Fluid flow patterns between 2 adjacent plates of a PHE.

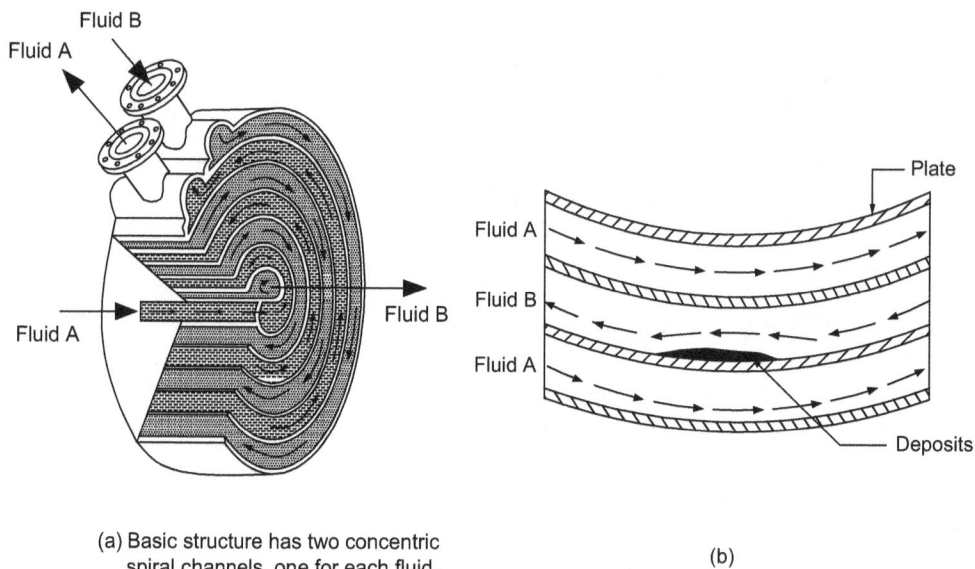

(a) Basic structure has two concentric
spiral channels, one for each fluid.   (b)

FIGURE 2.9   Fluid flow pattern in a SPHE.

Note: (b) shows scrubbing action of fluid flow.

2.7.15   EQUIPMENT DESIGN

Careful equipment design can contribute to reduction in fouling. Heat exchanger tubes that extend beyond tubesheet, for example, can cause rapid fouling and deposition.

2.7.16   HEAT EXCHANGER GEOMETRY AND ORIENTATION

Heat exchanger geometry influences the uniformity of flows on the shellside and tubeside. Orientation of heat exchangers eases cleaning [4]. Finned tube heat exchanger geometry and surface characteristics like tube layout, tube pitch, secondary surface density, etc., influence fouling.

STHE. In a shell and tube heat exchanger, the conventional segment baffle geometry is largely responsible for higher fouling rates. Uneven velocity profiles, back-flows and eddies generated on the shellside of a segmental baffled heat exchanger result in higher fouling and shorter run lengths between periodic cleaning and maintenance of tube bundles.

2.7.17   HEAT TRANSFER PROCESSES LIKE SENSIBLE HEATING, COOLING, CONDENSATION, AND VAPORIZATION

The fouling resistances for the same fluid can be considerably different depending upon whether heat is being transferred through sensible heating or cooling, boiling, or condensing [4].

2.7.18   TYPE OF HEAT EXCHANGER

A shell and tube heat exchanger consists of a shell with an internal tube bundle that is typically supported by tubesheets at the ends and intermittent tube support plates known as baffles. The baffle configuration shall be segmental or nonsegmental. Based on the baffles' configuration, the types of STHEs are classified as shown in Table 2.1.

TABLE 2.1
Shell and Tube Heat Exchanger with Segmental Baffles and Nonsegmental Baffles/No baffle

With Segmental Baffles	With Nonsegmental Baffles
1. Shell and Tube Heat Exchanger	1. Phillips RODbaffle Heat Exchanger
2. Disk and Doughnut Heat Exchanger	2. EMbaffle® Heat Exchanger
	3. HELIXCHANGER® Heat Exchanger
	4. Twisted Tube® Heat Exchanger (use no baffle)

More often, the fouling mechanism responsible for the deterioration of heat exchanger perform-ance is flow-velocity dependent. Maldistribution of flow, stagnant zones, wakes, and eddies caused by poor heat exchanger geometry can have detrimental effect on heat exchanger performance and reliability. Over two decades, different kinds of shells and tube heat exchangers have been developed to improve shellside performance. Many baffle types such as Philips RODbaffle heat exchanger, Twisted Tube® heat exchanger and Helixchanger® heat exchanger, and EMbaffle® heat exchanger have been designed to improve flow velocity, enhance heat transfer performance, and reduce fouling on shellside. Few of these types from fouling reduction point of view are discussed next.

2.7.18.1 Philips RODbaffle Heat Exchanger

RODbaffle heat exchangers, which replace baffle plates with an arrangement of rods that support heat exchanger tubes, have been successfully used in a variety of refining, chemical, and water-treatment applications around the world. Phillips Petroleum selected the RODbaffle heat exchanger design for its superior pressure drop performance and its ability to operate without tube vibration failures. Refer to Figure 1.13 of Chapter 1 for Philips RODbaffle heat exchanger concept.

2.7.18.2 EMbaffle® Heat Exchanger[1]

The EMbaffle heat exchanger is an innovation designed to improve performance and simultaneously reduce operating costs by reducing fouling losses. The patented EMbaffle design uses expanded metal baffles made of plate material that has been slit and expanded. In fouling services, for instance, in crude preheat trains, the dead zones typically found with conventional segmental designs reduce the performance of the heat exchanger with an increasing pressure drop during operation. However, with EMbaffle, due to the longitudinal flow, no low shellside velocity "dead zones" are created behind the EMbaffle, and there is a reduction in fouling. Consequently, the maintenance, cleaning, and refurbishment schedules can be optimized. Refer to Figure 1.14 of Chapter 1 for expanded metal concept and Figure 1.15 for a section of EMbaffle heat exchanger tube bundle cage.

2.7.18.3 HELIXCHANGER Heat Exchanger[2]

In a HELIXCHANGER heat exchanger, the conventional segmental baffle plates are replaced by quadrant-shaped baffles positioned at an angle to the tube axis creating a uniform velocity helical flow through the tube bundle. Figure 2.10 shows flow pattern in a helical baffle heat exchanger. Near plug flow conditions are achieved in a HELIXCHANGER heat exchanger with little back-flow and eddies. HELIXCHANGER heat exchangers have demonstrated significant improvements in the fouling behavior of heat exchangers in operation. Uniform velocities and near plug flow conditions achieved in a HELIXCHANGER provide low fouling characteristics offering much longer heat exchanger run-lengths between scheduled cleaning of tube bundles. Figure 2.11(a) shows a heavily fouled conventional segmental baffle shell and tube heat exchanger with a lightly fouled Helixchanger heat exchanger, Figure 2.11(b).

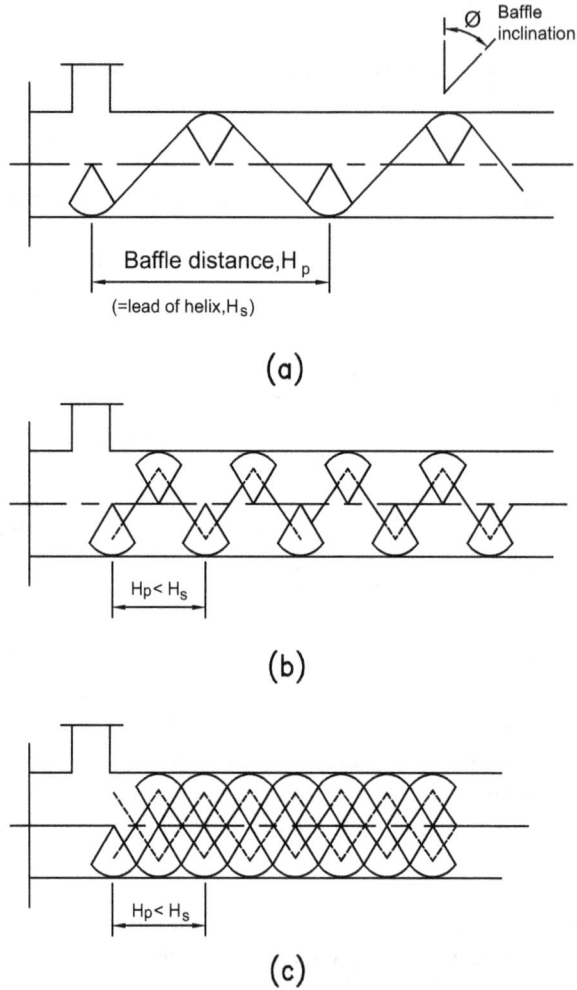

(a)

(b)

(c)

FIGURE 2.10 Orientation of helical baffles in a Helixchanger heat exchanger.

(a)   (b)

FIGURE 2.11 Comparison of level of fouling of conventional shell and tube heat exchanger and Helixchanger heat exchanger: (a) heavily fouled segmental baffle exchanger and (b) lightly fouled Helixchanger heat exchanger. (Courtesy of Lummus Technology Inc., Bloomfield, NJ.)

2.7.18.4 Koch Twisted Tube Heat Exchanger[3]

Sedimentary-type fouling is enhanced in areas of low-velocity or "dead" spots. Turbulent flow and the elimination of dead spots are needed to minimize this type of fouling. On the tubeside, higher velocities can be obtained with multiple tube passes and devices that enhance swirl flow to keep particles in suspension, such as Twisted Tube. On the shellside, elimination of dead spots is most effective. With traditional segmental baffles, fouling occurs in the dead zones on each side of a baffle up against the shell. Baffle free Twisted Tube bundles have no such dead spots and the constant and uniform shellside velocity ensures considerable reduction in fouling while in service. Refer to Figure 1.17 of Chapter 1, which shows an adjacent Twisted Tube contact pattern in a Twisted Tube heat exchanger.

2.8 GAS-SIDE FOULING IN INDUSTRIAL HEAT TRANSFER EQUIPMENT

If fouling occurs in the presence of a flue gas or combustion gas stream, the process is known as gas-side fouling. Gas-side fouling occurs when a fuel is burned and the products of combustion pass through a heat exchanger. Combustion gases typically contain ash, unburned hydrocarbons, and elements such as sodium, sulfur, chlorine, potassium, calcium, and magnesium, which are all potential foulants. Metallic elements contained in coal, oil, and gasoline are mobilized by combustion processes and emitted through the flue gases mainly as components of submicron particles [12].

2.8.1 Heat Transfer Equipment Prone for Gas-Side Fouling

The types of heat exchangers in which gas-side fouling can be a problem are grouped as follows:

(a) Recuperators like finned tube heat exchangers.
(b) Regenerators.
(c) Direct contract exchangers.
(d) Fluidized bed exchangers.
(e) Heat pipe exchangers.

Level of gas-side fouling on finned-tube heat exchangers depends on features such as

a. Tube layout pattern—in-line or staggered.
b. Tube pitch.
c. Fin density.
d. Fin surface characteristic such as plain fin, surface-modified fin.

Of the six categories of fouling, precipatation fouling, particular fouling, chemical reaction fouling, corrosion fouling, biological fouling and solidification fouling, particulate fouling is most likely to be assoicated with gas-side fouling but chemical reaction fouling, corrosion fouling, dew-point corrosion, etc., can also be improtant. In some cases, gas-side fouling may be accompanied by corrosion and/or erosion. Because it can increase pressure drop and decrease heat transfer, gas-side fouling can lead to increased capital and maintenance costs and major production and energy losses in many energy-intensive industries [12].

2.8.2 Dew-Point Corrosion

Dew-point condensation is a very important deposition mechanism in gas-side fouling. In low-temperature dew-point condensation, which takes place at temperatures below 500°F, the principal constituents of interest include H_2SO_4, HCl, and H_2O. Acid deposits tend to be highly corrosive and also to attract particulate matter from the gas stream because of the wet surface created by the condensate. The problem of low-temperature condensation and corrosion is frequently avoided by making certain that the surface temperature always remains above the acid dew-point temperature of the gas stream.

2.8.3   Gas-Side Fouling Prevention, Mitigation Techniques

A number of prevention, mitigation, and accommodation techniques are used by the industry in an attempt to deal with gas-side fouling. These techniques may be grouped here under [12]

    i.   fuel cleaning techniques;
    ii.   control of combustion conditions;
    iii.   fuel and gas additives;
    iv.   surface cleaning techniques;
    v.   quenching;
    vi.   control of operating conditions; and
    vii.   gas cleaning techniques.

2.8.3.1   Surface Cleaning Techniques

A number of surface cleaning techniques have been developed and used by industry, including steam and air soot blowing, sonic soot blowing, and water washing. Other approaches that are used to some extent include chemical, mechanical, and thermal cleaning. Each of these methods is discussed in this section.

2.8.3.2   Steam and Air Soot Blowers

The most widely used technique to clean gas-side fouled heat transfer surfaces is soot blowing using steam or air with both pressure and temperature of the blowing medium important parameters. The location and spacing of the soot blowers within the heat exchanger are very important. The important parameters associated with the soot blower nozzles are the number, type, size, and angle of attack. Finally, the speed and frequency of the soot blower operation depends on at least the type of fuel, the amount of excess air, and the operating conditions.

2.8.3.3   Water Washing

Water soluble deposit accumulations formed on heat transfer surfaces may be easily removed by washing, provided that a sufficient quantity of water is used. The standard water washing apparatus is a stationary, high-penetration, multinozzle device. The high-velocity jets produce a high fluid shear stress along with contractions caused by thermal shock to accomplish the removal of the deposits. Water washing devices are normally located at both the hot and cold ends of the exchanger, in the case of an air preheater, and are generally operated simultaneously in order to effectively remove the deposits. The actual washing operations may be carried out under the following conditions: (a) out-of-service; (b) in-service isolated; (c) in-service on-stream.

2.8.3.4   Other Cleaning Techniques

Other cleaning techniques may be grouped into the categories of chemical, mechanical, and thermal cleaning.

2.9   FOULING OF REFINERY PROCESS EQUIPMENT

Fouling as it pertains to petroleum is deposit formation, encrustation, deposition, scaling, slagging, etc., and sludge formation, which has an adverse effect on operations. Fouling of refinery process equipment is a common problem resulting in severe economic penalties, as well as significant safety and environmental concerns. Typical problem areas include crude preheat exchangers and furnaces, hydrotreater exchangers and reactor beds, slurry exchangers, and thermal cracking process exchangers and furnaces. Heat exchangers are used to recover heat from product streams to preheat the feedstock, minimizing the furnace fuel needed to raise the feedstock to the required process temperature. Fouling in the exchanger train and the related reduction of heat transfer can cause significant energy loss and increase operating costs. Fouling can also become so severe that unit capacity and production limits are reached [13a, 13b, 13c, 14].

2.9.1 ECONOMIC IMPACT OF FOULING

Fouling of refinery process heat exchange systems can cause significant economic penalties. Of the total processing cost, energy is a major part, particularly in distillation processes, which can directly influence refinery profitability. Controlling, or minimizing, heat exchanger fouling can dramatically lower overall operating costs. Minimizing operating costs, energy cost, safety, sustaining throughput, and maintaining equipment reliability are primary goals in refinery operation.

Safety: regular equipment opening for cleaning increases the risk of leakages, specifically during the start-up. Although difficult to quantify, the economic penalties associated with a fire or explosion, and moreover personnel injuries, can be enormously high.

2.9.2 FOULING MECHANISMS

Multiple factors impact fouling including equipment design; flow rates; temperatures; cracking units operational severity; fluid characteristics; caustic addition, can impact crude preheat fouling and coker furnace fouling; and upstream processes, such as desalter operation. Fouling deposits can be categorized into two major types, inorganic and organic. To properly control fouling, the differences between these two categories must be known and accounted for when identifying the fouling mechanisms involved and designing an appropriate chemical treatment program [13].

2.9.3 ANTIFOULANT CHEMICAL TREATMENT

Antifoulant chemical treatment consists of use of dispersants (high- and low-temperature stable, organic and inorganic fouling), polymerization inhibitors (free-radical, condensation), metal coordinators/deactivators and various corrosion inhibitors that exhibit dispersant properties.

2.10 FOULING DATA

Allowance for fouling is largely a matter of experience. There are tables of typical values for various services, e.g., TEMA Standards [15], and there is a proprietary correlation available for cooling water. However, fouling behavior is strongly dependent upon many variables and these interactions are very complex, so each problem needs to be examined for its special considerations. In choosing the fouling resistances to be used in a given heat exchanger, the designer has sources as follows:

1. Past experience of heat exchanger performance in the same or similar environments.
2. Results from fouling monitoring.
3. TEMA [15] values, which are overall values for a very limited number of environments.
4. Rules of thumb (i.e., 20%–25% overdesign).
5. Model analysis/bench scale measurements under accelerated conditions.
6. Empirical correlations based on laboratory experiments.
7. Numerical simulations (i.e., CFD).

2.11   HOW FOULING IS DEALT WHILE DESIGNING HEAT EXCHANGERS

2.11.1   Specifying the Fouling Resistances

Values of the fouling resistances are specified that are intended to reflect the values at the point in time just before the exchanger is to be cleaned. This implies that the exchanger is oversized for clean condition and barely adequate for conditions just before it should be cleaned. The fouling resistances result in higher heat transfer surface area. Planned fouling prevention, maintenance, and cleaning can justify lower fouling resistances, but at higher ongoing costs [4].

2.11.2   Oversizing

In this approach, the overall heat transfer coefficient is determined for clean conditions, and subsequently, the surface area required for clean condition is increased by a certain percentage. Based upon experience, the oversurface specified can range from 15% to 50% depending on the service [4]. In effect, the fouling for the exchanger is combined and no longer can be identified as belonging to one side or another.

2.12   TEMA FOULING RESISTANCE VALUES

The influence of the TEMA Fouling Resistance values [15] on design of heat exchangers has been enormous. In practice, some thought the TEMA values too high, others too low, since the tendency to fouling in a heat exchanger is dependent upon parameters such as local flow velocities, heat fluxes, etc., rather than overall values or point values as given for most cases in TEMA. Fouling types and effects of fouling are included in the RGP section of TEMA fouling resistance values taken from Ref. [4] is shown in Table 2.2.

TABLE 2.2
Fouling Resistance Values for Various Liquid Streams

Fluid	Fouling Resistance $r_s \times 10^4$ (M² K/W)
Industrial water streams	
Seawater	2.75–3.5
Brackish water	3.5–5.3
Treated cooling tower water	2.75–3.5
Artificial spray pond	2.75–3.5
Closed loop treated water	2.75
River water	3.5–5.3
Engine jacket water	2.75
Distilled water or closed cycle condensate	0.9–2.75
Treated boiler feed water	0.9
Boiler blow down water	3.5–5.3
Industrial liquid streams	
No. 2 fuel oil	3.5
No. 6 fuel oil	0.9
Transformer oil, engine lube oil	2.75
Refrigerants, hydraulic fluid, ammonia	2.75
Industrial organic HT fluids	2.75–3.5
Ammonia (oil bearing)	5.3
Methanol, ethanol, ethylene glycol solutions	3.5

TABLE 2.2 (Continued)
Fouling Resistance Values for Various Liquid Streams

Fluid	Fouling Resistance $r_s \times 10^4$ (M^2 K/W)
Process liquid streams	
MEA and DEA solutions	3.5
DEG and TEG solutions	3.5
Stable side draw and bottom products	2.75–3.5
Caustic solutions	3.5
Crude oil refinery streams	
Temperature (°C)	
120	3.5–7
120 to 180	5.3–7
180 to 230	7–9
>230	9–10.5
Petroleum streams	
Lean oil	3.5
Rich oil	2.75–3.5
Natural gasoline, liquefied petroleum gases	2.75–3.5
Crude and vacuum unit gases and vapors	
Atmospheric tower overhead	
Vapors, naphthas	2.7
Vacuum overhead vapors	3.5
Crude and vacuum liquids	
Gasoline	3.5
Naphtha, light distillates, kerosene, light gas oil	3.5–5.3
Heavy gas oil	5.3–9
Heavy fuel oil	5.3–12.3
Vacuum tower bottoms	17.6
Atmospheric tower bottoms	12.3
Cracking and coking unit streams	
Overhead vapors, light liquid products	3.5
Light cycle oil	3.5–5.3
Heavy cycle oil, light coker gas oil	5.3–7
Heavy coker gas oil	7–9
Bottoms slurry oil	5.3
Catalytic reforming, hydrocracking, and hydrodesulfurization streams	
Reformer charge, reformer effluent	2.6
Hydrocharger charge and effluent	3.5
Recycle gas, liquid product over 50°C	2.75
Liquid product 30°C to 50°C (API)	3.5
Light ends processing streams	
Overhead vapors, gases, liquid products	2.75
Absorption oils, reboiler streams	3.5–5.3
Alkylation trace acid streams	3.5
Visbreaker	
Overhead vapor	5.3
Visbreaker bottoms	17.5

(*continued*)

TABLE 2.2 (Continued)
Fouling Resistance Values for Various Liquid Streams

Fluid	Fouling Resistance $r_s \times 10^4$ (M^2 K/W)
Naphtha hydrotreater	
Feed	5.3
Effluent, naphthas	3.5
Overhead vapor	2.6
Catalytic hydro desulfurizer	
Charge	7–9
Effluent, HT separator overhead, liquid products	3.5
Stripper charge	5.3
HF alky unit	
Alkylate, depropanizer bottoms	5.3
Main fractional overhead and feed	5.3
Other process streams	3.5
Industrial gas or vapor streams	
Stream (non-oil-bearing)	9
Exhaust steam (oil-bearing)	2.6–3.5
Refrigerant (oil-bearing)	3.5
Compressed air	2.75
Ammonia	2.75
Carbon dioxide	3.5
Coal flue gas	17.5
Natural gas flue gas	9
Chemical process streams	
Acid gas	3.5–5.3
Solvent vapor	2.75
Stable overhead products	2.75
Natural gas processing streams	
Natural gas	2.75–3.5
Overheat products	2.75–3.5

Source: Adapted from Chenoweth, J.M., Final report of the HTRI/TEMA joint committee to review the fouling section of the TEMA Standards, Heat Transfer Research, Inc., Alhambra, CA, 1988 also reproduced from *Journal of Heat Transfer Engineering*, 11(1), 73–107, 1990.

2.12.1 RGP-T-2.5 Fouling Mitigation Design Method

i. Fouling may be mitigated for many services through proper heat exchanger design and operation.

ii. A small design margin may be added to the design to address design uncertainties. Rarely is this margin in excess of 30%.

iii. Some fluids like refrigerants, demineralized water, LNG, etc., are known to not foul during operation. For these fluids, the only reason to apply design margin is to address uncertainty in fouling prediction methods. This is normally on the order of 20% or less.

iv. Cooling water: for the case of those cooling-water streams, which are closely regulated in the plant for velocity control and are kept reasonably clean with a water maintenance program, fouling mitigation strategies apply. The cooling-water temperatures should be designed and operated to not exceed a maximum bulk temperature of 120°F (49°C) nor exceed a maximum wall temperature of 140°F (60°C). In addition, there must be sufficient velocity to maintain any particulate in suspension as it travels through the heat exchanger as well as to produce enough wall shear to stabilize any fouling, which does occur. There are many sources for information on minimum cooling-water velocities for design. An old rule of thumb for tubeside minimum water velocities has been 3 ft/sec (1 m/s).

2.13 RESEARCH IN FOULING

Much research is in underway for fouling. This research will some day enable us to understand the parameters responsible for fouling and hence to devise means to control or eliminate fouling.

2.14 FOULING MONITORING

Exchangers subject to fouling or scaling should be monitored for their efficient functioning. A light fouling on the tube greatly reduces its efficiency. A marked increase in pressure drop and/or reduction in performance usually indicates fouling. The unit should first be checked for air or vapor binding to confirm that this is not the cause for the reduction in performance.

2.14.1 FOULING INLINE ANALYSIS

Guidelines are available to assist the heat exchanger user in determining the extent of fouling. By using these industry guidelines, cleaning and maintenance tasks can be better assigned.

> *EPRI heat exchanger performance monitoring*: the EPRI *Heat Exchanger Performance Monitoring Guidelines*, EPRI NP-7552 lists several methods to evaluate the in-line heat transferring ability of heat exchangers. That document presents the details of each method and explains the strengths and weaknesses of each one. A few of the methods are given below [16]:
> *Heat transfer method*: in this method, the fouling resistance can be determined from the reduction in heat transfer during the deposition process. The data may be reported in terms of changes in overall heat transfer coefficient or temperature variation compared to the design values. This is a reliable method because it is a direct test of heat transfer ability.
> *Pressure drop method*: the pressure drop is increased for a given flow rate by virtue of the reduced flow area in the fouled condition and due to the increased roughness of the deposit. It is most challenging to measure small pressure drop changes and then make a determination regarding heat transfer ability.

In addition to the EPRI report, various heat exchanger monitoring techniques are presented in ASME OM-S/G-1994, *Standards and Guides for Operation and Maintenance of Nuclear Power Plants*, April 1995. A few of the techniques are given below:

1. *Functional test method*: temperature is measured and compared to the acceptance criterion. An advantage to this method is that the parameter of interest is directly measured.
2. *Heat transfer coefficient test method (without phase change)*: the fouling is computed directly based on conventional heat transfer analysis. An assumption is that the shellside flow is kept in the same flow regime.
3. *Heat transfer coefficient test method (with condensation)*.
4. *Transient test method*.
5. *Temperature effectiveness test method*: it is applicable to single-phase fluid stream monitoring. The result is a projected temperature that is compared to its acceptance criterion.
6. *Batch test method*.

7. *Temperature difference monitoring method*: the relationship between inlet cooling fluid and outlet process fluid temperatures is determined.
8. *Pressure loss monitoring method.*
9. *Visual inspection monitoring method.*

2.14.2 Tube Fouling Monitors

The ASME Condenser Performance Test Code (PTC 12.2–1998), released in September 1998, requires extensive use of a device to measure and determine the fouling characteristics of the tube bundle by monitoring individual tubes. A typical fouling monitors of Powerfect, USA, is shown in Figure 2.12 and its working principle is briefly discussed next.

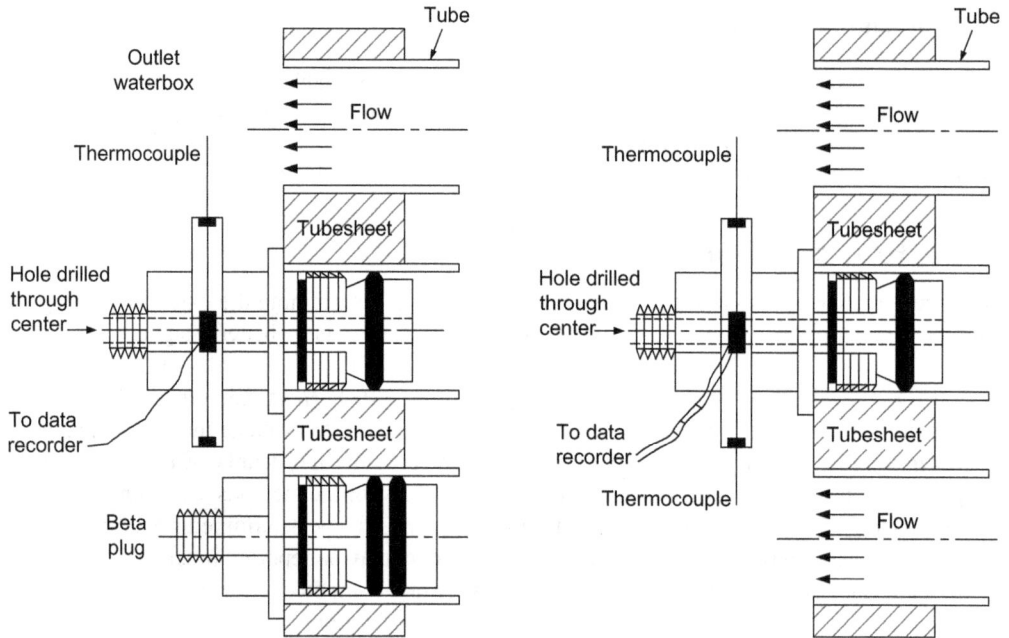

(b)

FIGURE 2.12 Tube fouling monitoring: (a) monitoring mechanism (illustration) and (b) monitoring plug. (Courtesy of Powerfect, Brick, NJ.)

2.14.3 FOULING MONITOR

1. After tube cleaning or retube, the Tube Fouling Monitor (TFM) is installed with one thermo-couple lead positioned to measure outlet temperature of unplugged tube.
2. A control tube adjacent to the TFM is plugged with removable plug on both ends.
3. Condenser is placed in service.
4. Outlet temperature data is recorded.
5. After a designated period of time, control tube plugs are removed.
6. A second thermocouple lead is installed to measure outlet temperature of the control tube.

2.14.3.1 Instruments for Monitoring of Fouling

Instruments have been developed to monitor conditions on a tube surface to indicate accumulation of fouling deposits and, in some cases, to indicate the effect on heat exchanger performance. Various fouling monitors are described in Ref. [17]. The following is a summary of the different fouling monitors [2].

1. Removable sections of the fouled surface, which may be used for microscopic examination, mass measurements, and chemical and biological analysis of the deposits.
2. Increase in pressure drop across the heat exchanger length. This method provides a measure of fluid frictional resistance, which usually increases with buildup of fouling deposits.
3. Thermal resistance monitors are used to determine the effect of the deposit on overall heat transfer resistance.

2.14.3.2 Gas-Side Fouling Measuring Devices

Marner and Henslee [18] carried out a comprehensive review of gas-side fouling measuring devices. They classified the devices into the five groups: heat flux meters, mass accumulation probes, optical devices, deposition probes, and acid condensation probes. A heat flux meter uses the local heat transfer per unit area to monitor the fouling. The decrease in heat flux as a function of time is thus a measure of the fouling buildup. A mass accumulation device measures the fouling deposit under controlled conditions. Optical measuring devices use optical method to determine the deposition rate. Acid condensation probes are used to collect liquid acid that accumulates on a surface that is at a temperature below the acid dew point of the gas stream.

2.15 EXPERT SYSTEM

A computerized, consultative expert system has been developed that can simulate human reasoning, perform water treatment diagnoses, and recommend procedures to minimize corrosion and fouling in cooling-water system [19]. In this work, a basic model adopted for knowledge representation is in the form of a fault tree. The expert system has been developed as a useful tool to accomplish the following:

1. Integrate myriad data from monitoring programs.
2. Provide a diagnosis.
3. Organize a logical stepwise approach to problem solving.
4. Serve as a valuable tool for training.

2.16 FOULING PREVENTION AND CONTROL

Specifying the fouling resistances or oversizing result in added heat transfer surface. The excess sur-face area can result in problems during startup and bring about conditions that can, in fact, encourage excess fouling due to low velocity [20]. There are a number of techniques that can overcome or mitigate the effects of fouling in heat exchangers, and they include the following:

1. Designing the plant or process in such a way that the condition leading to the fouling is limited or reduced.
2. Instituting an online mechanical cleaning system, or cleaning the equipment when the effects of the fouling can no longer be tolerated to restore its effectiveness by various offline cleaning techniques.
3. Using chemical additives or antifoulants in the fouling stream.

These aspects are discussed in detail next.

2.16.1 Measures to be Taken During the Design Stages

No hard and fast rules can be applied for heat exchanger design in relation to fouling, but the following points should be kept in mind during the conception and design of a heat exchanger:

1. Make the design simple.
2. Select the heat exchanger type with point 1 in mind. Heat exchangers other than shell and tube units may be better suited to fouling applications. Gasketed plate exchangers and spiral plate exchangers offer better resistance to fouling because of increased turbulence, higher shear, or other factors. Before commissioning a heat exchanger, carry out design checks and ensure that all constructional details and clearances conform to specification.
3. Prevent the possibility of corrosion and fouling during and subsequent to hydrostatic testing.
4. Startup conditions should avoid temperatures higher or velocities lower than the design values [3].
5. Maximize the flow velocities of process fluids to enhance the removal of the fouling deposits, provided that the fluid velocity is not high enough to cause excessive pressure drop or flow-induced vibration on the shellside. Ensure that velocities in tubes are in general above 2 m/s and about 1 m/s on the shellside [3]. Avoid stagnant areas where the flow velocities are less than those in the bulk of the core.
6. Assume nominal fouling resistance either from past experience or from published standards and design the heat exchanger with nominal oversizing. The oversizing may be of the order of 20%–40%. It is generally prudent to avoid large fouling factors, which result in larger equipment. Larger equipment generally results in lower velocities and hence may accelerate fouling.
7. To minimize fouling of finned tube or plate fin heat exchangers, use optimum fin density. Otherwise the initial benefit of increased heat transfer will be offset by fouling in the long run.
8. Fouling fluid on the tubeside: when the fouling fluid is on the tubeside, Mukherjee [10] recommends measures such as (1) using larger diameter tubes (a minimum of 25 mm OD); (2) maintaining high velocity (for cooling water, a minimum velocity of 2.5 m/s for mild steel, 2.2 m/s for nonferrous tubes, and as high as 5 m/s for titanium tubes is recommended); (3) leaving sufficient margin in pressure drop (for high fouling services, leave a margin of 30%–40% between the allowable and calculated pressure drop); (4) using a spare tube bundle or spare exchanger; (5) using two shells in parallel (each with 60%–70% of total capacity); (6) using wire-fin tube inserts; and (7) using online cleaning methods.
9. Fouling mitigation by installing hiTRAN® wire matrix elements (discussion follows). Fouling characteristics are dictated largely by the properties of the thermal and hydrodynamic boundary layers. hiTRAN wire matrix elements are a useful tool for controlling the conditions near the tube wall, especially in the laminar and transition flow regions and they have proven successful for direct fouling mitigation.

10. Fouling fluid on the shellside: when the fouling fluid is on the shellside, use a square or rotated square tube layout, minimize dead spaces by optimum baffle design, and maintain high velocity [10].

11. If severe fouling is inevitable, it is frequently better to install spare units. Installed spares will permit cleaning while the other unit is in service.

12. Proper selection of cooling medium can frequently avoid problems associated with fouling. For example, air cooling in place of cooling water solves many of the corrosion and fouling problems, such as scaling, biological growth, and many of the aqueous corrosion. The cleaning of bare tube or finned tube surfaces fouled by air is easier than surfaces fouled by water.

13. Particulate fouling, scaling, and trace-metal-catalyzed hydrocarbon reaction fouling can often be prevented by pretreatment of the feed streams to a heat exchanger by filtration, softening, and desalting, respectively [21].

14. Once the unit is on-stream, operate at the design conditions of velocity and temperature.

2.17 CLEANING OF HEAT EXCHANGERS

In most applications, fouling is known to occur in spite of good design, effective operation, and maintenance. Hence, heat exchangers and associated equipment must be cleaned periodically. The time between cleaning operations will depend upon the severity of the fouling problem. Each time the tube deposits, sedimentation, biofouling, and obstructions are removed, the surfaces are returned almost to bare metal, providing the most effective heat transfer and the tube itself with a new life cycle, the protective oxide coatings quickly rebuild themselves to re-passivate the cleaned tube. Great care must be taken to avoid damaging any tubesheet tube or tube coatings which may be present; otherwise the successful removal of fouling deposits may become associated with new tube leaks or increased tubesheet corrosion, only observable after the unit has been brought back online.

2.17.1 CLEANING TECHNIQUES

In general, the techniques used to remove the foulants from the heat exchanger surfaces, both on the shellside and on the tubeside, can be broadly classified into two categories: mechanical and chemical. The mechanical cleaning methods may be manual, automated, or semiautomated. The cleaning process may be employed while the plant is still operating, that is, *online*, but in most situations it will be necessary to shutdown the plant to clean the heat exchangers, known as *offline* cleaning. In some instances, combinations of these cleaning methods may be necessary. Each method of cleaning has advantages and disadvantages with specific equipment types and materials of construction.

2.17.2 DEPOSIT ANALYSIS

Information about the composition of fouling deposits through deposit analysis is extremely helpful to identify the source of the major foulants, to develop proper treatment, and as an aid in developing a cleaning method for a fouling control program [8]. The sample should represent the most critical fouling area. For heat exchangers and boilers, this is the highest heat transfer area [22]. Many analytical techniques are used to characterize deposit analysis. Typical methods include x-ray diffraction analysis, x-ray spectrometry, and optical emission spectroscopy.

2.17.3 SELECTION OF APPROPRIATE CLEANING METHODS

Before attempting to clean a heat exchanger, the need should be carefully examined. Consider the following factors for selecting a cleaning method:

Degree of fouling.

Nature of the foulant, known through deposit analysis.

In chemical cleaning the compatibility of the heat exchanger material and system components in contact with the cleaning chemicals.

Regulations against environmental discharges.

Accessibility of the surfaces for cleaning.

Cost factors.

2.17.3.1 Developing an Appropriate Cleaning Procedure

The selected cleaning procedure should remove the particular deposits that are present as effectively as is possible, and will render the unit out of service for the minimum amount of time. Some other major considerations in the selection process are as follows:

Removal of obstructions. Many tube-cleaning methods are ineffective when there are obstructions within tubes, or when various forms of macrofouling such as marine biofouling are present. When such obstructions are found it is inadvisable to proceed with the cleaning regimen as planned.

2.17.3.2 Precautions to be Taken while Undertaking a Cleaning Operation

Precautions to be taken while undertaking a cleaning operation are listed in TEMA [15] paragraph E-4.32 and Ref. [22]:

1. Individual tubes should not be steam blown because this heats the tube and may result in severe thermal strain and deformation of the tube, or loosening of the tube to tubesheet joint.
2. When mechanically cleaning a tube bundle, care should be exercised to avoid damaging the tubes. Tubes should not be hammered with a metallic tool.

Various cleaning methods are discussed next. (Cleaning of heat exchangers is also discussed in Chapter 5.)

2.18   OFFLINE MECHANICAL CLEANING

Offline mechanical cleaning is especially useful where fouling problems exist and are too severe to be handled by any of the other methods. Obviously, the tool selected has to be the most appropriate for removing a particular type of deposit. Molded plastic cleaners (pigs) are quite popular for some light silt applications. Brushes can also be used to remove these soft deposits as well as some microbiological deposits. Brushes are also useful for cleaning tubes with enhanced surfaces (e.g., spirally indented or finned), or those tubes with thin wall metal inserts or epoxy type coatings. The various offline mechanical cleaning methods are as follows:

1. Manual cleaning.
2. Jet cleaning.
3. Drilling and roding of tubes.
4. Turbining.
5. Hydro drilling action.
6. Passing brushes through exchanger tubes.
7. Scraper-type tube cleaners.
8. Blasting.
9. Soot blowing.
10. Thermal cleaning.

Table 2.3 shows a few mechanical fouling control techniques and their application areas.

TABLE 2.3
Mechanical Fouling Control Techniques and Their Application Areas

Technique	Application
Circulation of sponge rubber balls	Inside of water cooled condenser tubes
Brush and cage systems	Inside of tubes in shell and tube heat exchangers
Soot blowers	Outside of tubes in combustion spaces/boiler or waste heat recovery tubular heat exchanger
Sonic vibration	Tubular exchangers
Inserts technology	Three types of inserts for tubeside by Total- and Petroval, France are Turbotal®, Fixotal®, and Spirelf® .

2.18.1 Manual Cleaning

Where there is good access, as with a plate or spiral heat exchanger, or a removable tube bundle, and the deposit is soft, hand scrubbing and washing may be employed, although the labor costs are high. One of the most common and most economical methods for cleaning tubes is manual water blasting. There are a number of variations on this, but all generally require an operator to manually feed the nozzle into each tube that needs to be cleaned.

2.18.2 Jet Cleaning

Jet cleaning or hydraulic cleaning with high-pressure water jets can be used mostly on external surfaces where there is an easy accessibility for passing the high pressure jet. Jet washing can be used to clean foulants such as the following [23]: (1) airborne contaminants of air-cooled exchangers at a pressure of 2–4 bar; (2) soft deposits, mud, loose rust, and biological growths in shell and tube exchangers at a pressure of 40–120 bar; (3) heavy organic deposits, polymers, tars in condensers and other heat exchangers at a pressure of 300–400 bar, and (4) scales on the tubeside and fire side of boilers, preheaters, and economizers at a pressure of 300–700 bar. This method consists of directing powerful water jets at fouled surfaces through special guns or lances.

2.18.3 Automated Systems for External Cleaning

A range of methods to clean the external surface used such as dry cleaning, foam cleaning, and mechanized hydro-jetting. Automated systems clean several tubes at a time while the operator stands clear. Automated systems for cleaning tubes typically have multiple lances or hoses, a means of moving them across the face of the tube bundle, and a system for feeding them in and out of the tubes without manual labor. The operator typically controls the actions of the system from a remote position nearby. Hoses and lances are usually provided in different sizes to suit a variety of tube diameters; changeover should be simple to minimize downtime.

Jet cleaning of tube bundle of a shell and tube heat exchanger is shown in Figure 2.13 [24, 25]. A variety of nozzles and tips is used to make most effective use of the hydraulic force. The effectiveness of this cleaning procedure depends on accessibility, and care is needed in application to prevent damage to the tubes and injury to the personnel. Multiple lance systems are available for fast cleaning of large main condensers. Similar to water jet cleaning, pneumatic descaling is employed on the fire side of coalfired boiler tubes.

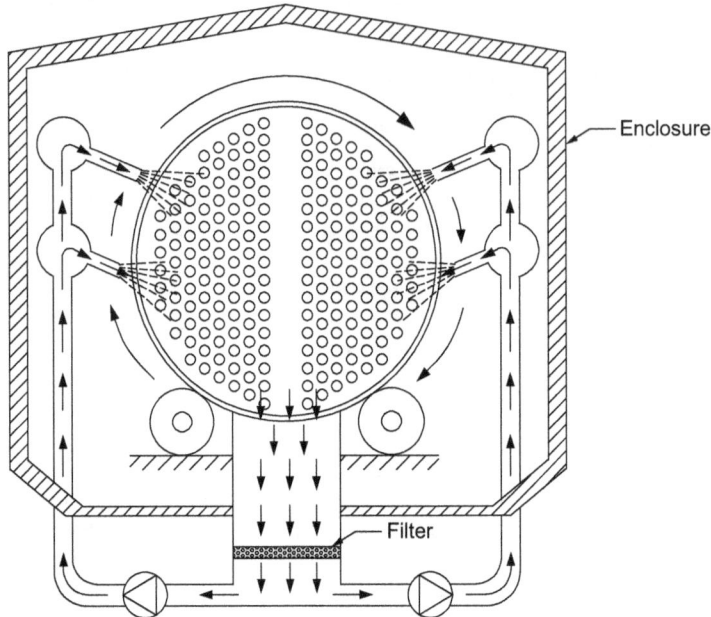

FIGURE 2.13 Jet cleaning of tube bundle of a shell and tube heat exchanger.

2.18.3.1 Safety and Ergonomics

Automated tube cleaning systems have benefits for operating personnel as well. Manual hose or lance feeding can be tiring over time (as previously noted) and can expose operators to backsplash and debris. With a semiautomated system, the operator has complete control without coming near the waterblasting, using a remote (even wireless) control.

2.18.3.1 Automated Systems for Internal Cleaning

Internal tube cleaning is performed by automated high-pressure water jetting techniques such as tube lancing equipment (TLE). Chemical cleaning can be used to remove hydrogen sulphide (H_2S), iron sulphide (FeS), and/or sticky pollutants that present challenges to hydro-jetting applications.

2.18.4 FOULING TENDENCIES OF AIR-COOLED CONDENSER

The external surfaces of the finned tubes on air-cooled heat exchangers are very prone to fouling from pollen, dust, insects, leaves, plastic bags, bird carcasses, etc. Not only is the air flow affected but also the heat transfer coefficient. To improve the heat removal capacity of an air-cooled heat exchanger under conditions of high ambient air temperature, operators will sometimes spray water on the heat exchanger to reduce surface temperature. Unfortunately, depending on the quality of water used, this sometimes leads to new scale formation on the tube fins and, again, reduces the heat transfer rate if the deposits are allowed to accumulate.

2.18.4.1 Cleaning Techniques for Air-Cooled Heat Exchangers

The three principal methods for cleaning the external surfaces of air-cooled condensers are hose, high-pressure hand lance and automated cleaning machine.

Other methods include dry cleaning, use soda blasting to clean induced and forced draft ACHE, *foam cleaning and mechanized hydro-jetting.*

(a)

(b)

(c)

FIGURE 2.14(a–c) Cleaning of air cooler/air cooled condenser by water jet showing unique top-side cleaning. (Courtesy of ACC-Team B.V., Nieuwdorp, the Netherlands).

Typical water jet cleaning of air cooled heat exchanger is shown in Figure 2.14. Traditionally, Fin-Fan Cooling Units were cleaned from the underside, which simply forces the dirt further into the tube bundle. ACC Team's new and unique system cleans from above and forces the dirt out from where it came resulting in a 25%–40% improvement in cooling performance.

2.18.4.2 Drilling and Roding of Tubes

Drilling is employed for tightly plugged tubes and roding for lightly plugged tubes. Drilling of tightly plugged tubes is known as bulleting. For removing deposits, good access is required, and care is again required to prevent damage to the equipment. A typical example is roding of radiator tubes plugged by solder bloom corrosion products.

2.18.4.3 Turbining

Turbining is a tubeside cleaning method that uses air, steam, or water to send motor-driven cutters, brushes, or knockers in order to remove deposits.

2.18.4.4 Hydro Drilling

Through a water flushing and rotary drilling action, the hydro-drilling system is the fastest and most effective way to remove difficult deposits from the inside of heat exchanger tubes, chemical reactors, condensers, reboilers, and absorbers. High torque drilling action removes obstructions from any type of tube ranging from 3/8″ to 6″ in diameter and up to 40′ in length. It can effectively clean hard deposits such as coke, calcium, sulfur, bauxite, asphalt, oxides, and baked-on hard polymers.

FIGURE 2.15 Hydro drilling system for dislodging fouling deposits. (Courtesy of Conco Systems, Inc., Verona, PA.)

The HydroDrill, shown in Figure 2.15 has been used in refineries, petrochemical plants, pulp and paper plants, electric power plants and other process industry plants throughout the world. The HydroDrill can effectively clean hard deposits, such as coke, calcium, sulfur, bauxite, asphalt, oxides, baked-on hard polymers, etc.

2.18.4.5 Projectile Cleaning

When a large number of tubes need to be cleaned quickly and effectively, projectile cleaning is an excellent choice. In the industry, this cleaning method is also known as "dart cleaning." Projectile tube cleaning is a method of cleaning condenser tubes and heat exchanger tubes using low water pressure and a "projectile" or "dart." Projectile's method works by loading patented mechanical devices, called projectiles, into each tube and then shooting the projectile down the length of the tube to scrape off adhering deposits. The unit consists of a long plastic dowel wrapped with nylon bristles. The brushes are propelled through the tube by a flexible shaft and the debris expelled by air/water shot right through the tube. The projectiles are propelled with safe, low-pressure water of approximately 350 psi (25 Bar).

Special projectiles. For example, Goodway Benelux offers various projectiles for different applications. For soft deposits, opt for a lighter type of projectile. When the deposits becomes harder and the tube material allows it, one can opt for heavier type projectiles. Every situation and deposit is different, therefore it is important to select the right type of projectile for a correct and effective cleaning result.

2.18.4.6 Scraper-Type Tube Cleaners

Scraper-type tube cleaners such as CONCO tube cleaner is an all-purpose tube cleaner for condensers and heat exchangers with tube sizes 1/2″–2.25″. This metal cleaner is effective on all types of deposits, including micro/macro fouling, organic-type scales, corrosion and pitting by-products, and all types of obstructions. The spring loaded tube cleaners are shot through the tubes with the Water Gun at 200–300 psi water pressure. The tube cleaner travels through the tube at 10–20 ft/s, plowing off deposits and corrosion and removing obstructions. Water flushes the removed matter out ahead of the cleaner, leaving a clean, polished bare metal inside diameter for optimum heat transfer efficiency. Typical wiper is shown in Figure 2.16 and CONCO tube cleaners is shown in Figure 2.17. Plastic or metal cleaner capable of navigating U-tube configurations in heat exchangers and feedwater heaters is also marketed by Conco Systems [26].

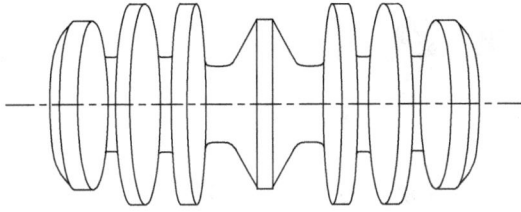

(a)   Four-wiper design is ideal for thicker, more viscous products or
      products that are difficult to clean and /or contain large particulates.

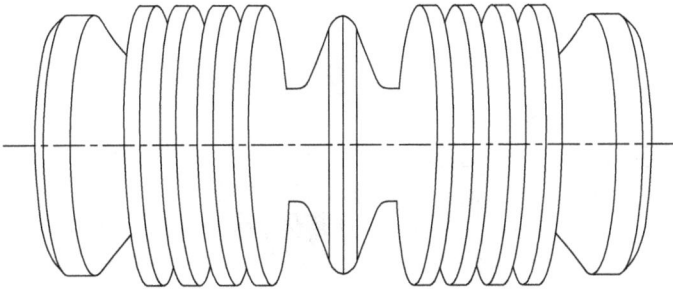

(b)   Eight-wiper design provides greater flexibility and efficiency, with a
      range of applications.

FIGURE 2.16   Foulant scrapper.

2.18.4.7   Blast Cleaning

Blast cleaning involves propelling suitable abrasive material at high velocity by a blast of air or water (hydroblasting) to impinge on the fouled surface. Hydroblasting is seldom used to clean tube bundles because the tubes are very thin. However, the technique is suitable to descale and clean tubesheet faces, shells, channel covers, bonnets, and return covers inside and outside.

2.18.4.8   Soot Blowing

Soot blowing is a technique employed for boiler plants, and the combustion or flue gas heat exchangers of fired equipment. The removal of particles is achieved by the use of air or steam blasts directed on the fin side. Water washing may also be used to remove carbonaceous deposits from boiler plants. A similar cleaning procedure is followed for air blowing of radiators on the fin side during periodical schedule attention.

2.18.4.9   Thermal Cleaning

The pyrolysis process. Pyrolysis is the thermal decomposition of organic materials in an oxygen-poor environment. At a temperature below 450°C (842°F), organic materials are converted into a homogeneous residue ready for further controlled processing. Aside from steady heating and cooling, a very important factor in the process is maintaining a constant temperature to prevent damage to the parts being cleaned [27].

Advantages of the thermal cleaning technique. It causes fouling in extremely inaccessible places to decompose into dust, which can then be removed easily using simple techniques. This is impossible if, for example, only high-pressure cleaning is used.

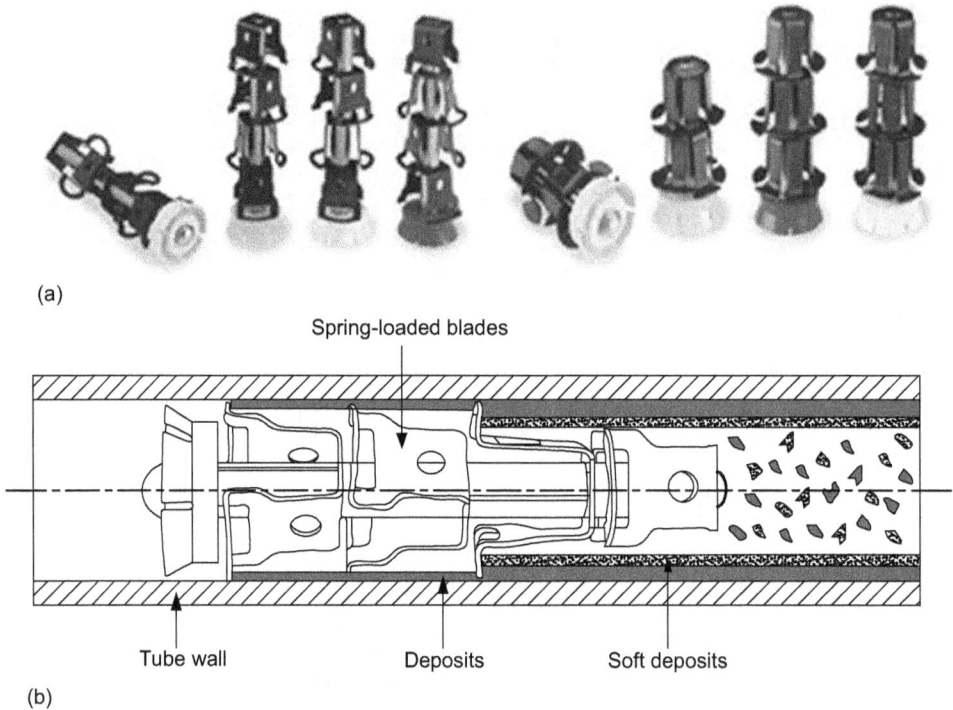

(a)

Spring-loaded blades

Tube wall   Deposits   Soft deposits

(b)

FIGURE 2.17 Scraper-type tube cleaners: (a) various models of tube cleaners and (b) tube cleaner in actionpositioned inside a fouled tube. (Courtesy of Conco Systems, Inc., Verona, PA.)

Steam Cleaning. Thermal cleaning involves steam cleaning, with or without chemicals. This method is also known as hydrosteaming. It can be used to clean waxes and greases in condensers and other heat exchangers.

2.18.5 MERITS OF MECHANICAL CLEANING

The merits of mechanical cleaning methods include simplicity and ease of operation, and capability to clean even completely blocked tubes. However, this method may damage the equipment, particularly tubes, it does not produce a chemically clean surface, and the use of high-pressure water jet or air jet may cause injury and/or accidents to personnel engaged in the cleaning operation—hence the personnel are to be well protected against injuries.

2.19 CHEMICAL CLEANING

The usual practice is to resort to chemical cleaning of heat exchangers only when other methods are not satisfactory. Chemical cleaning involves the use of chemicals to dissolve or loosen deposits. The chemical cleaning methods are mostly offline known as clean-in-place systems.

2.19.1 CLEAN-IN-PLACE SYSTEMS

Clean-in-place systems using chemicals such as detergents and sterilizing agents have come into broad use due to a number of advantages:

Time saving.

Cost savings resulting from the use of less chemical solution.

Elimination of unit openings on hazardous duties, etc.

2.19.2 Choosing a Chemical Cleaning Method

Chemical cleaning methods must take into account a number of factors as given in the following:

1. Compatibility of the system components with the chemical cleaning solutions. If required, inhibitors are added to the cleaning solutions.
2. Information relating to the deposit must be known beforehand.
3. Chemical cleaning solvents must be assessed by a corrosion test before beginning cleaning operation.
4. Adequate protection of personnel employed in the cleaning of the equipment must be provided.
5. Chemical cleaning poses the real possibility of equipment damage from corrosion. Precautions may be taken to reduce the corrosion rate to acceptable levels. Online corrosion monitoring during cleaning is necessary [22].
6. Disposal of the spent solution.

2.19.3 Chemical Cleaning Solutions

Chemical cleaning solutions include mineral acids, organic acids, alkaline bases, complexing agents, oxidizing agents, reducing agents, and organic solvents. Inhibitors and surfactant are added to reduce corrosion and to improve cleaning efficiency [22]. Common foulants and cleaning solvents are given in Table 2.4 and common solvents and the compatible base materials are given in Table 2.5.

2.19.4 General Procedure for Chemical Cleaning

The majority of chemical cleaning procedures follow the below steps [23]:

1. Flush to remove loose debris.
2. Heating and circulation of water.
3. Injection of cleaning chemical and inhibitor if necessary in the circulating water.

TABLE 2.4
Foulants and Common Solvents

Foulant	Cleaning Solvent
Iron oxides	Inhibited hydrofluoric acid, hydrochloric acid, monoammoniated citric acid or sulfamic acid, EDTA
Calcium and magnesium scale	Inhibited hydrochloric acid, citric acid, EDTA
Oils or light greases	Sodium hydroxide, trisodium phosphate with or without detergents, water–oil emulsion
Heavy organic deposits such as tars, asphalts, polymers	Chlorinated or aromatic solvents followed by a thorough rinsing
Coke/carbonaceous deposits	Alkaline solutions of potassium permanganate or steam air decoking

Sources: Compiled and modified from Lester, G. D. and Walton, R., Cleaning of heat exchangers, in *Practical Application of Heat Transfer*, C55/82, IMechE, London, UK, pp. 1–10, 1982; Garverick, L., Corrosion in the chemical processing industry, *Corrosion in the Petrochemical Industry*, ASM International, OH, 163–224, 1994.

TABLE 2.5
Solvent and Compatible Base Metals

Solvent	Base Metal/Foulant
Hydrochloric acid	Waterside deposits on steels. Inhibited acid can be used for cleaning carbon steels, cast iron, brasses, bronzes, copper-nickels, and Monel 400. This acid is not recommended for austenitic stainless steels, Inconel 600, Incoloy 800, and aluminum
Hydrofluoric acid	To remove mill scale
Inhibited sulfuric acid	Carbon steel, austenitic stainless steels, copper-nickels, admirality brass, aluminum bronze, and Monel 400. It should not be used on aluminum
Nitric acid	Stainless steel, titanium, and zirconium
Sulfamic acid	To remove calcium and other carbonate scales and iron oxides. Inhibited acid can be used on carbon steel, copper, admiralty brass, cast iron, and Monel 400
Formic acid with citric acid or HCl	To remove iron oxide deposit. Can be used on stainless steels
Acetic acid	To remove calcium carbonate scale
Citric acid	To clean iron oxide deposit on aluminum or titanium
Chromic acid	To remove iron pyrite and certain carbonaceous deposits that are insoluble in HCl on carbon steel and stainless steels. It should not be used on copper, brass, bronze, aluminum, and cast iron

Source: Compiled and modified from Deghan, T.F., Corrosion in the chemical processing industry, in *Metals Handbook*, Vol. 13, *Corrosion*, 9th edn., American Society for Metals, Metals Park, OH, pp. 1134–1185.

4. After sufficient time, discharge cleaning solution and flush the system thoroughly.
5. Passivate the metal surfaces.
6. Flush to remove all traces of cleaning chemicals.

It is suggested that one employ qualified personnel or a qualified organization for cleaning services.

2.19.4.1 CIP of PHE

Chemical cleaning of plate and frame heat exchangers works by circulating a specialized descaling agent through the cooling side or water side of the heat exchanger. The chemical descaler is typically an inhibited acid that is matched to the deposit inside the heat exchanger. The other leading method for the cleaning plate and frame exchangers is to first disassemble the heat exchanger and remove the plates to be cleaned one by one. This process can be very expensive and time consuming. For more details on CIP of PHE refer to Chapter 7, Heat Exchangers: Classification, Selection, and Thermal Design.

2.19.5 OFFLINE CHEMICAL CLEANING

Major offline chemical cleaning methods are as follows:

1. Circulation.
2. Acid cleaning.
3. Fill and soak cleaning.
4. Vapor-phase organic cleaning.
5. Steam injection cleaning.

Circulation: this method involves the filling of the equipment with cleaning solution and circulating it by a pump. While cleaning is in progress, the concentration and temperature of the solution are monitored.

Acid cleaning: scales due to cooling water are removed by circulating a dilute hydrochloric acid solution. This is discussed in detail with the discussion of cooling-water fouling.

Fill and soak cleaning: in this method, the equipment is filled with a chemical cleaning solution and drained after a period of time. This may be repeated several times until satisfactory results are achieved. However, this method is limited to small units only.

Vapor-phase organic cleaning: this method is used to remove deposits that are organic in nature.

Steam injection cleaning: this method involves an injection of a concentrated mix of cleaning solution and steam into a fast-moving stream. The steam atomizes the chemicals, increasing their effectiveness and ensuring good contact with metal surfaces.

2.19.5.1   Integrated Chemical Cleaning Apparatus

Generally, chemical cleaning is performed in four stages: degreasing, pickling, passivating, and flushing. Each of these stages ordinarily requires a separate and distinct setup of apparatus to complete the job adding considerable extra expense for additional man hours taken by pipefitters and other skilled labor. As per this invention, an integral chemical cleaning apparatus for supplying and recirculating chemical cleaning solutions used in the process of removing scale, rust, grease, and dirt coatings from internal metallic surfaces of closed industrial equipment where the stages of pickling, passivating, and flushing are performed consecutively without change-over to separate devices for each of such stages of the process and further it eliminates the need for virtually all auxiliary pumping requirements.

2.19.5.2   Merits of Chemical Cleaning

Chemical cleaning offers the following advantages over the mechanical cleaning:

1. Uniform cleaning and sometimes complete cleaning.
2. Sometimes chemical cleaning is the only possible method.
3. No need to dismantle the unit, but it must be isolated from the system.
4. Capable of cleaning inaccessible areas.
5. Moderate cleaning cost and longer intervals between cleaning.

2.19.5.3   Disadvantages of Chemical Cleaning Methods

Chemicals used for cleaning are often hazardous to use and require elaborate disposal procedures. Noxious gases can be emitted from the cleaning solution from unexpected reactions. Chemical cleaning corrodes the base metal and the possibility of excess corrosion cannot be ruled out. Complete washing of the equipment is a must to eliminate corrosion due to residual chemicals.

2.20   ONLINE MECHANICAL CLEANING METHODS

There is an obvious need for an industrial online cleaning procedure that can remove fouling deposits without interfering with a plant's normal operation. Online cleaning methods can be either mechanical or chemical. Online chemical cleaning is normally achieved by dosing with chemical additives. Various online mechanical cleaning methods to control fouling in practice are as follows:

1. Upstream filtration in a cooling-water system.
2. Flow excursion.
3. Air bumping of heat exchangers.
4. Reversing flow in heat exchangers.
5. Automatic/online tube cleaning systems such as the following:
   a. Sponge rubber balls cleaning system.
   b. Brush and cage system.
6. Insert technology.
7. Grit cleaning.
8. Self-cleaning fluidized-bed exchangers.

2.20.1 UPSTREAM FILTRATION (DEBRIS FILTER)

Power-station condensers are more vulnerable to the intake of debris and biological organisms. One solution to prevent the blockage of condenser tubes is the installation of an upstream filtration system [28]. Despite all of the prescreening technology employed at raw cooling-water intake sites, additional filtration is often required to avoid macro-fouling of the units served. The solution to this problem is provided by the Automatic Debris Filter System (ADFS). The ADFS is designed to effectively prevent macro-fouling by fibrous, course and marine life debris (larger than the selected filter perforation size), by capturing and ejecting all debris in an optimal manner. The optimal cleaning capability is achieved through the incorporation of dedicated suction ports for the two basic classes of water-borne debris (coarse and fibrous). Fouling control of cooling-water system by Automatic Debris Filter System (ADFS) is shown in Figure 2.18.

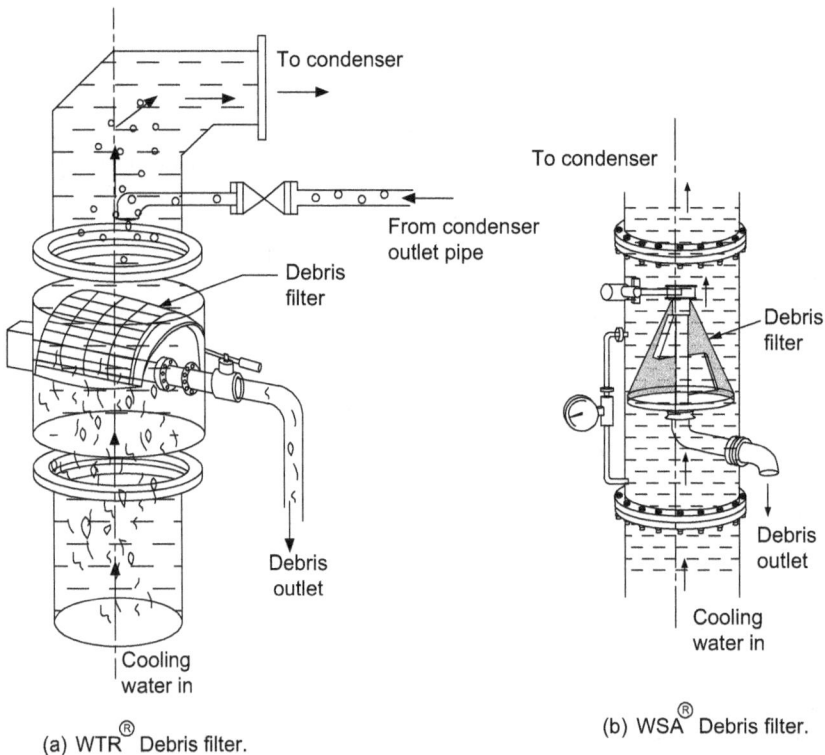

(a) WTR® Debris filter.

(b) WSA® Debris filter.

FIGURE 2.18 Fouling control of cooling water system by fitment of automatic debris filter system.

2.20.2 Flow Excursion

In this method the instantaneous flow is increased to remove the fouling deposits. This method is particularly applicable to a heat exchanger fouled badly due to the effects of low velocity either on the shellside or the tubeside.

2.20.3 Air Bumping

This technique involves the creation of slugs of air, thereby creating localized turbulence as slugs pass through the equipment. The technique has been applied to the liquid system on the shellside of heat exchangers. Care has to be taken to avoid the possibility of producing explosive mixtures of gases if the process fluid is volatile and flammable [3].

2.20.4 Reversing Flow in Heat Exchangers

Reversing of flow or backflushing or back wash system is followed on the water side of the cooling-water system by intermittent reversal of flow. The heart of the system includes (1) a backflushing valve and (2) control panel. To avoid "water hammer" the valves must be opened and closed in a certain sequence, hence a control system is absolutely necessary. A typical backflushing *valve and its working principle, i.e., reverse flow cleaning is shown in Figure 2.19.* The merits of backflushing include *energy savings because* the backflushing valve does not affect the standard operation, and hence no extra pumping energy costs, and it is *nonpolluting* and ecologically harmless. By regular backflushing of heat exchangers in open cooling systems, the use of chemical cleaners can be reduced considerably.

2.20.5 Online Tube Cleaning Systems

Tube cleaning systems are used to avoid sedimentation on the inner tube surface. Apart from high-pressure water lance cleaning and the brush method by MAN, the sponge ball cleaning system by TAPROGGE is the most common procedure applied. This system is an online continuous mechanical cleaning system utilizing specially engineered sponge rubber balls, which are cycled through the condenser tubes in the cooling water.

2.20.6 Flow-Driven Brushes

The flow-driven brush system is marketed by the MAN Corporation of West Germany. It consists of plastic baskets permanently fixed to the ends of each heat-exchanger tube. A brush consisting of a titanium core and plastic bristles is shuttled between the baskets by means of seawater flow reversal mechanisms. Experience with seawater-cooled power plants has shown that the success of this method depends strongly on the control of marine macro-fouling organisms from the inflow. The presence of macro-fouling organisms either on the catching basket surfaces or on the bristles would impede operation of the MAN system.

Recirculating rubber balls. This system is applicable to those heat exchanger designs that have water flowing within smooth, round tubes. It uses sponge rubber balls, which are injected at the upstream water and randomly driven by the water flow into heat exchanger tubes. The oversized balls wipe the inside tube surfaces. At the downstream water box, the balls are collected by special screens and recirculated to the upstream end of the tube bundle. Due to wear and contamination, the balls need to be counted, sorted, and replaced about once a month. The procedure can be fully automated.

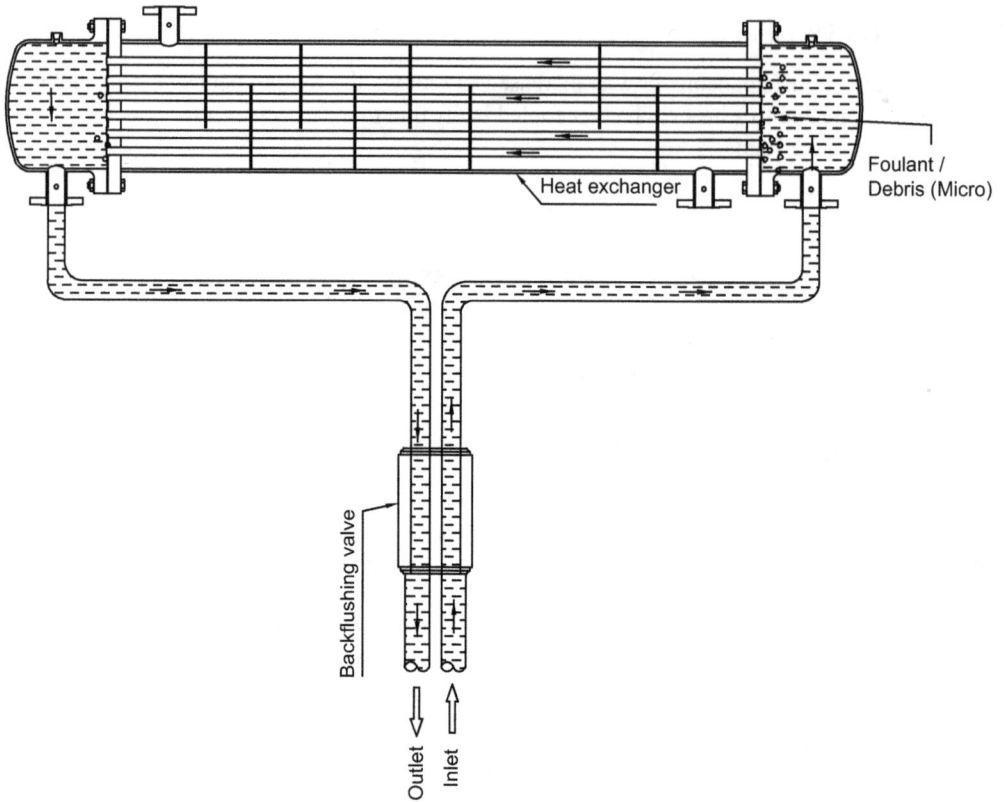

(a)

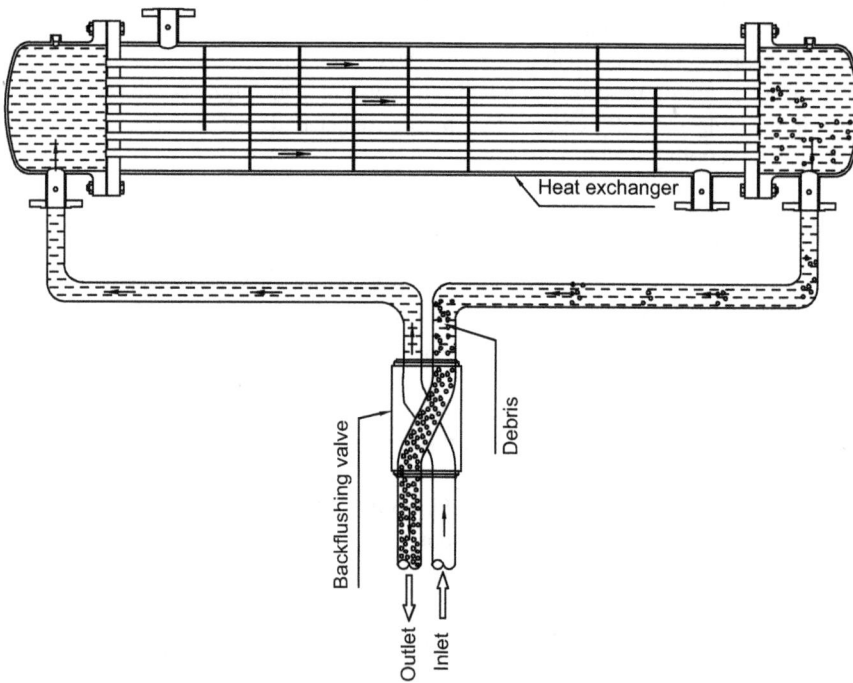

(b)

FIGURE 2.19 Flow reversing—working principle.

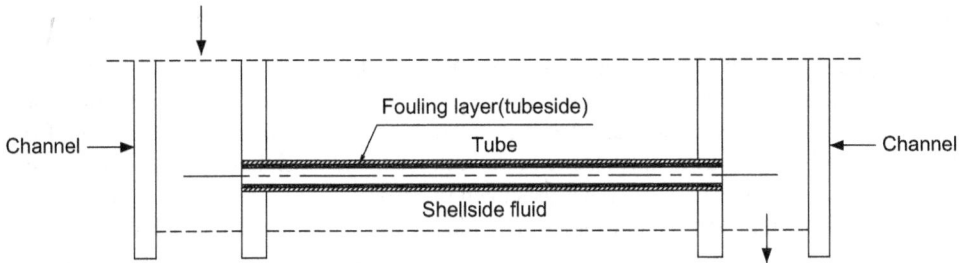

(a) Fouled tube.

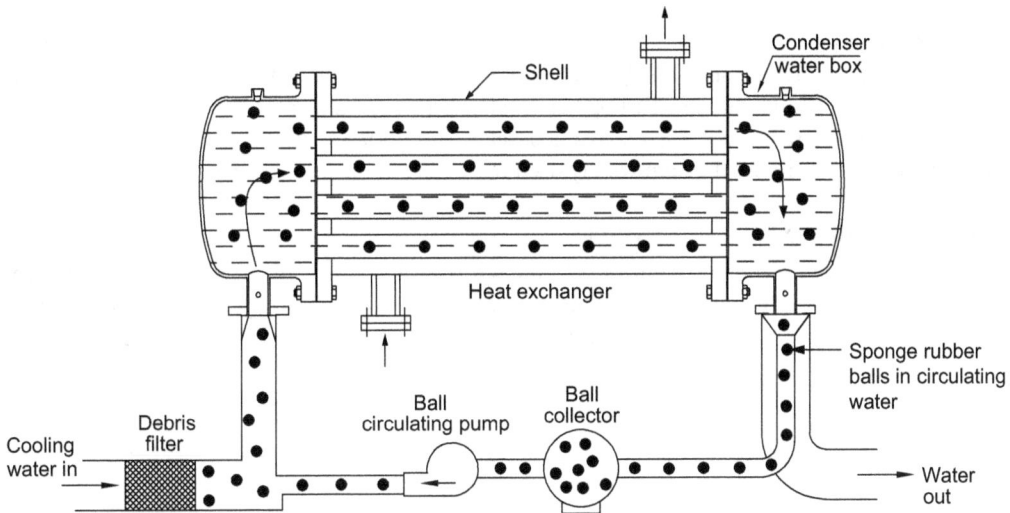

(b) Sponge rubber ball cleaning system of a water-cooled condenser / Heat exchanger.

FIGURE 2.20   Sponge rubber ball cleaning system of a water-cooled condenser.

2.20.7   SPONGE RUBBER BALL CLEANING SYSTEM

Major parts of a ball-type cleaning system include ball circulation pump, ball strainer, ball collector vessel, ball injector, control panel, and ball cleaning systems. The sponge ball diameter
is typically 2 mm above the inner tube diameter and having about the same specific gravity as
seawater are passed continuously into the inlet water box as shown in Figure 2.20a (for instance,
the Beaudrey or Taprogge system) [28] and a sponge rubber ball cleaning system with an inlet
denris filter is shown in Figure 2.20b. As each sponge ball passes through the tube it presses
slightly against the tube wall, wiping off deposits, organic matter, silt, and other accumulations.
The method may not be effective on longer runs once hard deposits are formed, or pitted. The
balls used for normal operation should have the right surface roughness to gently clean the tubes,
without scoring the tube surface. To remove heavy deposits, special abrasive balls that have a
coating of carborundum are available [29].

2.20.8   Brush and Cage System. Automatic Online Tube Cleaning System (ATCS)

The brush and cage or the Mesroc automatic online tube cleaning system used for cleaning heat exchangers consists of (1) a four-way flow diverter; (2) a control panel; and (3) one lot of brushes and basket sets. The four-way flow diverter is installed between the tubeside piping to and from the unit. A specifically selected brush is inserted into each tube, and catch baskets are semipermanently installed at both ends of each of the tubes as shown in Figure 2.21. By reversing the flow direction, every brush is being moved from one end of the tube to the other, where it is retained by the basket. It remains there until the next cleaning cycle. The brushes moving to and fro keep the inner walls clean. An actuator and a control system initiate the cleaning cycles.

2.20.9   Insert Technology

Inserts are devices installed in tubular heat exchangers as a means of heat transfer augmentation device. Few types of inserts continuously reduce fouling and improve heat transfer by means of mechanical effects. Three inserts resulting from research by Total are available: Turbotal®, Fixotal®, and Spirelf® (Spirelf, Fixotal, and Turbotal are registered trademarks of Total, the French oil company, and Petroval, France, is the exclusive licensee of Total for the design and the commercialization of these inserts). Today they are widely used in preheat trains in oil refineries.

2.20.9.1   Spirelf System

In the Spirelf system shown in Figure 2.22, fine wire springs are threaded through the tubes and held in place by straight wires at the ends of the tubes as fluid flows through the tubes. After installation, the insert stays under tension. The springs vibrate radially and axially under the

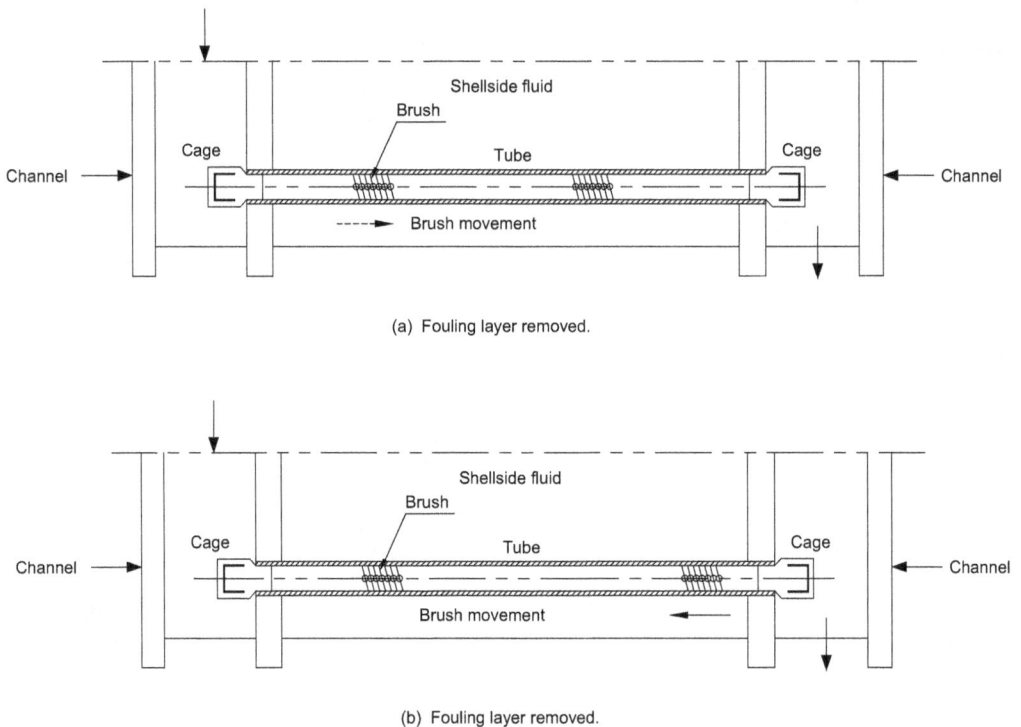

(a) Fouling layer removed.

(b) Fouling layer removed.

FIGURE 2.21   Brush and cage ATCS—schematic representation.

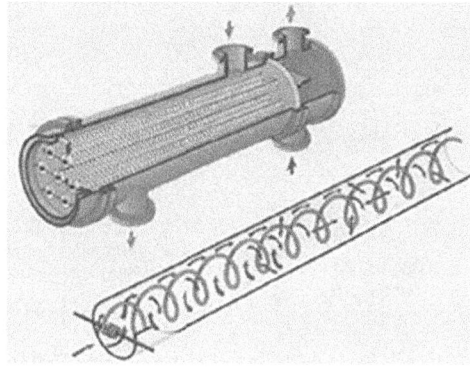

(a)

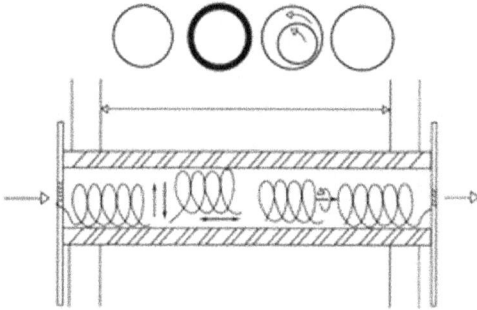

(b)

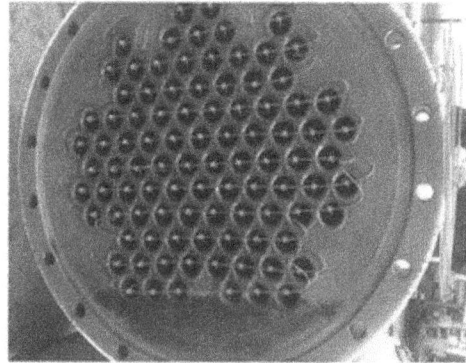

(c)

FIGURE 2.22 Spirelf: (a) schematic, (b) principle of working, and (c) spirelf placed inside tubes, application in reboilers. (Spirelf, Fixotal, and Turbotal are registered trademarks of Total, the French oil company, and Petroval, France, is the exclusive licensee of Total for the design and the commercialization of these inserts).

influence of fluid flow. This action reduces buildup of fouling inside the tube walls and break the boundary layer in the tubeside flow. These two effects achieve an improvement of the heat transfer rate. The effect of Spirelf system on thermohydraulic performance of a heat exchanger is shown in Figure 2.23.

2.20.9.2 Turbotal

The principle is based on the insertion of rotating metal devices, in rigid helicoidal form, into the tubes of shell and tube heat exchangers (Figure 2.24). Turbotal device is held at the inlet of tubes by a fixing device allowing the device to rotate around its axis by means of the fluid flow. This rotation causes a high turbulence in the flow and improves thus the internal heat transfer coefficient. Typical applications include crude preheat train, hydrotreaters, and feed-effluent exchangers.

2.20.9.3 Fixotal

The purpose of this fixed device is mainly to increase the rate of heat transfer. Fixotal consists of a wire coil, which is inserted inside every tube, with the wire in firm contact with the inside tube wall. Once in place, the device has no possibility of the slightest displacement (no vibration, no rotation, and no translation), the device can be easily removed if necessary.

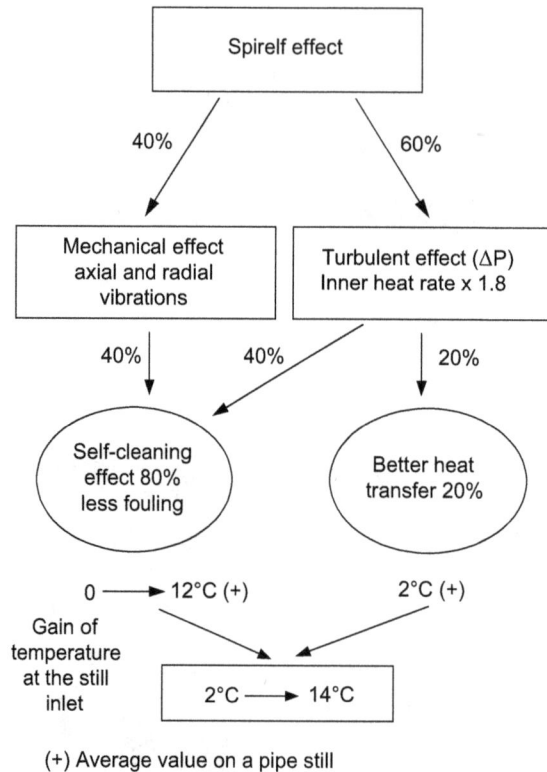

FIGURE 2.23   The effect of Spirelf system on thermohydraulic performance of a heat exchanger.

Fixotal acts as a source of turbulence in contact with the internal walls of the tube, preventing the stagnation of products in the layer adjacent to the tube. Typical applications include air coolers, cooling-water system, U-tubes, and mixed phase flow.

2.20.10   MERITS OF ONLINE CLEANING

The merits of online cleaning are as follows:

1. Increased heat transfer rate due to higher flow turbulence on tubeside.
2. Reduction in maintenance cost due to reduced tubeside fouling.
3. Extended run time between cleaning intervals.
4. Convenient to install.
5. Does not require any plant shutdown.
6. Can save time and labor.

However, the initial cost may be very high in certain cases.

2.21   GRIT CLEANING

In this method, abrasive materials, such as sand, glass, or metal spheres, are passed through the tubes. The scouring action removes the deposits from the inside of the tubes. The method has found application in cooling-water systems, but it could be used in conjunction with any fouling fluids. A special grit blasting nozzle accelerates the grit and causes it to follow a sinusoidal path through

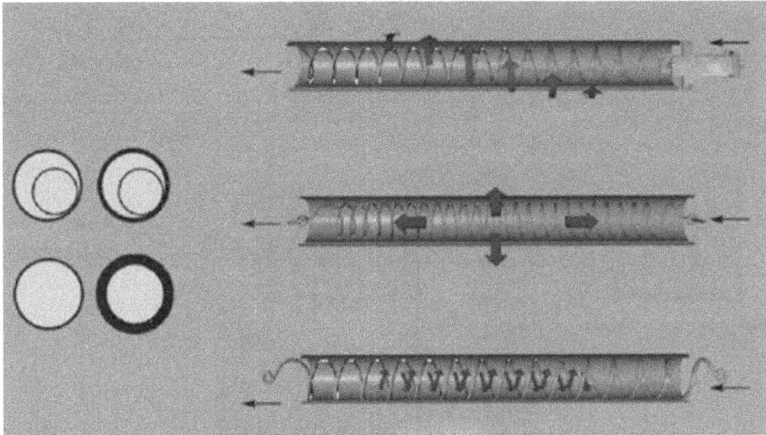

(a)

(b)

FIGURE 2.24 *Turbotal: (a) Turbotal effect and (b) turbotals placed inside a tube.*

the tube, dislodging the deposits [28]. Velocities more than 3 m/s are probably required for the technique to be effective.

2.22 SELF-CLEANING HEAT EXCHANGERS

The operating principle of the self-cleaning exchangers is shown in Figure 2.25.

The fouling liquid is fed upward through a vertical shell and tube exchanger that incorporates specially designed inlet and outlet channels. Solid particles are also fed at the inlet, where a flow distribution system uniformly provides the liquid and suspended particles throughout the internal surface of the bundle. At the top, inside the separator connected to the outlet channel, the particles disengage from the liquid and are returned to the inlet channel through an external downcomer, and the cycle is repeated. The process liquid fed to the exchanger is divided into a main flow and a control flow that transport the particles into the exchanger. By varying the control flow, it is possible to control the amount of particles in the tubes. This controls the efficiency of the cleaning mechanism, allowing the particle circulation to be either continuous or intermittent. It can be seen from Figure 2.25, the process liquid fed to the exchanger is divided into a main flow and a control flow that transport the particles into the exchanger.

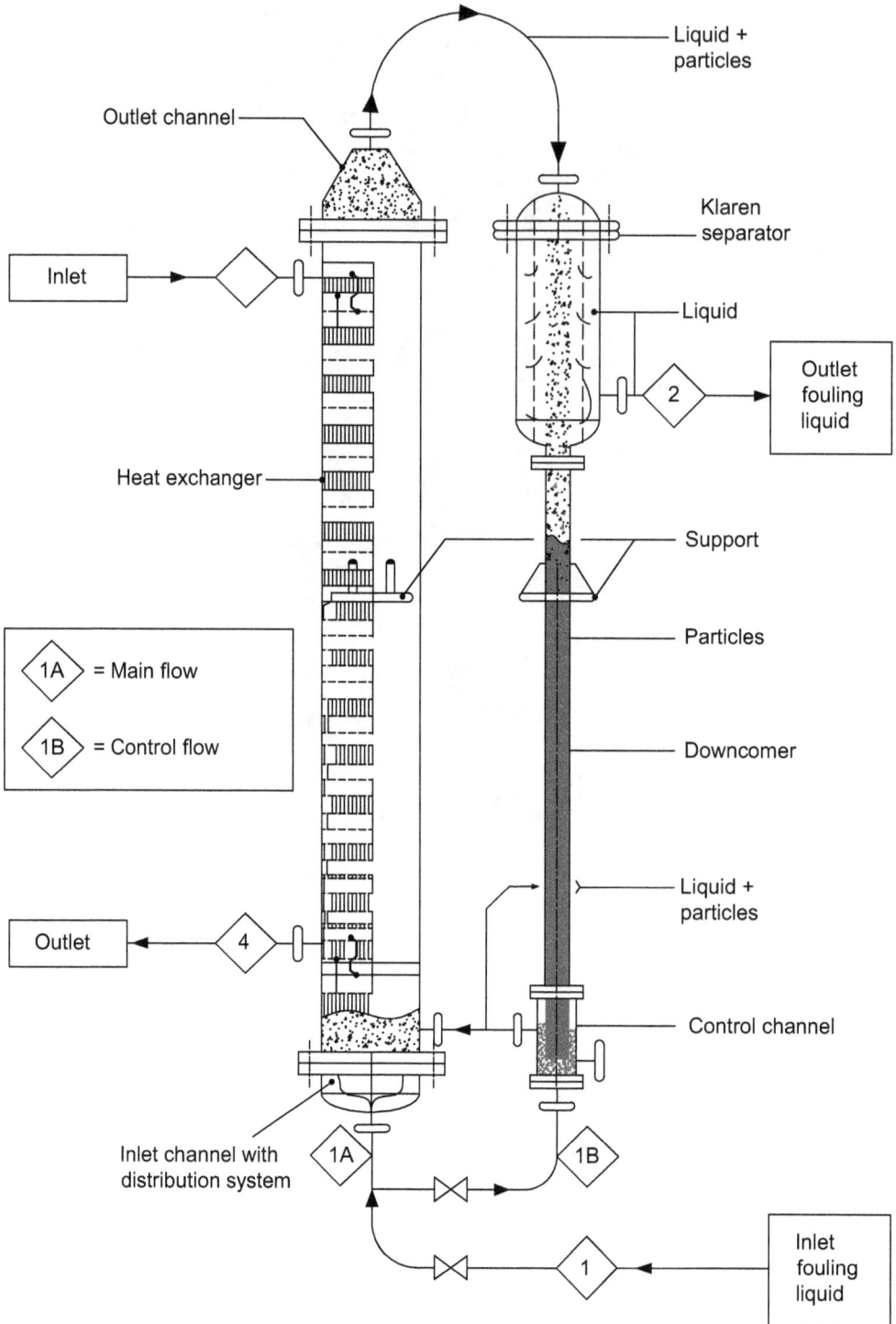

FIGURE 2.25   Principle of self-cleaning heat exchanger. (Courtesy of Klaren BV, Rotterdam, the Netherlands.)

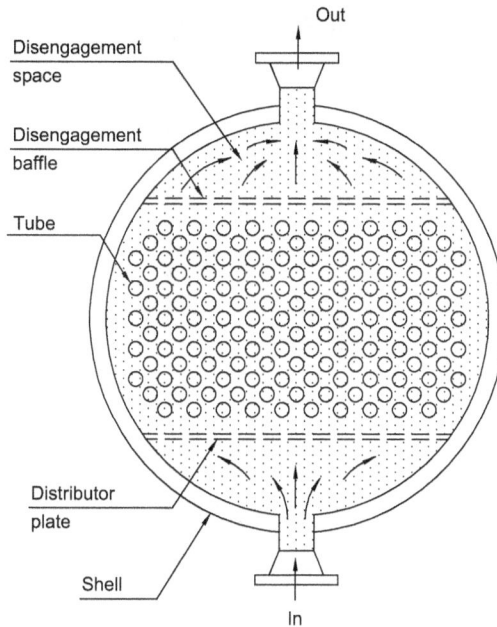

FIGURE 2.26 Liquid bed regenerator principle applied to shellside of a heat exchanger.

A variation of the abrasive cleaning method is to use a fluidized bed of particles to control fouling on the outside tubular exchangers; its working of the same is briefly discussed next.

2.23 LIQUID FLUIDIZED-BED TECHNOLOGY

Shellside application: liquid fluidized-bed technology offers the potential for scale control and increased heat transfer coefficients in heat exchangers. Fluidized beds consist of a bed of solid particles (e.g., sand) with fluid passing upward through them, Figure 2.26. The bed of particles will continue to expand as the velocity is increased and it will behave as a fluid until the terminal velocity is reached. At terminal velocity, the particles will be entrained in the fluid. This process of fluidization has been applied to liquid heat exchangers to eliminate the common problem of heat transfer surface scaling on shellside. The primary fluid may be used to fluidize a bed material such as sand. The fluidizing action of the bed creates two distinct advantages over conventional shell and tube flow arrangements: (1) the scouring action of the bed prevents scaling and limits corrosion on the tubes and (2) the heat transfer coefficient for the fluidized bed is almost double the coefficient for a conventional exchanger, Cole et al. [30].

Tubeside application: liquid fluidized-bed technology offers the potential for scale control and increased heat transfer coefficients on the tubeside of shell and tube heat exchangers. Figure 2.27 shows schematic of the fluidized-bed principle applied to tubeside of a heat exchanger.

2.24 FOULANT CONTROL BY CHEMICAL ADDITIVES

If fouling cannot be tackled adequately either by process design or equipment design, it can be further reduced by periodical online injection of additives into the process stream [21]. Chemical additives find wider use in cooling-water systems and to control fouling due to crystallization and

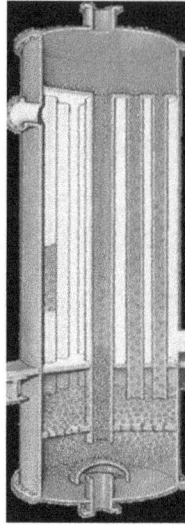

FIGURE 2.27 Schematic of the fluidized-bed principle applied to tubeside of heat exchanger. (Courtesy of Klaren BV, Rotterdam, the Netherlands.)

freezing, chemical reaction or polymerization, precipitation, particulate, and scaling. Various types of additives and their functions are as follows:

1. *Alkali or acid dosing*: by dosing either alkali or acid, pH may be controlled. This may be effective to control precipitation fouling. Acid dosing controls the hardness of hard or brackish water.

2. *Complexing agents*: agents such as chelants complex the metallic ion into a ring structure that is difficult to ionize.

3. *Chemical reactants*: these are used to complex or tie up the active foulants. They can solubilize or condition the foulants to prevent deposition.

4. *Sequestrant*: this additive complexes the metallic ions into a water-soluble structure, thus preventing its adhesion to the heat exchanger surface. A sequestrator physically surrounds and isolates particles (e.g., EDTA).

5. *Oxidizing agents*: these oxidize the deposits, making them suitable for dissolution (e.g., chromic acid, sodium nitrite, potassium permanganate).

6. *Reducing agents*: reduce the compounds in deposits and make them suitable for dissolution and to prevent the formation of hazardous by-products.

7. *Inhibitors*: control cooling-water corrosion. Corrosion inhibitors such as filming amines stifle chemical reaction fouling [3].

8. *Surfactant*: added to chemical cleaning solution to improve the wetting characteristics.

9. *Antiscalants*: in aqueous system, these additives chemically combine the scales to form soluble compounds.

10. *Distortion agents*: these interfere with the crystal structure so that it becomes more difficult for coherent crystal structures to form on surfaces.

11. *Dispersants*: dispersants impart an electrical charge to the particles so that they are held in suspension in the bulk of the liquid and the particles pass through the equipment without deposition on heat transfer surfaces. Dispersants are helpful to control chemical reaction fouling.

12. *Depressants*: depressants lower the freezing point of the solution such that the potential forming solids is brought down (e.g., glycols, alcohols).

TABLE 2.6
Fouling Control by Chemical Additives

Fouling method/problem	Additive
Scaling	Threshold inhibitors
	Dispersant to prevent atachment.
	Crystal modifer to weaken the deposit structure
	Chemicals that enhance the solubility of calcium carbonate and other mineral scale
	Cooling-water polymers.
Deposition	
Particle deposition	Dispersant to prevent deposition.
	Polymerization inhibitors
	Metal coordinator
	Flocculant followed by settling or filtration.
	Modifiers to reduce the strength of deposits in combustion systems.
Biological growth	Broad spectrum of oxidizing and nonoxidizing biocides
	Biocides to kill the living matter and algaecides.
	Biostat to redce biological activity.
	Biodispersant to prevent attachemtn to surfaces.
Chemical reaction	Antioxidants to prevent oxidation.
	Reaction chain terminators to remove free radicals.
Corrosion	Corrosion inhibitors to restrict the opportunity for corrosion reactions.
Freezing	Crystal modifiers to weaken the deposit structure.

13. *Flocculating agent*: causes the particles to agglomerate so that they may be settled out of the cooling water or suitably filtered.
14. *Threshold agents* that prevent the creation of crystal nuclei around which the larger crystals form. They also arrest the growth of nuclei (e.g., polyphosphates). In cooling-water systems, threshold agents retard the precipitation of scale-forming salts.
15. *Stabilizers*: on reaching the solubility limit, stabilizers are able to retard the nucleation of individual low-solubility compounds and prevent any existing crystals from forming adhesive deposits.
16. Metal coordinators react with the trace metals and prevent them from functioning as fouling catalysts in the case of chemical reaction or polymerization fouling.
17. Biocides kill the micro- or macro-organism. Biostats arrest the growth of microorganisms.

Table 2.6 shows fouling control by chemical additives.

2.25 ELECTRONIC WATER CONDITIONING

Electronic water treatment (EWT) is one of the methods of removing fouling deposits. It prevents crystals from sticking together and forming scales. Electronic water treatment is a noninvasive system utilizing a solenoid coil or coils wrapped around the pipework to be treated. A signal generator—of which the frequency is continuously changed supplies current to the coils. The pulse-shaped current creates an induced electric field concentric around the axis inside the pipe. With this arrangement, any charged particle or ion moving within the field experiences a so-called Lorenz force generated by the interaction between charged particles and magnetic and electric fields. Electronic water treatment also can be used for biological fouling control [31].

2.26 CONTROL OF FOULING FROM SUSPENDED SOLIDS

Methods of control of fouling from suspended solids include [3] the following:

1. Pretreatment of process fluids by means such as filtration, softening, and desalting to control precipitation fouling, particulate fouling, and scaling
2. Chemical treatment using dispersants and flocculating agents

2.27 COOLING-WATER MANAGEMENT FOR REDUCED FOULING

Water is by far the most common fluid subject to fouling. The quality of water used in the cooling system varies depending on its sources, like sea, river, ocean, lake, etc., and on the three forms of cooling-water systems: once through, open recirculating, and closed systems. Cooling-water quality factors that contribute to fouling include turbidity, salinity, dissolved solids and hardness, biological organisms, airborne contaminants, etc. The quality of raw water changes according to weather conditions also.

The primary types of fouling occurring in cooling tower water systems are crystallization of inverse solubility salts, sedimentation, corrosion of the heater surface, and biological growth.

Traditionally, the treatment of cooling water has often been oriented toward corrosion control followed by foulant control. In the cooling-water system, fouling occurs when insoluble particulates suspended in recirculating water form deposits on a surface. Fouling mechanisms are dominated by particle–particle interactions that lead to the formation of agglomerates. At low water velocities, particle settling occurs under the influence of gravity. Foulants enter a cooling system with makeup water, airborne contamination, process leaks, and corrosion. Most potential foulants enter with makeup water as particulate matter, such as clay, silt, and iron oxides. Airborne contaminants usually consist of clay and dirt particles but can include gases such as hydrogen sulfide, which forms insoluble precipitates with many metal ions. Process leaks introduce a variety of contaminants that accelerate deposition and corrosion. Cooling-water corrosion control measures are discussed in Chapter 3.

2.27.1 Forms of Waterside Fouling

2.27.1.1 Macro-Fouling

Fouling caused by large organisms, such as oysters, mussels, clams, and barnacles, is referred to as macro-fouling. Typically, organisms are a problem only in large once-through cooling systems or low cycle cooling systems that draw cooling water directly from natural water sources (rivers, lakes, coastal seas). Macro-fouling can occur despite common prescreening systems. This can lead to deposition on tubesheet surface and tube clogging, reduction of cooling surfaces, erosion–corrosion, etc. To avoid such consequences, it is recommended that, in addition to the common cleaning systems a self-cleaning filter, monitored automatically by the differential pressure caused by the collected debris [32].

2.27.1.2 Macro-Fouling Prevention: Use of a Debris Filter

Automatic debris filtering prevents macro-fouling of heat exchangers and condensers. Intake screens are often ineffective in protecting heat exchanger tubes from debris. These strainers can be sized to remove the carryover and fine debris that pass through intake screens to prevent buildup inside the condenser or exchanger tubes. Figure 2.28 illustrates a typical installation for such in-line debris filters in the inlet pipe to a heat exchanger or condenser combined with sponge rubber ball system (to be discussed later). Debris larger than the mesh size is trapped over the screen area. The debris cleaning cycle lasts only a few minutes. The debris is then discharged into the sewer or into the cooling-water discharge from the plant using a small quantity of water.

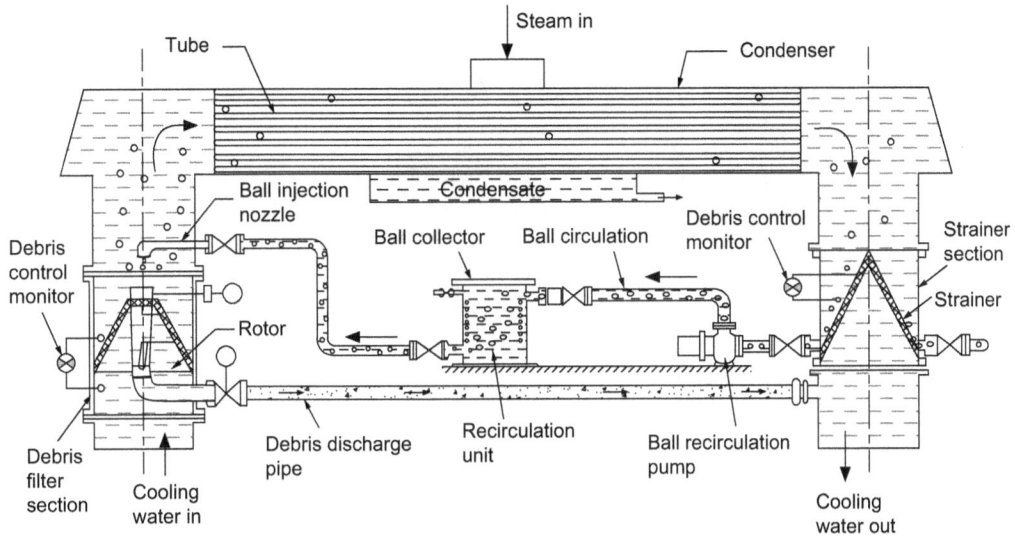

FIGURE 2.28 Installation of in-line debris filters in the inlet pipe to a heat exchanger or condenser combined with sponge rubber ball automatic cleaning system. (Not to scale).

For example, Brackett Green's Automatic Self-Flushing Debris Filters (www.glv.com) fit within the cooling-water lines. The filter housing actually replaces the segment of the cooling-water piping, eliminating any need for additional space or extra piping work. The filtration mechanism is designed to remove coarse as well as fibrous debris, including plastic, muscles and clams, fish, tree roots, seaweed, algae, rocks, wood chips, pine cones, leaves, etc. Foreign matter will first settle on the filter surface. As fouling builds up, differential pressure between filter inlet and outlet increases and the filter has to be cleaned by backwashing.

2.27.1.3 Use of Alfa Laval Backflushing Filter (ALF)

ALF-B is a pressure filter with an automatic flushing arrangement. A differential pressure control system monitors pressure drop over the filter basket and indicates when cleaning is necessary. ALF-B units can be cleaned either automatically, at predetermined intervals, or manually. Alfa Laval backflushing filter is shown in Figure 2.29 [33]. During normal operation, liquid passes through the inlet into the filter basket. The flow diverter valve is open and the flushing valve closed. The liquid passes through the filter basket prior to being discharged at the main outlet.

Regeneration. ALF units can be cleaned either automatically, using a timer, at predetermined intervals, or manually by pushing a button on the control panel. An optional differential pressure control system is available as a back-up and for monitoring the filter status.

2.27.1.4 Macro-Fouling Control

Macro-fouling control measures include the following [34]:

i.   Toxic coating of heat transfer surfaces.
ii. Use of toxic alloys as tube material. Copper alloys have a long and successful history of use in the marine environment as antifoulants. They are strongly resistant to macro-fouling and have low corrosion rates.
iii. Chlorination. Four types of chlorine-based compounds are available to control macro-fouling: calcium hypochlorite, liquid chlorine, liquid sodium ·hypochlorite, or on-site hypochlorite generation.

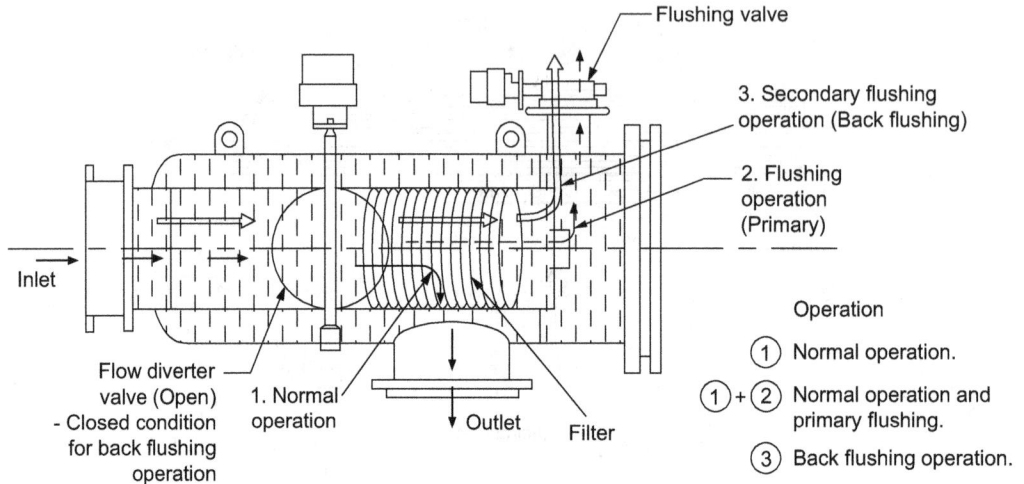

FIGURE 2.29   Alfa Laval backflushing filter (Schematic).

2.27.1.5   Oxidizing Antimicrobials

The application of an oxidizing antimicrobial such as chlorine for the control of undesirable organisms is a well-known and long-practiced procedure. Chlorine is toxic to all living organisms from bacteria to humans.

2.27.1.6   Nonoxidizing Antimicrobials

There are several categories of nonoxidizing antimicrobials that have proven to be very effective in controlling macrofouling organisms. Quaternary amine compounds and certain surfactants have been applied to infested systems for relatively short intervals (6–48 hr). The advantages of a nonoxidizing antimicrobial program include ease of handling, short application time, and relatively low toxicity to other aquatic organisms. In addition, some of these compounds can be readily detoxified [35].

2.27.1.7   Macro-Fouling Cleaning Methods

a.   Mechanical Cleaning

Once a cooling system becomes fouled, mechanical cleaning is required to restore capacity. The system is usually shut down and the walls scraped or hydroblasted to dislodge adhering organisms.

b.   Thermal Treatment (Thermal Backwash)

The organisms that cause macro-fouling can be killed by heated water. Some systems are designed to allow the heated water from the outlet of the condenser to be recirculated back to the intake. As the water recirculates, it is heated and improves macro-fouling control. Thermal treatment is not used extensively because most systems are not designed to recirculate water. Also, when heated water is recirculated the system cooling capacity is greatly diminished.

c.   Miscellaneous Methods

Critical areas of the cooling system may be protected by additional screens and strainers. Other methods that have been used include ultrasonic vibration and electrical shock. They, like thermal treatment, can be effective but require relatively expensive capital equipment and are difficult to maintain and use. Therefore, these methods are not widely used.

2.28 OTHER FOULING TYPES

In general, fouling associated with cooling water can be classified under the following mechanisms [3]:

Scaling due to crystallization of inverse solubility salts, mainly in cooling-tower water and sometimes in water drawn from a river, lake, or well.

Microbiological fouling of water drawn from a river, lake, sea, or the ocean; algae growth in open recirculating water.

Particulate fouling—deposition of silts, sediments, and suspended solids when water is drawn from a lake, river, or nearby seashore and due to airborne objects when water is exposed to the atmosphere.

Corrosion fouling due to cooling-water quality.

Crystallization of inverse solubility salts. One of the most common causes of fouling is due to crystallization of salts having inverse solubility, i.e., when the temperature of the system is raised, as by contact with a hot surface, their solubilities decrease. Common inverse solubility salts include $CaCO_3$, $CaSO_4$, and $Mg(OH)_2$. This type of fouling is referred to as scaling or precipitation fouling. An induction period of a certain time duration is normally present, during which only negligible fouling deposition is observed. At a certain point in the fouling process, the nucleation sites become so numerous they combine together and the fouling increases rapidly.

Sedimentation. This refers to the deposition of particulate matter such as rust and dust particles commonly contained in cooling tower waters. It is frequently superimposed on crystallization fouling processes.

Biological growth. Warm surfaces can provide suitable environments for bacteria, algae, and fungi, which can form layers and reduce the rate of heat transfer. Biofouling is usually prevented by adding chlorine. Fouling of condensers by microbiail slime is controlled in conventional power plants by chlorination and/or by mechanical cleaning. In the power industry, gas refineries, and other industries using large volumes of cooling water, chlorine is commonly used to prevent microbial slime formation on condenser surfaces. In the event that chlorine is unacceptable, other chemicals including detergents, acid, and basic cleaning solutions, are available.

2.28.1 FOULING INFLUENCING PARAMETERS

The parameters that appear to be the most important in effecting the fouling process are velocity/shear stress, surface temperature, water chemistry, and material of the heated surface.

Water chemistry. Generally, pH and the concentration of different mineral salt components have been used to characterize the water chemistry. These quantities can be related to fouling tendencies of water.

Velocity effects. Velocity affects the fouling process with respect to both deposition and removal. For the deposition term, velocity affects the transport of fouling material to the surface. The effect of velocity on removal is characterized by wall shear stress and the mechanical strength of the deposit.

Temperature effects. In cooling-water systems, the temperature of the surface is higher than the bulk fluid temperature. In such cases, the inorganic substances that are inverse solubility salts may deposit on the high-temperature surface. The surface temperature of the deposit is an important parameter and the deposition rate function is of the Arrhenius form, which is characteristic of chemical reactions.

The maximum tube surface temperature (the average between the temperature of the inlet fluid and the exchanger outlet water) on the water side at the tube–water interface is usually the critical concern. The recommended design surface temperature is 145°F (63°C) with 160°F (71°C) as a practical maximum. Heat exchangers with surface temperature above 160°F are prone to localized boiling. Boiling allows concentration of even the most soluble salts, with the threat of severe deposition and subsequent corrosion. For such a situation, water quality is to be improved by full softening or demineralization. (This material is based on Chenoweth [4].)

2.29   METHODS OF COOLING-WATER FOULING CONTROL

2.29.1   REMOVAL OF PARTICULATE MATTER

The amount of particulate entering a cooling system with the makeup water can be reduced by filtration and/or sedimentation processes. Particulate removal can also be accomplished by filtration of recirculating cooling water.

2.29.2   HIGH WATER VELOCITIES

The ability of high water velocities to minimize fouling depends on the nature of the foulant. Clay and silt deposits are more effectively removed by high water velocities than aluminum and iron deposits. To prevent floating particles from settling a minimum flow rate of 1.0 m/s should be maintained at every point of the equipment. Average water velocities of 1.5 to 2.0 m/s have been found to be successful.

2.29.3   DISPERSANTS

Dispersants are materials that suspend particulate matter by adsorbing onto the surface of particles and imparting a high charge. Electrostatic repulsion between like-charged particles prevents agglomeration, which reduces particle growth. The presence of a dispersant at the surface of a particle also inhibits the bridging of particles by precipitates that form in the bulk water.

2.29.4   SURFACTANTS

Surface-active or wetting agents are used to prevent fouling by insoluble hydrocarbons. They function by emulsifying the hydrocarbon through the formation of microdroplets containing the surfactant. The hydrophobic (water-hating) portion of the surfactant is dissolved within the oil drop, while the hydrophilic (water-loving) portion is at the surface of the droplet. The electrostatic charge imparted by hydrophilic groups causes the droplets to repel each other, preventing coalescence. Through a similar process, surfactants also assist in the removal of hydrocarbon-containing deposits

2.30   FOULANT CONTROL VERSUS TYPE OF COOLING-WATER SYSTEM

Fouling problems and their control in each of the three basic types of cooling systems, (once-through, open recirculating, and closed) are often quite different and thus require different techniques. Cooling system operation and specific characteristics should be analyzed constantly as a continuing part of the foulant control program.

2.30.1   COOLING-WATER SYSTEMS

There are three main categories of industrial cooling-water systems:

   i.   once-through where the water is used only once before being returned to the water source;
   ii. recirculating where the warm cooling water is cooled and reused;
   iii. closed recirculating systems.

The above three cooling-water system is shown in Figure 2.30.

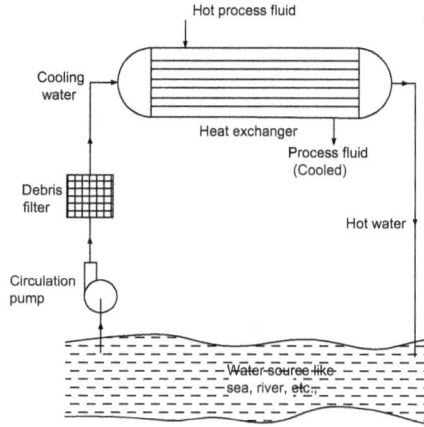

(a) Once through cooling water system.

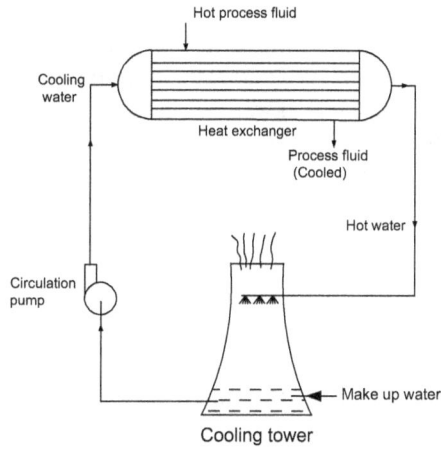

(b) Open recirculating water system.

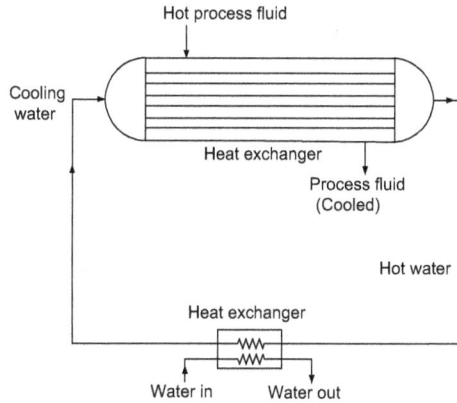

(c) Closed cooling water system.

FIGURE 2.30 Three types of industrial cooling water system.

2.30.1.1   Once-Through System

In once-through cooling-water systems, usually the major foulants are biological organisms, mud, silt, debris, or other suspended matter, and pollutants [8]. This type of cooling-water system generally needs only upstream filtration and a mechanical online cleaning method of passing plugs or sponge rubber balls as in condensers cooled by seawater. Generally, economics favor the chemical treatment. The chemical additives most generally used are mud fluidizers, dispersants, and biocides, particularly chlorination. Environmental regulations may restrict certain chemical additives.

2.30.1.2   Open Recirculating Cooling Systems

An open recirculating cooling system uses the same water repeatedly to cool process equipment. Heat absorbed from the process must be dissipated to allow reuse of the water. Cooling towers, spray ponds, and evaporative condensers are used for this purpose. In a cooling tower the warm water enters at the top and spreads down over numerous vertical panels. The large surface area facilitates evaporation, which lowers the temperature of the water that remains behind. The cooled water collects in a basin at the bottom of the tower, from where it is recirculated to again perform its cooling function. As water evaporates, dissolved solids remain behind and increase in concentration. A continuing increase in dissolved solids can lead to salts of calcium, magnesium, or silica precipitating out of solution and forming scale deposits on cooling system surfaces. To dilute the water and minimize scaling, the concentrated water of the cooling tower is discharged and is then replaced by an equivalent volume of fresh makeup water. The discharge is referred to as "bleed off," or "blowdown." A cooling tower operating at relatively high cycles of concentration will save water compared to a similar one operating at lower cycles.

2.30.1.3   Fouling in Open Recirculating Cooling Systems

In an open recirculating system with cooling towers or spray ponds, foulants originate in the makeup water, air, or process contamination, and from lack of good corrosion control [8]. Hence, this system is frequently treated for foulant control. Though open recirculating cooling systems save fresh water compared to the alternative method, once-through cooling, open recirculating cooling systems are inherently subject to more treatment-related problems than once-through systems:

  i.   cooling by evaporation increases the dissolved solids concentration in the water, raising corrosion and deposition tendencies and periodical blowdown of cooling tower water is necessary;
  ii. the relatively higher cooling-water temperatures significantly increase corrosion potential and the tendency for biological growth;
  iii. airborne gases and nutrients and potential foulants can also be absorbed into the water across the tower.

2.30.1.4   Closed Recirculating Systems

The source of foulants is usually the corrosion products from heat exchangers and piping components. Corrosion control is mostly by inhibitors.

2.30.2   Fouling Mechanisms in Cooling-Water Systems

The primary types of fouling occurring in cooling tower water systems are scaling, i.e., crystalization of inverse solubility salts, deposition/sedimentation, corrosion of the heater surface and biological growth.

2.30.2.1   Crystalization of Inverse Solubility Salts

One of the most common causes of fouling is due to crystallization of salts having inverse solubility, i.e., when the temperature of the system is raised, as by contact with a hot surface, their solubilities decrease. Common inverse solubility salts include $CaCO_5$, $CaSO_4$, and $Mg(OH)_2$. This type of fouling is referred to as scaling or precipitation fouling. An induction period of a certain time duration is normally present, during which only negligible fouling deposition is observed. At a certain point in the fouling process, the nucleation sites become so numerous they combine together and the fouling increases rapidly.

2.30.2.2   Sedimentation

This refers to the deposition of particulate matter such as rust and dust particles commonly contained in cooling tower waters. It is frequently superimposed on crystallization fouling processes.

2.30.2.3   Water Chemistry

Generally, pH and the concentration of different mineral salt components have been used to characterize the water chemistry. These quantities can be related to fouling tendencies of water.

2.30.3   Deposition Monitoring and Control

Deposition presents the most serious barrier to the transfer of heat through a surface and can be divided into two forms:

1. fouling and
2. scaling.

Fouling, due to suspended solids in the water, is the accumulation of water suspended materials on tower fills or heat exchanger surfaces. Scale is a dense coating of inorganic materials and results from the precipitation of soluble minerals from supersaturated water.

2.30.3.1   Fouling Control

Both mechanical and chemical methods are used singly or jointly to overcome fouling problems. Remove solid particles through sedimentation ponds and/or continuous filtration such as automatic debris filter system (ADFS). Mud fluidizers, dispersants, flocculating chemicals and foulant solubilizers, biocides, and other chemicals are added to control fouling.

2.31   ONLINE CHEMICAL CONTROL OF COOLING-WATER FOULANTS

Foulant control through some of the chemical additives include the following:

Threshold chemicals.
Sequestrant and chelating chemicals, such as ethylenediamine tetraacetic acid (EDTA) and derivatives, nitrilotriacetic acid (NTA), organic phosphate esters, and organic phosphonates [8].
Dispersants (e.g., lignins, tannins, alginates, cellulose, starch products, sodium polymethacrylate, and polyvinyl pyridinium butyl bromide) [8].
Sludge fluidizers.
Biocides.

2.32 SCALING TENDENCIES AND CONTROL

The formation of scale most often occurs when the water becomes oversaturated or when water temperature increases. For some scale species, such as calcium compounds, solubility decreases with increasing temperature. As such, these deposits usually occur first at the outlet end of the condenser where the temperature is the highest. The rate of formation of the scale will depend on temperature, alkalinity, or acidity, and the amount of scale forming material in the water. Deposition monitoring can be carried out by measuring the Scaling Indices and/or visual inspections.

2.32.1 SCALING INDICES

A formation prediction can be calculated with the use of indices such as the Langelier Saturation Index (LSI). This index is described as follows:

$$LSI = pH_s - pH$$

where
pH_s is the measured water pH,
pH is the pH at saturation in calcite or calcium carbonate.

If the LSI calculation produces a positive value, the potential to form scale exists; and if a negative value is produced, no scale potential exists. At no time should an LSI value be greater than 2.2. If the values reach 2.2, changes in the treatment program need to be made immediately to decrease the potential for deposition. It is recommended that an LSI calculation be performed once a day, preferably in the afternoon when temperatures are warmest and conditions are most favorable for deposition.

2.32.2 CONTROL OF SCALE FORMATION AND FOULING RESISTANCES FOR TREATED COOLING WATER

The following values for cooling water assume that corrosion is under control and that biological growth does not represent a significant portion of fouling [4]:

1. Since scaling in cooling-water system is due to inverse solubility phenomena, which normally take place about 140°F (60°C), keep the surface temperature below this temperature.
2. The cooling-water velocity is at least 4 ft/s on the tubeside for most nonferrous alloy tubes and 6 ft/s for carbon steel tubes. Velocities as high as 15 ft/s have been used inside titanium tubes.
3. The velocity of cooling water on the shellside is at least 2 ft/s. Higher velocities are permitted if erosion can be tolerated or flow-induced vibration is not possible.
4. Under the preceding conditions, a reasonable design value for the cooling-waterside fouling resistance is 0.001 h · ft^2 °F/Btu.
5. The fouling resistance for a clean heat exchanger should be taken as 0.0005 h · ft^2 · °F/Btu.

The concept of the basic cooling-water treatment program for foulant control is shown in Figure 3.31

2.32.2.1 Chemical Means to Control Scaling

Calcium carbonate formation can be controlled by adding acids or specific chemicals, including sulfuric acid, polymeric inorganic phosphates, phosphonates, and organic polymers like polycarboxylates [36].

```
┌─────────────────────────────────┐
│   Basic cooling water treatment │
│            program              │
└─────────────────────────────────┘
                 │
                 ▼
┌─────────────────────────────────┐
│ 1.  Cooling water system audit and │
│     assessment of water parameters │
│     that influence corrosion, scaling │
│     and fouling.                │
└─────────────────────────────────┘
                 │
                 ▼
┌─────────────────────────────────┐
│ 2.  Cooling water pretreatment. │
└─────────────────────────────────┘
                 │
                 ▼
┌─────────────────────────────────┐
│ 3.  Inhibitors / Chemicals / Biocides │
│     additions.                  │
└─────────────────────────────────┘
                 │
                 ▼
┌─────────────────────────────────┐
│ 4.  Monitoring and control and  │
│     periodical replenishment of │
│     additives.                  │
└─────────────────────────────────┘
```

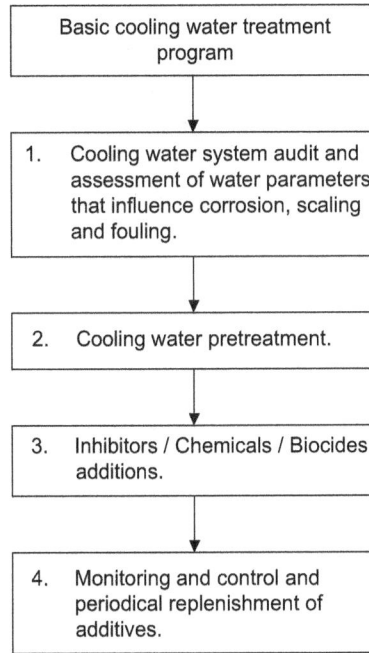

FIGURE 2.31 Basic cooling water treatment program for fouling control.

Removal of the hardness salts: removal of the scaling chemical species such as magnesium and calcium from water prior to use in the system, by ion exchange and lime softening procedures, is effective in scaling control. The large volumes of water usually encountered in many cooling-water systems are likely to make the treatment costs of these methods very high. The modern trend toward very small volumes and high recirculation rates may offer cost restrictions [3].

Conversion of hardness salts to a more soluble form: control of scale formation involves conversion of hardness salts to a more soluble form. Since the solubility of scale-forming species in cooling water generally increases with decreasing pH, the addition of acid to maintain the pH in the range 6.5–7.5 (with 6.5 as the recommended minimum) may reduce the scaling. These effects are shown schematically in Figure 2.32. However, corrosion may be involved if the treatment is not carried out carefully. If large concentrations of sulfate are present, the use of sulfuric acid may cause sulfate scale to appear. Hydrochloric acid may be preferable under these circumstances [3].

Alkaline cooling-water operation for scale control: alkaline cooling-water treatment can also help in scale control. When the organic phosphorous compounds, including the well-known phosphates and polyol esters, and low-molecular-weight acrylic-acid-based polymers and copolymers are used properly, they completely prevent calcium carbonate deposition under alkaline conditions [37].

2.32.2.2 Electrostatic Scale Controller and Preventer

This is online equipment designed as a one-time fitment that performs the dual function of scale prevention and scale removal without using chemical additives. The equipment enables the use of untreated water. By subjecting the hard water to a carefully controlled electrostatic field, hardness ions are kept in suspension and prevented from depositing onto heat transfer surfaces. The merits of this system include (1) elimination of the use of expensive chemicals for descaling; (2) absence of

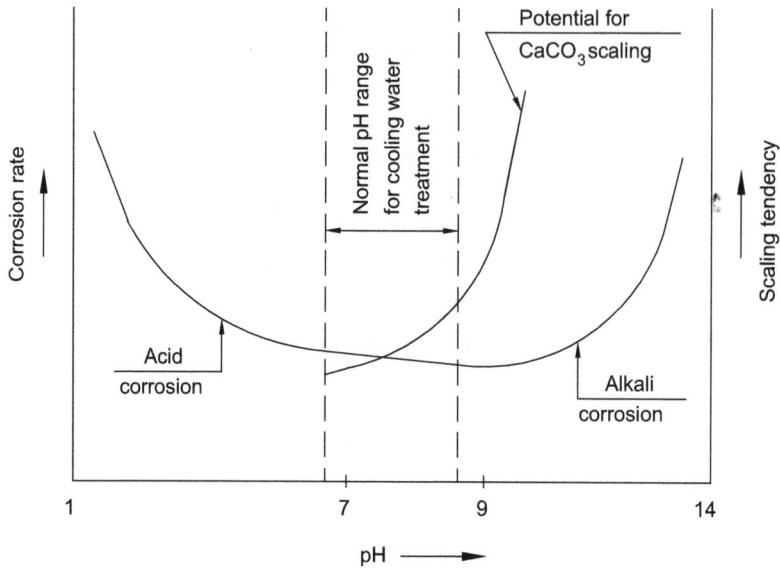

FIGURE 2.32 The effect pH on cooling water corrosion and scaling control.

corrosion of heat transfer surfaces from the use of chemicals for cleaning; and (3) eliminating the need to shut down the equipment for scale cleaning.

2.32.3 CLEANING OF SCALES

2.32.3.1 Chemical Cleaning

Chemical cleaning has been the most common method of scale removal. Mineral acids such as inhibited hydrochloric, ammonium bifluoride, sulfamic acid, and phosphoric acid are strong scale dissolvers. Organic acids are much weaker. They are often used in combination with other chemicals to complex scales. Refer to Table 2.5 for solvents and compatible base metals for scale removal.

2.33 IRON OXIDE REMOVAL

Conventional cleaning methods: two common procedures generally followed in the industries for iron deposit removal are mechanical cleaning and chemical cleaning [38]:

1. Mechanical cleaning methods include water hydroblasting, lancing, and passing abrasive sponges, which remove most of the soft deposits but can leave hard, baked-on deposits.
2. Chemical cleaning with strong mineral acids or high concentrations of chelants up to 10% is used for temperatures up to 180°F (82°C).

Refer to Tables 2.4 and 2.5 for solvents and compatible base metals for iron oxide removal.

With the chemical cleaning methods, the possibility of corrosion of underlying metal surfaces as they dissolve away from the iron deposits cannot be ruled out. To overcome this problem, a new online procedure for removing iron-based deposits from cooling-water systems has been developed. This is explained next.

Online removal of iron deposits [38]: the first step of the online cleaning process is the addition of a tannin-based, iron conditioning agent, which penetrates and softens the deposits. Later, a mild organic acid and dispersants are added and cause sloughing of the conditioned deposits. Postcleaning passivation of all metal surfaces can be accomplished by the normal corrosion inhibitor or by addition of online passivators to prevent flash corrosion. The process cleans transfer lines as well as heat exchangers and usually can be completed in 3–5 days.

2.34 MONITORING OF FOULING

In common practice, assessing the condition of fouling consists of calculating the overall heat transfer coefficient from temperature and flow-rate measurements at a specific point in time and comparing this value with that calculated under clean conditions. The difference between the two is the fouling resistance. The fouling of heat exchanger and other piping or vessel surfaces can be monitored by measuring pressure, flow, temperature, and heat flux, then using the results to calculate fluid frictional resistance. Results can be employed to evaluate treatments for fouling control and study the effect of equipment configurations and operating conditions on containment buildup. The monitoring methods demand for measuring and analyzing systems in heat exchangers several experimental (pressure drop, temperature, heat transfer parameters, electrical parameters, acoustics) and computational methods are adopted including the following [39]:

i. Pressure drop—pressure between inlet and outlet is measured.
ii. Temperature—product outlet/heating medium temperature is measured, and
iii. Heat transfer parameters—heat flux, heat transfer coefficient, thermal resistance is measured

NOMENCLATURE

A_{cr}	rate constant
A_i	surface area on the tube inside, m^2 (ft^2)
A_m	tube wall surface at the mid plane, m^2 (ft^2)
A_o	surface area on the tube outside, m^2 (ft^2)
A_r	chemical reaction fouling rate constant
A_w	total wall area for heat conduction, m^2 (ft^2)
d	tube outside diameter, m (ft)
d_i	tube inside diameter, m (ft)
E	activation energy for chemical reaction fouling
h_i	heat transfer coefficient on the tube inside, W/m$^2 \cdot$ °C (Btu/h $\cdot$ ft$^2 \cdot$ °F)
h_o	heat transfer coefficient on the tube outside, W/m$^2 \cdot$ °C (Btu/h $\cdot$ ft$^2 \cdot$ °F)
L	tube length, m (ft)
k_f	thermal conductivity of fouling deposit, W/m $\cdot$ °C (Btu/h $\cdot$ ft $\cdot$ °F)
K_R	constant for crystallization rate
K_r	chemical reaction fouling (polymerization) rate
K_{sp}	solubility product of CaCO$_3$, (mol/m^3)^2
k_w	thermal conductivity of the conduction wall material, W/m $\cdot$ °C (Btu/h $\cdot$ ft $\cdot$ °F)
$\dot{m}_d$	rate of deposition of fouling mass on the heat transfer surface, kg/m$^2 \cdot$ s (lbm/ft$^2 \cdot$ h)
m_f	net rate of fouling mass deposition per unit area, kg/m$^2 \cdot$ s (lbm/ft$^2 \cdot$ h)
m_f*	asymptotic value of m_f
$\dot{m}_r$	rate of removal of fouling mass from the heat transfer surface, kg/m$^2 \cdot$ s (lbm/ft$^2 \cdot$ h)
m_s	CaCO$_3$ scaling rate, kg/m$^2 \cdot$ s (lbm/ft$^2 \cdot$ h)
R	gas constant
R_f	thermal resistance due to fouling on both sides of a heat exchanger surface, °C/W (°F $\cdot$ h/Btu)
R_f	thermal resistance of fouling deposit for a unit surface area, m^2/W (ft$^2 \cdot$ °F $\cdot$ h/Btu)
R_f*	asymptotic value of thermal resistance due to fouling,°C/W (°F $\cdot$ h/Btu)
R_i	fouling resistance on tube inside surface, m$^2 \cdot$ °C/W (ft$^2 \cdot$ °F $\cdot$ h/Btu)

R_o	fouling resistance on tube outside surface, $m^2 \cdot °C/W$ ($ft^2 \cdot °F \cdot h/Btu$)
R_T	total thermal resistance to heat transfer, $°C/W$ ($h \cdot °F/Btu$)
R_w	thermal resistance of the separating wall, $°C/W$ ($h \cdot °F/Btu$)
t	time, s
t_c	time constant
t_d	time delay period, s
T_s	heat transfer temperature, $°C$ ($°F$)
T_w	conduction wall temperature, $°C$ ($°F$)
U_o	overall heat transfer coefficient based on tube outside surface, $W/m^2 \cdot °C$ ($Btu/h \cdot ft^2 \cdot °F$)
$U_{o.c}$	overall heat transfer coefficient based on tube outside surface in clean condition, $W/m^2 \cdot °C$ ($Btu/h \cdot ft^2 \cdot °F$)
$U_{o.f}$	overall heat transfer coefficient based on tube outside surface in fouled condition, $W/m^2 \cdot °C$ ($Btu/h \cdot ft^2 \cdot °F$)
x_f	thickness of fouling deposit, m (ft)
β	constant $= 1/t_c$
Δt_w	tube wall thickness, m (ft)
ρ_f	density of the foulant, kg/m^3 (lbm/ft^3)

NOTES

1   EMbaffle is a registered trademark of EMbaffle B.V., Alphen a/d Rijn, the Netherlands.
2   HELIXCHANGER is a registered trademark of Lummus Technology Inc.
3   Twisted Tube is a registered trademark of Koch Heat Transfer Company.

REFERENCES

1.   Collier, J. G., Heat exchanger fouling and corrosion, in *Heat Exchangers: Thermal-Hydraulic Fundamentals and Design* (S. Kakac, A. E. Bergles, and F. Mayinger, eds.), Hemisphere, Washington, DC, 1981, pp. 999–1012.

2.   Zelver, N., Roe, F. L., and Characklis, W. G., Monitoring of fouling deposits in heat transfer tubes: Case studies, in *Industrial Heat Exchangers Conference Proceedings* (A. J. Hayes, W. W. Liang, S. L. Richlen, and E. S. Tabb, eds.), American Society for Metals, Metals Park, OH, 1985, pp. 201–208.

3a. Bott, T. R., *Fouling Notebook*, Institution of Chemical Engineers, London, UK, 1990.

3b. Bott, T. Reg, FOULING, DOI: 10.1615/AtoZ.f.fouling, 2011. FOULING (thermopedia.com).

4.   Chenoweth, J. M., Final report of the HTRI/TEMA joint committee to review the fouling section of the TEMA standards, Heat Transfer Research, Inc., Alhambra, CA, 1988 also reproduced in *J. Heat Transfer Eng.*, 11(1), 73–107, 1990.

5a. Epstein, N., Fundamentals of heat transfer surface fouling: With special emphasis on laminar flow, in *Low Reynolds Number Flow Heat Exchangers* (S. Kakac, R. K. Shah, and A. E. Bergles, eds.), Hemisphere, Washington, DC, 1982, pp. 951–964.

5b. Epstein, N., Fouling in heat exchangers, in *Proceedings of the Sixth International Heat Transfer Conference*, Toronto, Canada, August, Vol. 4, Hemisphere, Washington, DC, 1978, pp. 279–284.

6.   O'Callaghan, M., Fouling of heat transfer equipment: Summary review, in *Heat Exchangers: Thermal-Hydraulic Fundamentals and Design* (S. Kakac, A. E. Bergles, and F. Mayinger, eds.), Hemisphere, Washington, DC, 1981, pp. 1037–1047.

7.   Kern, D. Q. and Seaton, R. E., A theoretical analysis of thermal surface fouling, *Br. Chem. Eng.*, 4, 258–262 (1959).

8.   Hasson, D., Precipitation fouling, in *Fouling of Heat Transfer Equipment* (E. F. C. Somerscales, and J. G. Knudson, eds.), *Proceedings of the International Conference on the Fouling of Heat Transfer Equipment, 1979*, Hemisphere, Washington, DC, 1981, pp. 569–586.

9.   Puckorius, P. R., Controlling deposits in cooling water systems, *Mater. Protect. Perform.*, 77, 19–22 (November, 1972).

10. Mukherjee, R., Conquer heat exchanger fouling, *Hydrocarbon Processing, 75*, 121–127 (January, 1996).

11. Silvestrini, R., Waste heat recovery: Heat exchanger fouling and corrosion, *Chem. Eng. Prog., 25*, 29–35 (December, 1979).

12. W.J. Marner J. W. Suitor, "A Survey of Gas-Side Fouling in Industrial Heat-Transfer Equipment" Final Report, Prepared for US Department of Energy Pacific Northwest Laboratory Battelle Memorial Institute Through an Agreement with National Aeronautics and Space Administration by Jet Propulsion Laboratory California Institute of Technology Pasadena, California.

13a. www.suezwatertechnologies.com/products/hydrocarbon-processing/antifoulant-chemical-solutions

13b. Refinery Process Fouling Control, Water Technologies & Solutions Technical Paper, Suez, pp. 1–8.

13c. www.suezwatertechnologies.com/sites/default/files/documents/TP1194EN.pdf

14. Refinery process fouling control, TP1194EN.doc, Feb-12, 2012, General Electric Company, pp. 1–8.

15. Standards of the Tubular Exchanger Manufacturers Association (TEMA), 10th edn., 2019, Tubular Exchanger Manufacturers Association, Inc., Tarrytown, NY.

16. *Heat Exchanger Performance Monitoring Guidelines*, Electric Power Research Institute, Palo Alto, CA, December 1991, Report NP-7552.

17. Knudsen, J. G., Apparatus and techniques for measurement of fouling of heat transfer surface, in *Fouling of Heat Transfer Equipment Proceedings of the International Conference on the Fouling of Heat Transfer Equipment, 1979* (E. F. C. Somerscales and J. G. Knudson, eds.), Hemisphere, Washington, DC, 1981, pp. 57–82.

18. Marner, W. J. and Henslee, S. P., A survey of gas side fouling measuring devices, in *Industrial Heat Exchangers Conference Proceedings* (A. J. Hayes, W. W. Liang, S. L. Richlen, and E. S. Tabb, eds.), American Society for Metals, Metals Park, OH, 1985, pp. 209–226.

19. Feltzin, A. E., Garcia, H., and Alberto, I. L., Fouling and corrosion in open recirculating cooling water systems: The expert system approach, *Mater. Perform., 44*, 57–61 (June, 1988).

20. Knudsen, J. G., Conquer cooling-water fouling, *Chem. Eng. Prog., 87*, 42–48 (April, 1991).

21. Epstein, N., On minimizing fouling of heat transfer surfaces, in *Low Reynolds Number Flow Heat Exchangers* (S. Kakac, R. K. Shah, and A. E. Bergles, eds.), Hemisphere, Washington, DC, 1982, pp. 973–972.

22. Deghan, T. F., Corrosion in the chemical processing industry, in *Metals Handbook*, Vol. 13, *Corrosion*, 9th edn. (L. J. Korb and D. L. Olson, eds.), American Society for Metals, Metals Park, OH, pp. 1134–1185, 1987.

23. Lester, G. D. and Walton, R., Cleaning of heat exchangers, in *Practical Application of Heat Transfer*, C55/82, IMechE, London, U.K., pp. 1–10, 1982.

24. www.nationalheatexchange.com/services/shell-side-cleaning/

25. www.chemicalprocessing.com/assets/wp_downloads/pdf/091215_NLP_wpa.pdf

26. George E. Saxon, Jr. and Richard E. Putman, The practical application and innovation of cleaning technology for heat exchangers, Heat Exchanger Fouling and Cleaning: Fundamentals and Applications, Art. 40 [2003], Conco Systems, Inc., Verona PA., pp.1–8. gsaxonjr@concosystems.net; http://dc.engc onfintl.org/heatexchanger/40

27. Mol, R. A., Thermal Cleaning Of Heat Exchangers, No Alternative But A Better Way To Clean, Thermo-Clean Group, Dellestraat 45, B-3550 Heusden-Zolder ISBN: 978-0-9984188-1-0; Published online www.heatexchanger-fouling.com, Heat Exchanger Fouling and Cleaning, 2019.

28. Woodward, A. R., Howard, D. L., and Andrews, E. F. C., Condensers, pumps and cooling water plant, in *Modern Power Station Practice*, Vol. C, *Turbines, Generators, and Associated Plant* (D. J. Littler, E. J. Davies, H. E. Johnson, F. Kirkby, P. B. Myerscough, and W. Wright, eds.), 3rd edn., Pergamon Press, New York, 1992.

29. Stegelman, A. F. and Renfftlen, R., On line mechanical cleaning of heat exchangers, *Hydrocarb. Process., 82*, 95–97 (1983).

30. Cole, L. T., Allen, C. A., Liquid-fluidized bed heat exchanger flow distribution models, Icp 1152. 1972. National Technical Information Service, Alexandria, VA,

31. Jan De Baat Doelman, Electronic water treatment reduces fouling in heat exchangers Jan De Baat Doelman, Chemical Industry Digest. November 2013, pp. 66–69.

32. www.westech-inc.com/products/debris-filter
33. www.alfalaval.cn/globalassets/documents/products/separation/automatic-back-flushing-filters/Alf-filters-pee00007en.pdf
34. Chapter 28—Macrofouling Control www.watertechnologies.com/handbook/chapter-28-macrofouling-control
35. www.watertechnologies.com/handbook/chapter-28-macrofouling-control
36. Boffardi, P., Control of environmental variables in water-recirculating systems, in *Metals Handbook*, Vol. 13, *Corrosion*, 9th edn., American Society for Metals, Metals Park, OH, pp. 487–497.
37. Freedman, A. J., Cooling water technology in the 1980s, *NACE Mater. Perform.*, *11*, 9–16 (1984).
38. Kaplan, R. I. and Ekis, E. W., Jr., The on-line removal of iron deposits from cooling water systems, *NACE Mater. Perform.*, *40*, 40–44 (1984).
39. Wallhäußer, E., Hussein M.A. and Becker, T., "Clean or not clean—Detecting fouling in heat exchangers," AMA Conferences 2013—SENSOR 2013, OPTO 2013, IRS 2013 Technische Universität München, Lehrstuhl für Brauund Getränke technologie, Weihenstephaner, Freising, Deutschland, pp. 121–125.

BIBLIOGRAPHY

Branch, C. A. and Muller-Steinhagen, H. M., Influence of scaling on the performance of shell and tube heat exchangers, *Heat Transfer Eng.*, *12*, 37–45 (1991).
Characklis, W. G., Microbial fouling: A process analysis, in *Fouling of Heat Transfer Equipment* (E. F. C. Somerscales and J. G. Knudson, eds.), Proceedings of the International Conference on the Fouling of Heat Transfer Equipment, 1979, Hemisphere, Washington, DC, 1981, pp. 251–292.
de Deus, J. and Pinheiro, R. S., Fouling of heat transfer surfaces, in *Heat Exchangers: Thermal-Hydraulic Fundamentals and Design* (S. Kakac, A. E. Bergles, and F. Mayinger, eds.), Hemisphere, Washington, DC, 1981, pp. 1013–1035.
Epstein, N., Fouling models: Laminar flow, in *Low Reynolds Number Flow Heat Exchangers* (S. Kakac, R. K. Shah, and A. E. Bergles, eds.), Hemisphere, Washington, DC, 1982, pp. 965–972.
Fassbender, L. L., Industrial fouling data base development, in *Industrial Heat Exchangers Conference Proceedings* (A. J. Hayes, W. W. Liang, S. L. Richlen, and E. S. Tabb, eds.), American Society for Metals, Metals Park, OH, 1985, pp. 227–238.
Gudmundsson, J. S., Particulate fouling, in *Fouling of Heat Transfer Equipment* (E. F. C. Somerscales and J. G. Knudson, eds.), Proceedings of the International Conference on the Fouling of Heat Transfer Equipment, *1979*, Hemisphere, Washington, DC, 1981, pp. 357–388.
How Automated Water Jetting Improves Tube Bundle Cleaning Efficiency "The Leader in Water Jet Productivity" 29830 Beck Road Wixom, MI.
www.projectiletube.com/heat-exchanger-tube-cleaning/
Jasner, M., Hecht, M. and Beckmann W., Heat exchangers and piping systems from copper alloys—commissioning, operating and shutdown, *KME; 1998*.
Klaren, D. G., Fluid bed heat exchangers—A new approach in severe fouling heat transfer, *Resour. Conserv.*, *7*, 301–314 (1981).
Marner, W. J., Progress in gas-side fouling of heat transfer surfaces, in *Compact Heat Exchangers—A Festschrift for A. L. London* (R. K. Shah, A. D. Kraus, and D. Metzger, eds.), Hemisphere, Washington, DC, 1990, pp. 421–490.
Pritchard, A. M., Fouling—Science or art? An investigation of fouling and antifouling measures in the British Isles, in *Fouling of Heat Transfer Equipment* (E. F. C. Somerscales and J. G. Knudson, eds.), Proceedings of the International Conference on the Fouling of Heat Transfer Equipment, 1979, Hemisphere, Washington, DC, 1981, pp. 513–527.
Sheikholeslami, R. and Watkinson, A. P., Scaling of plain and externally finned heat exchanger tubes, *Trans. ASME J. Heat Transfer, 108*, 147–152 (1986).
Smith, S. A. and Dirks, J. A., Costs of heat exchanger fouling in the U.S. industrial sector, in *Industrial Heat Exchangers Conference Proceedings* (A. J. Hayes, W. W. Liang, S. L. Richlen, and E. S. Tabb, eds.), American Society for Metals, Metals Park, OH, 1985, pp. 339–344.

Somerscales, E. F. C., Introduction and summary, in *Fouling of Heat Transfer Equipment* (E. F. C. Somerscales and J. G. Knudson, eds.), Proceedings of the International Conference on the Fouling of Heat Transfer Equipment, 1979, Hemisphere, Washington, DC, 1981, pp. 1–30.

Suitor, J. W., Precipitation fouling, in *Fouling of Heat Transfer Equipment* (E. F. C. Somerscales, and J. G. Knudson, eds.), Proceedings of the International Conference on the Fouling of Heat Transfer Equipment, 1979, Hemisphere, Washington, DC, 1982.

3 Corrosion

3.1 BASICS OF CORROSION

Most common metals and their alloys are attacked by environments, such as the atmosphere, soil, water, or aqueous solutions. This destruction of metals and alloys is known as corrosion. It is generally agreed that metals are corroded by an electrochemical mechanism. With practically all commercial processes engineered on a continuous basis of operation, premature failure from corrosion of various types of equipment, including heat exchangers, piping, and others, may mean costly shutdowns and expensive maintenance operations. It is especially troublesome in oil refining, chemical industries, and electric power plants on land and sea, as well as in food and liquor processing, paper manufacture, refrigeration, air-conditioning, etc. Therefore, an understanding of corrosion principles and corrosion control should be of great interest to industry and the general public.

3.1.1 REASONS FOR CORROSION STUDIES

There are two main reasons for concern about and study of corrosion: (1) economics and (2) conservation of materials. Of these, the economic factors mostly favor study and research into the mechanisms of corrosion and the means of controlling corrosion. Economic reasons for corrosion study include the following:

1. Loss of efficiency: corrosion can result in the buildup of corrosion products and scale, which can cause a reduction in heat transfer as well as an increase in the power required to pump the fluid through the system.
2. Loss of product due to leakage: high fuel and energy costs as a result of leakage of steam, fuel, water, compressed air, or process fluid that absorbed energy.
3. Possible impact on the environment: if the leaking fluid is corrosive in nature, it will attack its surroundings, and if lethal or poisonous, it will create hazards and environmental problems. Discharge of copper- and chromate-treated water is severely regulated to conserve aquatics and biosphere.
4. Lost production as a result of a failure.
5. High maintenance costs.
6. Warranty claims on corroded equipment and the consequent loss of customer confidence, sales, and reputation.
7. Contamination and loss of product quality, which can be detrimental to the product, such as foodstuffs, soap products, discoloration with dyes, etc.
8. Extra working capital to carry out maintenance operations and to stock spares to replace corroded components.

DOI: 10.1201/9781003352068-3

9. Overdesign: in many instances, when the corrosive effect of a system is known, additional thickness to components is provided for in the design. This is known as corrosion allowance and involves additional material cost and extra weight of new units.
10. Highly corrosive fluids may require the use of expensive materials such as titanium, nickel-base alloys, zirconium, tantalum, copper–nickels, etc. The use of these materials contributes to increased capital cost.
11. Damage to adjacent equipment and the system components.

3.1.2 CORROSION MECHANISM

According to electrochemical theory, the combination of anode, cathode, and aqueous solutions constitutes a small galvanic cell, and the corrosion reaction proceeds with a flow of current in a manner analogous to the way current is generated by chemical action in a primary cell or in a storage battery on discharge. Due to the electrochemical action, the anode is dissolved. For a current to flow, a complete electrical circuit is required. In a basic corroding system as shown in Figure 3.1, the circuit is made up of four components:

1. Anode.
2. Electrolyte.
3. Cathode.
4. External circuit.
1. Anode: the anode is the electrode at which oxidation (corrosion) takes place and current in the form of positively charged metal ions enters the electrolyte. At the anode, the metal atom loses an electron, oxidizing to an ion.
2. Electrolyte: the electrolyte is the solution that surrounds, or covers, both the anode and the cathode. The conductivity of the solution is the key to the speed of the corrosion process. A solution with low conductivity produces a slow corrosion reaction, while a solution with high conductivity produces rapid corrosion [1]. In the total absence of an electrolyte, little or no corrosion takes place. For example, iron exposed to dry desert air remains bright and shiny since water necessary to the rusting process is not available, and in arctic regions, no rusting is observed because ice is a nonconductor [2]. The electrolyte need not be liquid. It can be a solid layer also. For example, at elevated temperatures, corrosion of a metal can occur in the absence of water because a thick metal oxide scale acts as the electrolyte (Figure 3.2). The surface metal oxide is the cathode and the metal oxide and metal interface are the anode [2].
3. Cathode: the cathode is the electrode at which reduction takes place and current enters from the electrolyte.
4. External circuit: if there are two pieces of metal, they must either be in contact or have an external connection in order for the corrosion to take place. The external circuit is a metallic path between the anode and cathode that completes the circuit. Where the anode and cathode are on the metal surface, as shown in Figure 3.3, the metal itself acts as the external circuit.

3.1.2.1 Basic Corrosion Mechanism of Iron in Aerated Aqueous System

The corrosion principle is explained by the basic corrosion mechanism of iron in an aerated aqueous system (Figure 3.4). In its simplest form, this reaction consists of two parts: (1) the dissolution of iron at the anode and (2) cathodic reaction in the absence of oxygen or the reduction of oxygen to form hydroxyl ions at the cathode. Hence, the overall theoretical corrosion reaction becomes

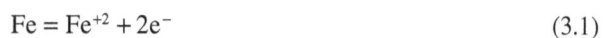

$$Fe = Fe^{+2} + 2e^- \tag{3.1}$$

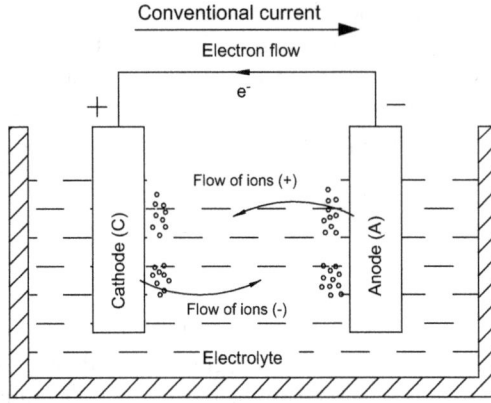

(a) Galvanic cell.

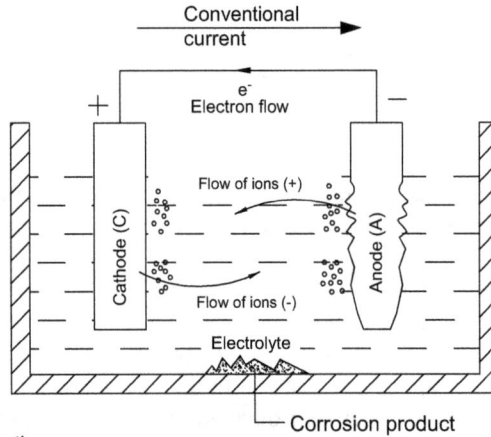

Reactions

At the anode : $Fe \rightarrow Fe^{++} + 2e^-$ (Current)

At the cathode : $2e^- + 2H+ \rightarrow H_2$

$H_2O \leftrightarrow OH^- + H^+$

$Fe^{++} + 2\ OH^- \rightarrow Fe(OH)_2$ (Corrosion product)

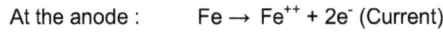

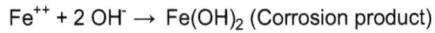

(b) Corrosion cell.

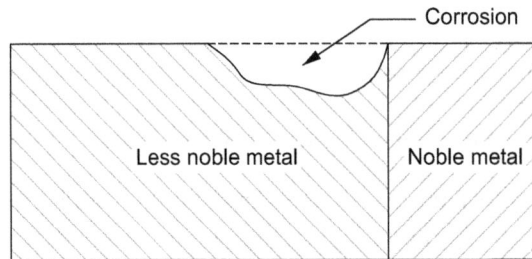

(c) Preferential galvanic corrosion of the more active metal.

FIGURE 3.1   Galvanic cell—basic corroding system. (Electrolyte is not shown).

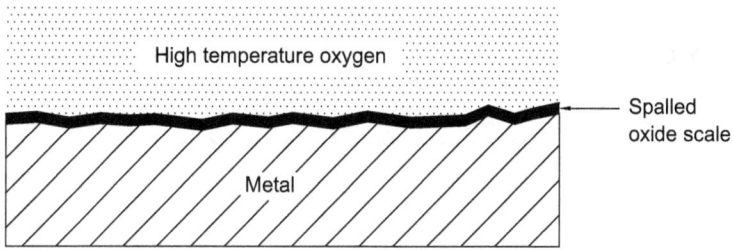

FIGURE 3.2 Thick metal oxide scale as the electrolyte.

Source: Adapted from Pludek, R., *Design and Corrosion Control*, The Macmillan Press Ltd., London, U.K., 1977.

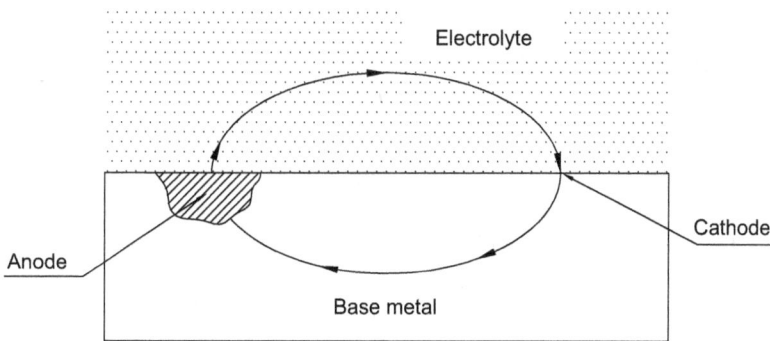

FIGURE 3.3 Anode and cathode on the same metal surface constituting a galvanic cell.

$$2H^+ + 2e^- \rightarrow H_2 \tag{3.2}$$

$$\frac{1}{2}O_2 + H_2O + 2e^- = 2OH^- \tag{3.3}$$

$$Fe + \frac{1}{2}O_2 + H_2O = Fe^{+2} + 2OH^- \tag{3.4}$$

followed by

$$Fe^+ + 2OH^- \rightarrow Fe(OH)_2 \tag{3.5}$$

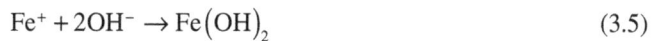

In practice, however, these reactions are much more complex. The rate of corrosion is governed by the rate of either the cathodic reaction or the anodic reaction or less frequently by the electrical resistance of the electrolyte [2]. When the anodic reaction is severely limited by films that form on the metal surface, as in the case of stainless steels, the metal is called passive.

3.1.3 Forms of Electrochemical Corrosion

Various forms of electrochemical corrosion are (1) bimetallic cell or dissimilar electrode cell; (2) concentration cell; and (3) differential temperature cells.

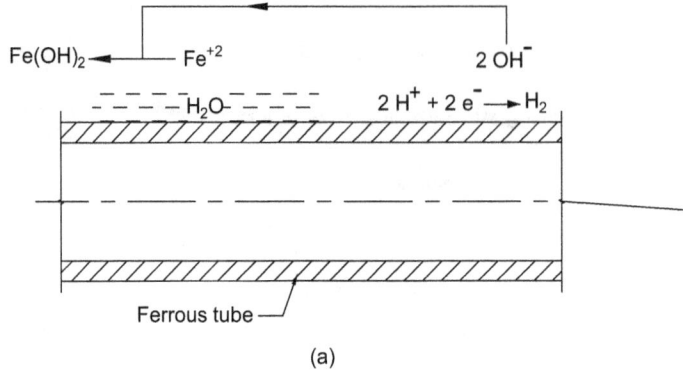

(a)

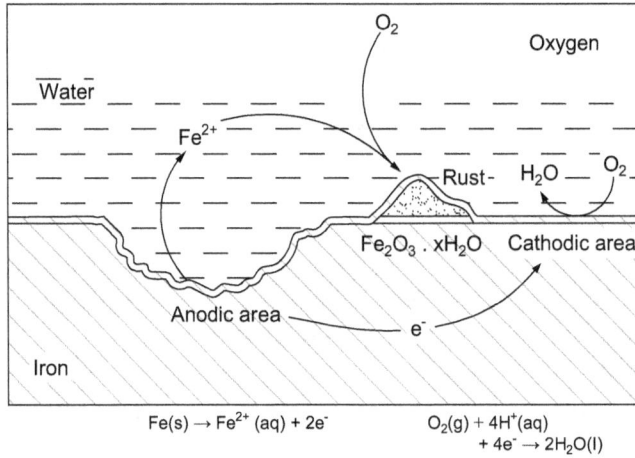

(b) Principle of rusting.

FIGURE 3.4   Corrosion of iron in water.

3.1.3.1   Bimetallic Cell

In a bimetallic cell, two dissimilar metals are in contact and are immersed in an electrolyte. The far-ther the two metals are apart in the electromotive series, the more severe is the corrosion. Bimetallic corrosion also takes place when a metal is nonhomogeneous. One part of a metallic structure may be anodic to another part if it is not exactly the same alloy. A bimetallic cell is also known as a dis-similar electrode cell.

3.1.3.2   Concentration Cell

A concentration cell is produced in an identical electrode, in contact with an electrolyte of a differing concentration. The area of metal in contact with the dilute solution will be anodic and it will corrode. It is often called crevice corrosion because a crevice acts as a diffusion barrier, and corrosion occurs most often within a crevice. There are mainly two kinds of concentration cell, as shown in Figure 3.5: (1) the salt concentration cell, formed due to differences in concentration of the salt in the electrolyte (Figure 3.5a), and (2) the differential aeration cell, where the oxygen con-centration varies on the electrode surface, with the anodic area being the one having lower oxygen concentration (Figure 3.5b). Such cells account for localized corrosion of metals at crevices formed by overlapping joints, threaded connections, or by microorganisms growing on metal surfaces [2]. The shielded area tends to be low in dissolved oxygen concentration and hence become anodic to the outside area with higher oxygen concentration.

(a)

(b)

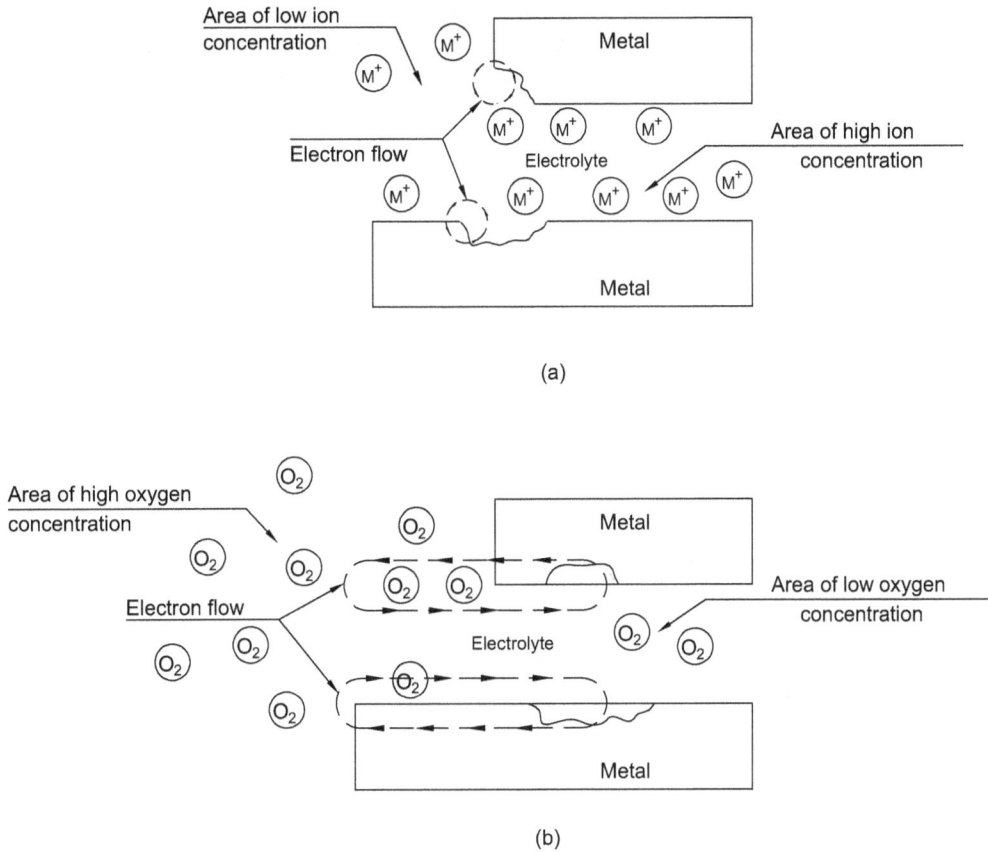

FIGURE 3.5 Concentration cell. (a) Salt concentration cell and (b) oxygen concentration cell.

Source: Adapted from *Resistance to Corrosion*, 4th edn., Inco Alloys International, Huntington, WV, 1985.)

3.1.3.3 Differential Temperature Cells

Components of these cells are electrodes of the same metal, each of which is at a different temperature, immersed in an electrolyte of the same initial composition. The surfaces at higher temperatures are generally anodic with respect to cooler ones.

3.1.4 CORROSION POTENTIAL AND CORROSION CURRENT

The driving force for current and corrosion is the potential developed between the metals. When no current flows between the anode and the cathode, the potential difference between them is at a maximum known as the open circuit potential. The current that flows between the anode and the cathode when a metal corrodes in an electrolyte is called the corrosion current, and the net potential of the corroding surface is the corrosion potential.

3.1.5 CORROSION KINETICS

The preceding discussion deals with the criteria leading to the formation of an electrolytic cell, which is the essential step in the corrosion process. However, the potential difference of the galvanic couple can change with time. As corrosion progresses, corrosion products may accumulate at the anode, cathode, or both. This reduces the speed at which corrosion proceeds. The phenomena affecting the corrosion kinetics are referred to as polarization and passivation. These two phenomena are also extremely important in the preventive measures that can be used for corrosion control.

3.1.5.1 Polarization Effects

The phenomenon that controls the rate of corrosion reaction is known as polarization, which is the ease with which anodic and cathodic reactions take place. In simple terms, due to polarization, the potentials of the metals in a corrosion cell tend to approach each other. The decrease in anode potential is anodic polarization, and the decrease in cathode potential is cathodic polarization. This reduced voltage can drive less additional current through the cell. It is not always true that both anodic and cathodic polarization will take place to the same extent. In some cases, greater polarization is at the anode and in other cases at the cathode. In the former case, the reaction is said to be anodically controlled. In the latter case, it is said to be cathodically controlled.

Polarization in iron–water system: in the iron–water system, the reaction is cathodically controlled because hydrogen ions are available in small quantity. In other words, cathodic polarization limits the rate of reaction [3]. Oxygen is a depolarizer because it decreases the slope of one of the lines, thereby increasing the corrosion current and, in this reaction, the amount of corrosion. A little consideration of this will indicate the importance of polarization in limiting the corrosion rate without it reaching an infinite value, which would have been the case in the absence of any polarization. It can take the form of slow ion movement in the electrolyte, slow combination of atoms to form gas molecules, or slow solvation of ions by electrolyte [1].

Factors affecting polarization: the degree of polarization is variable; some corrosion reactions proceed rapidly owing to high spontaneity and low polarization, and others proceed very slowly owing to high polarization even though they have a pronounced tendency to corrode as shown by reversible EMF of the corrosion cell [4].

Polarization diagrams: plots of anode and cathode potential versus current flow (E_p vs. I) are called polarization diagrams. By plotting current density on a logarithmic scale, the polarization lines will be linear, in accordance with the Tafel equation. These diagrams are also called Evans diagrams, after one of the founders of corrosion science, Ulick Evans. Idealized polarization diagrams are shown in Figure 3.6. Figure 3.6a shows a polarization diagram for a cathodically controlled corrosion cell and Figure 3.6b shows a polarization diagram for an anodically controlled corrosion cell. A polarization diagram for a passive metal anode is shown schematically in Figure 3.6c, and it can be seen from this figure that it does not polarize along a straight line as shown in an idealized diagram, but follows an S-shaped curve. The electrochemical behavior of active–passive transitions is illustrated by such curves. Such diagrams can be used to show the effects of cathodic polarization by hydrogen and anodic polarization by accumulated metal ions and corrosion products. In actual practice, the polarization curves would not be straight lines. The shapes of these curves will depend on the particular process responsible for the polarization [4].

Polarization measurement: polarization measurements on the members of a galvanic couple can provide precise information regarding their behavior, particularly the prediction of localized corrosion. Polarization techniques and critical potentials are used to measure the susceptibility to pitting and crevice corrosion of metals and alloys in a chloride solution [5].

3.1.5.2 Passivation

Sometimes material corrodes, producing an adherent corrosion product that protects it from further corrosion. Such (passivated) material corrodes very little in a specific environment, even though it

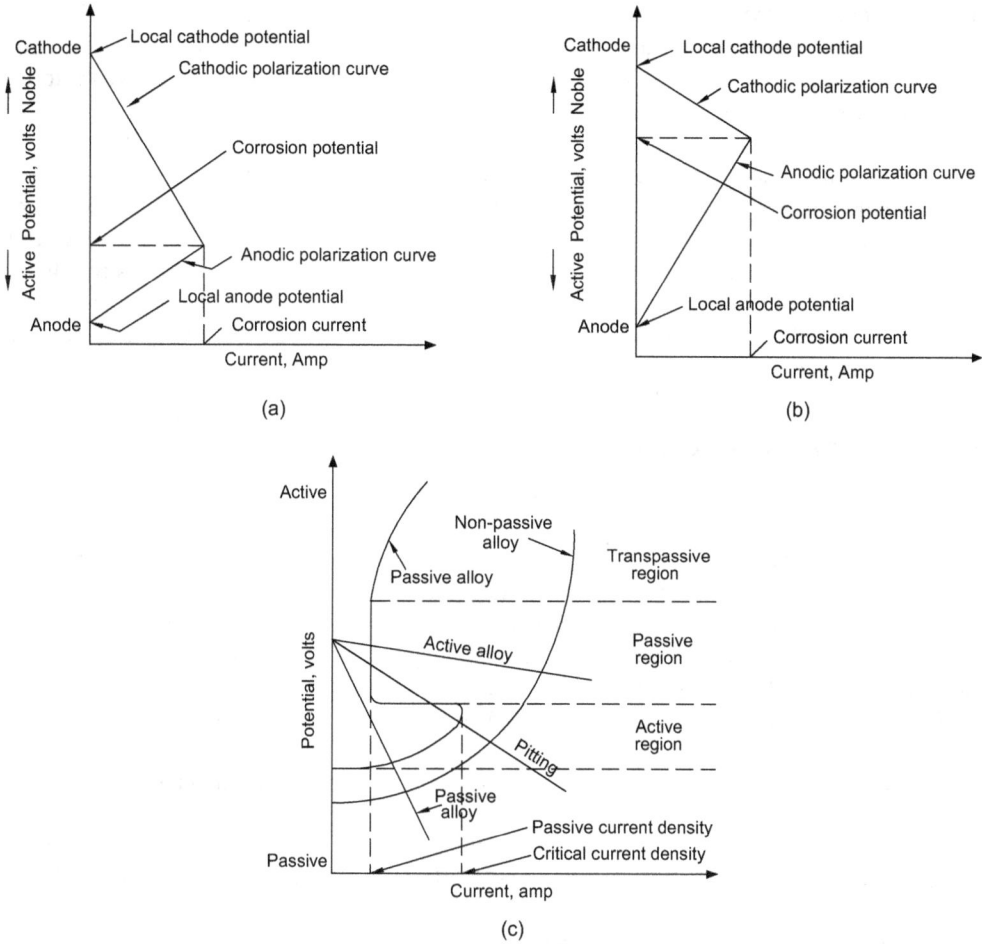

FIGURE 3.6 Polarization diagrams. (a) Cathodically controlled corrosion, (b) anodically controlled corrosion, and (c) comparison of anodic polarization curves for passive and nonpassive materials.

would otherwise corrode considerably [6]. For example, a look at the galvanic series will indicate that aluminum should corrode at a high rate. In practice, however, it is found that aluminum is highly resistant to attack in most of the media except halides. This phenomenon is known as passivation. Materials such as nickel, titanium, zirconium, chromium, and stainless steel owe their corrosion resistance to natural passivation.

Passivity can be understood through a study of polarization diagrams (schematics) presented by Roser et al. [7]. The anodic polarization curves of passive alloys shown in Figure 3.6b are distinctly different from those of nonpassive alloys. Comparison of anodic polarization curves for passive and nonpassive materials is shown in Figure 3.6c. Passivation is a result of marked anodic polarization whereby a barrier of thin protective film, either metal oxide or chemisorbed oxygen, is formed between the metal and the environment, preventing further contact with the electrolyte. In the case of iron, when more oxygen reaches the metal surface than that can be used in the cathodic reaction, a protective passive film is able to form [3]. Thus, the attainment of passivity is thus most important in avoiding accelerated corrosion. Whether a given alloy will be passive in a given situation depends on both the anodic and the cathodic polarization effects.

Passive alloys are widely used as corrosion-resistant materials for the construction of heat exchangers. The corrosion resistance of passive alloys depends on the chromium content, chloride and oxygen content in the environment, and the temperature [7]. Attainment of passivity in a given situation depends on the relative value of all factors rather than on any one of them. For example, high chromium aids passivity, low temperature aids passivity, depassivating ions, such as chlorides, hinders passivity, and oxygen aids passivity.

Behavior of passive alloys: passive material corrodes very little in a specific environment, even though it would otherwise corrode considerably [6]. Conversely, alloys that commonly exhibit passivity are invariably quite active in the nonpassive state. Some elements break down passive films, causing the metal to corrode where the film is discontinuous. Chlorine ions, e.g., destroy the passivity of aluminum, iron, and the stainless steels, causing pitting corrosion. Therefore, the users of passive alloys should be particularly on guard for pitting, stress corrosion cracking (SCC), sensitization, and oxygen starvation-type corrosion [7].

3.1.6 Factors Affecting Corrosion of a Material in an Environment

The corrosion process is affected by various parameters:

1. Environment factors, such as concentration of chemicals, pH, velocity, impurities and suspended matter, and temperature of the medium.
2. Source of heat, if any. If the environment is heated through the material being selected, the effects of heat transfer and surface temperature may be the controlling factors.
3. Material factors like composition, alloying elements, passivity, tendency for fouling.
4. Design conditions and geometry of the joints, like gasketed surfaces, crevices, stagnant areas, and U-bends.
5. Fabrication techniques: corrosion due to welding, brazing, soldering, and heat treatment.

Factors influencing corrosion are shown schematically in Figure 3.7. Only the environmental factors are discussed next. The other factors are discussed while discussing various forms of corrosion.

3.1.6.1 Environmentally Assisted Cracking

Environmentally assisted cracking (EAC) or environmental cracking is a form of corrosion, in which the combined action of a tensile stress and a corrosive environment causes cracks to form in the metal. The likely damages include the following:

i. Stress corrosion cracking (SCC).
ii. Corrosion fatigue.
iii. Hydrogen embrittlement, hydrogen-induced cracking (HIC), etc.

Stresses that cause environmental cracking arise from residual cold work, welding, grinding, thermal treatment, or may be externally applied during service and, to be effective, must be tensile (as opposed to compressive). Forms of environmental cracking are discussed later.

3.1.6.2 ISO 12944—The Corrosive Environment

Corrosive Environment

The corrosive environment describes the environment in which the asset to be protected is situated. The environmental factors that determine an environment's corrosivity are as follows:

i. Climate (the weather, which is established by reference to historical data).
ii. Atmosphere (the gases—including aerosols and particles—that surround the asset to be protected against corrosion).

Corrosivity is dependent on the corrosive agents present in the environment, especially gases such as sulfur dioxide, and salts such as chlorides and sulfates.

3.1.6.3 Surface Layer

Formation of a surface layer, whether it is based on an oxide, carbonate, sulfate, or any other compound, is a major factor in corrosion resistance, particularly if the layer effectively isolates the metal substratum from the environment. Some oxides of copper, aluminum, and chromium typically form very slowly and coat the underlying metal. Some metals, such as SS, titanium, or aluminum, are frequently left unpainted when exposed to the atmosphere since oxygen in the air helps develop a protective oxide layer on the metallic surface.

3.1.6.4 Environments and Corrosion Effects

Some environments are more corrosive than others. Although there are exceptions, the following statements are generally accepted as facts [8]:

i   Moist air is more corrosive than dry air.
ii   Hot air is more corrosive than cold air.
iii   Polluted air is more corrosive than clean air.
iv   Hot water is more corrosive than cold water.
v   Salt water is more corrosive than fresh (low chloride content) water.
vi   Acids are more corrosive than bases (alkalis) to steels.
vii   SS will outlast ordinary steel.
viii   No corrosion will occur in a vacuum, even at very high temperatures.

3.1.6.5 Details of Environmental Factors that Cause Corrosion

Presence of impurities: impurities or contaminants in the corrosive environment can cause either general corrosion or localized attack within the system, or both. The presence of even minor amounts of impurities can alter the corrosion rate significantly. For example, chlorides above 30 ppm will increase the corrosion rate of austenitic stainless steel drastically.

Temperature of the corrodent: as a rule, the degree of corrosion increases with increase in temperature, but increasing temperature also tends to drive dissolved gases out of solution so that a reaction that requires dissolved oxygen can often be slowed down by heating [2]. There are numerous cases where metals satisfactory for cold solutions are unsuitable for the same solutions at elevated temperatures. For instance, refrigerant-quality brine can be handled by a plate heat exchanger (PHE) in 18Cr-12N-2.5Mo (AISI 316) stainless steel, provided the surface temperature does not exceed 10°C (50°F). At higher temperatures, plate failure due to pitting and/or stress corrosion is inevitable.

Degree of aeration and oxygen content: the design of the plant and equipment selection, in particular, can influence the amount of air introduced into a process stream, which in turn may have an influence on corrosion. Oxygen can behave as a depolarizer and increase the rate of corrosion by speeding up the cathodic reaction. It can also act as a passivator because it promotes the formation of a stable passive film [2]. It also must be understood that the major contributor to the corrosion of all metals in the atmosphere is oxygen.

Velocity of corrodent: velocity of the corrodent affects both the type and the severity of the corrosion and removal of fouling deposits. Corrosion is favored by too low or too high velocities. Uniform and constant flow of process fluids past heat exchanger favors less fouling and hence less corrosion. High velocity helps to prevent the accumulation and deposition of corrosion products, which might create anodic sites to initiate corrosion, to maintain clean surfaces free from fouling deposits, and to avoid

Environment factors

Degree of aeration
Flow velocity
Impurities, pH
Temperature,
concentration of corrodent, etc.

Material factors

Heat treatment
Microstructure
Alloying elements
Composition
Surface conditions
Passivity

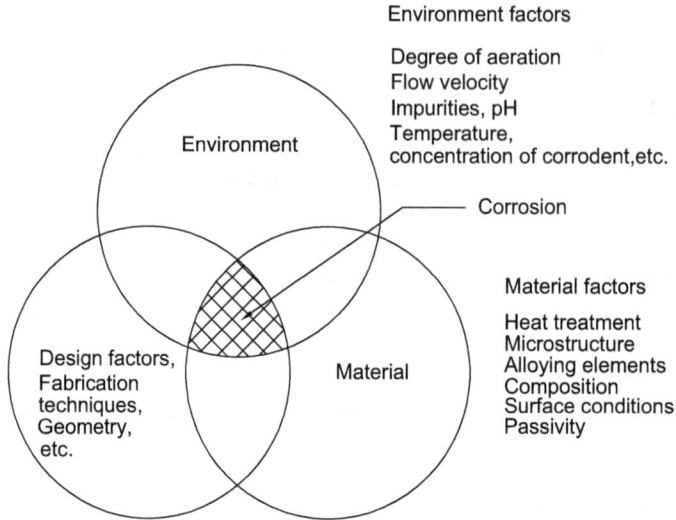

FIGURE 3.7 Factors influencing corrosion.

crevices and stagnant areas. On the other hand, too high a velocity can destroy the protective surface film and result in erosion–corrosion, especially on metals such as copper and aluminum alloys.

Adherent deposits: deposits on the metal surface cause crevices. These act as sites for accelerated corrosion. Adherent deposits cause localized hot spots, which in turn contribute to high-temperature corrosion.

Concentration of corrodent: in general, the corrosion rate increases with increasing concentration, including the concentration of the aggressive chemical species, such as chloride ions. However, there are exceptions also. For example, iron is attacked vigorously in dilute nitric and sulfuric acids, but the corrosion rate is drastically reduced in concentrated acids due to passivation of the metal surfaces when once corrosion has begun.

Effects of pH: increasing the acidity of a solution can result in a very large increase in general corrosion rate below a critical value or a range for a given alloy and environment. Since the reaction of a metal in an aqueous environment can be expressed as a simple displacement reaction

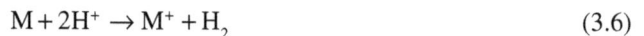

$$M + 2H^+ \rightarrow M^+ + H_2 \tag{3.6}$$

it is evident that where a greater number of hydrogen ions (low pH) are available, the corrosive reaction should occur more rapidly. Corrosion rate versus pH is presented schematically in Figure 3.8. The general shape of this curve may be considered as fairly typical for copper-base alloys. Such curves prepared for various metals and alloys could have their minima broadened or shortened, the slopes varied considerably, and the whole curve displaced in any direction, depending upon the characteristics of the alloy and the composition of the solution.

3.1.6.6 High-Temperature Corrosion

High-temperature corrosion is a form of corrosion that does not require the presence of a liquid electrolyte. It is an attack from gases, solid, or molten salts, or molten metals, typically at temperatures

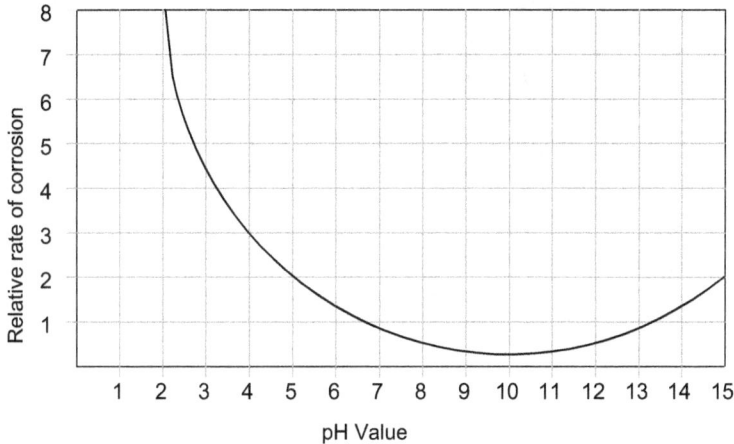

FIGURE 3.8 Corrosion rate versus pH (schematic).

above 750°F (400°C). Different types of high-temperature corrosion include carburization, including metal dusting, chloride attack, flue gas and deposit corrosion, nitridation, oxidation, and sulphidation. Oxidation is by far the most common form of high-temperature corrosion—almost all useful metals and alloys will oxidize above a certain temperature, leading to scaling, loss of material and changes in physical properties. Gaseous attack is not limited to oxygen however, with sulphur-bearing gases, carbon oxides, nitrous oxides, halogens, and many more all attacking materials in a different manner [9, 10].

3.2 UNIFORM CORROSION VERSUS LOCALIZED CORROSION

Corrosion attack on the metal surfaces can be either uniform or localized, where the major part of the surface of the metal remains almost unaffected while certain localized areas are attacked at a very high rate with rapid penetration into the section of the metal. Uniform corrosion occurs during the corrosion of a metal in an acid, alkali, and during the exposure of certain metals to natural environments like air and soil. In general, uniform corrosion takes place when the metal and environment system are homogeneous; that is, the metal is uniform in composition and structure and the nature of the environment (composition, oxygen concentration, acidity/alkalinity), temperature, velocity, etc., are the same at all parts of the metal surface [4]. Conversely, in many metals and environment systems, due to heterogeneities in the metal or variations in the environment or both, corrosive attack may be localized. Various forms of localized corrosion are pitting corrosion, crevice corrosion, intergranular corrosion, dealloying, erosion–corrosion, etc.

3.2.1 FACTORS THAT FAVOR LOCALIZED ATTACK

For metals and alloys, grain boundaries, intermetallic phases, inclusions, impurities, regions that differ in their mechanical or thermal treatments, discontinuities on metal surface such as cut edges or scratches, discontinuities in oxide or passive films or in applied metallic or nonmetallic coatings, and geometrical factors such as crevices favor localized attack [4]. Localized corrosion of a metal surface due to irregularities in metals is shown in Figure 3.9 [11].

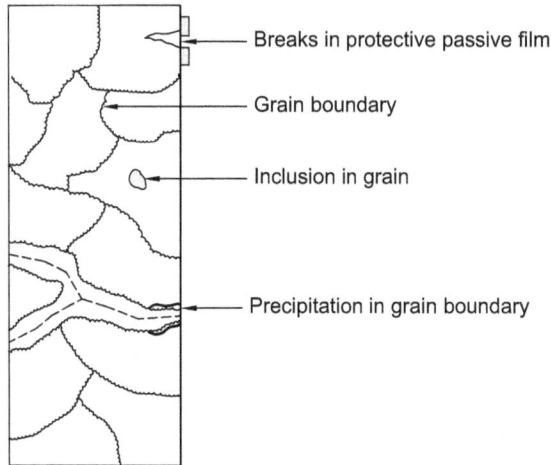

FIGURE 3.9 Localized corrosion of a metal surface due to irregularities in metals.

3.3 FORMS OF CORROSION AS DEFINED BY FONTANA AND GREENE [12]

Over the years, corrosion scientists and engineers have recognized that corrosion manifests itself in forms that have certain similarities and therefore can be categorized into specific groups. The most familiar and often used categorization of corrosion is probably the following eight forms presented by Fontana and Greene [12]:

1. Uniform corrosion.
2. Galvanic corrosion.
3. Pitting corrosion.
4. Crevice corrosion.
5. Intergranular corrosion.
6. Dealloying or selective leaching.
7. Erosion–corrosion.
8. SCC.

Hydrogen damage, although not a form of corrosion, often occurs indirectly as a result of corrosive attack [12]. This classification of corrosion was based on visual characteristics of the morphology of corrosion attack. Other forms of corrosion classified based on the mechanisms of attack rather than the visual characteristics are as follows:

Fretting corrosion.
Corrosion fatigue.
Microbiologically influenced corrosion (MIC).

General corrosion takes place uniformly over a broad area. In galvanic corrosion, two dissimilar metals form a galvanic couple, and the less noble metal in the couple corrodes. Pitting is the localized attack that produces small pits, which may penetrate the metal thickness. Crevice corrosion is another form of localized corrosion and takes place under crevices or deposits. Intergranular corrosion takes place at grain boundaries in weld metal or in heat-affected zones (HAZs) of sensitized metals. In dealloying, an alloying element is preferentially corroded over others from the parent alloy, leaving behind a weak structure. Erosion–corrosion is a localized corrosion that occurs mostly on the tubeside, in areas where the turbulence intensity at the metal surface is high enough to remove the protective surface film. SCC results when the corrosive action of a susceptible metal and the tensile stress combine in a particular environment. Corrosion may combine with other forms of attack, such as fatigue, to produce severe damage. Corrosion fatigue is the reduction in the fatigue strength of a metal exposed to a corrosive environment. Various forms of corrosion are shown schematically in Figure 3.10, and forms of corrosion, factors influencing corrosion, susceptible metals, and corrosion control measures, are shown in Table 3.1. Forms of corrosion are discussed in detail in the following sections.

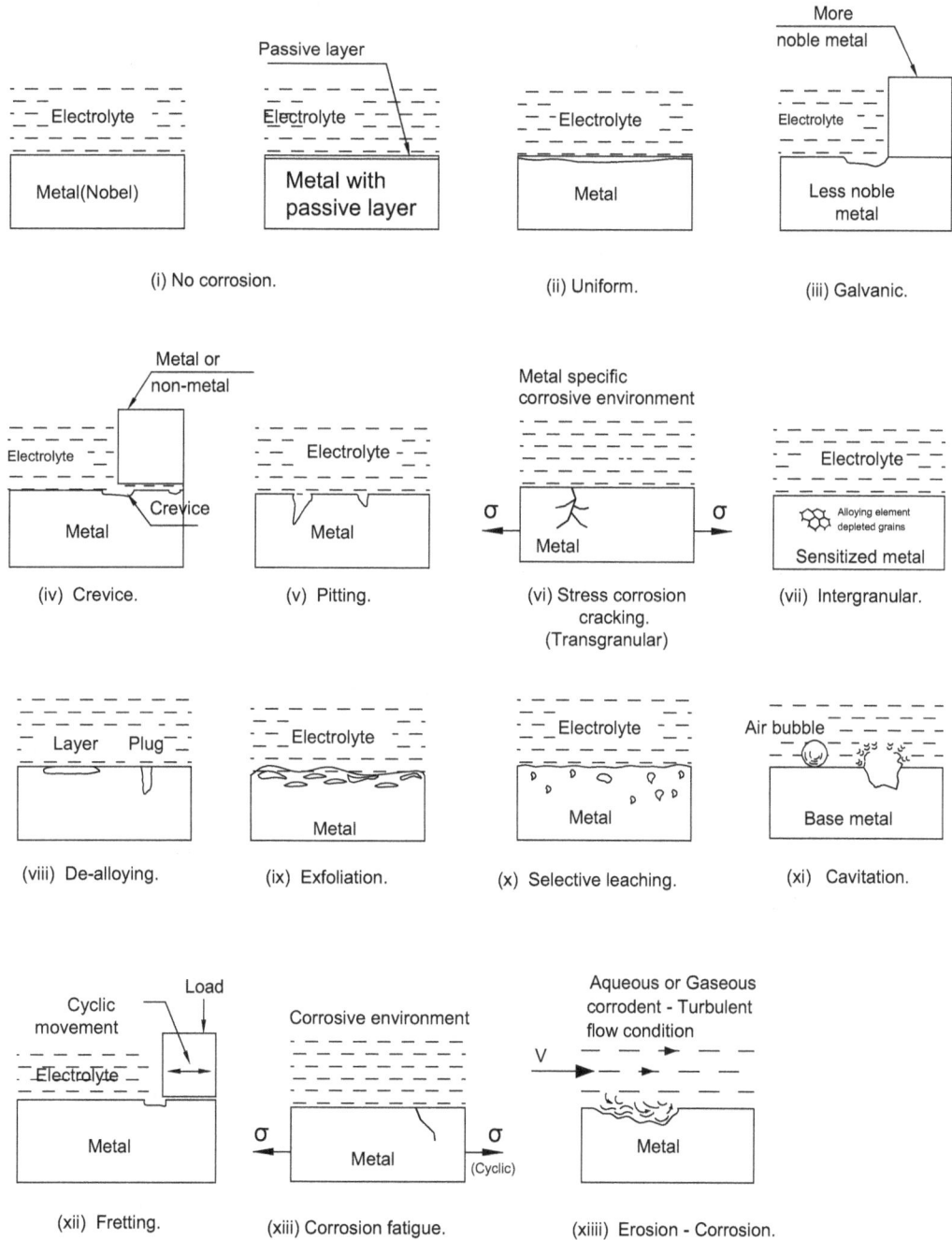

FIGURE 3.10 Schematic of various forms of corrosion.

TABLE 3.1
Forms of Corrosion, Factors Influencing Corrosion, Susceptible Metals, and Corrosion Control Measures

Forms of Corrosion	Factors Influencing Corrosion	Susceptible Metals	Corrosion Prevention/Control Measures
Uniform corrosion (or general corrosion).	Corrosion process dominated by uniform thinning and proceeds without appreciable localized attack. Uniform corrosion results from prolonged contact with environments such as atmosphere, water, acids, alkalies, soil, etc.	Though all metals corrode/ tarnish in atmosphere, rusting of iron is most important	Proper material selection, alloying additions to the base metal, cathodic protection using sacrificial anodes, use of inhibitors, surface coatings, and corrosion allowances
Galvanic corrosion	When two dissimilar metals or alloys placed at different potentials in electromotive series are in contact with each other in a electrolyte, a galvanic couple is formed and results in corrosion of one of the metals (less noble metal), known as the anode of the couple	In seawater cooled condensers, tube materials such as copper–nickels, stainless steels, or titanium are more noble than tubesheet materials such as Muntz metal, naval brass, or aluminum bronze; consequently, the tubesheets may suffer galvanic attack	Choose the combination of metals as close together as possible in the galvanic series and avoid the unfavorable area effect
Pitting corrosion	Instead of a uniform corrosion in a corrosive environment, a metal very often suffers localized attack resulting in pitting. Metallurgical and structural factors, environmental factors, and polarization phenomena, breakdown of passive film, Mill scale or applied coating. Compositional heterogeneity; inhomogeneities in the alloys	Aluminum, austenitic stainless steels, nickel, and their alloys, titanium in which surface film develops	Surface cleanliness and selection of materials resistant to pitting in the given environment: (a) For aqueous solutions of chlorides, choose AISI 316 or 317, Hastelloy G-3, Inconel alloy 625, and Hastelloy C-22, copper–nickel, Monel, or titanium. (b) In seawater and stagnant natural water applications, use copper–nickel, aluminum bronze, inhibited admiralty brass, titanium, superferritics, and duplex stainless steels. (c) Reduce the aggressiveness of the environment (the control of acidity, temperature, oxidizing agents, and chloride ions concentration)
Crevice corrosion	Localized form of corrosion that takes place at localized environments/areas that are distinctly different from the bulk environments	Metals or alloys that depend on protective surface film for corrosion resistance such as aluminum, stainless steels, and titanium	Materials can be alloyed to improve resistance to crevice corrosion. Design to minimize crevices and maintenance practices to keep surfaces clean and avoid deposit build up

TABLE 3.1 (Continued)
Forms of Corrosion, Factors Influencing Corrosion, Susceptible Metals, and Corrosion Control Measures

Forms of Corrosion	Factors Influencing Corrosion	Susceptible Metals	Corrosion Prevention/Control Measures
Inter granular corrosion (IGC)	It takes place as the result of local cell action in the grain boundaries due to potential difference between second-phase micro-constituents and the depleted solid solution from which the constituents are formed. The high cathode to anode area ratio results in rapid corrosion of the grain boundary material and the metal disintegrates	Austenitic stainless steels 304 (18-8), ferritic, superferritic, and duplex stainless steels. Nonferrous metals such as nickel 200 and nickel-base alloys like Inconel alloys 600 and 601, Incoloy alloys 800 and 800H, Hastelloys B and C and aluminum base alloys containing copper	Employ a suitable high-temperature solution heat treatment, commonly known as quench annealing or solution annealing. Ensure that the steel contains insufficient carbon to form $Cr_{23}C_6$ and resulting alloy depletion. Such steels contain less than 0.03% C and are called extra-low-carbon (ELC) steels, signified by the suffix L (e.g., type 304L and 316L). Use stabilized steels such as type 347 (columbium stabilized) and type 321 (titanium stabilized) instead of type 304
Dealloying or selective leaching	Corrosion process in which an alloying element is preferentially corroded over others from the parent alloy, leaving behind a weak structure due to interaction of metallurgical, and environmental factors	Brasses-dezincification, cupronickels denickelification, aluminum bronze-dealuminification, silicon bronze destannification or desiliconification	Material substitution- use admiralty brass, inhibited admiralty brass, aluminum brass (inhibited), cupronickel, and titanium. Surface cleanliness and avoid stagnant fluid conditions
Erosion–corrosion	A form of localized corrosion, which takes place due to the movement of a high velocity fluid over a material surface. Both mechanical and chemical factors such as turbulence, fluid velocity, impingement attack, level of suspended particles, aeration, bubble level, local partial pressure, cavitation, etc., affects	Normally restricted to copper and certain copper-base alloys and aluminum alloys.	Use filtered water. Use titanium or stainless steel or aluminum brasses or copper–nickel tubes, which are resistant to erosion–corrosion. Installation of tube inserts made of wear-resistant materials; on the shellside use impingement plate protection beneath the inlet nozzle
Stress corrosion cracking (SCC)	The corrosion attack on a susceptible alloy due to combined and synergistic interaction of tensile stress, conducive environment and its temperature and susceptible material	Chloride cracking, of austenitic stainless steel and aluminum. Caustic cracking of carbon steel. Ammonia cracking of admiralty brass. Polythionic acid cracking of austenitic stainless steel. Sulfurous acid cracking of austenitic stainless steel	To avoid the metal and environment combinations that are favorable for SCC. To resist chloride SCC of Austenitic SS, use Inconel 600, Incoloy alloys 800 and 825; ferritic-, superferritic-, and duplex stainless steels or titanium

(continued)

TABLE 3.1 (Continued)
Forms of Corrosion, Factors Influencing Corrosion, Susceptible Metals, and Corrosion Control Measures

Forms of Corrosion	Factors Influencing Corrosion	Susceptible Metals	Corrosion Prevention/Control Measures
Corrosion fatigue	Combined action of a cyclic stress, corrosive environment and metallurgical factors	Aluminum alloys exposed to chloride solution, Cu–Zn and Cu–Al alloys exposed to aqueous chloride solutions, and high-strength steel in a hydrogen atmosphere	Use of protective coatings, adding inhibitors to the environment, cathodic protection, and introduce residual compressive stresses by methods such as shot peening
Hydrogen attack or hydrogen damage. Other forms of hydrogen attack are hydrogen-induced cracking (HIC), hydrogen embrittlement (HE), hydrogen sulfide stress cracking (HSCC), or sulfide stress corrosion cracking (SSCC)	Degradation of steel by hydrogen sulfide (H_2S) in pressure vessels handling sour crude oil or gas in petroleum refinery equipment. Diffusion of mono atomic hydrogen into the steel and recombine to form molecular hydrogen at voids, laminations, discontinuities or around inclusions, generating extremely high pressure. It stiffens the metal, impairing its ductility. It also gives rise to blistering and cracking	High strength steels, titanium alloys and aluminum alloys. Steels with weldments are susceptible to hydrogen embrittlement. Use pressure vessel quality steels with 0.25%–0.35% C, in plate thickness 1–2 in (25.4–50.8 mm), such as ASTM A515 and A516	Selection of steel to resist high-temperature and high-pressure hydrogen/hydrogen sulfide environment. Follow the operating limits on temperature, partial pressure of hydrogen, and alloy composition set forth by Nelson curves. Cr-Mo steels are more resistant to hydrogen attack than plain carbon steel
Microbiologically influenced corrosion (MIC)	MIC occurs directly or indirectly as the result of metabolic activity of microorganisms on heat transfer surfaces	Common engineering metals and alloys such as carbon steel, lined steel, stainless steels, aluminum alloys, and copper alloys, 6% Mo stainless steels, and high–nickel alloys (Monel 400 and Alloy B-2)	Uptake filtration. Systematic cleaning and elimination of stagnant areas. Proper use of a biocide. Thermal shock treatments

3.3.1 UNIFORM OR GENERAL CORROSION

Uniform corrosion results from prolonged contact with environments such as atmosphere, water, acids, alkalies, soil, etc. It is manifested by a chemical or electrochemical reaction. Uniform corrosion of the metal surface, as in the rusting of iron, results from anodes and cathodes rapidly interchanging sites at random; should the anode become fixed on the surface, a localized pitting corrosion will result instead [2]. Like other forms of corrosion, many factors influence general corrosion. The corrosive media is the most important factor governing corrosion. Environmental factors such as acidity, temperature, concentration, motion relative to metal surface, degree of oxidizing power and aeration, and presence or absence of inhibitors influence general corrosion [3].

Forms of uniform corrosion: most general types of uniform corrosion are atmospheric corrosion and high-temperature (gaseous) corrosion. Although high-temperature attack in gaseous environments may manifest itself as various forms of corrosion, it has been incorporated into the category "general corrosion" because it is often dominated by uniform thinning [13].

Atmospheric corrosion: atmospheric corrosion is defined as the corrosion of a material exposed to the air and its pollutants rather than immersed in a liquid. Many variables influence the atmospheric corrosion. Relative humidity, temperature, contents like sulfur dioxide, hydrogen sulfide, and chloride, amount of rainfall, dust, etc., influence corrosion behavior. Therefore, in an arid atmosphere, free of contaminants, only negligible corrosion would be expected. It also must be understood that the major contributor to the corrosion of all metals in the atmosphere is oxygen.

Types of corrosive atmospheres: a common practice is to divide atmospheres into categories such as rural, industrial, marine, and indoor [14]. Most of them are mixed and no clear lines of demarcation are present, and the type of atmosphere may vary with the wind pattern, particularly where corrosive pollutants are concentrated. Except for severe marine environments, there is no need to protect against atmospheric corrosion except with protective coatings [15].

Protection against atmospheric corrosion: important considerations to protect against atmospheric corrosion should involve [15] (1) specific plant environment around the equipment; (2) operating temperature; and (3) protection against ingress of rain water into crevices. Merrick discusses these factors in detail [15].

The specific plant environment around the equipment that involves releases of corrosive gases, such as hydrogen sulfide, sulfur dioxide, or ammonia, from nearby process plant will significantly increase atmospheric corrosion of equipment.

The operating temperature of the pressure vessels and heat exchangers has a significant effect on atmospheric corrosion. Units that operate below the dew point experience more corrosion than those operating at higher temperatures.

For protection against ingress of rain water into crevices, saddles and reinforcing pads for supports should be continuously welded to the vessel. Partial welding or stitch welds will act as a crevice. This will permit the ingress of rain water between the saddle and the vessel and can accelerate crevice corrosion.

Control of uniform corrosion: uniform corrosion can be easily predicted and controlled by

Proper material selection.
Small alloying additions to the base metal.
Cathodic protection using sacrificial anodes.
Use of inhibitors.
Surface coatings.
Adding extra material known as corrosion allowances.

Rating of metals subject to uniform corrosion: in contrast to localized forms of corrosion, uniform corrosion is rather predictable. The uniform attack on an entire area exposed to a corrosive environment is usually expressed in terms of an average loss of metal thickness for a given period of time in units of mils (1 mil = 1/1000 in.) per year (mpy), inches per month (ipm), or millimeter per year (mm/yr). Corrosion rate less than 5 mpy is excellent (for very critical application) and 5–20 mpy is satisfactory [16].

Determination of general corrosion: general corrosion is determined by immersion tests conducted according to ASTM G-31. Eight different boiling acid and alkali solutions—20% acetic acid, 45% formic acid, 10% oxalic acid, 20% phosphoric acid, 10% sodium bisulfate, 50% sodium hydroxide, 10% sulfamic acid, 10% sulfuric acid—are used to compare the performance of different alloys in a variety of solutions rather than to simulate a particular process industry environment. Duplicate samples are exposed for five 48 h periods, and an average corrosion rate is determined.

Material selection for general corrosion: of the various forms of corrosion, general corrosion is the easiest to evaluate, and hence the material selection is straightforward; if a material shows only a general and uniform attack, a corrosion rate, 0.25 mm/yr (10 mils/yr) or less for low-cost material such as carbon steel, a negligible contamination of the process fluids, and availability, ease of fabrication etc., then that material favors the choice of selection. For more costly materials such as the 300 series austenitic stainless steels and copper- and nickel-base alloys, a maximum corrosion rate of 0.1 mm/year (4 mils/year) is generally acceptable [17]. The corrosion rate is multiplied by the nominal design life of the vessel (normally 20 years) to determine the corrosion allowance.

3.3.1.1 Corrosion Rate—Units of Measurement

Corrosion rate is normally expressed as weight loss per unit area per unit time (weight loss/area/time) or as depth of penetration of the metal in unit time. The common units are given below [18]:

1. weight loss: mdd—milligrams per square decimeter per day
2. penetration: ipy—inches per year
   mpy—mils (1/1000 inch) per year
   mpd—mils per day
   micrometer per year: μm/y
   millimeter per year: mm/y

Weight loss is the basic measurement. Corrosion, assumed to be uniform is then calculated from weight loss, using the metal density in the calculation.

3.3.2 GALVANIC CORROSION

When two dissimilar metals or alloys placed at different positions in electromotive series are in contact with each other in an electrolyte, a galvanic couple is formed and results in the corrosion of one of the metals, known as the anode of the couple. In other words, galvanic corrosion does not affect the cathode, which is known as a noble metal. This form of corrosive attack is known as galvanic corrosion since the entire system behaves as a galvanic cell. Galvanic corrosion can also take place even within the same group of metals due to local imperfections or heterogeneities on the metal surfaces or due to variation in local solution chemistry. For initiation of galvanic corrosion, four essential components are required: the anode, the cathode, the electrolyte, and a metallic path between the anode and the cathode, which completes the circuit. These four basic components were already discussed.

3.3.2.1 Heat Exchanger Locations Susceptible to Galvanic Corrosion

Components such as tubesheets, water box, bolts and flanges, and supports made of less noble metals will corrode at the following locations:

Interfaces between the tube and baffle plates.
Between the tubes and tubesheet areas.
Welded joints, brazed joints, and soldered joints.

In seawater-cooled condensers, tube materials such as copper–nickels, stainless steels, or titanium are more noble than tubesheet materials such as Muntz metal, naval brass, or aluminum bronze; consequently, the tubesheets may suffer galvanic attack when fitted with more noble tube materials.

3.3.2.2   Galvanic Corrosion Sources

Two important sources of galvanic corrosion are (1) metallurgical sources and (2) environmental sources.

Metallurgical sources: metallurgical sources are within the metal and/or in relative contact between dissimilar metals. Such sources include difference in potential of dissimilar materials, distance apart in galvanic series, relative areas of anode and cathode, oxide or mill scales, strained metal (cold work), inclusions in metal, and differences in microstructure, HAZ, and sensitization [19,20].

Environmental sources: environmental sources include conductivity of the fluid, concentration differences in solution, changes in temperature, velocity, and direction of fluid flow, aeration, and ambient environment (seasonal changes) [19].

3.3.2.3   Types of Galvanic Corrosion

Among the various forms of electrochemical corrosion; (1) bimetal corrosion; (2) differential aeration corrosion; (3) differential concentration corrosion; and (4) work area corrosion belong to galvanic corrosion [21]. Items (1)–(3) have been discussed earlier. Only work area corrosion is defined here.

When a metal is cold worked so that it is denser in one place than another, a corrosion cell with the stressed area as an anode and the remaining area as a cathode is set up if these two areas are immersed in an electrolyte [21].

3.3.2.4   Magnitude of Galvanic Effects

The discussion given earlier deals with the criteria leading to the formation of an electrolytic cell, which is the essential step in the corrosion process. However, the extent of corrosion attack taking place is dependent on many factors as follows:

1. The polarization behavior of the metals or alloys.
2. Passivation of the alloys.
3. Potential difference between the metals or alloys, i.e., the distance effect.
4. Area effect, i.e., the geometric relationships such as relative surface area.

Items 1 and 2 have been discussed earlier, and the remaining points are discussed next.

Distance effect: enhanced corrosion takes place when the metals in galvanic couple are placed further apart in the galvanic series. This is due to high electrochemical current density. In some cases, the separation between the two metals or alloys in the galvanic series gives an indication of the probable magnitude of the corrosive effect.

Area effect: another important factor in galvanic corrosion is the area effect. As the ratio of the cathode-to-anode area increases, the corrosion rate of the anode metal rapidly accelerates (Figure 3.11). On the other hand, if the area of anode is large compared to the cathode area, the corrosion of the anode is so widely distributed that the amount of metal loss in terms of its thickness may be so small that it may be ignored. For example, the nuts and bolts that are critical to the flanged joints must always be noble (cathodic) to the larger area of the flange. The nuts and bolts (Figure 3.12a) corrode at a rate well below normal at the negligible expense of the flange, which corrodes at only slightly

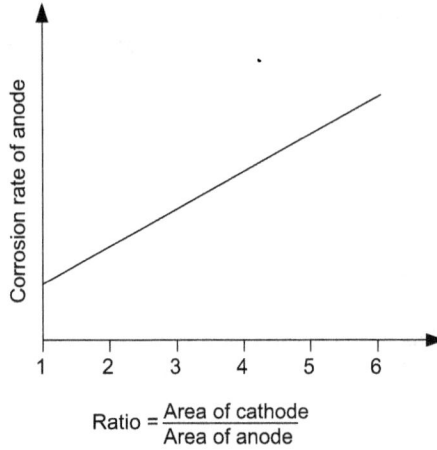

FIGURE 3.11   Galvanic corrosion due to area effect.

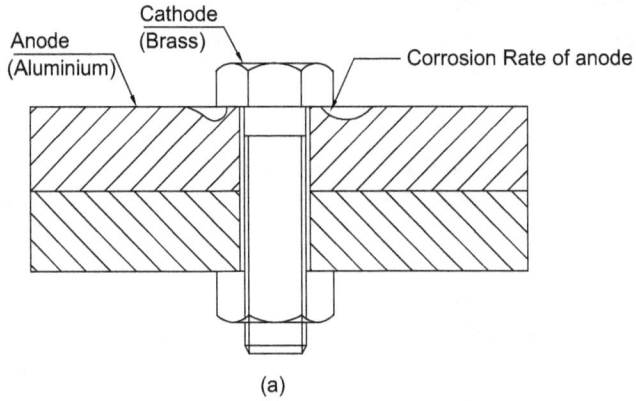

(a)

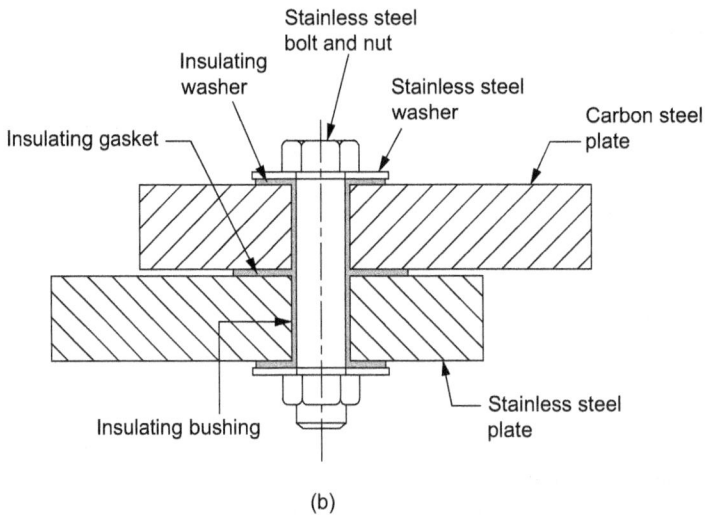

(b)

FIGURE 3.12   Bolt and Nut (cathodic)in a flanged joint (anodic) (a) Crevice corrosion and (b) Bolt and nut with insulator.

above-normal rate over a large area and Figure 3.12b shows dissimilar metals joint with an insulating layer separating the metal parts.

3.3.2.5 Tools to Determine the Degree of Galvanic Corrosion

The degree of galvanic corrosion is normally known from the following two tools. They are the electromotive force (EMF) series and the galvanic series.

Electromotive force series: the EMF series is a list of elements arranged according to their standard electrode potentials, with the sign being positive for elements whose potentials are cathodic to hydrogen and negative for those anodic to hydrogen. In this listing, hydrogen is used as an arbitrary reference element, and metals such as gold and platinum have large positive values indicating little tendency for corrosion attack. Since the EMF series is of little value to the practical corrosion scientist or technologist, it is necessary to develop some alternative system by which the relative corrodibility of a galvanic couple may be assessed [4]. Such a listing is known as the galvanic series.

TABLE 3.2
Galvanic Series in Seawater at 25°C (77°F)

Anodic (least noble) end or active end	Cartridge or yellow brass C27000
Magnesium	Admiralty brass C44300, C44400, C44500
Zinc	Aluminum bronze C60800, C61400
Galvanized iron	Red brass C23000
Aluminum alloy 5052H	ETP copper C11000
Aluminum alloy 3004	Silicon bronze C65100, C65500
Aluminum alloy 3003	Copper–nickel, 10%
Aluminum alloy 1100	Copper–nickel, 30%
Aluminum alloy 6053T	Nickel 200 (passive)
Alclad aluminum alloys	Inconel alloy 600 (passive)
Aluminum alloys, 2117	Monel alloy 400
Aluminum alloys, 2017T	Stainless steel type 410 (passive)
Aluminum alloys, 2024T	Stainless steel type 430 (passive)
Cadmium	
Low-carbon steel	Stainless steel type 304 (passive)
Low-alloy steel	Stainless steel type 316 (passive)
Cast iron	E-Brite alloy
Stainless steel type 410 (active)	AL-29-4C alloy
Stainless steel type 430 (active)	AL-6XN alloy
50–50 Lead-tin solder	Inconel alloy 825
Stainless steel type 304 (active)	Inconel alloy 625, alloy 276
Stainless steel type 316 (active)	Hastelloy alloy C
Lead	Silver
Tin	Titanium
Muntz metal C28000	Graphite
Manganese bronze A—C67500	Zirconium
Naval brass: C46400, C46500, C46600, C46700	Tantalum
Nickel 200 (active)	Gold
Inconel alloy 600 (active)	Platinum
Hastelloy alloy B	Cathodic (most noble) end

Sources: Adapted from *Resistance to Corrosion*, 4th edn., Inco Alloys International, Huntington, WV, 1985; *AL-6XN*ᴿ *Alloy*, Allegheny Ludlum Steel Corporation, Pittsburgh, PA; Steven, P.L., General corrosion, in *Metals Handbook*, 9th edn., Vol. 13, Corrosion, American Society for Metals, Metals Park, OH, pp. 80–103, 1987.

Galvanic series: whether a given metal or alloy is naturally anodic or cathodic with respect to another metal is judged by the galvanic series, which consists of an arrangement of metals and alloys in accordance with the measured potentials in flowing seawater at velocities ranging from 2.5 to 4 m/s. A typical tabulation is given in Table 3.2. The metals and alloys that are next to each other have little tendency to galvanic corrosion when connected together, so it is relatively safe to use such combinations in jointed assemblies [22].

3.3.2.6 Galvanic Corrosion Control

Design is a major factor in preventing or minimizing galvanic corrosion. Typical measures to control galvanic corrosion are as follows:

1. Choose the combination of metals as close together as possible in the galvanic series, unless the more noble metal is easily polarized.
2. Avoid the unfavorable area effect, i.e., small area of anode and a large cathode area. Otherwise small components such as bolts and fasteners should be of more noble metal.
3. Insulate or break the circuit between the two metals by applying coatings, introducing gaskets, nonmetallic washers, etc., and make sure that metal-to-metal contact is not restored in service.
4. Add corrosion inhibitors to decrease the aggressiveness of the environment or to control the rate of cathodic and/or anodic reaction.
5. *Maintain coatings*: coating is the most common method for combating corrosion.
6. Cathodic protection is one of the recommended methods of protecting the anode. Use a sacrificial anode such as Zn, Al, or Mg that is anodic to both the structural metals. Examples include sacrificial cladding (Alclad) applied in the internal aluminum tube surfaces of radiators to give protection to the tube core alloy and planting of Zn anode on the water box of condensers to protect both the tubes and the tubesheet. It is important to note that overprotection of the titanium tubesheet may result in the hydriding of the titanium tubes.

3.3.3 PITTING CORROSION

Instead of a uniform corrosion in a corrosive environment, a metal very often suffers localized attack resulting in pitting. Pitting usually occurs on metals that are covered with a very thin adherent protective surface film that formed on the metal surface during a surface treatment process or produced by reaction with an environment [23]. Thus, pitting corrosion occurs on aluminum, titanium, stainless steels, nickel, and their alloys, in which surface film develops. Pitting takes place when there is a breakdown of protective surface film. Pitting is the most aggressive form of corrosion and leads to premature failure due to perforation of the surfaces. Pitting corrosion is shown schematically in Figure 3.13.

3.3.3.1 Parameters Responsible for Pitting Corrosion

Pitting corrosion is caused by factors, such as [24] (1) metallurgical and structural factors; (2) environmental factors; and (3) polarization phenomena. Other causes for pitting are the attachment of microorganisms, presence of corrosion products, deposits, etc. These parameters are explained next.

Metallurgical and structural factors: the following metallurgical and structural factors act as nucleation sites for initiation of pitting corrosion:

1. The breakdown of passive film, Mill scale [25], or applied coating.
2. Defect structures.

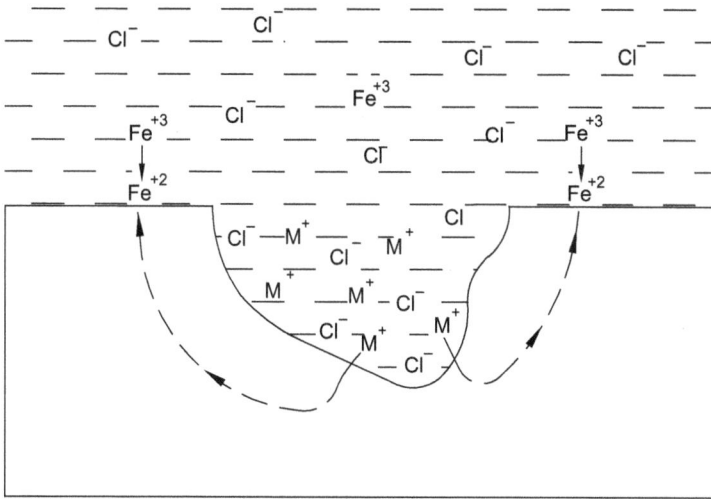

FIGURE 3.13 Schematic illustration of a growing pit. (Metal or alloy M is being attacked by a solution of ferric chloride. The dissolution reaction, $M \rightarrow M^+ + e^-$ is localized within a pit and cathodic reduction of the ferric ion, $Fe^{+3} + e^- \rightarrow Fe^{+2}$, occurs on the rest of the exposed metal surface.)

Source: Adapted from *Resistance to Corrosion*, 4th edn., Inco Alloys International, Huntington, WV, 1985.

3. A compositional heterogeneity; inhomogeneities in the alloys caused by segregation of alloys or by cold working [7].
4. Weld-related parameters [6]: inclusions, multiple phases, compositional differences within the same phase, sensitization, arc strikes, spatter, and inhomogeneities in the base materials can act as potential pitting sites.

Pits develop if there is a breakdown of passive film, Mill scale, or applied coatings due to high turbulence in the flow, chemical attack, or mechanical damage or under deposits. Mill scale is cathodic to steel and is found to be one of the more common causes of pitting. Figure 3.14a illustrates the pit action due to breakdown of Mill scale. A pit caused by broken Mill scale becomes deeper, an oxygen concentration cell is formed by the depletion of oxygen in the pit, and this will accelerate the rate of penetration.

Studies show that the sites for initiation of pits on passive metal surfaces may be generally related to defect structures of the underlying metal such as dislocations, grain boundaries, or nonmetallic inclusions [24].

In terms of compositional heterogeneity, nonmetallic inclusions like sulfide inclusions, particularly manganese sulfides (MnS), are potential nucleation sites in austenitic stainless steels and ferritic stainless steels.

Environmental factors: certain chemicals, mainly halide salts and particularly chlorides, are well-known pit producers. The passive metals including austenitic stainless steels are particularly susceptible to pitting in chloride environments. The chloride ions accumulate at anodic areas and either penetrate or dissolve the passive film at these points.

Influence of polarization on passive alloys, after Roser et al. [7]: pitting can occur even in a relatively homogeneous alloy due to electrochemical causes. This is explained by polarization curves as shown schematically in Figure 3.6c. If the cathodic polarization curve crosses the anodic polarization curve

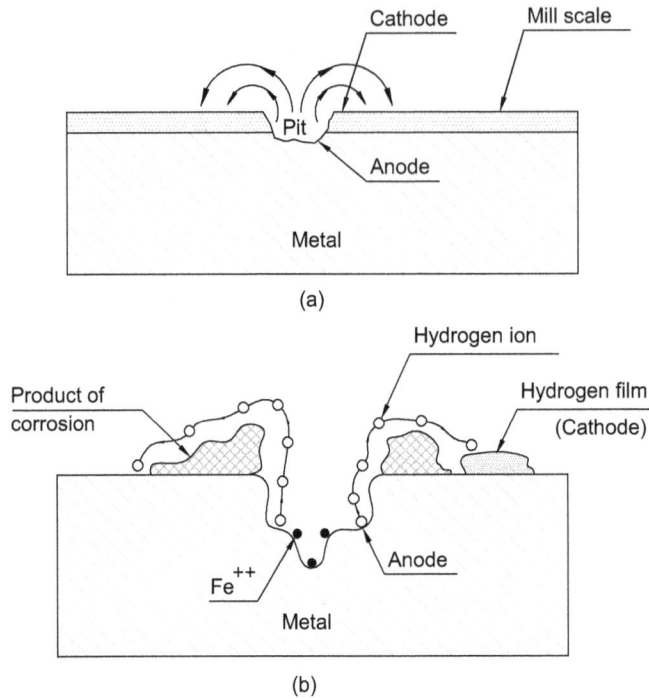

FIGURE 3.14 Pitting corrosion. (a) Pitting due to breakdown of mill scale and (b) pitting due to deposit on the base metal (galvanic cell).

in the active region, the alloy will be active; if it crosses in the passive region, it will be passive; and if it passes in the intermediate region, the alloy will be partly active and partly passive. If the active regions are small and passive regions are large, the alloy can pit. Pitting can be avoided by either increasing or decreasing the polarization of the cathode to avoid the pitting region. Changes in alloy composition or structure can also be useful. For example, molybdenum bearing steels are resistant to chloride pitting.

Attachment of microorganisms, presence of corrosion products, and deposits: costly pitting failures of austenitic stainless steel components and weldments can take place by the attachment of microorganisms. Austenitic stainless steels form small tubercles from microbial action, under which severe pitting occurs. Or the sulfides produced by the dying microorganisms depassivate stainless steel, and pitting takes place underneath the fouling. Figure 3.14b illustrates the pit action due to the galvanic cell formed between the corrosion deposit and the base metal.

Basic condition for the initiation and propagation of pitting corrosion: a basic condition must be fulfilled for the initiation and propagation of pitting corrosion: "Pitting takes place when the anodic breakdown potential of the metal surface film is equal to or less than the corrosion potential under a given set of conditions."

3.3.3.2 Mechanisms and Theories of Pitting Corrosion

Nucleation and growth: the modern theory of pitting presupposes the formation of a pit at a minute area of a metal surface that suffers a breakdown in passivity [22]. This is known as the pit nucleation stage. The passive film breakdown is followed by formation of an electrolytic cell, which leads to growth and propagation of a pit rather than to spreading along the entire surface. The anode of this cell is a minute area of active metal, and the cathode is a considerable area of passive metal [26, 27]. The large cathode-to-anode area ratio accounts for the considerable flow of current with rapid corrosion at the anode. Pits once initiated can propagate inside the pits. They stop propagating if the metal is polarized to (or below) the potential of metal inside the pits, which in the extreme is that of the active (nonpassive) state [28].

Growth and propagation of pit: a pit develops in stages: original attack, propagation, termination, and reinitiation [23]. Termination will occur with increase in internal resistance of the local cell.

3.3.3.3   Pitting Potential

For passive metals, pits are initiated at or above a specific potential. The potential at which pit initiation occurs is called the pitting potential. Resistance to pitting increases with the pitting potential. The value of pitting potential depends on the material and its composition, the environment, and the concentration of aggressive ions, pH of the solutions, temperature, and history of heat treatment operation [28].

Determination of pitting potential: pitting potential is determined by electrochemical techniques, which consist of measuring current and potential potentiostatically either stepwise or by applying a constant potential sweep rate in a standard chloride-containing solution. The recorded values of the current and potential are plotted. A theoretical curve obtained by the electrochemical method is shown schematically in Figure 3.15 [16]. From such a curve, the following values are obtained and used to characterize alloys with respect to pitting and crevice corrosion: (1) pitting potential E_p where pits start to grow; (2) repassivation potential E_{rpp} below which already growing pits are repassivated; and (3) critical current densities characterizing the active/passive transition.

Morphology of pits: while the shapes of pits vary widely, they usually are roughly saucer shaped, conical, or hemispherical. If appreciable attack is confined to a relatively larger area and is not so deep, the pits are called shallow, whereas if the pit is confined in a small area, it is called deep pit. The depth of pitting is sometimes expressed by the term "pitting factor" [29]. This is the ratio of the deepest metal penetration to average metal penetration as determined by the weight loss of the specimen. This is shown schematically in Figure 3.16.

Detection: pitting is usually a slow process (taking several months or years to become visible) but still can cause unexpected failures. However, the small size of a pit and the small amount of metal dissolution make its detection difficult in the early stages.

3.3.3.4   Prevention of Pitting Corrosion

Surface cleanliness and selection of materials known to be resistant to pitting in the given environment are usually the safest ways of avoiding pitting corrosion. Details of these measures are as follows:

1. Reduce the aggressiveness of the environment, which includes the control of acidity, temperature, oxidizing agents, and chloride ion concentration [29].
2. Modify the design to avoid crevices, circulate/stir to eliminate zero velocity regions, and ensure proper drainage.
3. Systematic cleaning and elimination of stagnant areas: since the presence of microorganisms, corrosion products, deposits, etc., stimulates pitting and, in particular, crevice corrosion in the tubes, keep the tubes clean [30].
4. Upgrade the materials of construction; chromium and nickel reduce pitting tendency very effectively, and these are often given a considerable boost with an alloy addition of molybdenum [6]. The resultant alloys are many superaustenitics highly resistant to pitting. Nitrogen

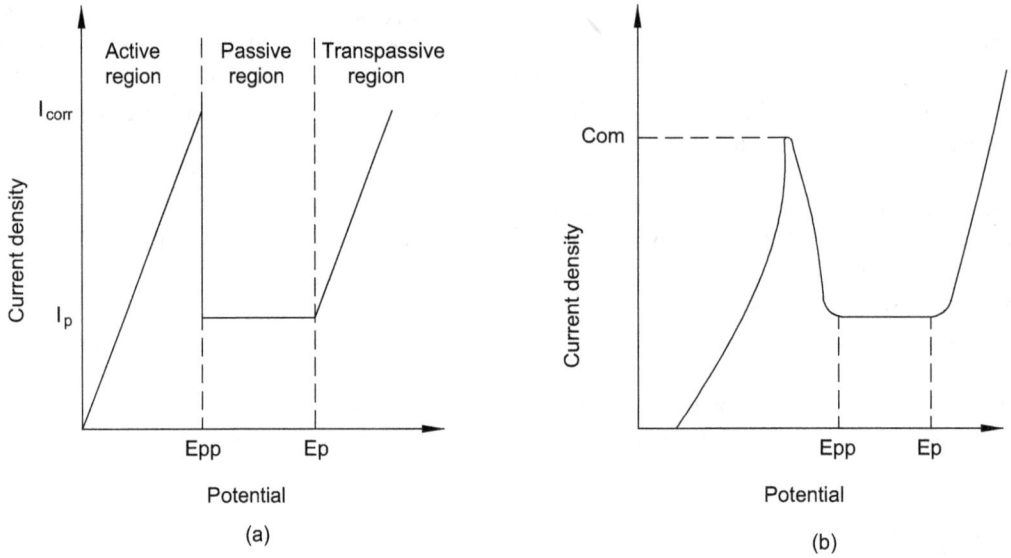

FIGURE 3.15 Pitting potential determined by electrochemical method. (a) Theoretical curve and (b) actual curve.

improves the pitting resistance of wrought stainless steels but has the opposite effect in the weld metal, although either way the effect is fairly small compared to molybdenum [31]. Overlay with lining resistant to corrosion. Typical alternative materials that are resistant to pitting are (a) for aqueous solutions of chlorides, choose molybdenum-containing steels such as AISI 316 or 317, or alloys containing greater amounts of chromium and molybdenum such as Hastelloy G-3, Inconel alloy 625, and Hastelloy C-22 [17], copper–nickel, Monel, or titanium. (b) In seawater and stagnant natural water applications, use materials such as copper–nickel, aluminum bronze, inhibited admiralty brass, titanium, superferritics, and duplex stainless steels. (c) Proven lower-cost nonmetallic coatings, linings, or cladding can be helpful. (d) The rating of certain stainless steel materials for pitting corrosion is discussed in Chapter 2, Heat Exchangers: Mechanical Design, Materials, Nondestructive Testing, and Manufacturing.

Avoid the metals and corrosive combinations that have the pitting tendencies [17]:

1. A1 and A1 alloys: electrolytes containing ions of heavy metals such as copper, lead, and mercury.
2. Plain carbon and low-alloy steels: waters containing dissolved O_2 or sulfate-reducing bacteria (SRB).
3. Austenitic stainless steel weldment: exposed to stagnant natural waters that are infected with iron and/or manganese bacteria.

3.3.3.5 How to Gauge Resistance to Pitting

The susceptibility of passive metals to pitting corrosion is usually investigated by experimental techniques: (1) simple immersion tests; (2) electrochemical methods; and (3) using empirical correlation for stainless steels. The methods are described next.

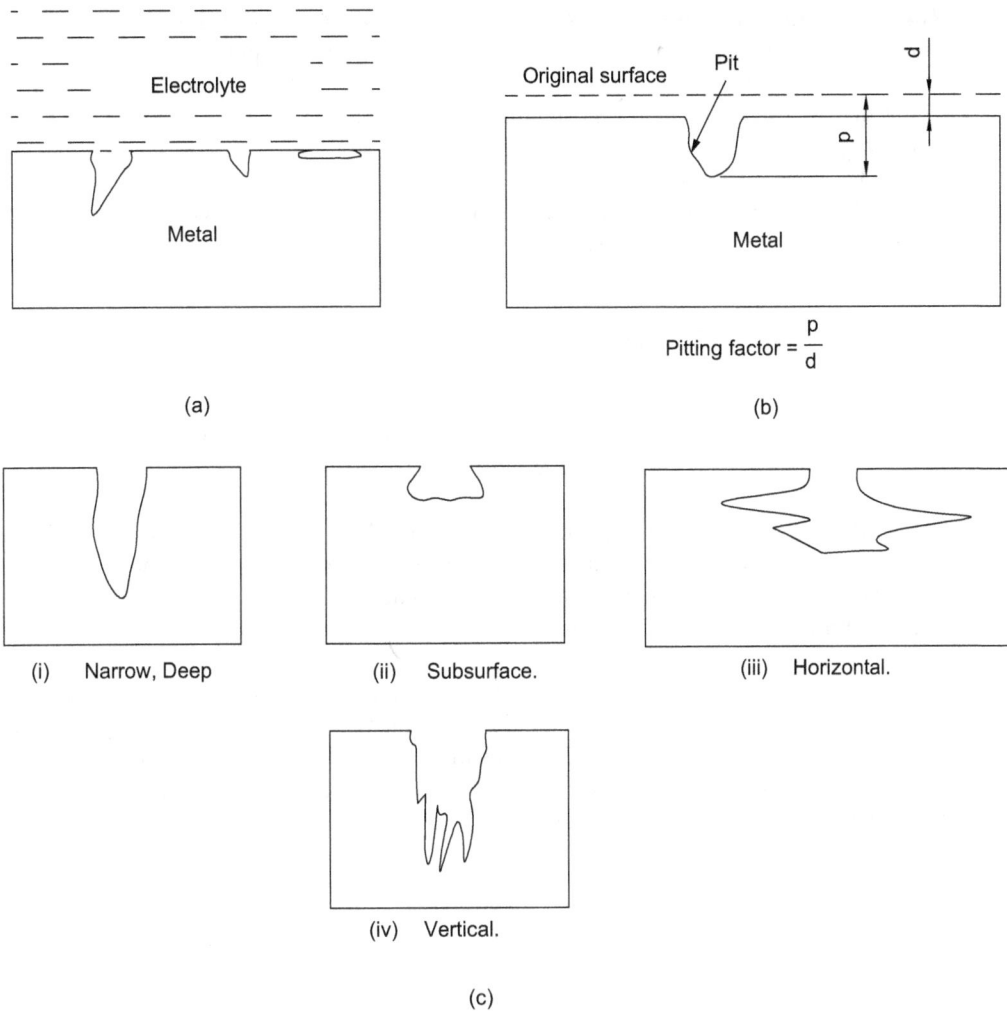

FIGURE 3.16 Pitting corrosion (a) Pitting corrosion, (b) Pitting factor and (c) Few forms of pits(schematic) (Refer to ASTM G 46-2021 for more pit forms).

Critical pitting corrosion temperature: the formation of visible pits in specimens exposed to aqueous chloride containing solutions, at different temperatures, is frequently used as a measure of pitting resistance. The temperature at which the pits begin to form is known as the critical pitting temperature (CPT). This is determined as per ASTM G48.

ASTM G48 is a laboratory test method for determining the resistance of stainless steels and related alloys to pitting and crevice corrosion.

The method uses a ferric chloride ($FeCl_3$) solution. The concentration of ferric chloride is usually 6%, but sometimes 10% is used instead and sometimes in an acidic mixture of chlorides and sulfates [4% $NaCl$ + 1% $Fe_2(SO_4)_3$ + 0.01 M HCl]. The resulting indices are known as critical pitting corrosion temperature (CPT or CPCT) and critical crevice corrosion temperature (CCT or CCCT). These are the minimum temperatures at which these types of localized attack start in the $FeCl_3$ solution. CCT is determined with crevices.

ASTM G46-202—Standard Guide for Examination and Evaluation of Pitting Corrosion
Significance and use. It is important to be able to determine the extent of pitting, either in a service application where it is necessary to predict the remaining life in a metal structure, or in laboratory test programs that are used to select the most pitting-resistant materials for service.

ASTM G48-03
Standard Test Methods for Pitting and Crevice Corrosion Resistance of Stainless Steels and Related Alloys by Use of Ferric Chloride Solution
Six procedures are described and identified as *Methods A, B, C, D, E*, and *F*.

1. *Method A*—ferric chloride pitting test.
2. *Method B*—ferric chloride crevice test.
3. *Method C*—critical pitting temperature test for nickel-base and chromium-bearing alloys.
4. *Method D*—critical crevice temperature test for nickel-base and chromium-bearing alloys.
5. *Method E*—critical pitting temperature test for stainless steels.

Pitting potential: another laboratory test that is frequently used for ranking stainless steel is pitting potentials as measured using an electrochemical apparatus in a standard chloride-containing solution. The pitting potential indicates the relative susceptibility of an alloy to localized corrosion. Resistance to pitting increases with the pitting potential.

Pitting index number (PRE_N): because the resistance of a stainless steel against pitting and crevice corrosion is primarily determined by the amount of chromium, molybdenum, and nitrogen in it, an index for comparing the resistance to these types of attacks is often evaluated in terms of these elements. The index is called the pitting resistance equivalent number (PRE or PRE_N). It is defined, in weight percent, using the following equation:

$$PRE_N = \%Cr + 3.3\%Mo + 16\%N \quad \text{for austenitic and duplex stainless steel}$$

$$= \%Cr + 3.3\%Mo \quad \text{for ferritic stainless steel}$$

$$= \%Cr + 3.3\%Mo + 30\%N \quad \text{for 6\% Mo superaustenitic stainless steel}$$

The higher the PRE_N number, the better the performance of an alloy in chloride environments. For example, the PRE number for 6% Mo superaustenitic alloys containing nitrogen is 43, Alloy 2205 has 35, while for type 316L it is 24.

3.3.4 CREVICE CORROSION

Crevice corrosion, similar to pitting, is a localized form of corrosion that takes place at localized environments/areas that are distinctly different from the bulk environments. Such localized environments include metal-to-metal joints, metal-to-nonmetal joints such as pipe support (Figure 3.17), gasketed joints, shielding by corrosion products and fouling deposits, beneath biological growth, stagnant areas, and sharp corners. At these locations, the crevice corrosion is usually attributed to one or more of the following [32]: (1) changes in acidity in the crevice; (2) lack of oxygen in the crevice; (3) buildup of a detrimental ion species; (4) for passive metals, the loss of passivity within the crevices; and (5) depletion of inhibitors. This leads to concentration cells or aeration cells.

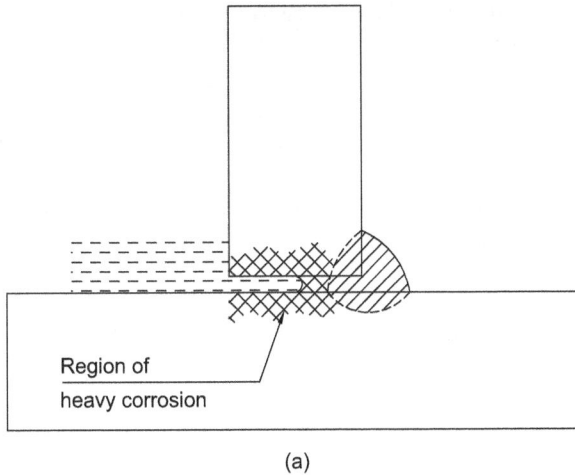

(a)

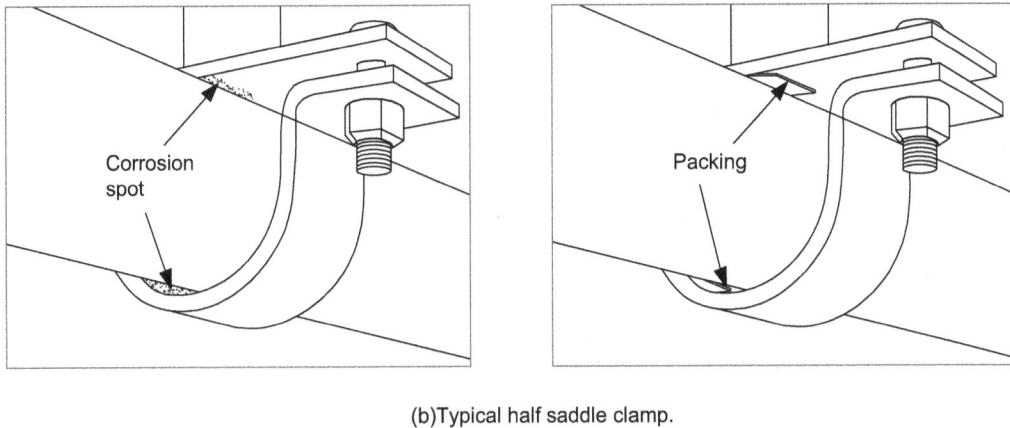

Corrosion
spot

Packing

(b)Typical half saddle clamp.

FIGURE 3.17   Crevice corrosion –(a) Metal-to-metal welded joint and (b) a Pipe support.

Susceptible alloys: similar to pitting corrosion, metals, or alloys that depend on protective surface film for corrosion resistance are particularly susceptible to crevice corrosion. Typical metals affected by crevice corrosion include aluminum, stainless steels, and titanium. In the case of passive metals like aluminum or the stainless steels, oxygen starvation within the crevice usually destroys the passive film responsible for the corrosion resistance and forms a passive–active cell. The passive–active cell exhibits a greater potential difference and hence induces higher current than that occurring within crevices formed by nonpassive metals like iron and copper [2].

Mechanism of crevice corrosion: crevice corrosion takes place when a small volume of solution gets into a crack or a small opening. It stays there, stagnant, and its composition changes by the corrosion process so that its composition is different from the bulk solution. It may be oxygen depleted (oxygen concentration cell), enriched in metal ions (metal ion concentration cell), or enriched in chloride ions or at a lower pH than the rest of the solution. Crevices are particularly detrimental when alternate wetting and drying occurs, because corrosive liquids retained in the crevices are concentrated by evaporation [31, 32]. To avoid crevice corrosion, the crevice must be wide open to allow the free movement of the electrolyte. It usually occurs in gaps just a few micrometers wide, not in wide gaps or grooves [6].

Heat exchanger locations prone to crevice corrosion: areas prone to crevice corrosion in a shell and tube heat exchanger and PHEs are as follows:

Shell and tube heat exchanger: clearance between the rolled tubes and the tubesheet, open welds at tubesheet, beneath deposits, water box gaskets, bolt holes, nuts, washer, disbonded water box linings, etc.

Plate heat exchanger: beneath gaskets, plate contact points, and beneath deposits.

Crevice corrosion versus pitting corrosion: the mechanism of propagation of pits and crevice corrosion is identical; however, the mechanisms of initiation differ [28]. Crevice corrosion is initiated by differential concentration of oxygen or ions in the electrolyte, whereas pitting is initiated (on plane surfaces) by metallurgical factors and structural factors only [33]. These may include discontinuities in a protective film or coating, or compositional variations such as inclusions. The level of crevice corrosion occurring at crevices such as under deposits or gaskets or at joints between two metals is significantly greater than that of pitting on open surfaces. These two forms of corrosion are compared with reference to austenitic stainless steels in Chapter 2, Heat Exchangers: Mechanical Design, Materials, Nondestructive Testing, and Manufacturing.

Crevice corrosion control: various practices recommended for safeguards against the occurrence of crevice corrosion include the following:

1. Structural designs should avoid any and all crevices. This is especially true for passive metals like aluminum, stainless steels, and various nickel-base alloys. Unavoidable crevices should be filled by weld metal or with nonconducting sealants or cements [2].
2. In a new equipment, specify butt welding joints and emphasize the necessity for complete penetration of the weld metal to guard against even minute crevices.
3. While designing the heat exchanger, avoid sharp corners, stagnant areas, or other sites favorable for the accumulation of precipitates or wherein the solute or O_2 concentration ion cell takes place.
4. Modify the design to avoid crevices. For example, projection of tubes beyond the tubesheet as shown in Figure 3.18 can lead to crevice corrosion.
5. For heat exchanger tubing, a minimum velocity of 5 ft/s is recommended to keep free of deposition. Also, regular cleaning will reduce susceptibility to crevice corrosion, as well as to pitting.
6. During the design stage and operation of tubular heat exchangers, endeavor to provide uniform velocity throughout the exchanger.
7. Upgrade the materials of construction. Use a higher alloy, which offers resistance to a broader range of conditions. Nitrogen combined with molybdenum has a beneficial effect on crevice corrosion resistance in chloride-bearing, oxidizing, and acid solutions. Use materials composed of these elements. High-molybdenum steels, particularly the superaustenitic stainless steels, give good corrosion resistance. Overlaying susceptible alloys with alloy that is resistant to crevice corrosion may also help.
8. Keep the crevices wide open or shallow to allow continued entry of the bulk fluid.
9. Reduce the aggressiveness of the environment, such as acidity, chloride ions' concentration, oxidizing agents, or cathodic reactants.

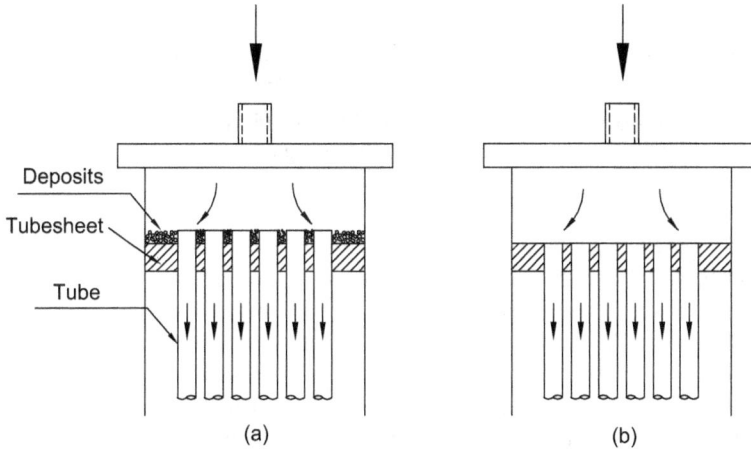

FIGURE 3.18 Projection of tubes beyond the tubesheet. (a) Original design with protrusion of tubes and hence prone for accumulation of deposits and (b) absence of crevices due to flush tube-to-tubesheet joint.

10. Wherever possible, use solid nonabsorbent gaskets like Teflon for gasketed metal joints [9]. This prevents the entry of moisture.
11. Crevice corrosion is not initiated at a specific externally applied potential. Its propagation can be avoided by polarizing metal outside the crevice (cathode) to the potential of metal inside the crevice (anode) [28].

Prediction of crevice corrosion—critical crevice corrosion temperature: tests that establish the temperature of a ferric chloride solution at which crevice corrosion is first observed on mill-produced alloy samples with crevices, which is referred to as the CCCT, are often used to compare the crevice corrosion resistance of various alloys. CCCT is determined as per ASTM Practice G 48 B (6% or 10% $FeCl_3$, for 72 h with crevices).

3.3.4.1 ASTM Standard Guide for Crevice Corrosion Testing

ASTM G78-20-Standard Guide for Crevice Corrosion Testing of Iron-Base and Nickel-Base Stainless Alloys in Seawater and Other Chloride-Containing Aqueous Environments.

This guide covers information for conducting crevice-corrosion tests and identifies factors that may affect results and influence conclusions.

3.3.5 INTERGRANULAR CORROSION

A localized and preferential form of corrosion attack in a narrow region along the grain boundaries or closely adjacent regions without appreciable attack on the grains is called intergranular corrosion (IGC). Due to this form of corrosion, the metal loses its strength and metallurgical corrosion. Intergranular corrosion generally takes place because the corrodent preferentially attacks the grain boundary phase or a zone adjacent to it that has lost an element necessary for adequate corrosion resistance. The depletion of a particular alloying element along the grain boundaries is usually caused by improper heat treatment or heat from welding or any other high-temperature operation that causes the precipitation of certain alloying element at the grain boundary. Conversely, alloys that do not form second-phase microconstituents at grain boundaries, or those in which the constituents have corrosion potentials similar to the matrix, are not susceptible to intergranular corrosion. It should be noted that the problem of sensitization seldom occurs in thin sheet metal [31].

3.3.5.1 Susceptible Alloys

In austenitic stainless steels (18-8), this form of attack is most common. Other susceptible alloys include ferritic, superferritic, and duplex stainless steels. Nonferrous metals such as nickel 200 and nickel-base alloys like Inconel alloys 600 and 601, Incoloy alloys 800 and 800 H, Hastelloys B and C, and aluminum-base alloys containing copper are susceptible to intergranular corrosion [17].

Sensitization of austenitic stainless steels: during heating of austenitic stainless steels between 800°F and 1500°F (450°C and 815°C) while the metal is subjected to welding, heat treatment, or high-temperature exposure, chromium carbides ($Cr_{23}C_6$) are precipitated along the grain boundaries. This precipitation causes the steel to lose chromium below 11% and makes the zone susceptible to corrosion, and this is known as sensitization. Compared to the rest of the grain, the chromium-depleted region is anodic, and severe attack occurs adjacent to the grain boundary if the metal comes into contact with an electrolyte. In the extreme case, whole grains become detached from the materials, which are considerably weakened.

Intergranular corrosion mechanism: intergranular corrosion is an electrochemical corrosion that takes place as the result of local cell action in the grain boundaries. A galvanic cell is formed due to the potential difference between second-phase microconstituents and the depleted solid solution from which the constituents are formed. The carbide precipitate and the grain matrix are cathodic to the locally depleted grain boundary region. The high cathode-to-anode area ratio results in rapid corrosion of the grain boundary material and the metal disintegrates [31].

Weld decay: during welding, there will be a region of the HAZ at either side of the weld bead, which is inevitably sensitized, and these regions are susceptible to intergranular corrosion. Reheating a welded component during multipass welding is a common cause of this problem. This phenomenon is invariably termed weld decay, which is an unfortunate choice of words since, as discussed earlier, welding alone does not cause sensitization of stainless steels [31]. Factors influencing IGC is shown in Figure 3.19.

3.3.5.2 Control of Intergranular Corrosion in Austenitic Stainless Steel

There are three basic remedies for combating intergranular corrosion of austenitic stainless steel:

1. *Heat treatment:* employ a suitable high-temperature solution heat treatment, commonly known as quench annealing or solution annealing. This involves heating the steel to 1976°F (1080°C) followed by rapid cooling. High-temperature heating causes decomposition of the $Cr_{23}C_6$ and homogenization of the chromium by diffusion. Rapid cooling is necessary to prevent the reformation of the carbide.
2. *Low-carbon steel:* ensure that the steel contains insufficient carbon to form $Cr_{23}C_6$ and resulting alloy depletion. Such steels contain less than 0.03%C and are called extra-low-carbon steels, signified by the suffix L (e.g., type 304L and 316L). Type 304L (0.03%C max), the low-carbon version of type 304, is now used extensively in applications calling for resistance to intergranular attack in the welded condition. In a similar manner, type 316L is the low-carbon version of type 316. Lowering the carbon content below 0.03% is normally achieved by argon oxidation process (AOD) and other modern steel melting/refining processes.
3. *Stabilization:* it is possible to stabilize an austenitic stainless steel, such as the 18Cr–8Ni, by adding a potent carbide-forming element such as niobium (also known as columbium) or niobium plus tantalum or titanium. These elements fix the carbon so that it is unable to

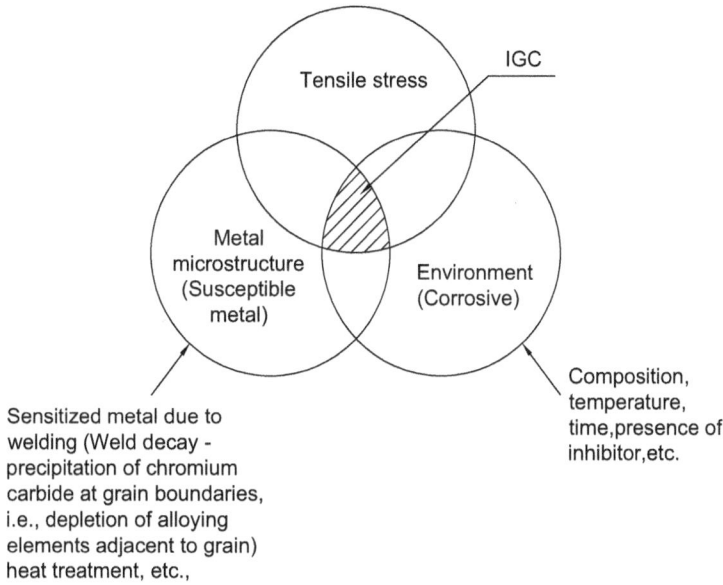

FIGURE 3.19 Factors influencing IGC.

form $Cr_{23}C_6$. The added elements are called stabilizers. It is usual to add titanium or niobium at about 5–10 × %C in order to ensure that no chromium carbides are formed. Typical stainless steels produced by adding stabilizers are type 347 (columbium stabilized) and type 321 (titanium stabilized).

Intergranular corrosion in ferritic stainless steels: ferritic stainless steels are susceptible to intergranular corrosion after being heated to 1700°F–1800°F (925°C–982°C) due to welding or improper heat treatment. It appears that sensitization of ferritic stainless steel occurs under a wider range of conditions than for austenitic steels. This problem is overcome by alloying with titanium and/or columbium to form the carbides of these elements.

Test for detecting susceptibility of austenitic stainless steels to intergranular corrosion: ASTM A262, Standard Practices for Detecting Susceptibility to Intergranular Attack in Stainless Steels, is followed for evaluating austenitic stainless steel HAZ for sensitization.

3.3.6 DEALLOYING OR SELECTIVE LEACHING

Dealloying or selective leaching (sometimes called parting) is a corrosion process in which an alloying element is preferentially corroded over others from the parent alloy, leaving behind a weak structure. The corrosion is detrimental largely because it leaves a porous metal with poor mechanical properties. Dealloying takes place in specific alloy and environment combinations [34]. Combinations of alloys and environments subject to dealloying and elements preferentially removed are given in Table 3.3. The most common example is dealloying of zinc in brass (copper–zinc alloys), known as dezincification. Specific categories of the dealloying process normally carry the name of the alloying element that is selectively leached out in their titles, such as Brasses-dezincification, cupronickels-denickelification, aluminum bronze-dealuminification, silicon bronze-destannification, or desiliconification.

TABLE 3.3
Combinations of Alloys and Environments Subject to Dealloying and Elements Preferentially Removed

Alloy	Environment	Element
Brasses (includes beta-phase attack of Muntz metal and Naval brass)	Stagnant water, chlorides, under deposits, high temperature	Zinc (dezincification)
Aluminum bronzes	Hydrofluoric acid, acids containing chloride ions	Aluminum (dealuminification)
Silicon bronzes	High-temperature steam and acidic species	Silicon (desiliconification)
Copper–nickels	High heat flux and low water velocity (in refinery condenser tubes)	Nickel (denickelification)
Monel	Hydrofluoric acids and other acids	Copper in some acids and nickel in others
Nickel–molybdenum alloys	Oxygen at high temperature	Molybdenum

3.3.6.1 Factors Influencing Dealloying

Dealloying is influenced by many critical factors; in general, factors that increase general corrosion will promote dealloying. However, specific accelerating factors may be further classified into one of three categories: (1) metallurgical; (2) environmental; and (3) water chemistry. These factors are explained while discussing dezincification of brasses.

3.3.6.2 Dezincification

Dealloying of zinc from brass containing more than 15% Zn (e.g., yellow brass 70% Cu: 30% Zn) under conditions of slow-moving water or stagnant water or under deposits is known as dezincification. Zinc is a chemically active element. Its standard electrode potential is very low (-0.763). Standard electrode potential of copper is much higher ($+0.337$).

The difference between the potentials is the driving force of dezincification. Alloys susceptible to dezincification include C 23000, C 26000, C 26800, C 27000, C 28000, C 36500, C 44300 (uninhibited), and C 46400.

Dezincification Types. There are two major types of dezincification, viz., (1) layer type and (2) plug type. The names are taken from the characteristic corrosion product morphologies. In layer-type dezincification, the component surface is converted to corrosion product to roughly uniform depth. The alloying generally increases with increasing temperature and with increasing chloride content of the cooling water. Since the corrosive wear is normally slight, visual observation is difficult; only microscopic examination will reveal the damage.

When dealloying is generally restricted to localized areas such as beneath deposits, at hot spots, or in stagnant regions, it is called "plug-type" dealloying. Plug-type dezincification produces small pockets of plugs of almost pure copper.

3.3.6.3 Factors Influencing Dezincification

Metallurgical: metallurgical factors cover the classes of copper alloys like brass, aluminum bronzes, and cupronickel susceptible to dealloying and the alloying elements parted from them. This has been given in Table 3.3. One important metallurgical factor influencing dezincification is known as beta-phase attack.

For beta-phase attack, in the two-phase (alpha and beta) alloys such as Muntz and naval brasses, dezincification may be concentrated initially on the beta phase and may be sufficient to weaken the metal. If the attack spreads to both alpha and beta phases, complete dezincification may result with the formation of a layer of porous copper [35].

Environment: stagnant conditions, deposits, high heat flux, crevice condition, and stresses accelerate dealloying. Porous or granular deposits further enhance the attack.

In common with many other types of corrosion and chemical reactions, an increase in temperature accelerates the rate of dezincification of brasses. At lower temperature, the rate of dezincification is low, and this explains the long life obtained from naval brass and Muntz metal piping, tubing, plates, sheets, etc., used in contact with cooling brine [35].

In terms of water chemistry, soft water, high concentration of dissolved carbon dioxide, acidic or high-pH conditions, and high chlorine concentrations promote dezincification.

Waters that deposit magnesium, calcium or iron silicates, or silica scale are protective to the underlying brass. Dense adherent calcium and iron carbonate scales are also effective in reducing or preventing dezincification. But thick scales are objectionable for well-known reasons [35].

Prevention of dealloying—general: dealloying is overcome by material substitution, surface cleanliness, and chemical treatment. The most common method of preventing dealloying is to ensure that the tubes are kept clean and free of deposits by screening methods and cleaning procedures. Stagnant conditions should also be avoided.

Prevention of dezincification: though dezincification can be minimized by reducing the aggressiveness of the environment (e.g., oxygen removal), or by cathodic protection [9], it is mostly overcome by substituting alloys immune to or less susceptible to this form of attack. The resistance of brass is greatly increased by the addition of a little arsenic, antimony, and phosphorus, and the resulting alloy is known as inhibited admiralty brass. As a result, inhibited grades of brasses are routinely used in condensers. Typical alloys resistant to dezincification include the following:

Admiralty brass, developed by addition of tin.
Inhibited admiralty brass, developed by the addition of inhibitors, such as phosphorus, antimony, or arsenic.
Alternative alloys like aluminum brass (inhibited), cupronickel, and titanium. Among the copper alloys resistant to dealloying, Cu–Ni alloys are considerably more resistant than Cu–Zn alloys.

3.3.7 Erosion–Corrosion

Erosion–corrosion, a form of localized corrosion, takes place due to the movement of a fluid over a material surface, as shown in Figure 3.20. It takes place mostly on the tubeside with water flowing through it. The corrosion damage involves both mechanical and chemical factors that allow the corrosion to proceed unhindered. The relative importance of mechanical wear (erosion) and corrosion is often difficult to assess and varies greatly from one situation to the other [32]. The role of erosion is usually attributed to the removal of protective surface film or adherent corrosion products by the fluid shear stress under high turbulence conditions.

Erosion–corrosion is usually accelerated when the fluid is entrained with air or abrasive solid particles, such as sand, but erosion–corrosion can also occur in filtered, bubble-free water [30]. The nature and properties of the protective films that form on some metals or alloys are very important from the standpoint of resistance to erosion–corrosion. A hard, dense, adherent, and continuous film would provide better protection than one that is easily removed by mechanical means or hydraulic force [12]. Erosion–corrosion is normally restricted to copper and certain copper-base alloys and aluminum alloys. In this section, erosion–corrosion is discussed in two different forms: (1) erosion-corrosion and (2) erosive wear.

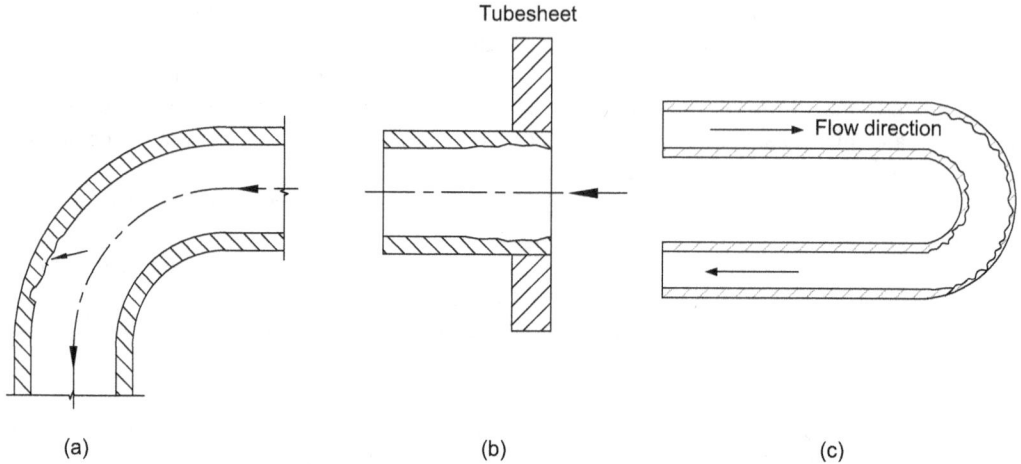

FIGURE 3.20   Tubeside erosion–corrosion.

3.3.7.1   Parameters Influencing Erosion–Corrosion

Erosion–corrosion is influenced by two parameters: (1) turbulence and parameters related to process fluids, such as fluid velocity, impingement attack, level of suspended particles, aeration, bubble level, local partial pressure, cavitation and (2) flow geometry. Turbulence increases with increasing velocity so that higher velocities favor the initiation of erosion–corrosion. Turbulence intensity is higher at tube inlets than downstream, resulting in the phenomenon of inlet-end erosion–corrosion. Similarly, on the shellside, the peripheral tubes located beneath the inlet nozzle without impingement plate protection are most affected by turbulence. Erosion–corrosion also occurs adjacent to a partial blockage in a tube where local velocities through the restricted opening are high or in the turbulent region just downstream of the blockage.

Pattern of erosion–corrosion: erosion–corrosion mostly exhibits a directional property and is characterized by directional grooves, waves, valleys, gullies, holes, etc. [12], or pits in the shape of horseshoes at the site where the protective surface film is damaged [36], or crescent-shaped indentations facing upstream of the water flow that are often influenced by the local flow conditions; consequently, this form of attack has been sometimes referred as "horseshoe," "star," "crescent," and "slot" attack [30]. The increased turbulence caused by pitting on the internal surfaces of a tube can result in rapidly increasing erosion rates and eventually a leak.

Condenser tube failure due to erosion–corrosion: the most common condenser tube failure is due to erosion–corrosion from impingement attack, which develops from a number of causes, such as (1) general impingement attack; (2) lodged debris; (3) localized impingement attack due to blocking by iron oxide scale; (4) waterborne debris; and (5) mussel fouling [37]. Probability of impingement attack of various condenser tube materials in seawater in decreasing order is titanium < cupronickel 70-30 < aluminum brass < cupronickel 95-5 < copper.

3.3.7.2   Erosive Wear

Four distinct forms of erosive wear have been listed by Paul Crook [38]:

1. Solid particle impingement erosion.
2. Slurry erosion.
3. Liquid droplet impingement erosion.
4. Cavitation erosion.

They have a common mechanism of attack, namely, damage to the surface film by mechanical action followed by localized corrosion.

Solid particle impingement erosion: solid particle impingement erosion is associated with solid particle and surface interaction in gaseous environments. Relative velocities range from 2 to 500 m/s; particle sizes range in average diameter from 5 to 500 μm [38].

Slurry erosion: slurry erosion refers to wear due to particle-laden fluid streams.

Liquid droplet impingement erosion: impingement attack has been defined as localized erosion–corrosion caused by turbulence or impinging flow. Entrained solids and air bubbles accelerate this action. Liquid droplet impingement erosion or impingement attack takes place in pumps, valves, pipelines, elbows, heat exchanger tubes, and on the shellside underneath the inlet nozzle. The solution to impingement attack is the use of more resistant alloys like cupronickel containing 0.4%–1.0% Fe, titanium, and stainless steels [39]. Impingement attack has become one of the most frequently reported failure modes in condenser tubes primarily located beneath the inlet nozzle that receive the direct impact of the exhaust steam. This is discussed next.

Impingement attack or steam-side erosion of condenser: impingement attack, so-called steam-side erosion, has been discussed in detail in Ref. [30]. The problem arises when water droplets entrained in the steam enter the condenser and impact on the tubes at high velocity. While corrosion may play a small part in the overall process, purely mechanical processes are considered more important. To resist impingement attack, stainless steel and titanium tubes are superior to copper alloys. For this reason, it is common practice to install stainless steels or titanium tubes, at least in the peripheral sections of condensers where high erosion resistance is required. When thin-walled titanium tubes are used in the peripheral sections, adequate tube support must be provided to avoid the flow-induced vibration (FIV) problems. Properly designed condenser necks will reduce the collection and development of big droplets, which are especially harmful. Protection against impingement attack tubes below the shellside inlet nozzle and clip-on or angle tube protectors made from a more erosion-resistant material fixed over tubes below the inlet nozzle, as suggested by Yokell [40], is shown in Figure 3.21.

TEMA [41] guidelines to limit impingement attack: on the shellside, an impingement plate or other means to protect the tube bundle against impinging fluids shall be provided when entrance line values of ρV^2 exceed the following (where V is the velocity of the fluid in ft/s and ρ is the fluid density in lb/ft^3):

1. Noncorrosive, nonabrasive, single-phase fluids, 1500.
2. All other liquids, including a liquid at its boiling point, 500.

In no case shall the shell or bundle entrance or exit area produce a value of ρV^2 in excess of 4000.

On the tubeside, consideration shall be given to the need for special devices to prevent erosion of the tube ends under the following conditions:

1. Use of an axial inlet nozzle.
2. Liquid ρV^2 is in excess of 6000.

3.3.7.3 Cavitation

Cavitation damage, sometimes referred to as cavitation corrosion or cavitation erosion, is a form of localized corrosion combined with mechanical damage that occurs in turbulent flow or high-velocity

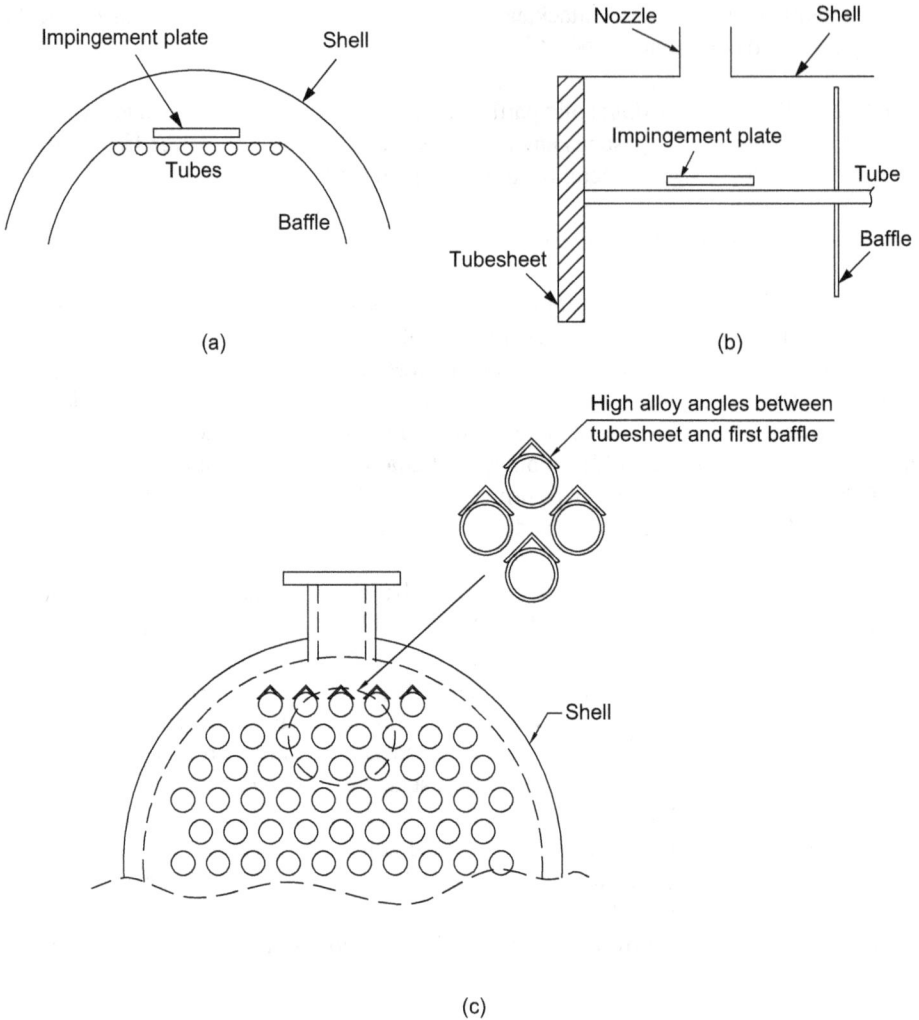

FIGURE 3.21   Impingement protection—angle tube protectors.

fluids. It takes the form of areas or patches of pitted or roughened surface [42]. Cavitation is caused by rapid formation and collapse of vapor bubbles (i.e., voids or cavities), which exert high-pressure forces at the metal surface; these high-pressure forces can deform the underlying metal and remove protective surface films and form pits on surfaces. The occurrence of cavitation damage is shown schematically in Figure 3.22. The bubbles are created as a result of turbulence or when pressure in liquid falls below its vapor pressure. Collapse is caused by subsequent pressure increase. Among various factors, the surface finish plays an important role in the formation of bubbles. Smooth surfaces are beneficial since they reduce the number of sites for bubble formation [32].

Control of cavitation erosion: cavitation damage involves both physical and electrochemical processes. Cavitation resistance increases as the grain size of a metal becomes smaller. Among the alloys, stainless steels is more resistant to cavitation damage because of its ductility, toughness, high corrosion fatigue limit, homogeneity, fine grain size, and ability to work harden [42]. Preventing cavitation damage requires the use of the most resistant alloys and designing the system to avoid turbulence and cavitation.

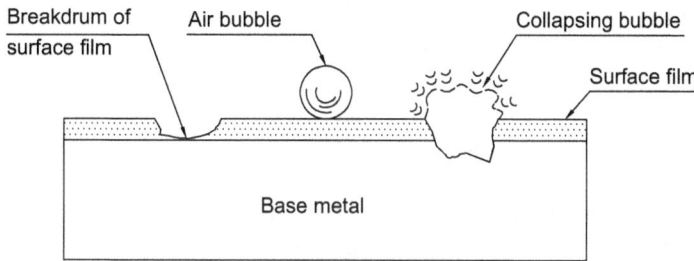

FIGURE 3.22   Cavitation damage.

3.3.7.4   Control of Erosion–Corrosion

Current engineering practice limits fluid velocities in tubes, pipes, and heat exchangers to some arbitrary value based on empirical tests or field experience [43]. The maximum allowable velocity is known as the threshold velocity or critical velocity. Below the critical velocity, the impingement attack does not occur, and above it, it increases rapidly. For example, the critical velocity for the protective film on aluminum brass and 70Cu–30Ni copper–nickel are 2.5 and 4 m/s, respectively [37]. However, not only is the fluid hydrodynamics important, but also the corrosiveness of the process stream and the use of inhibitors, if any, to control corrosion affects corrosion; therefore, a simple prediction based on velocity is not always valid [43]. Recent research demonstrates that it is possible to quantify and model flow-enhanced corrosion and erosion–corrosion phenomena in terms of hydrodynamics, electrochemical corrosion kinetics, and film growth/removal phenomena [44]. Erosion–corrosion is mostly overcome by these measures [30]:

Use filtered water.

If susceptibility to erosion–corrosion is unusually high, it may be necessary to avoid the use of copper and certain copper alloys like brasses, and aluminum alloys, and to use titanium or stainless steel or aluminum brasses or copper–nickel tubes, since these alloys are essentially immune to erosion–corrosion.

Installation of tube inserts made of wear-resistant materials like stainless steel or plastics at the tubes inlets. This aspect is further discussed in detail later. An alternative to inserts, sprayed-on epoxy coatings, is also preferred.

Inlet-end erosion–corrosion can sometimes be prevented by streamlining the flow, redesigning the water box or inlet nozzle, or installing vanes and diffusers to reduce turbulence in the inlet region or eliminate low-pressure pockets.

An important and even newer application of protective coatings is for the inlet end of tubes to prevent inlet-end erosion–corrosion [36].

Install a cathodic protection system in the water box.

On the shellside, use impingement plate protection beneath the inlet nozzle.

Periodic injection of ferrous sulfate into the cooling water has proven to form protective surface film on brass that can be an effective method of controlling erosion–corrosion of power-plant condenser tubes.

It has been shown for several multiphase alloys that if the ratio of surface hardness to abrasive hardness is less than 0.6, wear resistance is low, independent of microstructural conditions [38].

Use alternative material; for brackish or sea water applications, use cupronickel (90:10) or Sea-Cure since both the materials are good resistance for erosion–corrosion [45].

Erosion–corrosion that takes place due to partial blockage of condenser tube ends can be overcome by [46].

Installing an upstream filtration in the intake point.

Introducing an online cleaning method like sponge rubber ball cleaning to remove the fouling and corrosion deposits.

Periodical reversal of the flow.

Prevention of biological fouling by injection of biocides or subjecting to thermal shock.

Periodical offline cleaning program.

3.3.7.5 Ferrules or Sleeves or Tube Inserts

To overcome the tube inlet erosion–corrosion, metallic or plastic inserts, also known as sleeves, have been used in many applications. Some sleeves are leak limiting and others are designed to be sealable sleeves that can serve as a new pressure boundary should a through-wall penetration develop in the tube. Sleeves are typically 150–300 mm long, but are occasionally longer. They are fabricated from erosion-resistant materials like stainless steel, copper–nickels, plastics, etc. One important area where inserts are used is in steam generators that have caustic intergranular attack of tubing in the tubesheet or sludge pile, primary-side SCC of tubing in the tubesheet roller-expanded region, and pitting attack of tubing in the sludge pile [47]. While there is some loss of thermal transfer efficiency, a sleeved tube does little to diminish the overall effectiveness of the exchanger.

Sleeving has long been used in power generation heat exchangers to reduce the inlet end erosion. Because of the need to keep existing equipment operating as long as possible, sleeving the inlets of closed feedwater heaters and steam surface condensers is increasingly widespread in the power generation industry. The process of sleeving involves insertion of thin tubes (sleeve material) into the existing inlet tubes of a heat exchanger and then expanding the inserts into the tubes. Typical sleeve thicknesses are in the range 0.01–0.03 in., depending upon the material of construction and the thickness of the original tubes.

Roller expanding: for many years, the principal method of fastening sleeves into tubes has been roller expanding. Some sleeves were also welded to the tubes at their sleeve outlet ends. Torque controls on rolling equipment produce less precise information than do controls on hydraulic sleeving equipment, and hydraulic methods have been developed to surmount this problem.

Hydraulic expanding: control of hydraulic expanding pressure is far more precise than control of rolling torque. Consequently, there is less risk of overexpanding and causing the tube to bulge between supports and baffles. Hydraulic sleeve expanding pressure is best determined experimentally using mock-ups.

3.3.7.6 Sleeving (Expansion) Process

The HydroproSleevePro® system is an electronically controlled means of hydraulically expanding sleeves into tubes, tubes into baffles, tubes into fins, and full-length liners into tubes by monitoring and controlling the growth or strain of the sleeve. Once the sleeve comes into contact with the parent tube, the SleevePro system controls additional tube growth to limit the radial expansion of the sleeve and parent tube, thus creating a tight interfacial fit by utilizing the elastic capacity of the parent tube. Figure 3.23a shows a sleeve placed inside a tube, and Figure 3.23b shows the pressure path during sleeving with the SleevePro system.

Seals for plugging leaking tubes: patented seals known as Pop-A-Plug for plugging leaking tubes can serve as a new pressure boundary should a through-wall penetration develop in the heat exchanger; they have been patented by M/s. Expando Seal Tools Inc., PA, and marketed by Martin Engineering & Equipment Sales, NJ. The salient features of Pop-A-Plug are discussed next.

The patented Pop-A-Plug System uses a 6600-lb or 17.5-ton hydraulic ram to expand a two-piece mechanical plug inside a leaking heat exchanger tube. The ram pulls a tapered center pin through an externally serrated ring. As the ring expands, the serrations compress against the tube wall. At a precise, predetermined pressure, a breakaway "pops," sealing the tube and separating the plug from the ram. The plugs are available in brass, carbon steel, stainless steel, Monel, copper–nickel,

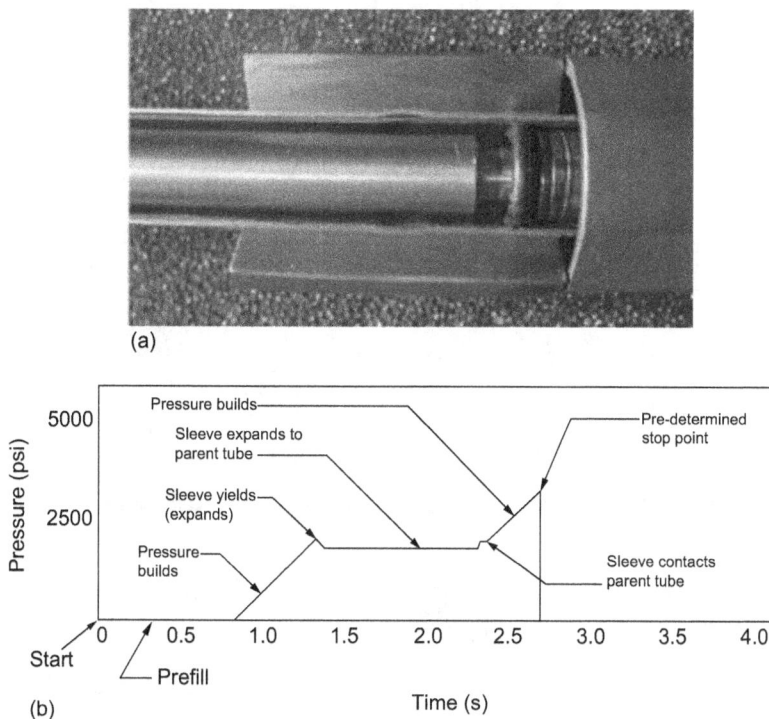

FIGURE 3.23 Sleeving of leaking/corroded tube. (a) Sleeve (SleevePro) and (b) pressure path during sleeving with the SleevePro system. (Courtesy of HydroPro, Inc., San Jose, CA.)

aluminum bronze, and Inconel. Sizes fit tube inner diameters from 0.400 to 1.250 in. Pop-A-Plug and its installment is shown schematically in Figure 3.24.

Considerations while using inserts: while using metallic inserts, care should be exercised to avoid formation of galvanic couple between the tube metal and the insert metal. Also ensure a smooth transition between the end of the tube insert and the tube surface at the downstream to avoid turbulence, which will further enhance the erosion–corrosion in the downstream.

3.3.8 STRESS CORROSION CRACKING

SCC is defined as the corrosion attack on a susceptible alloy due to combined and synergistic interaction of tensile stress and conducive environment. The stress required to cause SCC is normally low and usually below the yield stress, and it can be an applied or residual stress, but it is always a tensile stress. In simple terms, SCC requires the simultaneous occurrence of the following three conditions:

1. A susceptible material.
2. A corrosive environment.
3. Tensile stress.

These conditions are shown schematically in Figure 3.25 and also refer to Figure 3.7.

Time for cracking ranges from few minutes under highly accelerated laboratory conditions to months or even years for resistant materials exposed to a mildly corrosive environment. SCC is much

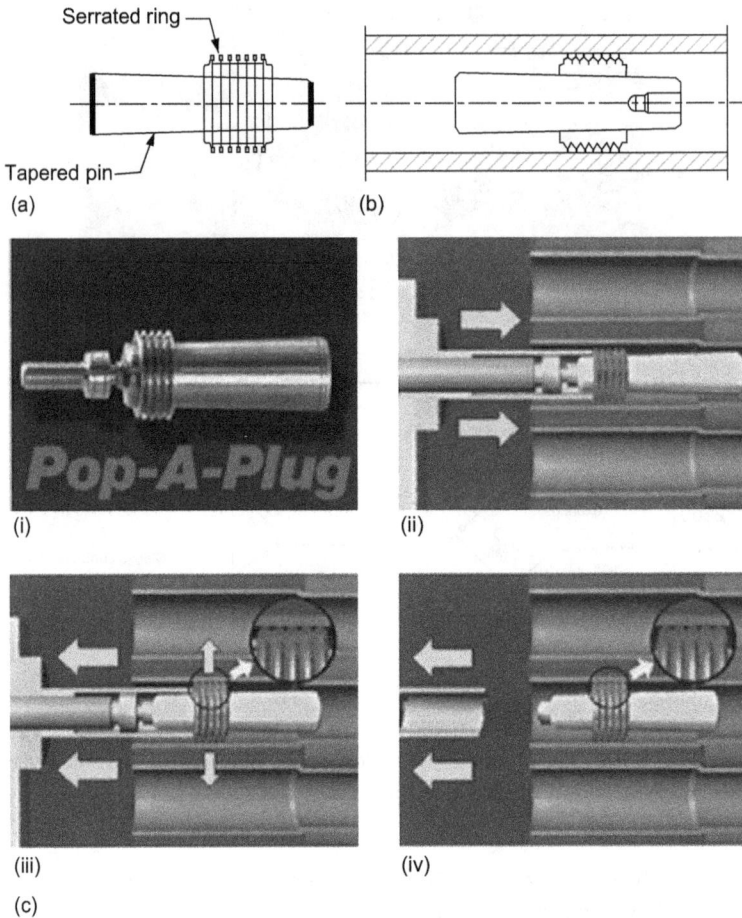

FIGURE 3.24   (a) Pop-A-Plug and (b) Pop-A-Plug positioned inside a tube (schematic). (c) Photo of a pop-a-plug. (i) Three mechanical parts: a threaded adapter, a sealing ring, and a tapered pin, (ii) insertion of Pop-A-Plug in to a tube, (iii) pulling of the Pop-A-Plug, and (iv) Pop-A-Plug positioned. (Courtesy of Maus Italia F. Agostino &C.s.a.s., BagnoloCremasco (Cr), Italy.)

more insidious; the cracks propagate perpendicular to the tensile stress and may be transgranular (Figure 3.26a), common in stainless steels or intergranular (Figure 3.26b) as found in many aluminum alloys with little or no evidence of telltale corrosion products. There is little plastic deformation so that the material appears to behave in a brittle fashion. The characteristic appearance of SCC includes the lack of deformation and relatively small amounts of general corrosion [48]. SCC is the result of the interaction of a number of variables—mechanical, metallurgical, and environmental.

Environmental effects on properties of materials: certain environments drastically alter the strength and stability of engineering structures, whereas the same structures perform satisfactorily in air and other environments [49]. For example, a deep drawn brass fails spontaneously by cracking in air containing traces of ammonia, a stressed mild steel may crack when exposed to condensates of gaseous combustion products containing nitrates, and 18-8 stainless steel above room temperature on exposure to moist environment containing traces of chlorides. These are examples of what is called SCC.

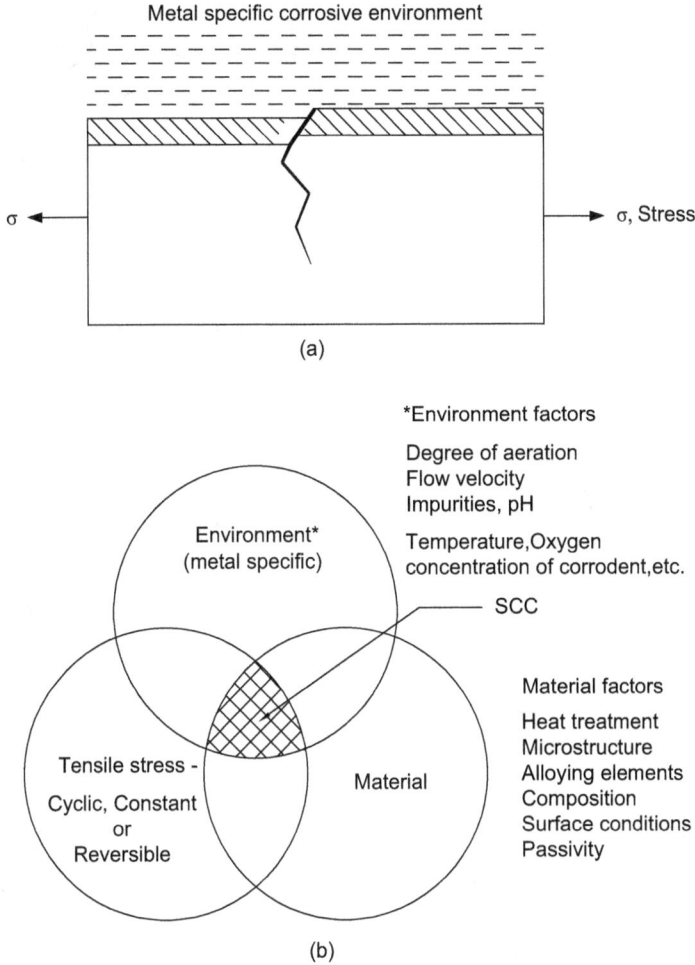

FIGURE 3.25 SCC. (a) Schematic and (b) factors influencing SCC.

3.3.8.1 Heat Exchanger Components Susceptible to SCC

The most common places wherein SCC takes place include at (1) tube-to-tubesheet expanded joints; (2) U-tube bends; (3) indented locations by the baffle plates, and (4) bellows-type expansion joints.

Tube-to-tubesheet expanded joint: all tube-expanding methods leave residual stresses in the tube wall. To minimize residual stress, the tube expansion should not exceed a certain percentage of the tube wall thickness, which varies from metal to metal (carbon steel 5%–8%, stainless steel 3%–5%, alloy steel 4%–6%).

U-bend: cold forming of U-bend tubes induces severe residual tensile stress in the outer portion of the bend, which subsequently fails by SCC. Annealing of the bend tubes will prevent SCC. Other areas prone to SCC are near the U-bend apex for tubes of steam generators where the tube's legs were brought closer together via the denting forces and at nonsmooth transition region of U-bends [47]. Various regions of U-tube heat exchanger susceptible to SCC are shown in Figure 3.27.

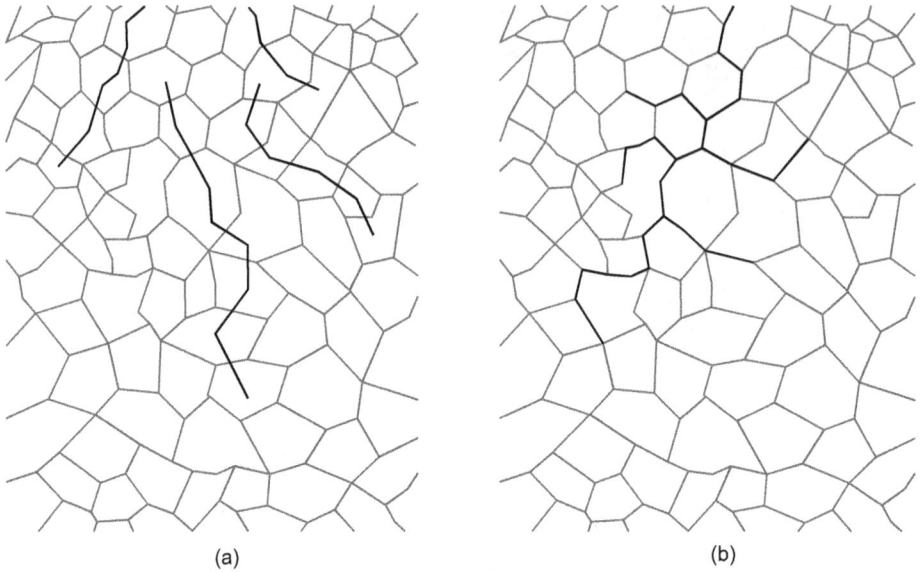

FIGURE 3.26   SCC cracking patterns. (a) Transgranular and (b) intergranular (schematic).

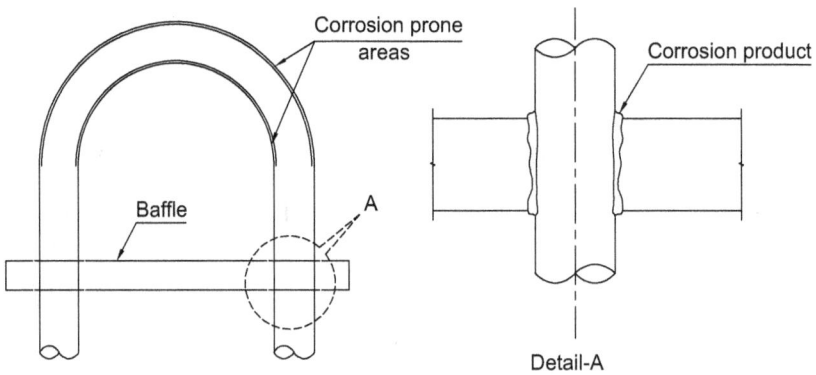

FIGURE 3.27   Some regions of U-tube heat exchanger susceptible to SCC. (a) Regions of U-bend and (b) tube indented locations by the baffle plate.

Thin-walled expansion joints: SCC occurs in thin-walled expansion joint elements made of austenitic stainless steel materials on the bottom of horizontal units that are not drainable.

Classification of SCC failures: SCC failures may involve an electrochemical mechanism, attack by a molten phase, hydrogen embrittlement (HE), or some other factors. According to the environment causing cracking of metals and metal alloys, SCC can be categorized as follows:

Chloride cracking, e.g., austenitic stainless steel, aluminum.
Caustic cracking, e.g., carbon steel.
Ammonia cracking, e.g., admiralty brass.
Polythionic acid cracking, e.g., austenitic stainless steel.
Sulfurous acid cracking, e.g., austenitic stainless steel.

TABLE 3.4
Environments That Cause SCC

Metal	Environment
Al and Al-base alloys	NaCl solution, seawater, H_2O_2, chloride solution, and other halide solutions
Cu and Cu-base alloys	Ammonia and ammonium hydroxide, amines, mercury, sulfur dioxide, H_2S, steam
Carbon steel	Sodium hydroxide solutions, ammonia and sodium nitrate solutions, carbonate/bicarbonate
Austenitic stainless steel	Aqueous chlorides, seawater, sulfurous, and polythionic acid
Nickel and nickel-base alloys	Caustic above 315°C, fused caustic soda, hydrofluoric acid, polythionic acid
Titanium and titanium alloys	Reducing acids such as chlorides, iodides, fluorides, red fuming nitric acid, nitrogen tetroxide, methanol

3.3.8.2   Discussion of Conditions Responsible for SCC

Susceptible alloys and the environment: SCC, like most forms of corrosion, is electrochemical and involves metals in contact with an electrolyte. The environments that cause SCC are specific for each metal. Except for ferritic stainless steel, virtually all metals and alloy systems are susceptible to SCC by a specific corrodent under conditions of temperature, stress level, etc. Chloride ions lead to SCC in 304 and 316 stainless steels when oxygen is present. Typical hostile environments for few of the most widely used metals are given in Table 3.4. Tables of environments and alloy combinations known to result in SCC are published by the National Association of Corrosion Engineers (NACE) [50], the Materials Technology Institute [51], and others for the chemical process.

Stress: tensile stress at the surface of the metal is an essential factor in SCC. Cracking has never been found in metals under compression. The tensile stresses may be due to internal stress caused by metal deformation near welds and bolts, deformation caused by shrink fit, unequal cooling from high temperature, or volume changes in the material caused by phase change or rearrangement of crystal structure [52], or residual stress from some prior cold work or metal-forming operation or caused by an applied stress (i.e., service-induced external load). Residual and applied stresses are additive to evaluate the effect on cracking, and both must be known. Welding often leaves residual stresses that lead to SCC in susceptible environments.

SCC of welded joints: welded joints are particularly prone to SCC for three reasons [31]: (1) the welding operation will leave a residual tensile stress in the weld area unless effective post-weld stress relief is carried out; (2) stress concentrations are usually present; and (3) the thermal cycle can produce a susceptible microstructure. SCC is overcome by shot peening after welding to induce compressive stress on the surface, stress-relief heat treatment, and use of alloys resistant to such cracking.

How to detect SCC: some cracks may be visible to the naked eye and thus require no special detection means. Others may be seen after surface deposits are removed, particularly under magnification. Sophisticated microanalytical instruments such as the electron beam microprobe analyzer, scanning electron microscope, and mass spectrometer are being increasingly applied to failure analysis of power plant components [53]. Crack detection methods like magnetic particle testing, dye penetrant testing, x-rays, ultrasonics, and eddy current testing can also be used [52].

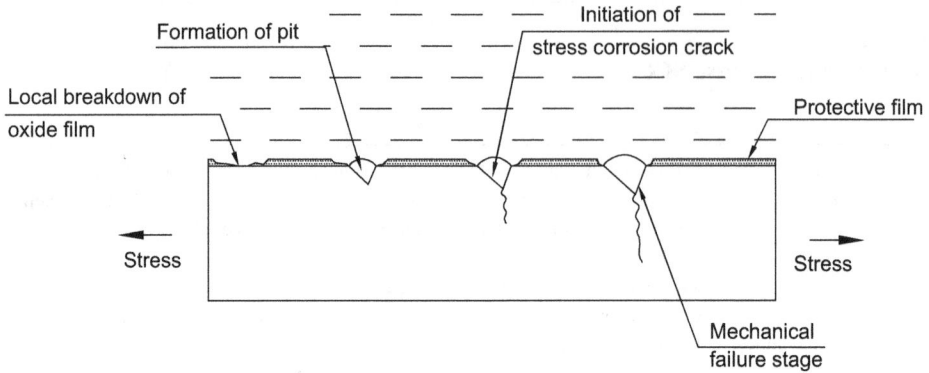

FIGURE 3.28   Growth and propagation of SCC.

3.3.8.3 Theory of SCC

No unified theory for SCC is at present accepted. Theories attributed the failure to mechanical, chemical, fracture mechanics, surface energy, etc. The sequence of events involved in the SCC process is usually divided into three stages [54]:

Crack initiation and propagation.
Steady-state crack propagation.
Crack propagation or final failure.

Metallurgists at the Naval Research Laboratory, Physical Metallurgy Branch, have applied a fracture mechanics theory to SCC [55]. According to this theory, SCC occurs by nucleation and growth at the site of discontinuity in a protective surface oxide or passive film or at the site of preexisting cracks or defects. Under stress or chemical action, pitting initiates at these sites, and pitting continues until the crack extends to its critical length and subsequent fracture. Growth and propagation of SCC are shown schematically in Figure 3.28.

3.3.8.4 Avoiding Stress Corrosion Cracking

Since SCC is due to the interaction of variables associated with (1) mechanical; (2) metallurgical; and (3) environmental factors, adjustment of these variables should minimize cracking [56]. To resist chloride SCC, use higher austenitic nickel-base alloys such as Inconel 600, Incoloy alloys 800 and 825, ferritic and superferritic stainless steels, duplex stainless steels such as ferralium 255, or titanium [57, 58]. Another important measure to prevent SCC is to avoid the metal and environment combinations that are favorable for SCC that are shown in Table 3.4.

Test for SCC: the sensitivity of a steel to SCC is determined by boiling a stressed sample in concentrated magnesium chloride or sodium chloride solutions. Stresses may come from a bead-on-plate weld or from plastic deformation of the specimen. The chlorine ions depassivate the steel and lead to SCC and eventual failure of the stressed specimen. ASTM B 154, mercurous nitrate test, is the standard test method for detecting the SCC of copper and copper alloys.

3.3.9 FRETTING CORROSION

Fretting corrosion is the combined wear and corrosion process that takes place at locations where there is a relative movement between two components and the movement is restricted to a very small amplitude. At the mating surfaces, the degree of deterioration increases because of repeated corrosion of the freshly abraded surface and the accumulation of abrasive corrosion products between these surfaces. A typical example is the wear of the heat exchanger tubes at the tube-baffle contacts. Various metallurgical, geometrical, and environmental factors influence the fretting wear of heat exchanger tubes. Factors that affect fretting wear include [59] contact load, amplitude, frequency, number of cycles, and temperature. Fretting wear of heat exchanger tubes has been covered in Chapter 1, Heat Exchangers: Operation, Performance and Maintenance.

3.3.10 CORROSION FATIGUE

Corrosion fatigue: corrosion fatigue is the reduction in the fatigue strength of a metal as a result of exposure to a corrosive environment. When a metal is subjected to cyclic stress in a corrosive environment, a marked drop in or elimination of the endurance limit may occur even in a mildly corrosive environment. Surface film-protected alloy is especially susceptible to corrosion fatigue [60]. Corrosion fatigue is shown schematically in Figure 3.29a and the drop in endurance strength in Figure 3.29b. In film-protected alloys, stress reversals cause repeated cracking of the otherwise protective surface film, and this allows access of the corrodent to the unprotected metal, with resultant corrosion. The reduction in fatigue strength has a consequence on the life and reliability of a component. The reduction may be so severe that a detail that has been ignored at the design or inspection stage can lead to catastrophic failure in service [31]. In heat exchangers, corrosion fatigue can occur in any tube material; it is caused by steam buffeting in the case of the condenser or FIV in association with inadequate tube support [37]. Generally, failure occurs at tube midspan due to collision between adjacent tubes above critical velocity.

Factors influencing corrosion fatigue and crack growth: corrosion fatigue occurs in metals as a result of the combined action of a cyclic stress, a corrosive environment, and metallurgical factors. The factors that influence corrosion fatigue include [59] stress intensity range, load frequency, stress state, environment, electrode potential, and metallurgical variables. The environmental factors include pH, concentration of corrosive species, dissolved oxygen content, conductivity, pressure, temperature, and flow conditions. Corrosion fatigue cracks are always initiated at the surface unless there are nearby surface defects that act as stress concentration sites and facilitate subsurface crack initiation.

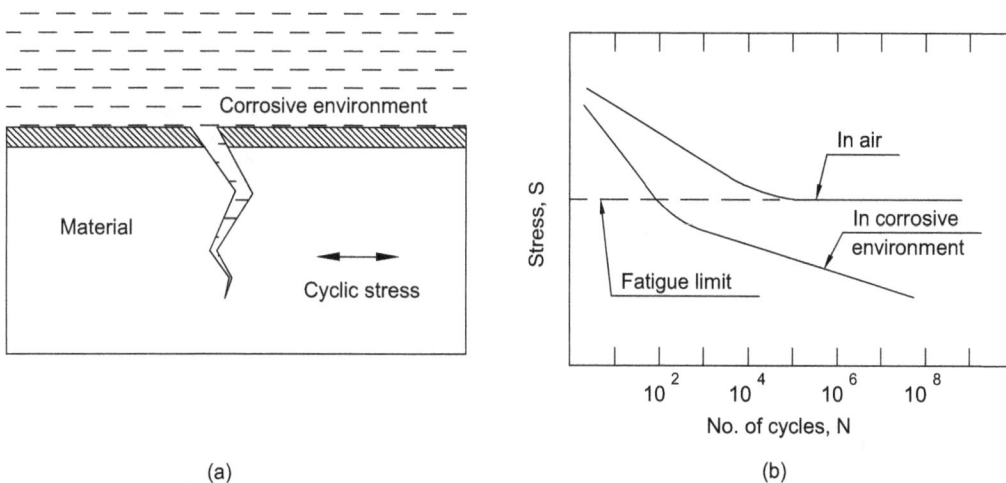

FIGURE 3.29 Corrosion fatigue. (a) Diagrammatic representative and (b) S–N curve for steels subjected to cyclical stress showing drop in endurance strength due to corrosion fatigue.

Intensity of corrosion fatigue: at any given time prior to failure, damage due to corrosion fatigue will be greater than the sum of corrosion damage plus fatigue damage [6]. Localized corrosion, such as pitting or intergranular corrosion, has a greater accelerating effect than the uniform corrosion [60]. The reduction in fatigue strength due to corrosion is sometimes expressed in terms of damage ratio, which is the ratio of corrosion fatigue strength in a particular environment divided by the air fatigue strength. The damage ratio for aluminum in seawater is 0.4, stainless steel 0.5, and mild steel 0.2 [31].

3.3.10.1 Corrosion Fatigue of Various Metals

The alloys that are generally affected by corrosion fatigue include aluminum alloys exposed to chloride solution, Cu–Zn and Cu–Al alloys exposed to aqueous chloride solutions, and high-strength steel in a hydrogen atmosphere.

Ferrous alloys: in a corrosive environment, ferrous alloys lose their fatigue limit, and hence the bottom of the standard stress-cycle curve (S-N) dips down from horizontal. As per theory, the drop in fatigue strength takes place as the cyclic stress causes progressive slip within metal grains, constantly producing clean metal surfaces that are anodic in nature and hence dissolve continuously [6].

Relationship between corrosion fatigue and SCC: corrosion fatigue is mostly interrelated to environmentally induced corrosive attack forms such as SCC and HE [59]. The relationship between corrosion fatigue and the two other environmental cracking mechanisms, SCC and HE, is shown in Figure 3.30.

Prevention of corrosion fatigue: corrosion fatigue can be prevented by several methods such as (1) using protective coatings; (2) adding inhibitors to the environment; (3) cathodic protection; and (4) introducing residual compressive stresses by methods such as shot peening.

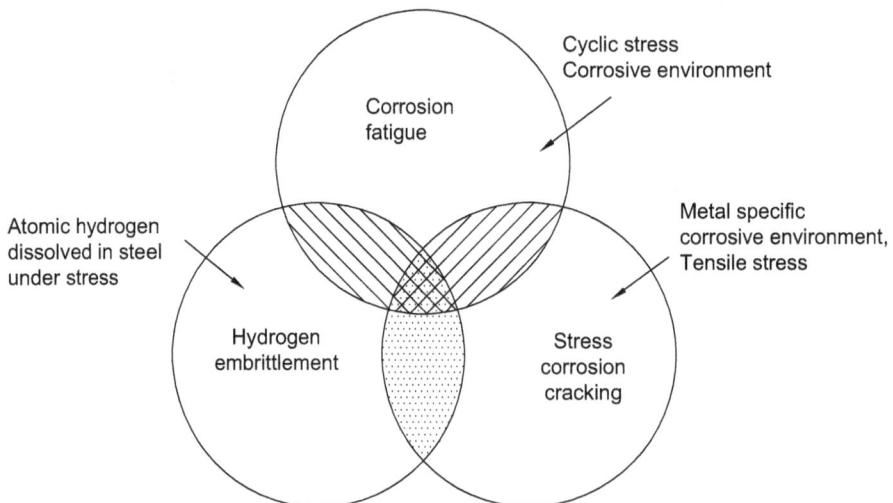

FIGURE 3.30 Relationship between corrosion fatigue, SCC, and HE.

Source: Adapted from Glaser, W. and Wright, I. G., Forms of corrosion—Mechanically assisted degradation, in *Metals Handbook*, 9th edn., Vol. 13, Corrosion, American Society for Metals, Metals Park, OH, 1987, pp. 136–144.

3.3.11 MICROBIOLOGICALLY INFLUENCED CORROSION (MIC)

MIC is the deterioration of metal by a corrosion process that occurs directly or indirectly as a result of metabolic activity of microorganisms on heat transfer surfaces. It also implies, however, that the corrosion would not have taken place in the absence of these organisms [61]. MIC is not a new form of corrosion, but some of the conditions created by microbes can lead to electrochemical reactions that make an environment much more corrosive. Certain microbes can metabolize nutrients (e.g., oxygen) and generate corrosive agents (e.g., organic acids) and other chemical compounds (e.g., sulfur and iron) or create a living crevice or active–passive cell due to biofouling [16]. Natural waters may contain several classes of microorganisms. Attack can be caused by SRB, biological slimes, mates, and tubercles by fungi and other microorganisms, including the iron bacterium [62]. Other synonyms for MIC are microbial corrosion, biodeterioration, and biocorrosion. Since the concept of anaerobic MIC was introduced in 1934, scientists and engineers have recognized that MIC can be a significant contributor to component failures.

Susceptible metals: all but the higher nickel–chromium alloys and titanium have been found to be subject to microbiological corrosion. Common engineering metals and alloys such as carbon steel, lined steel, stainless steels, aluminum alloys, and copper alloys, 6% Mo stainless steels, and high–nickel alloys (Monel 400 and Alloy B-2) have exhibited susceptibility to MIC [63].

Sources of microorganisms: microorganisms are widely distributed in nature, being present in streams, rivers, lakes, seawater, and coastal estuaries. Such microorganisms include bacteria, fungi, algae, and yeasts, depending on the sources of water.

3.3.11.1 Classification of Microbiological Organisms

Microbiological organisms, or microbes, known to cause corrosion of metals may be classified in four general flora groups [64]: (1) bacteria; (2) fungi; (3) algae; and (4) yeasts. The three major types of microbes commonly associated with MIC are (1) SRB; (2) iron and manganese bacteria; and (3) sulfuroxidizing bacteria [16].

Sulfate-reducing bacteria: the SRB seem to be involved in at least some form of the MIC of most of the susceptible alloys like iron and mild steels, aluminum alloys, and copper alloys [65]. They reduce sulfates to sulfides and depolarize cathodic sites on metal surfaces by consuming hydrogen.

Iron and manganese bacteria: iron- and manganese-oxidizing bacteria, known as metal ion concentrators/oxidizers, are most often associated with the corrosion of stainless steels [16]. Such bacteria produce iron and manganese metabolites that form deposits that in turn create concentration cells or harbor other corrosive microbes.

Sulfur-oxidizing bacteria or acid producer: some microbes can oxidize sulfur compounds to sulfuric acid; a pH as low as 2 has been recorded where sulfur-oxidizing microbes are active.

Attachment, growth, and influence of microorganisms on metal surfaces: on attachment to metal surfaces, microorganisms begin colonizing and produce a biofilm, known as slime or mat. Certain microbes can metabolize nutrients and other chemical compounds (e.g., sulfur and iron) to create corrosive environments or they can directly participate in electrochemical reactions. The metabolic processes of the microorganisms can influence the corrosion behavior of materials by [64,66,67] (1) destroying protective surface films; (2) producing a localized acid environment; (3) creating corrosive deposits; (4) creating corrosion cells, notably differential aeration and ion concentration cells; and (5) altering anodic and cathodic reactions, depending on the environment and organisms involved. These processes may occur either alone or in combinations. The microorganisms tolerate elevated pressures and a wide temperature range, and oxygen content from 0 to almost 100%. Temperatures greater than 40°F (4°C) but less than 140°F (60°C) tend to promote MIC.

Categories of MIC: MIC can be divided into two categories: aerobic and anaerobic. Aerobic microbes exist in the presence of air. Anaerobic microbes can exist and multiply in the absence of air. SRB are the most important microorganisms in the anaerobic category, whereas slime-forming bacteria, sulfur bacteria, and iron bacteria are important microorganisms in the aerobic category.

Detection, Testing, and Evaluation of MIC
To properly diagnose MIC, investigations should comprise a combination of chemical, metallurgical, and microbiological analyses. NACE TM0212-2018 presents information on sample collection, testing methods, corrosion monitoring, and how to relate the results to MIC.

NACE TM0212-2018-SG
NACE TM0212-2018-SG Detection, Testing, and Evaluation of Microbiologically Influenced Corrosion on Internal Surfaces of Pipelines.

Genetic MIC Testing Service
Gas Technology Institute (GTI), IL offers a comprehensive corrosion testing service that directly detects and quantifies corrosion-causing microorganisms typically found in pipes, production wells, and other equipment used by the natural gas, petroleum, chemical, water, produced water, and wastewater industries [68].

3.3.11.2   MIC of Industrial Alloys
Mild steel: due to MIC, carbon steels have experienced random pitting, general corrosion, and formation of tubercles on surfaces. MIC of carbon steels can be prevented by chemicals, coatings, or cathodic protection [65].

Austenitic stainless steel and weldments: pitting failures of austenitic stainless steel components and weldments by MIC commonly result when residual natural water is left in stainless steel equipment after hydrotesting. MIC failures of austenitic stainless steel welds have been reviewed by Pope et al. [65] and in Refs. [67,69]. Measures to overcome MIC in stainless steel include solution annealing and pickling, and a temporary measure, spraying the heat exchanger tube interior by a quick set epoxy to a distance of 5–6 ft instead of plugging [70].

Copper alloys: copper and copper alloys are more resistant to the attachment of biofouling organisms than steel and most of the other common materials of construction [71]. This is due to the inherent characteristic of copper alloys and appears to be associated with copper ion formation within the corrosion product film. Therefore, coupling of steel or less noble materials or cathodic protection, which suppresses copper ion formation, allows biological fouling to occur [72]. Copper alloys are particularly sensitive to SRB and ammonia-producing bacteria and have exhibited failures by pitting, plug-type dealloying, SCC, and erosion–corrosion [65].

Titanium's resistance to MIC: titanium and its alloys exhibit excellent resistance to MIC under both anaerobic and aerobic conditions. More than 30 years of extensive titanium alloy use in biologically active process and raw cooling waters, especially seawater, appears to substantiate titanium's resistance to MIC [73]. Nevertheless, titanium alloys are not biotoxic and permit growth or attachment of any micro or macro-biofilm or organism on metal surfaces. However, in no case has localized corrosion ever been observed beneath these biofilms. Periodical cleaning to avoid the buildup of the biofouling is necessary or the water velocity should be sufficiently high.

3.3.11.3 Control of MIC

Prevention of MIC in most of the metals involves trying to prevent the occurrence, growth, and metabolic activities of MIC causing microbes in the vicinity of the metals. The important means of microbial control are as follows:

1. Uptake filtration.
2. Systematic cleaning and elimination of stagnant areas.
3. Proper use of a biocide.
4. Thermal shock treatments

Biocides: the most practical and efficient method of controlling microbiological activity in cooling waters is through chemicals known as biocides. These biocides kill the organism or inhibit their growth and reproductive cycles. Biocides used in cooling-water system include chlorides, chlorine dioxide, bromine, organo-bromide, methylene bisthiocynate, isothiazolinone, quaternary ammonium salts, organo-tin/quaternary ammonium salts [74], copper salts, chlorine donors (e.g., phenates), thiocyanates, and acrolein [75].

Limitations of biocides:

1. Even though these chemicals can do a good job of killing organisms in the bulk water phase, they are much less effective in penetrating biofilms (slime) and killing the organisms therein.
2. While selecting biocides, consideration should be given for toxicity to plant and animal life exposed to the discharged water.
3. Hard-shelled organisms such as barnacles, mussels, and others possess the ability to tightly close their shells when first sensing a toxic substance such as chlorine in the water, to remain closed until it passes, and then to reopen and resume feeding [71].

Thermal shock: the thermal shock treatment involves recirculating the cooling water to allow the temperature to increase to 120°F (49°C), a condition that ensures the death of most organisms.

Microbiologically influenced corrosion failure analyses: MIC can be diagnosed by techniques such as in situ bacterial sampling of residual water, bacterial analysis of corrosion products using analytical chemistry, culture growth, and scanning electron microscopy, as well as nondestructive examination using ultrasonics and radiographic techniques. Metallographic examination can reveal MIC characteristics such as dendritic corrosion attack in weld metal [76]. If special techniques are not followed, they can be misdiagnosed as attack caused by conventional chloride crevice/pitting corrosion attack.

3.4 CORROSION OF WELDMENTS

Weld metal versus parent metal: corrosion of weldments occurs in spite of proper selection of base metal and filler metal, codes and standards followed, and post-weld heat treatment (PWHT) carried out. In welded joints, the weld metal is usually of matching composition to the parent material or overmatched, with its choice normally being dictated by the need to obtain a sound weld of mechanical properties comparable with those of the parent material and resistance to galvanic corrosion [77]. A survey of weld failures, reported at Corrosion 82, a forum on corrosion sponsored by the National Association of Corrosion Engineers, showed that poor welding practices such as poor fitups, misalignments, and incompletely fused root beads have caused many weld failures in process

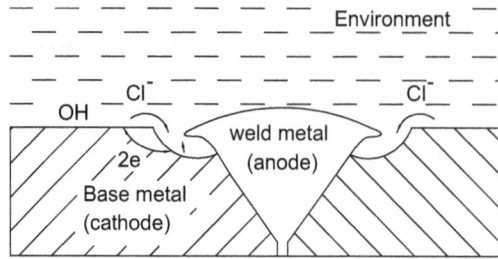

FIGURE 3.31   Crevice corrosion due to weld defect.

vessels. Incomplete fusion, particularly in root passes, is a common source of notches and crevices [6]. The factors that control the corrosion resistance of the weldments and how to optimize the weld quality are discussed in this section.

Causes of corrosion of weldments: in addition to the material composition, the following welding design/process-related features contribute to weld material corrosion [78]:

1. Weldment design.
2. Fabrication technique.
3. Heat input and welding practice.
4. Protection of welding environment, carbon and nitrogen pickup during welding [6].
5. Formation of oxide film and scale.
6. Weld slag and spatter.
7. Welding defects like incomplete weld penetration or fusion, porosity, cracks, crevices, HAZ, and high residual stresses.
8. Metallurgical factors like microsegregation, precipitation of intermetallic phases, formation of unmixed zones, recrystallization and grain growth in the weld HAZ, volatilization of alloying element(s) from the molten weld pool.
9. PWHT.
10. Post-weld cleaning.

The thermal cycle due to the welding process affects the microstructure and surface composition of welds and adjacent base metal. Often both crevice corrosion (Figure 3.31) and pitting corrosion are associated with the weld defects. The presence of secondary phases having different oxidation potential, and the attack on the phases of highest potential can be quite common. Weld deposits containing nonmetallic inclusions and porosity can accelerate the corrosion rate. Sometimes the weld metal may be deficient in corrosion resistance compared to the parent metal due to microsegregation of an important alloying element of the parent metal. This is especially true for welding 6-Mo superaustenitic stainless steel. This situation is overcome by overalloying the filler metal. Weld deposits containing entrapped slag particles or residues from fluxes act as sites for cathodic reaction and can lead to increased weld metal corrosion.

3.5   HYDROGEN DAMAGE

Hydrogen damage or hydrogen embrittlement (HE) or hydrogen attack is the generic name given to a large number of metal degradation processes due to interaction of certain metals with hydrogen atoms. HE is a low-temperature phenomenon, seldom encountered above 200°F, and most often occurs as a result of hydrogen evolved from aqueous corrosion reactions.

Hydrogen embrittlement (HE) is the process by which steel loses its ductility and strength due to tiny cracks that result from the internal pressure of hydrogen (H_2) or methane gas (CH_4), which forms at the grain boundaries or it can be used to encompass all embrittling effects that hydrogen has on metals. Atomic hydrogen will diffuse into and pass through the crystal lattices of metals causing damage, called hydrogen attack. Metals subject to hydrogen attack are as follows: the carbon, low-alloy, ferritic, and martensitic stainless steels even at subambient temperatures and atmospheric pressure. The different forms of hydrogen damage are as follows:

1. hydrogen embrittlement;
2. hydrogen blistering;
3. hydrogen SCC;
4. sulfide stress cracking;
5. wet H_2S cracking;
6. hydrogen-induced cracking (HIC); and
7. stress-oriented hydrogen-induced cracking (SOHIC);
8. stepwise cracking.

3.5.1 Effects of Hydrogen in Steel

Hydrogen sulfide below dew point forms complex sulfides on metallic surfaces. Monoatomic hydrogen can then diffuse into the material and recombine to form molecular hydrogen at voids, laminations, microcracks, or discontinuities around inclusions, generating extremely high pressures [79]. As an interstitial element in the body-centered cubic (bcc) ferrite lattice, hydrogen can cause a number of mechanical problems for steels. It gives rise to blistering and cracking. It stiffens the metal, impairing its ductility. Spontaneous cracking of materials is possible if the material hardness is not limited. The effect of hydrogen absorbed by carbon steels is well known and extensively reported in the literature. The primary effects are the following:

1. Hydrogen-induced cracking (HIC) producing stepwise cracks to the metal surface.
2. HE due to loss of ductility under a slow tensile loading.
3. Cracking in high-strength or high-hardness steels, known as hydrogen-induced stress corrosion cracking (HSCC) or sulfide stress corrosion cracking (SSCC).
4. Blister formation originating at nonmetallic inclusions, known as hydrogen blistering.

Figure 3.32 shows the effects of hydrogen on steel except HE. Requirements for the manufacture of pressure vessels for wet H_2S source are discussed in WRC Bulletin 374 [80]. It has been shown that the sound attenuation in hydrogen-damaged steel can be used to quantify the level of degradation of the material's mechanical property and knowing this, the remaining life of an affected plant can be estimated.

3.5.2 Sources of Hydrogen in Steel

HE and other phenomena influenced by hydrogen require a susceptible material and a fabrication method or a process capable of promoting the entry of hydrogen into the steel. Nascent or newly created atomic hydrogen can result from various operational processes [79]: initial melting, electroplating, acid pickling, high-pressure hydrogen environment, decomposition of gases (e.g., ammonia), hydrogen pickup during welding, corrosion with cathodic liberation of hydrogen, and hydrogen from cathodic protection. Acid corrosion, particularly in the presence of a "poison" such as sulfide, cyanide, arsenic, selenium, or antimony ions, promotes the entry of the hydrogen into the

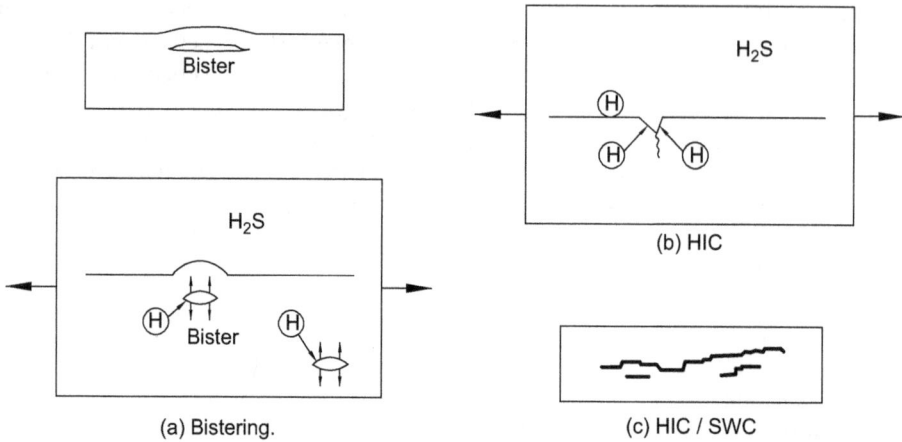

FIGURE 3.32 Schematic of hydrogen damage. (a) Blister (b) HIC and (c) HIC/stepwise cracking.

steel [3]. The nascent atomic hydrogen ($H°$) charged into the steel as a by-product of the corrosion reaction rather than gas evolution as H_2 is shown by the following equation [79]:

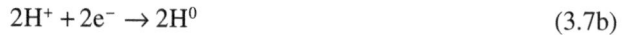

$$Fe \rightarrow Fe^+ + 2e^- \tag{3.7a}$$

$$2H^+ + 2e^- \rightarrow 2H^0 \tag{3.7b}$$

3.5.3 HYDROGEN-INDUCED CRACKING

Some of the atomic hydrogen evolved by the corrosion reaction diffuses into the steel and forms molecular hydrogen, which accumulates at discontinuities, principally inclusion metal interfaces [81]. When the internal pressure due to the molecular hydrogen buildup exceeds a critical level, HIC is initiated. It is recognized that HIC can occur in sour environments in the absence of external stress. Cracking can proceed along segregated bands containing lower-temperature transformation products such as bainite and martensite. Cracking tends to be parallel to the surface but can be straight or stepwise.

3.5.3.1 Stress-Oriented Hydrogen-Induced Cracking

When the small hydrogen-induced cracks or stepwise cracks coalesce perpendicular to the axis of an applied tensile stress on a relatively low-hardness steel having a ferrite–pearlite microstructure, the phenomenon is known as stress-oriented hydrogen-induced cracking (SOHIC) [82]. It occurs around weldments in fabricated steel vessels and process equipment [83]. According to Kane et al. [83], the combination of stress concentration, residual tensile stress, hardness, and microstructure makes these locations prone to SOHIC.

3.5.3.2 Susceptibility of Steels to HIC

Metallurgy of steels: the susceptibility of steels to cracking is influenced strongly by inhomogeneities such as the nonmetallic inclusions, their shape and distribution, and segregation of alloying elements [81]. Voids, MnS stingers, oxide inclusions, and laminations present in the steels act as nucleation sites for blistering, stepwise cracking, and SOHIC [82].

Environmental factors: environmental factors such as H_2S partial pressure, pH of the solutions, and other phenomena relevant to the absorption of hydrogen by the steel influence HIC [81].

Strength and alloying elements: the risk of cracking is increased by increasing the strength of steel as well as the partial pressure of H_2S.

Microstructure: in general, a nonhomogeneous microstructure such as pearlite banding can give high susceptibility to HE.

3.5.3.3 Prevention of HIC

Steels with resistance to HIC embody the following features:

1. The effects of segregation can be minimized by restricting the levels of elements such as C, S, Mn, and P and controlled rolling of plates, which enables high strengths to be obtained at low carbon levels. For wet H_2S systems, there are recommendations to use low-sulfur steels that have been qualified for HIC and SOHIC resistance using laboratory testing. HIC-resistant steels are fine-grain pressure vessel steels manufactured via electric arc furnace with desulfrization, dephosphorization, ladle refining, and vacuum degassing to provide an ultraclean and homogeneous steel.
2. Addition of calcium or rare earth metals to globularize the sulfide inclusions.
3. Low levels of oxide inclusions.
4. Mn is beneficial in improving toughness by refining the ferrite grain size, and therefore a minimum level of Mn must be maintained. Long elongated inclusions such as type II MnS are particularly favorable sites for crack initiation. Spheroidal inclusions are less susceptible to crack initiation except when they are in clusters [81].
5. Freedom from segregation so as to avoid bainitic or martensitic bands.
6. Reduced corrosion effects.
7. Periodic inspection of the wet H_2S refinery equipment even if it is made from HIC-resistant steel, as these steels may not provide total resistance to HIC and SOHIC in actual refinery service [83].
8. Use of stainless steel and stainless steel clad material for severe wet H_2S service as an alternative to reduce inspection and reinspection requirements. Clad products can be used for repair, replacement, or new construction [83].

3.5.4 HYDROGEN EMBRITTLEMENT

3.5.4.1 Mechanism of Hydrogen Embrittlement

Hydrogen embrittlement (HE) causes a reduction in the ductility of a metal due to absorbed hydrogen. HE involves two issues: (1) the crack has been formed by hydrogen and (2) the toughness changes because the material has been embrittled. Atomic hydrogen dissolved in steel can interfere with the normal process of plastic deformation. HE does not affect all metallic material. The most vulnerable are (i) high-strength steels; (ii) titanium alloys; (iii) zirconium alloys—hydrogen embrittlement is caused by zirconium hydriding; and (iv) aluminum alloys. If a steel of initially limited notch ductility (e.g., a steel hardened to 175,000 psi) is charged with atomic hydrogen and strained at the appropriate rate and temperature, it cannot plastically deform to locally accommodate a notch or stress raiser. Consequently, a crack is initiated and brittle fracture can take place. Severely cold-worked ferritic steels and welded steels of higher carbon content are similarly vulnerable to HE [109]. Plain carbon or alloy steels heat-treated to tensile strengths greater than 120,000 psi or hardness >Rc 22–23 are susceptible to HE. However, the other forms of hydrogen damage such as hydrogen attack or hydrogen blistering are associated with unhardened low-alloy or carbon steel [17]. The susceptibility of steels to HE increases with [79] the following:

1. Increasing strength level (hardness) of the steel.
2. Increasing amounts of cold work (plastic deformation).
3. Increasing residual or applied stress.

The risk of cracking initiation should be reduced by avoiding notch effects, whether geometrical or "metallurgical." Metallurgical notches include the formation of hard zones in the parent metal and in the weld areas. Crack propagation can be controlled by limiting operating and/or residual stresses and also by PWHT.

3.5.4.2   Hydrogen Embrittlement of Steel Weldments

Susceptible steels: steels with weldments susceptible to HE include pressure vessel quality steels with 0.25%–0.35% C, in plate thickness 1–2 in (25.4–50.8 mm), such as ASTM A515 and A516 [79].

Prevention of hydrogen embrittlement and cracking of welds: welding-related measures for the prevention of hydrogen cracking, such as use of dry electrode, joint surface cleanliness, sufficient preheat, and PWHT, will be helpful to prevent HE of welds. Hardness testing of weldments is a necessary requirement for most typical operating refinery or chemical plant environments. Limiting the weld hardness of carbon and low-alloy steels as indicated in NACE Standard MR-01-75 [84] can generally avoid this problem. Hardness of completed P-l welds should be 200 BHN or less.

3.5.5   Hydrogen-Assisted Cracking

Hydrogen-assisted cracking, also known as hydrogen sulfide stress corrosion cracking (HSCC), is a special case of localized HE where the nascent atomic hydrogen is supplied to the steel as a by-product of the corrosion reaction. Acid corrosion of steel in the presence of a "poison" such as sulfide, cyanide, arsenic, antimony, or selenide ions is a prime example [79]. Examples for HSCC include sulfide stress cracking and cyanide stress cracking of hardened steel.

3.5.5.1   Prevention of HSCC

HSCC can be controlled or prevented by (1) selecting a material that is resistant to process corrosion and (2) avoiding the formation of hard and brittle microstructures. These two aspects are discussed next.

Material selection: HSCC can be controlled or prevented by selecting a material that is resistant to process corrosion; that is, if there is no corrosive attack on the material, there can be no cathodic liberation of hydrogen and, hence, no localized HE.

Avoiding the formation of hard and brittle microstructures: HSCC poses problems in weld zones. It can occur especially in HAZs with high hardness [81]. Attention should be given to composition, hardenability, and welding procedures to avoid the formation of hard and brittle microstructures. The occurrence of HSCC observed typically in high-strength steels having yield strength higher than 90,000 psi (620 MPa) is currently mitigated by imposing a restriction of maximum hardness of 22 Re or 225–235 BHN. Nevertheless, countermeasures similar to those described for the prevention of HIC are necessary to prevent HSCC in HAZ even with relatively low hardness [81].

3.5.6   Hydrogen Blistering

Hydrogen blistering is a mechanism that involves hydrogen damage of unhardened steels near ambient temperature [17]. The blister originates at nonmetallic inclusion or other internal defects such as voids, laminations, or microcracks. Blistering requires a vulnerable material and a fabrication method or process condition that promotes the entry of atomic hydrogen into the steel as a byproduct of the corrosion reaction rather than gas evolution as H_2.

3.5.6.1 Susceptible Materials

Blistering is usually confined to the lower strength grades of steel, with tensile strength 70,000–80,000 psi, and Re 22 or less hardness, rimmed, or semikilled steels because of laminations, and free machining steels with extensive sulfide or selenide stringers that can serve as blister sites.

3.5.6.2 Prevention of Blistering

Some of measures suggested by Warren [79] for the prevention of blistering include the following:

1. Selection of steel: (a) where blistering is a concern, calcium-treated and argon-blown steels with 0.010% S maximum offer optimum resistance; (b) steels fully killed by silicon additions are preferable to aluminum-killed steel; (c) avoid susceptible steels such as rimmed, semikilled, and free machining steels; (d) hot-rolled or annealed material is preferable to cold-rolled material; and (e) apply cladding, lining, or protective coating with a more corrosion-resistant material.
2. Quality control and inspection of steel: to minimize blistering at planar defects, such as laminations and inclusion aggregates, steel plate should be ultrasonically inspected as per ASTM standards E114 and A578.
3. Venting of multilayer vessels.

3.5.6.3 Detection of Blisters in Service

Surface blisters may be detected by (1) visual inspection or by hand touch; (2) wet fluorescent magnetic particle inspection; (3) ultrasonic inspection to detect internal blisters within the steel; and (4) acoustic emission testing for locating active sites of SSC, HIC, and SOHIC [83]. Inspect both internal and external surfaces, including the less obvious items such as repair patch plate and unvented attachments (e.g., lifting lugs, and support legs).

3.5.6.4 Correction of Blistered Condition in Steel Equipment

If blisters have jeopardized the vessel integrity, cut out the blistered areas and replace with ultra-sonically inspected plate. Vent the blisters by careful drilling from the outside of the vessel. Avoid ignition of hydrogen within the blister; do not torch cut [79].

3.5.7 PRESSURE VESSEL STEELS FOR SOUR ENVIRONMENTS

Plates for Pressure Vessels in sour gas service are normally constructed of low-carbon steel ASME SA 516-70 normalized to ASTM Specification A300.

3.5.8 HIC TESTING SPECIFICATION

HIC testing is performed as per the standards of NACE TM-02-84 [85]. Refer to Figure 3.33 for the various parameters evaluated by this test method. Testing frequency by the manufacturer is one plate of each thickness rolled from each heat of steel. Melting practices by Lukens, one of the leading manufacturers of A 516-70, are mentioned here [86]. All "HIC-Tested A 516" steels are produced to the Lukens Fineline Double-O Two or more restrictive requirements, which utilize various maximum sulfur levels and calcium treatment for inclusion shape control. Phosphorus and oxygen level maximums may also be accepted. Two classes of chemistry control and HIC test guarantee levels may be specified for crack length ratio (CLR), crack thickness ratio (CTR), and crack sensitivity ratio (CSR).

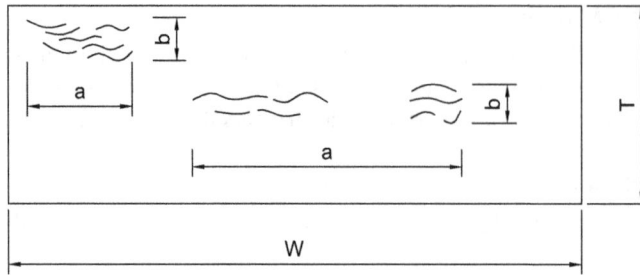

FIGURE 3.33 Determining hydrogen-induced cracking (HIC) resistance (NACE specification TM-02-84). Test specimen and crack dimensions to be used in calculating the following: crack sensitivity ratio (CSR)

$$= \frac{\Sigma(a \times b)}{(W \times T)} \times 100\%; \text{ crack length ratio (CLR))} = \frac{\Sigma a}{W} \times 100\%; \text{ crack thickness ratio (CTR)} = \frac{\Sigma b}{T} \times 100\%$$

where a = crack length; b = crack thickness; W = section width; and T = test specimen thickness.

Note: in the past CSR has been calculated by some investigators as $(\Sigma a \times \Sigma b)/(W \times T)$, which is simply the product of CLR × CTR (i.e., $\Sigma a/W \times \Sigma b/T$). This does not give the same value as $\Sigma(a \times b)/(W \times T)$.

3.5.9 DETECTING HYDROGEN DAMAGE

Detecting hydrogen damage in components is important for monitoring the state of any equipment made of a metal or alloy that might be susceptible to hydrogen damage. The following methods can be used to adequately quantify and measure hydrogen damage [87]:

   i.   Ultrasonic echo attenuation.
   ii.   Amplitude-based backscatter.
   iii. Velocity ratio (of shear-to-longitudinal wave velocity).
   iv.   Creeping wave velocity.
   v.   Advanced ultrasonic backscatter techniques (AUBT).
   vi.   Pitch-catch mode shear wave velocity.
   vii. Time-of-flight diffraction (TOFD).
   viii. In situ metallography.

3.5.10 MATERIAL SELECTION FOR HYDROGEN SULFIDE ENVIRONMENTS

Degradation of steel by hydrogen sulfide (H_2S) is now recognized as a serious problem in pipelines handling sour crude oil or gas in petroleum refinery equipment. Sour environments are defined as fluids containing water as a liquid and hydrogen sulfide. Hydrogen sulfide-resistant steels for pressure vessels or pipelines in sour environments are often required to have both resistance to H_2S cracking and low-temperature toughness. Because iron-base alloys are principal materials of construction, these alloys have been the focus of study for many years. An important reference on hydrogen effects on steels is Warren [79].

3.5.10.1 NACE MR-0175

MR0175 is globally recognized as ISO 15156. MR0175/ISO 15156 addresses requirements and recommendations for selection and qualification of materials for H_2S service in oil and natural gas production. MR0175 addresses all forms of cracking caused by H_2S and applies to equipment using conventional elastic design criteria.

3.5.11 High-Temperature Hydrogen Attack

High-temperature hydrogen attack (HTHA), also called hot hydrogen attack, is a problem, which concerns steels operating at elevated temperatures (typically above 400°F or 204°C) in hydrogen environments, in refinery, petrochemical and other chemical facilities and, possibly, high-pressure steam boilers. It is different from hydrogen embrittlement or other forms of low-temperature hydrogen damage. In this form of attack, atomic hydrogen forms a reaction with unstable carbides, producing methane gas that is responsible for gas pockets and further degradation [25, 26]. At room temperatures, even high-pressure hydrogen can be handled in steel cylinders. At higher temperatures, hydrogen diffused in a steel can react with carbon atoms to form methane gas as follows:

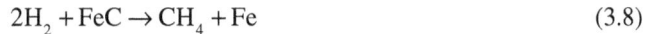

$$2H_2 + FeC \rightarrow CH_4 + Fe \tag{3.8}$$

At elevated temperature/pressure, hydrogen can react with carbon in interstitial solid solution, resulting in initial decarburization, followed by destabilization of the carbides and the formation of methane bubbles at grain boundaries. Since methane cannot diffuse out of steel, its accumulation causes fissuring and blistering, resulting in loss of ductility, and in ultimate failure. Stable carbide compositions, morphologies, and distributions are required to reduce this problem [27].

HTHA damage may occur in welds, weld HAZs, or in the base metal. Even within these specific areas, the degree of damage may vary widely. Consequently, if damage is suspected, then a thorough inspection of representative samples of these areas should be conducted. HTHA affects carbon and low alloy steels, but is most commonly found in carbon steel and carbon-1/2 Mo steel that is operating above its corresponding Nelson Curve limits. Welds often suffer from HTHA degradation as well.

3.5.11.1 Description of Damage

When steel is exposed to atomic hydrogen at high temperatures and pressures, hydrogen atoms may dissolve into methane, accumulating in bubbles that connect to create microfissures at steel grain boundaries. These microfissures reduce the strength of the metal and cause cracks to form in the steel. HTHA cracking can cause complete asset failure for critical steel component such as welds, piping, and exchangers, catalytic equipment, and more [88,89].

3.5.11.2 Affected Materials

(a) In order of increasing resistance: as-welded (non-PWHT) carbon steel, nonwelded carbon steel and carbon steel that has received PWHT, C-0.5Mo, Mn-0.5Mo, 1Cr-0.5Mo, 1.25Cr-0.5Mo, 2.25Cr-1Mo, 2.25Cr-1Mo-V, 3Cr-1Mo, 5Cr-0.5Mo, and similar steels with variations in chemistry.

(b) 300 series SS, as well as 5Cr, 9Cr, and 12Cr alloys, are not susceptible to HTHA at conditions normally seen in refinery units [90].

3.5.11.3 Nelson Curves

In 1949, Nelson gathered and rationalized a number of experimental observations on different steels. In the Nelson diagram, boundaries are placed in a temperature/hydrogen partial pressure graph, which delineates the region of safe use for carbon steels, 1.25Cr-0.5Mo steels, etc. This diagram has been updated a number of times by the American Petroleum Institute (API) and published in the API recommended practice 941.

API RP 941, 8th Edition, February 2016—Steels for Hydrogen Service at Elevated Temperatures and Pressures in Petroleum Refineries and Petrochemical Plants.

This RP does not address the resistance of steels to hydrogen at lower temperatures [below about 400 °F (204 °C)], where atomic hydrogen enters the steel as a result of an electrochemical mechanism.

3.5.11.4   HTHA Damage Detection and Inspection

HTHA damage may occur in welds, weld HAZs, or in the base metal. If damage is suspected, then a thorough inspection of representative samples of these areas should be conducted [88, 89]. Of all the inspection methods for base metal examination, UT techniques and high sensitivity wet fluorescent magnetic particle testing (HSWFMT) are the most sensitive techniques and have the best chance of detecting HTHA damage while still in the fissuring stage, prior to the onset of significant cracking. The most recent approach is a combination of time of flight diffraction (TOFD), phased array UT (PAUT), and/or full matrix capture/total focusing method (FMC/TFM).

Advanced ultrasonic backscatter technique (AUBT) is a collection of techniques used to evaluate frequency dependence, velocity ratio, and several other factors within an asset's material to determine which pieces have been affected by HTHA.

Velocity ratio measurements. By analyzing the difference in ratio between sound velocities in materials both affected and unaffected by HTHA, can determine the presence and extent of HTHA damage.

Phased array is an ultrasonic testing technique that uses specialized multielement "array" transducers and pulses those elements separately in a patterned sequence called "phasing." This phasing sequence allows wave steering, focusing, and scanning. This is all performed electronically. The examination can be tailored for each application, increasing speed and reliability of the inspection.

Radiographic testing (RT). Can detect late-stage HTHA damage in the form of cracks. Cannot detect early-stage HTHA damage.

Visual testing (VT). Surface blisters are readily apparent. HTHA damage has been detected below blistered or damaged cladding.

Acoustic emission testing (AET). Capable of detecting discontinuities with high-stress concentration factors and has a higher probability of detection for late-stage HTHA damage.

3.6   OTHER TYPES OF CORROSION

3.6.1   Touch Point Corrosion

Touch point corrosion (TPC) is a nonuniform type of corrosion that occurs in piping systems in areas where the pipes are in contact with other metal and nonmetallic materials that offer support; this form of corrosion is observable at the base of a pipe. Generally, proper insulation is required in these points that touch the pipes to prevent moisture and chemicals from creating acidic and alkaline conditions that accelerate the corrosion of metal. It is the beam supports and the saddle clamps that have historically caused the majority of the problems. They have the following undesirable features in common [91–93]:

i. crevice forming;
ii. water trapping;
iii. poor inspectability and maintainability;
iv. galvanic couple forming.

3.6.1.1 Design of Supports

When designing pipe supports, avoid the use of saddle clamps wherever possible. Never use a rubber pad between a pipe and a pipe support if the area is exposed to a corrosive environment. When using U-bolts to stabilize piping, always use polyolefin-sheathed bolts. The half-round rod solution has proven to be very effective in controlling pipe-support corrosion over a long period of years on

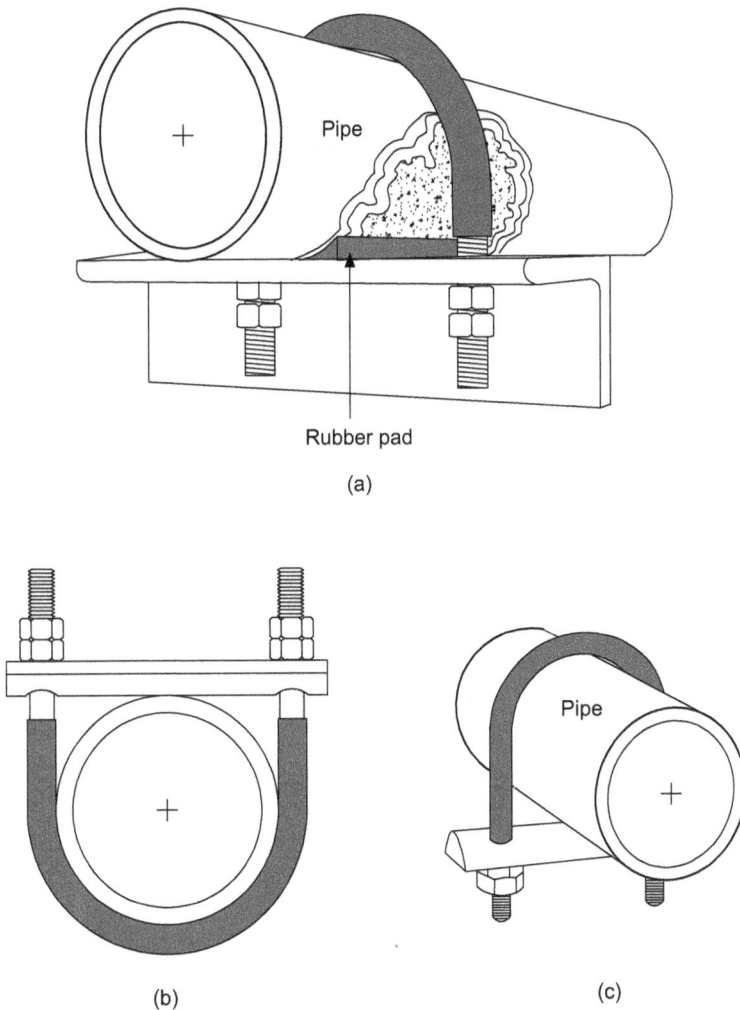

FIGURE 3.34 Touch point corrosion and methods of pipe support to control TPC.

offshore structures. Some operators still use rubber pads of varying types in an attempt to solve this problem, despite industry knowledge that they are counter-productive. The metal-to-metal contact is eliminated, and if used with an insulated bolt, the pipe can be totally isolated from the support structure [92].

Figure 3.34 shows TPC and methods of pipe support to control TPC.

3.6.2 Corrosion Under Insulation

Incidence of corrosion of piping, pressure vessels, and structural components resulting from water trapped under insulation or fireproofing is known as corrosion under insulation. Corrosion under insulation (CUI) is a form of external corrosion that can be widespread or localized, caused by trapped water/moisture on surfaces covered with insulation. Because these surfaces are not generally available/accessible for visual examination, the onset of corrosion cannot be easily identified, and in extreme cases, severe corrosion with consequential impairment of system integrity can occur. Corrosion under insulation is shown in Figure 3.35.

Intruding water is the key problem causing CUI. Corrosion may attack the jacketing, the insulation hardware, or the underlying piping or equipment. Chloride, galvanic, acidic, or alkaline

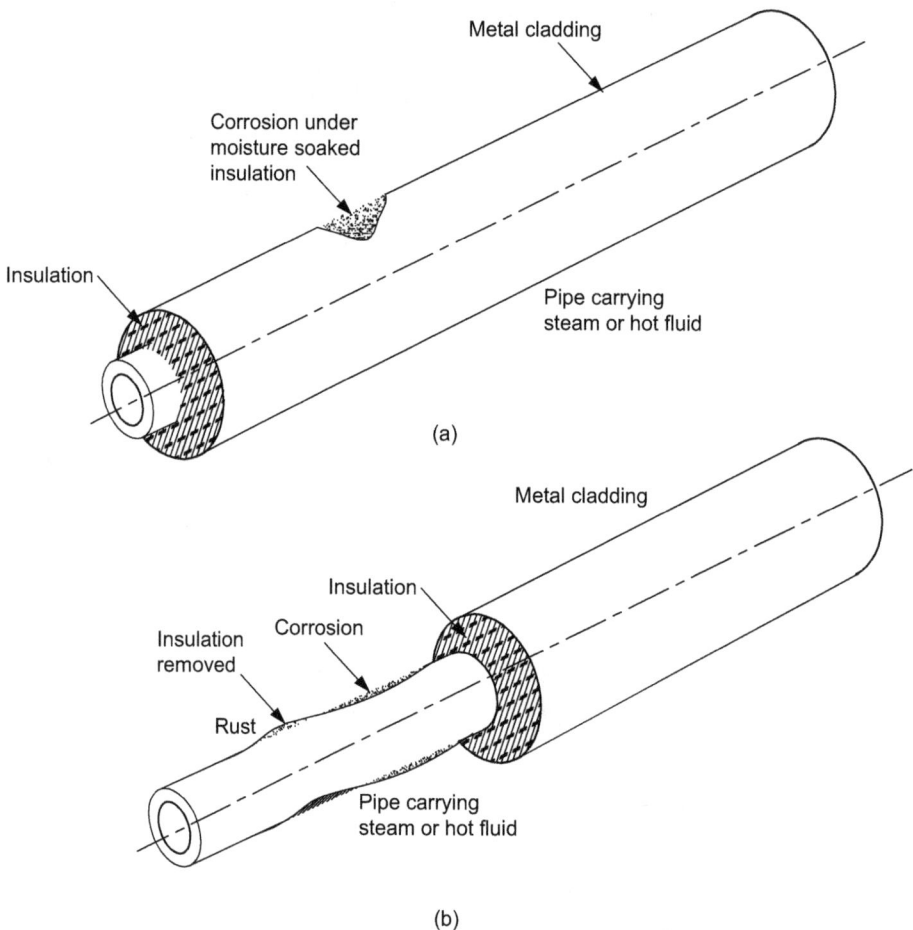

FIGURE 3.35   Corrosion under insulation (CUI).

corrosion can cause corrosion under insulation (CUI), but an understanding of these corrosion types and preventative steps can mitigate the problem [94–101].

3.6.2.1 Affected Materials

Carbon steel, low-alloy steels, 300 series SS, 400 series SS, and duplex stainless steels.

3.6.2.2 Critical Factors

Temperature, duration of wetting, design of the insulation system, insulation type, and environment are critical factors. Corrosion rates increase with increasing metal temperature up to the point where the water evaporates quickly.

3.6.2.3 Description of CUI

According to the National Board of Boiler and Pressure Vessel Inspectors [94]:

1. Galvanic corrosion generally results from wet insulation with an electrolyte or salt present that allows a current flow between dissimilar metals (i.e., the insulated metal surface and the outer jacket or accessories).
2. Alkaline or acidic corrosion results when an alkali, or acid, and moisture are present in certain fibrous or granular insulations.
3. Chloride corrosion can be caused by the combination of insulation containing leachable chlorides with the 300 series austenitic stainless steel surfaces, when moisture is present and temperatures are above 140°F.

3.6.2.4 Detection of CUI

The inspection of plant and pipework for CUI can be extremely expensive if all insulation material has to be removed. Often windows are cut into the insulation for localized inspection. (NDT) techniques can be employed to detect corrosion under insulation or to identify areas that may be susceptible to corrosion without the need to remove insulation. Other techniques that are available include special eddy current techniques, x-ray, remote TV monitoring, and electro-magnetic devices. Here are a few of the best inspection methods for CUI.

i. conventional and advanced radiography,
ii. pulsed eddy current,
iii. guided-wave ultrasonics, and ultrasonic thickness measurements from the internal surface of the equipment.

3.6.2.5 Prevention/Mitigation of CUI

There are five factors in preventing CUI [94]:

(1) insulation selection—careful selection of insulating materials is important. Both water absorption properties and water retention characteristics are important and should be considered;
(2) equipment design;
(3) apply protective paints and coatings;
(4) weather barriers—insulation, insulation jacketing, sealants, and vapor barriers should be properly maintained to prevent moisture ingress; and
(5) maintenance practices.

3.6.3 PIPELINE CORROSION

Pipeline corrosion occurs on both the inside and outside of any pipe and related structures, exposed to corrosive elements. Different types of corrosion can develop in pipelines. Reactions to the substances carried by pipelines as well as external conditions such as weather all contribute [102–106].

3.6.3.1 Pipe Material

The most common materials used in pipe manufacturing are galvanized steel, iron, copper, polybutene, PVC, chlorinated polyvinylchloride (CPVC), and polyethylene. The type of material used for a pipeline depends largely on its intended use. Galvanized steel is the material of choice for conveying chemicals and gasses such as oil, petrol, and gas.

3.6.3.2 Types of Pipeline Corrosion

Pipelines suffer from a number of corrosion forms such as pitting corrosion, uniform corrosion, galvanic corrosion, erosion–corrosion, crevice corrosion, and microbiologically influenced corrosion, etc. Corrosion can also develop in nonmetallic pipes, such as plastic or even carbon fiber.

3.6.3.3 Internal Corrosion

Internal corrosion refers to corrosion occurring on the inside of a pipeline. Internal corrosion often results from the presence of carbon dioxide (CO_2), hydrogen sulfide (H_2S), water, chlorides, organic acids, and others. The rate of internal corrosion depends on the concentration of these corrosive species, the temperature, the flow velocity and the surface material, and many other factors.

3.6.3.4 NACE SP0106-2018, Control of Internal Corrosion in Steel Pipelines and Piping Systems

This standard presents recommended practices for the control of internal corrosion in steel pipelines and piping systems used to gather, transport, or distribute crude oil, petroleum products, or gas.

3.6.3.5 Preventive/Mitigative Measures for Internal Corrosion

Dehydration

Dehydration is the most commonly applied measure to protect against internal corrosion in gas pipelines (and also in liquid pipelines that contain oil with free water or other electrolytes).

Use of inhibitors

Coatings internal coatings have been used on some gas transmission pipelines to improve product flow by reducing drag and eliminating dust.

Buffering in principle

Buffering agents that change the chemical composition of fluids that remain in the pipeline can be utilized to prevent internal corrosion.

Use cleaning pigs

The frequent use of cleaning pigs to scour the internal surfaces of a pipeline is another viable preventive measure.

Use biocides

Biocides can be used in the pipeline to inhibit the corrosive actions of the microbes that cause MIC and thereby reduce or eliminate MIC.

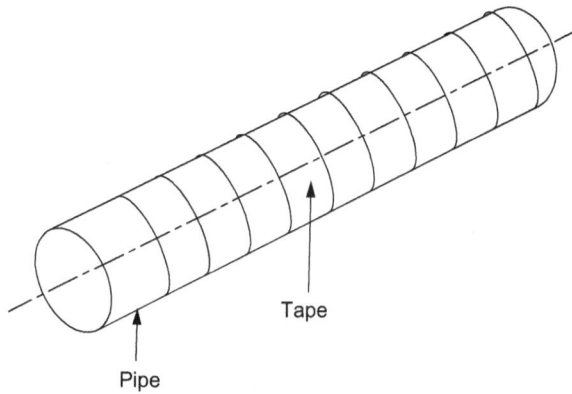

FIGURE 3.36 A tape applied to a pipeline.

Robotic investigation

To assess segments of pipe that are difficult to access, the use of robots with various technologies such as MFL, ultrasonic testing (UT), and photography currently is being investigated.

3.6.3.6 PHMSA

The Pipeline and Hazardous Materials Safety Administration (PHMSA) mission is to protect people and the environment by advancing the safe transportation of energy and other hazardous materials that are essential to our daily lives. To do this, the agency establishes national policy, sets and enforces standards, educates, and conducts research to prevent incidents.

1. Use coatings to prevent chemical corrosion.
2. Use cathodic protection to prevent electrochemical corrosion.

Cathodic protection of pipelines are shown in Figure 3.44 to 3.48 (to be discussed later).

3. Add tapes, mastics, and wraps to prevent abrasions. Figure 3.36 shows a tape applied to a pipeline.

3.6.3.7 Fusion-Bonded Epoxy Coating

Fusion-bonded epoxy coating (FBE), a powder-based coating, possesses the same properties as traditional epoxy and is used to coat and protect steel pipelines, piping connections, and valves. It earns its name "fusion bonding" from the process by which it adheres to a substrate—when the hardener and epoxy react and the coating takes solid form, the chemicals involved are cross-linked—the process is irreversible, even with severe heat applications. FBE coatings can be applied in layer thicknesses of between 350 μm and 1000 μm. The maximum permissible service temperature in continuous operation is 90°C [107]. Many pipelines operate at temperatures above ambient temperature. Depending on the circumstances and environment, the pipeline coating may be classified into [107]:

i.   Single-layer fusion-bonded epoxy (FBE).
ii. Dual-layer FBE.
iii. Three-layer polyolefin (PE/PP).

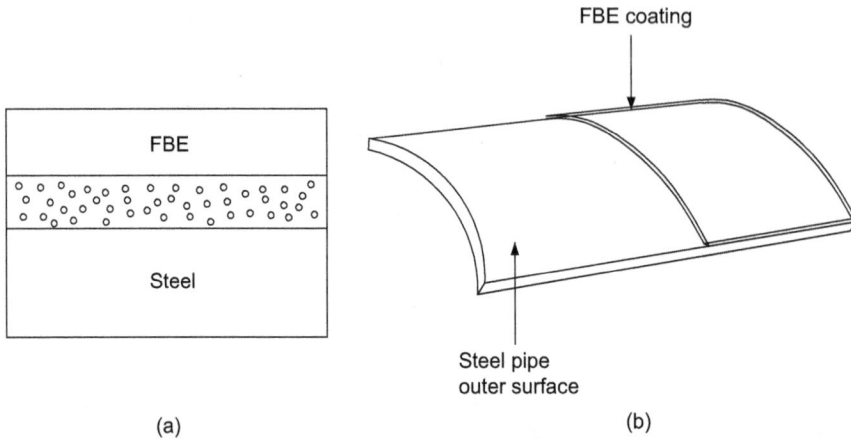

FIGURE 3.37   Fusion bonded epoxy coated pipeline.

Figure 3.37 shows fusion-bonded epoxy coated pipeline. References [108–111] give information on pipeline corrosion issues and surface coating methods.

3.6.3.8   Pipeline Internal Coating

The internal coating equipment consists of self-contained robots that travel inside the pipe, find the weld, blast clean, vacuum, and coat.

Field joint coating

The application of internal field joint coating requires sandblast precleaning of the internal cut-backs prior to welding.

3.6.4   Soil Corrosion

Soil corrosion is a geologic hazard that affects buried metals and concrete that is in direct contact with soil or bedrock. Soil corrosion is a complex phenomenon, with a multitude of variables involved. In soil corrosion, soil acts as an electrolyte. Therefore, properties of soils play a crucial role in accelerating corrosion. Properties of soils such as electrical resistivity, pH, moisture content, porosity, sulfate and chloride content, redox potential, presence of microorganism, temperature are important for evaluating the corrosion potential of soils. Pitting corrosion and stress-corrosion cracking (SCC) are a result of soil corrosion, which leads to underground oil and gas transmission pipeline failures. Three common indicators of a soil's tendency to corrode ferrous metals are as follows [112]:

   i.   electrical resistivity;
  ii.   chloride content;
 iii.   pH level.

Principle of soil corrosion is shown in Figure 3.38.

3.6.5   Concrete Corrosion

Corrosion of reinforcing steel and other embedded metals is the leading cause of deterioration in concrete. When steel corrodes, the resulting rust occupies a greater volume than the steel. This expansion creates tensile stresses in the concrete, which can eventually cause cracking, delamination, and

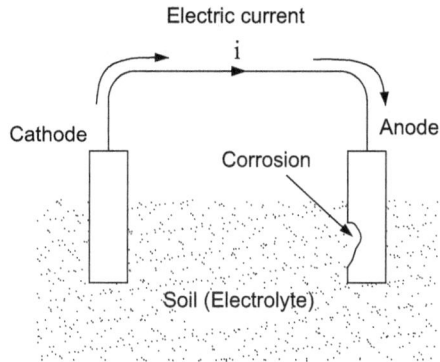

Note : Soil is equivalent electrolyte.Therefore properties of
solids play a crucial role in accelerating corrosion.

FIGURE 3.38   Principle of soil corrosion.

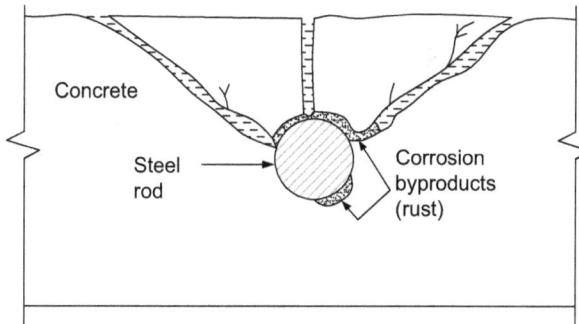

FIGURE 3.39   Concrete corrosion (Illustration).

spalling. Corrosion is initiated when materials that are harmful to steel, such as CO_2 and chloride from de-icing salt, start to penetrate concrete and reach the structure's steel reinforcement [113–115]. Concrete corrosion is shown schematically in Figure 3.39

3.6.5.1   Concrete Corrosion Control

Few concrete corrosion control methods include the following:

   a. Anticorrosion treatments such as interruption or restriction of any of these four corrosion components will slow down the rate of corrosion.
   b. Surface treatments—application of penetrating sealers or breathable coatings, to prevent further water ingress.
   c. Inhibitors may be added to the concrete either as admixtures when the concrete is placed or surface-applied to the steel reinforcing bar.
   d. Coatings may also be applied to the reinforcing steel to create a barrier between the steel and the concrete.

3.7   CORROSION PREVENTION AND CONTROL

The majority of the corrosion control techniques employed center around the ideas of either isolating the corroding metal from the environment or modifying the environment so that either the anodic or the cathodic reaction is brought under control [31]. Alternately, use corrosion-resistant metals in equipment construction. These include coatings and cladding of alloy steels and minor alloy additions to the 300 series stainless steels, chromizing of internal surfaces of 2.25Cr–1Mo [116], and alonizing steel tubes. Corrosion control measures have been mentioned already briefly while discussing various forms of corrosion. However, some generalized procedures are explained in this section.

3.7.1 PRINCIPLES OF CORROSION CONTROL

The most essential technical activities in the practice of corrosion control include the following:

1. Assessing the applicable corrosion threat mechanisms.
2. Selecting and applying appropriate prevention and mitigation control measures.
3. Monitoring equipment and conditions to determine whether the control measures are effective.

3.7.2 CORROSION CONTROL TECHNIQUES

There are proven methods to prevent and mitigate corrosion, thereby reducing or eliminating its negative consequences to the environment, economy, and safety. Corrosion control can be divided into two principal approaches: (1) corrosion prevention and (2) corrosion protection. Corrosion prevention is based on the idea of designing equipment so that corrosion cannot occur, whereas corrosion protection aims at minimizing the corrosion attack. Since corrosion of a material takes place due to electrochemical reaction with its environment, in most practical situations, this attack cannot be prevented; it can only be controlled so that a useful life is obtained from the equipment. Various techniques for corrosion control are as follows:

1. Use of proper design.
2. Changing the characteristics of the corrosive environment.
3. Use of corrosion-resistant materials.
4. Bimetal concept involving cladding and bimetallic tubes.
5. Application of barrier coats, surface treatment.
6. Providing electrochemical protection by cathodic or anodic protection.
7. Passivation.
8. Use of chemical inhibitors.
9. Corrosion management.

There are advantages, disadvantages, and areas of the most economical use for each of these methods. No single method is a universal cure for all corrosion problems [1]. Each problem must be individually studied before implementing the corrosion control measure. The technical solution most suitable to these problems should be decided through cost–benefit analysis by the plant user.

3.7.3 CORROSION CONTROL BY PROPER ENGINEERING DESIGN

Design is an important aspect in the prevention of corrosion. In fact, in many engineering structures, the "weakest spot" is the lack of consideration given to corrosion control during the design stage.

3.7.3.1 Design Details

The design details of pressure vessels and heat exchangers can have a significant effect on corrosion. Design details to minimize corrosion are discussed in Ref. [19]. During the design stage, give consideration to crevices, galvanic couples, drainage, and ventilation. Vessels should allow complete drainage. Inclined heat exchanger installation permits a dead space that may allow overheating if

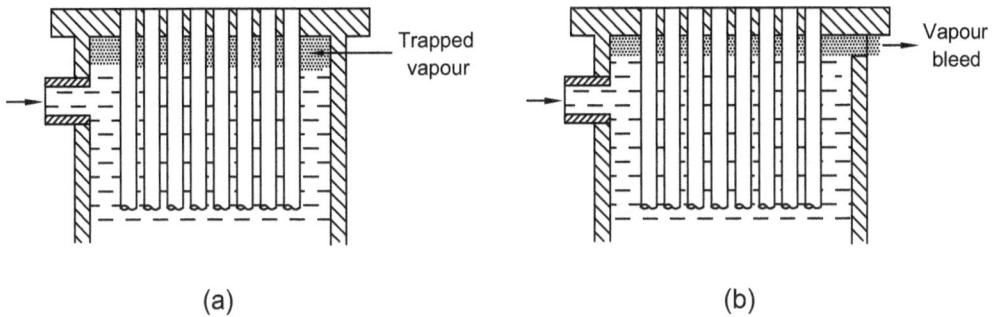

FIGURE 3.40 Crevice corrosion due to concentration of vapor phase and remedial measure. (a) Poor design and (b) correct design.

very hot gases or vapors are allowed on the tubeside (Figure 3.40). Disturbances to flow can create turbulence and cause impingement attack, and direct impingement should be avoided—introduce deflectors or protection devices. Nonaligned assembly distorts the fastener, which forms crevices.

3.7.3.2 Preservation of Inbuilt Corrosion Resistance

The design details should preserve the inbuilt corrosion resistance, including the passivity of the materials. For metals like carbon steels, alloy steels, stainless steels, aluminum, titanium, zirconium, etc., corrosion resistance is inbuilt by means of an adherent protective surface film that separates the metal from the surrounding media. The designer must consider potential corrosion problems from fabrication practices and PWHT.

3.7.3.3 Design to Avoid Various Forms of Corrosion

Crevice and galvanic corrosion, erosion–corrosion, and SCC can be controlled by proper design. Figure 3.40 illustrates the insidious nature of crevices occurred in the tubesheet of a vertical condenser handling vapors of formic and acetic acid [117,118]. If there is a dead space (air pocket) at which chlorides are allowed to concentrate by alternate wetting and drying of tubing surfaces, the tubes will be attacked by SCC. This problem is overcome by air venting the dead space or allowing complete flooding of all tubing surfaces. Another example is shown in Figure 3.41 that permits a dead space at the top of the shell zone due to inclination of a heat exchanger with reference to horizontal direction and design to avoid stagnation of fluid at bottom.

3.7.3.4 Weldments, Brazed, and Soldered Joints

Corrosion in weldments and brazed and soldered joints is also controlled through proper design, fabrication, and joining techniques. Give attention to joint design, surface continuity, and concentration of stress and complete corrosive flux removal.

3.7.3.5 Plant Location

Nearby sources of exhaust and atmospheric pollution due to local industries have an important bearing on the materials of construction [76]. Plant structures should be located in the upwind direction. For example, placing stainless steel equipment just downwind of a hydrochloric acid scrubber can make the condenser susceptible to SCC [15,32].

3.7.3.6 Startup and Shutdown Problems

The construction materials should withstand not only the requirements at the design conditions but also the process upsets, startups, shutdowns, and standby. During these conditions, corrosion

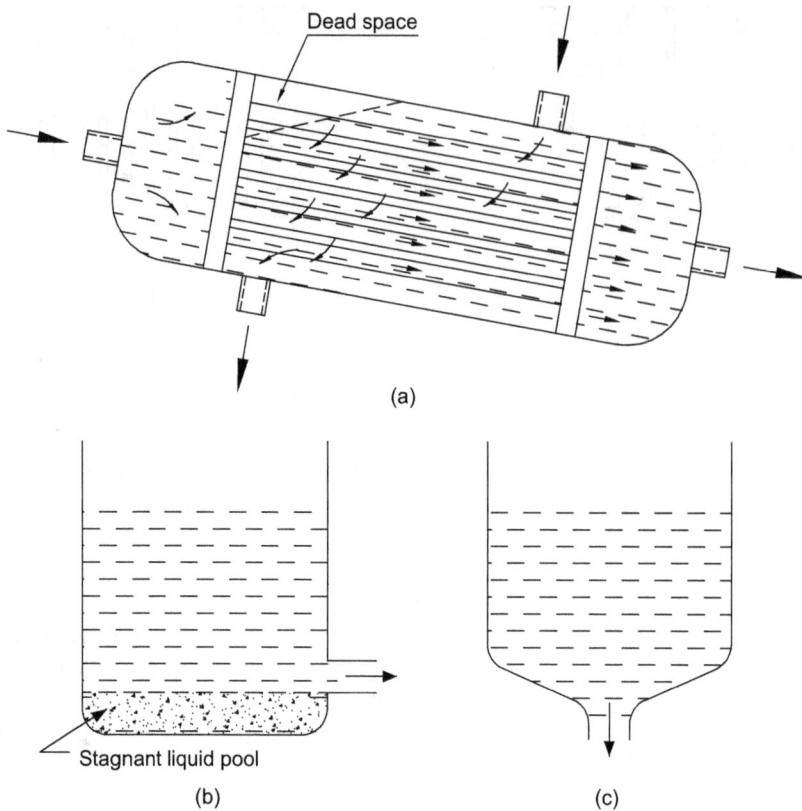

FIGURE 3.41 Inclined heat exchanger that permits a dead space at the top of the shell zone and design to avoid stagnation of fluid at bottom.

conditions can change significantly. Many corrosion problems do not originate when a plant is onstream. Rather, they can be traced to irregularities during startup or shutdown. Startup problems are often related to high temperature, difference in corrodent concentration, inadequate distribution of inhibitors, or incomplete oxygen removal [32]. Downtime problems are usually caused by residual process fluids or dry residues that result from prolonged shutdown periods or decaying of organic matter during shutdown periods. Residual fluids can promote localized corrosion such as pitting or crevice corrosion of stainless steel. Typical cases for shutdown problems are polythionic acid corrosion of stainless steel vessels in refinery applications, and SCC and sulfide attack of copper alloys due to ammonia liberated from putrefying organisms during shutdown periods.

3.7.3.7 Overdesign

In many instances, when the corrosive effect of a system is known, additional thickness to various components will be provided for in the design. This is known as a corrosion allowance. Because this thickness is in addition to that required for the design conditions, an extra cost is involved. Highly corrosive fluids may require the use of expensive corrosion-resistant metals such as stainless steel, titanium, nickel-base alloys, zirconium, tantalum, and nonmetals such as ceramics, graphite, glass, Teflon, etc. However, with the use of the special materials, normally the corrosion allowance is neglected or minimized.

Effect of the service environment—code and regulatory requirements: the codes recognize that corrosion can occur and provide rules for including corrosion allowances to be specified by the designer in the calculation of required thickness.

3.7.4 Corrosion Control by Modification of the Environment (Use of Inhibitors)

Since corrosion is the reaction between a metal and its environment, any modification to the environment that makes it less aggressive will be beneficial in limiting the corrosion attack upon the metal. The corrosivity of a corroding medium can be changed by using various methods such as the following [119]:

1. Use a chemical additive, known as an inhibitor, that will have an effect on the electrochemical reaction to stifle corrosion.
2. Change the corrosivity of the environment by removing the active corrosive constituents. Some examples are removal of oxygen and carbon dioxide from boiler feed water, softening, demineralization of cooling waters, removal of chloride ions from a solution, and deaeration of acid solutions in contact with copper and nickel alloys.

3.7.4.1 Inhibitors

An inhibitor can be generally defined as a material that, when added to a corrodent (liquid), interferes with or retards the electrochemical reaction. To control corrosion, inhibitors are commonly added in small amounts to acids, cooling waters, steam, and other environments, periodically or continuously [120]. Inhibitors can be roughly classified according to the way in which they retard the electrochemical reaction to stifle corrosion. On this basis, if they inhibit the anodic reaction, they are called anodic inhibitors; those that inhibit the cathodic reaction are cathodic inhibitors; and if they inhibit both the anodic and cathodic reactions, they are called mixed inhibitors. The effect of an inhibitor in most cases is to form a barrier between the metal surface and the environment. Inorganic compounds such as chromates and nitrites interfere with the anodic reaction while the polyphosphates suppress the cathodic reaction. The underlying principles of anodic and cathodic inhibitors are discussed next.

a. Anodic Inhibitors

During electrochemical reaction, a metal is dissolved at anodic areas to give metal ions and electrons that flow to cathodic areas. This may be represented as follows for a divalent metal:

$$M \rightarrow M^{++} + 2e^- \tag{3.9}$$

Typical anodic inhibitors include chromates, nitrites, phosphates, molybdates, orthophosphates, silicates, and some organic materials.

There are of three general types of anodic inhibitors:

1. Passivators.
2. Oxidizing inhibitors.
3. Nonoxidizing inhibitors such as orthophosphate.

The passivators function by converting an anodic area to cathodic area, usually by means of impermeable oxide films. Oxidizing inhibitors function by increasing the oxidation potential at the anodic surfaces to increase the rate of formation of gamma-iron oxide [121] or are effective in repairing discontinuity in the passive film by quickly oxidizing iron wherever the surface is exposed. Chromate is the best known oxidizing anodic inhibitor in cooling-water systems. Its merits and demerits are discussed next.

Chromate: chromate is probably the most effective corrosion inhibitor for water systems. Protection is afforded by a film consisting of alpha-ferric oxide and chromic oxide [11]. Chromates are the least expensive for use in water systems and are widely used in the recirculating cooling-water systems of internal combustion engines. However, chromates and heavy metals have recently become ecologically unacceptable in many places. To avoid toxic chemicals and meet regulations on effluent discharge, a large number of substitutes were developed, including zinc, poly- and orthophosphates, phosphonates, and a variety of polymers.

b. Cathodic Inhibitors

Cathodic inhibitors work by increasing the degree of cathodic polarization, thereby reducing the overall corrosion rate and current density, or they reduce corrosion by interfering with any of the steps of the oxygen reduction reaction. By preventing absorption of the electrons released by the anodic reaction, corrosion must stop. Zinc is a well-known cathodic inhibitor [122]. Typical cathodic inhibitors are calcium bicarbonate, polyphosphates, phosphonates, metal cations, and organics. Cathodic inhibitors are often termed "safe" because they do not usually cause localized pitting attack [24]. Cathodic inhibitors are somewhat less effective than anodic inhibitors [123].

Adsorption inhibitors or organic (nonchromate) corrosion control polymers: some organic compounds are effective as inhibitors because of adsorption mechanisms. Oil inhibitors are effective because of the physical adsorption process [124]. Organic inhibitors used in automotive diesel engine cooling-water systems include amines, benzoates, organic phosphates, mercaptans, triazoles, and polar-type oils. Organic inhibitors such as starch quinoline and its derivatives and thiourea and its derivatives are commonly used for inhibition in acid media.

Multicomponent systems: single-component inhibitors include chromates, sodium nitrite, silicates, sodium molybdate, and sodium phosphate. Multicomponent systems with many combinations have pronounced synergism in controlling steel corrosion in recirculating cooling-water systems compared with the individual components. Multicomponent systems can be either heavy metal treatments or nonheavy metal treatments (no zinc or chromate). Typical heavy metal multicomponent treatments include zinc chromate, zinc chromate/phosphonate, zinc polyphosphate, and zinc phosphonates. Nonheavy metal multicomponent systems (no zinc or chromium) in current use are [74,123] (1) combination of the phosphates (AMP/HEDP); (2) polyphosphate–phosphonate mixtures; and (3) the polyphosphate–orthophosphate.
  Nonheavy metal treatment programs are receiving increased attention because of environmental regulations against discharge.

Passivation inhibitors: passivation inhibitors, also known as film formers, work by depositing protective films over the entire surface, which provides a barrier to the dissolution of the metal in the corrosive environment. Typical passivation inhibitors include (1) chemical oxidizing substances, such as chromate and nitrate, and (2) organic substances such as tannin, gelatin, saponin, and betadiketones, used in alkaline solutions.

Precipitation inhibitors: precipitation inhibitors produce insoluble films on the cathode under conditions of locally high pH and isolate the cathode from the environment [69]. Sodium polyphosphate and zinc salts such as zinc sulfate and zinc chloride are examples of precipitation inhibitors.

Copper inhibitors: though the previously mentioned ferrous inhibitors exert some control over corrosion of copper-base alloys, three specific inhibitors are extensively used on to protect copper-base alloys [74]: (1) mercaptobenzothiazole (MBT); (2) benzotriazole; and (3) polytriazone.

Requirements for effectiveness of inhibitors: the key to the success of an inhibitor is the maintenance of clean metal surfaces, which are essential for effective inhibitor film formation [125]. The
inhibitors used should be compatible with the process fluids being used. Consideration should be
given to avoid adverse effects such as foaming, decrease in catalytic activity, degradation of another
material, and loss of heat transfer [120].

3.7.4.2 Inhibitor Evaluation

To determine the effectiveness of an inhibitor for use in a specific application, a comparison is made
by using any of the corrosion testing techniques to determine the corrosion rate of the medium
without inhibitor, and the test is repeated with each inhibitor present in the medium. The effectiveness of each inhibitor can be calculated from the following equation:

$$\eta_i = \frac{m_0 - m_i}{m_0} \times 100 \tag{3.10}$$

where

η_i is the inhibitor efficiency,
m_0 is the corrosion rate without inhibitor,
m_i is the corrosion rate with inhibitor.

Electrochemical techniques such as the galvanostatic pitting potential test can be used to determine the effectiveness of various corrosion inhibitors in the automobile cooling-water system. This
method correlates the effectiveness of inhibitors in the prevention of pitting corrosion by measuring
the inhibitor's effect on the pitting potential (E_p).

3.7.5 Corrosion-Resistant Alloys

Corrosion of certain metals is due to impurities in them. By controlling the impurities, the corrosion
resistance can be improved markedly. Before the advent of the advanced refining techniques, ferritic stainless steels had relatively high levels of interstitials like carbon and nitrogen and, as a
result, had serious limitations with respect to fabricability, toughness, and corrosion resistance. The
newer ferritics, especially those with high chromium content and very low carbon content, have
become possible through argon oxygen decarburization (AOD), vacuum induction melting, and
vacuum oxygen decarburization [126]. On the other hand, a small addition of alloying element
may have a profound effect on an alloy's resistance to certain types of corrosion. Addition of small
amounts of copper to steel increases its atmospheric corrosion resistance; small amounts of phosphorus, antimony, and arsenic added to brasses inhibit dezincification; and small amounts of arsenic
added to copper, and iron added to copper–nickel, increase the resistance to impingement attack and
erosion–corrosion.

Besides changes in composition, metallurgical variations and mechanical effects may have an
effect on reducing corrosion damage. Introduction of compressive stresses on the surface by shot
peening will reduce corrosion fatigue or SCC. Annealing to reduce the internal stresses and solution annealing to remove the grain boundary segregation help in eliminating SCC and intergranular
attack, respectively.

If corrosion of a metal in an environment is inevitable, use an alternative material, such as boro-
silicate glass, impervious graphite, zirconium, tantalum, Teflon, etc., that shows chemical inertness
to most of the chemicals.

A recent focus to tackle severe corrosion is on three alloy families [127]: (1) duplex stainless
steels, which exhibit good corrosion resistance to both pitting and SCC, and have about twice the

yield strength of the typical austenitic stainless steel; (2) 6-Mo superaustenitic stainless steels; and (3) high-nickel alloys such as UNS NO 6625, UNS NO 4400, and UNS NO 7716.

Duplex stainless steels offer several advantages over the common austenitic stainless steels. The duplex grades are highly resistant to chloride stress corrosion cracking. Because the chromium content of the ferritic SSs ranges from 11% to 29%, the general corrosion resistance can vary from moderate to excellent. An important feature of the ferritic SSs is their resistance to chloride SCC. Additions of molybdenum increase the resistance to pitting corrosion by producing a more stable passive film.

The corrosion resistance of duplex alloys depends primarily on their composition especially the amount of chromium, molybdenum, and nitrogen content. For these steels, the chromium content may vary from 22% to 27% and molybdenum 2% to 4%, depending upon corrosion resistance required. The PRE number is calculated using the formula

$$PRE = \%Cr + 3.3(\%Mo) \tag{3.11}$$

3.7.6 Bimetal Concept

The term bimetal is applicable to any combination of two metals or alloys having nearly matching or widely dissimilar physical and chemical properties and bonded by one of a variety of processes. The bimetal concept is employed in (1) cladding and (2) duplex or bimetallic tubes.

3.7.6.1 Cladding

Cladding has been used for many years in pressure vessels, piping systems, and heat exchangers (Figure 3.42). Nuclear and petrochemical pressure vessels are weld overlay clad with 300 series stainless steels or high-chromium nickel-base alloys. The severe corrosion problems in various coal combustion and incineration environments have renewed interest in cladding technology for both code-approved and enhanced-strength developmental alloys [116]. It is well known that various grades of austenitic SSs of types 304, 304L, 308, 316, and 347, nickel, Monel, Inconel, cupro-nickel, aluminum, copper, zirconium, titanium, etc., exhibit excellent corrosion resistance in many corrosive environments. However, the construction of large assemblies such as pressure vessels and heat exchangers in corrosion-resistant metal involves costs. Consequently, increasing use is being made of clad materials to achieve the optimum balance of strength and surface properties to overcome corrosion by an economical means. For example, clad steels satisfy severe service conditions at less cost than full-thickness sections of the costly corrosion-resistant material used for the cladding. Various cladding methods are discussed in detail in a separate section in Chapter 2, Heat Exchangers: Mechanical Design, Materials, Nondestructive Testing, and Manufacturing.

3.7.6.2 Bimetallic or Duplex Tubing

Two dissimilar corrodents in a heat exchanger can cause havoc, with material having a propensity to corrosion to one fluid and virtually immune to the other fluid, or the problem of severe corrosion of different natures that sometimes attacks both sides of a tube can often be solved by treating such a condition as two separate corrosion problems. By selecting the specific kind of metal that is best for each condition, it is possible to combine two such metals into one tube, known as duplex tubing or bimetallic tubing. Duplex tubing is shown schematically in Figure 3.43. Specific information on duplex tubing is given in Ref. [128].

The tube manufacturing method, mostly hot coextrusion, ensures close mechanical contact between the two metals without affecting the heat transfer properties. Duplex tubing is made up of various combinations of ferrous and nonferrous materials. Bimetallic tubes are available in a variety

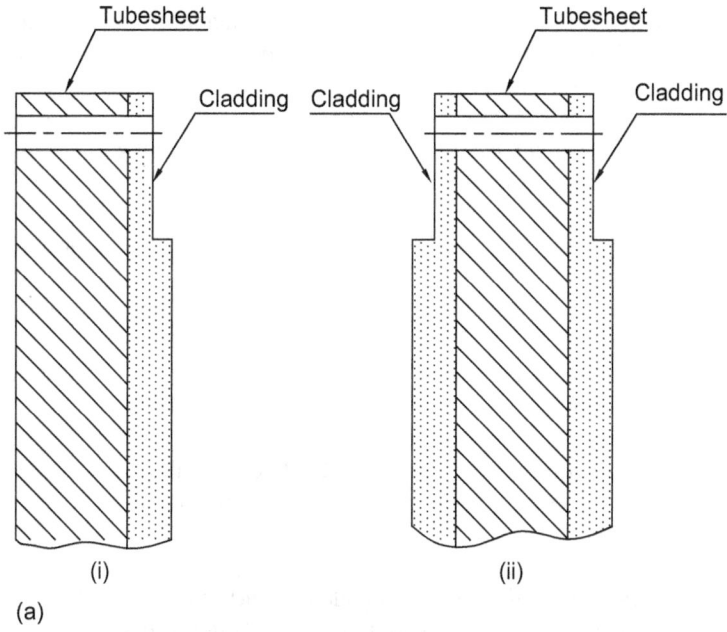

FIGURE 3.42 Cladded heat exchanger component. (a) Tubesheet and (b) heat exchanger heads. (Courtesy of VoestalpineGrobblech GmbH, Linz, Austria.)

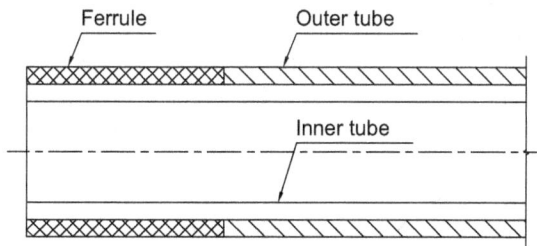

FIGURE 3.43 Duplex tubing.

of stainless steel–carbon steel, Cu, brass, Al, and other combinations. Typical metal combinations of bimetallic tubes used in refinery applications, high-temperature boiler corrosion, and condenser applications are discussed later.

Duplex tubing for refinery applications: for oil refining and in the natural gas industry, duplex tubing with the following combinations is used [35]:

1. Steel outside to resist various corrosive petroleum vapors and copper or copper alloy inside toward the fresh water.
2. Steel outside toward the oil and admiralty, aluminum brass, or cupronickel inside toward circulating saltwater.
3. Other applications in these industries call for the combinations of alloy steels with copper alloys or of aluminum with copper or brass either inside or outside.

Duplex tubing in high-temperature boiler corrosion: C-Mn or low-alloy steels, clad with a stainless steel or superalloys, are used in parts of the boiler where loss is excessive due to high-temperature corrosion [129].

Duplex tubing in a condenser application: in a condenser with stainless steel tubing, hot corrosive vapors caused little trouble within the tubes. But brackish cooling water on the shellside caused a problem. This situation was overcome by installing Carpenter bimetallic tubing consisting of type 304 welded stainless on the inside and deoxidized copper tube on the outside. Ferrules at the tube ends facilitated expansion of the tubes into type 304 stainless tubesheets.

3.7.7 PROTECTIVE COATINGS

Coatings are generally relatively thin films separating the two reactive materials or a metal from an environment. Applying a protective barrier between a corrosive environment and the material to be protected is a fundamental method of corrosion control [130]. It is most widely used for the protection of steel and other metals. Coatings can be metallic, plastic, paints, or organic. Typical metallic coatings include nickel coatings, lead coatings, zinc coatings, cadmium coatings, tin coatings, and aluminum coatings. Mechanisms by which coatings protect can be summarized as follows [131]:

1. They isolate the metal substrate and the environment, as with nickel electroplating.
2. They limit contact between the environment and the substrate, as with most organic coatings.
3. They release substances that are protective or inhibit attack, as with chromate primers.
4. They produce an electrical current that is protective, as with galvanizing.

3.7.7.1 Zinc-Rich Coatings

A zinc-rich coating is an anticorrosive coating for iron and steel incorporating zinc dust in a concentration sufficient to enable the zinc metal in the dried film to corrode preferentially to the ferrous substrate, i.e., to give galvanic protection. Some zinc-rich coatings are used as prefabrication primers or shop primers, where they are applied to freshly blast cleaned steel plates and sections.

3.7.7.2 Thermal Spray Coating

Thermal spray coating is a process that sprays coating materials using the pressure of a hot gas. The material then bonds to the surface, forming a coating that grants properties such as corrosion protection, thermal resistance, environmental protection, lubricity, wear resistance, and electrical/thermal conductivity [132].

3.7.7.3 Phosphating

Phosphating is the most widely used metal pretreatment process for the surface treatment and finishing of ferrous and nonferrous metals due to its economy, speed of operation and ability to afford excellent corrosion resistance, wear resistance, adhesion and lubricative properties.

3.7.7.4 Plastic Coatings

Plastic coatings often represent the ideal solution to a cooling-water problem. Typical plastic materials used for coatings are polyethylene, polyvinyl chloride, epoxy resins, and polyamides. Under certain conditions, they allow a cheap and corrosion-susceptible metal to be covered with a high-grade corrosion-resistant plastic coating. To ensure satisfactory protection, these must be nonporous and sufficient thickness (0.150–0.250 mm) [133].

3.7.7.5 Effectiveness of Coatings

To be effective, the coating film must be completely continuous. Any breakdown of the coating film leads to corrosion. Resistance to water is the most important requirement of the coating, since all coatings will come into contact with water or moisture in one form or another [130, 131].

3.7.7.6 Surface Treatment

Surface treatment is resorted to on certain metals to improve their corrosion resistance in specific applications. For example: (1) mild steel or low-alloy steel heat exchanger tubes are alonized or aluminized (aluminum vapor diffused) for high-temperature oxidation resistance and protection from sulfide corrosion in refinery applications [134]; and (2) titanium tubes are anodized or thermally oxidized to form an inert surface oxide film for its corrosion resistance, especially where hydrogen uptake is of concern.

3.7.8 ELECTROCHEMICAL PROTECTION (CATHODIC AND ANODIC PROTECTION)

3.7.8.1 Principle of Cathodic Protection

Cathodic protection is defined as the reduction or elimination of corrosion by making a metal cathode by means of an impressed current or attachment to a more anodic metal (sacrificial anode) than the metal in the galvanic couple. For pipelines, a cathodic protection system is designed in accordance with regulatory requirements and pipeline industry codes and practices giving consideration to length of pipeline to be protected, thickness of the pipeline coating, soil characteristics like corrosive nature of soil and soil resistance to passage of electrical current, water table characteristic, etc. Components of a cathodic protection system include, rectifiers, anode ground beds, conductive wire and test leads [135]. Cathodic protection system based on sacrificial anode and impressed current methods are shown schematically in Figure 3.44 and these two methods are discussed next.

a. Sacrificial Anode

A cathode is the electrode where practically no corrosion takes place. If follows, then, that if all anodic areas can be converted to cathodic areas, the entire structure will become a cathode, and corrosion will be eliminated. By coupling the structure with a less noble metal, it is allowed gradually to corrode but is replaced after an interval of time. Corrosion control by this principle is known as cathodic protection by sacrificial anode. Zinc or magnesium is often coupled with iron for this purpose and is termed a sacrificial anode. This method is mostly followed in heat exchangers for protecting the end closures.

Sacrificial anodes are relatively cheap pieces of metal. They are designed specifically to corrode instead of the more expensive metal parts. For marine applications, the most common sacrificial

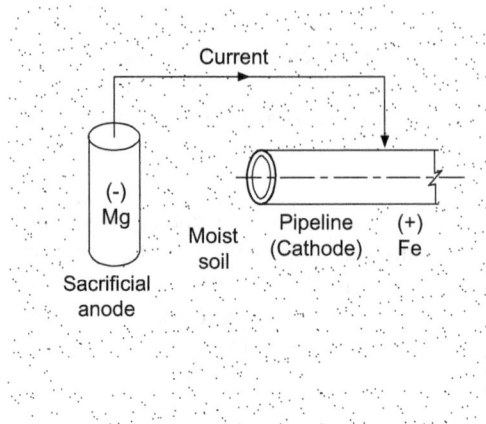

(a) Basic concept of sacrificial anode.

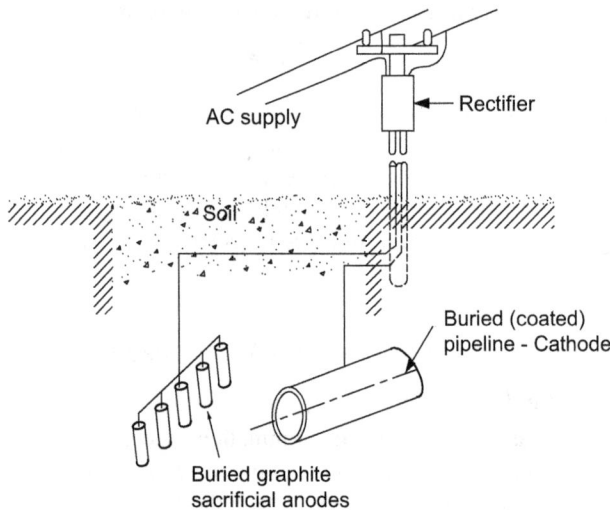

(b) Impressed current method of cathodic protection of a buried pipeline.

FIGURE 3.44   Cathodic protection system based on sacrificial anode and impressed current methods.

anodes are aluminum and zinc with aluminum preferred due to longer-term reliability. The three most active materials used in sacrificial anodes are zinc, aluminum, and magnesium and each have different properties and uses.

Galvanic combinations such as zinc or cadmium or magnesium coatings on steel, often used to avoid corrosion, are known as sacrificial anodic coatings, because they corrode in preference to steel and protect the latter from rusting. Galvanizing a steel heat exchanger of a tube bundle did not improve the life expectancy, and galvanizing is not sufficiently effective to be worth even the slight extra cost involved [136]. Figure 3.45(a) [137] shows protection of an offshore structure and Figure 3.45(b) shows a liquid holding tank with sacrificial anodes.

Impressed current method: because the corrosive action is electrochemical, electric current can be applied to effectively stop the attack of metals exposed to corrosive media. This technique is called cathodic protection by impressed current. In this method, an impressed current from a dc

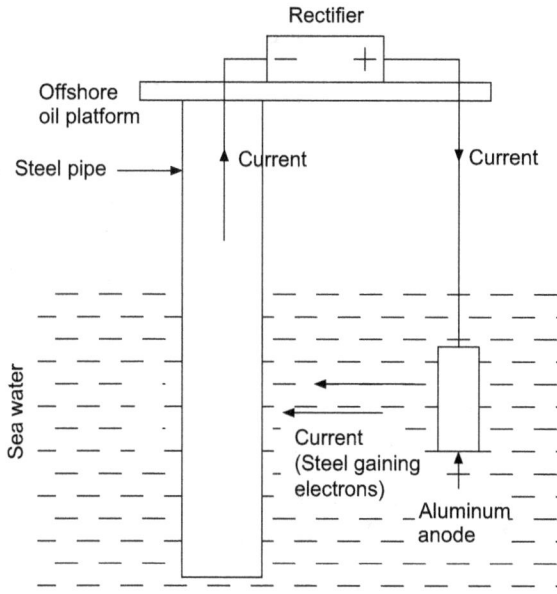

(a)

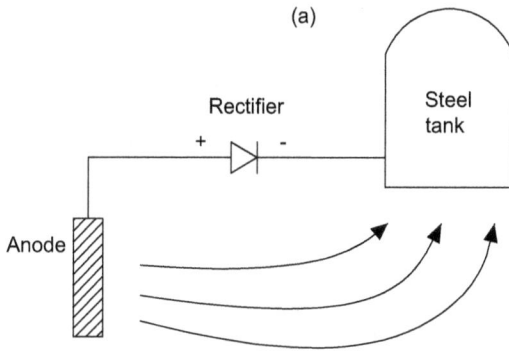

(b)

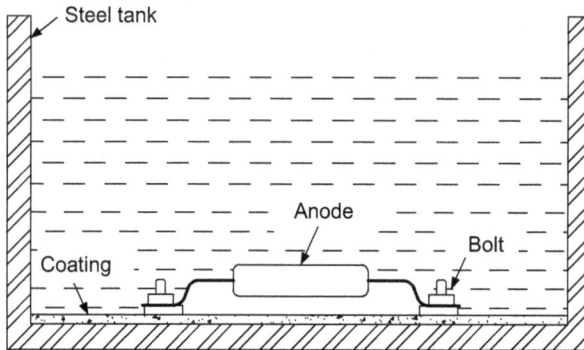

(c)

FIGURE 3.45 Cathodic protection (a) Protection of an offshore structure and (b) A liquid holding tank with sacrificial anodes.

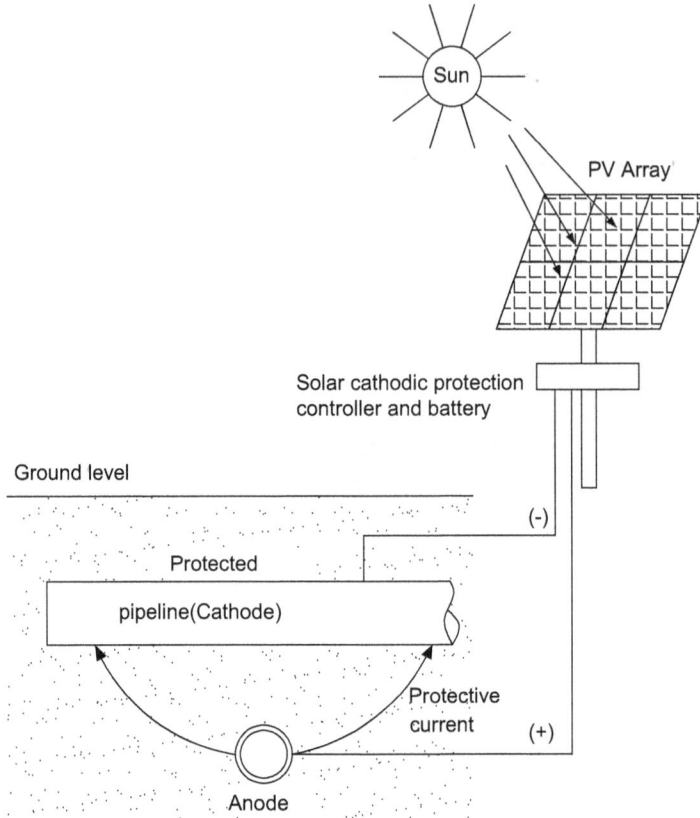

FIGURE 3.46 Solar powered cathodic protection system.

source is passed on to the metal to be protected and coupled to an insoluble anode such as platinum, graphite, or ferrosilicon alloy or an expendable anode such as zinc, aluminum, or magnesium, thus forcing the structure that has to be protected to have a large cathode so that it will not suffer attack. This method is presently employed in practice to protect buried pipelines carrying gas, oil, and water from corrosion by the surrounding soil. Cathodic protection uses a rectifier to convert alternating current (AC) power to direct current (DC). On one side, the rectifier output is electrically connected to the pipe, and on the other side, it is connected to anodes (metal rods). The rectifier is usually sited adjacent to existing power lines in the area. Anodes are buried in groups (referred to as ground beds) along the pipeline and are backfilled with a carbon-based conductive material to improve their effectiveness. A solar powered cathodic protection system is shown in Figure 3.46.

3.7.8.2 Flexible Anode Based on MMO/Ti Coating Technology

The flexible anode based on the technology of the mixed metal oxide (MMO) coating composed of MMO/Ti wire, cable, coke breeze, and woven acid-resistant jacket. The flexible anode embraces four parts as given below [138,139]:

1. continuous MMO/Ti wire;
2. continuous inner copper conductor;
3. continuous fabric jacket;
4. coke breeze in place around the anode cable.

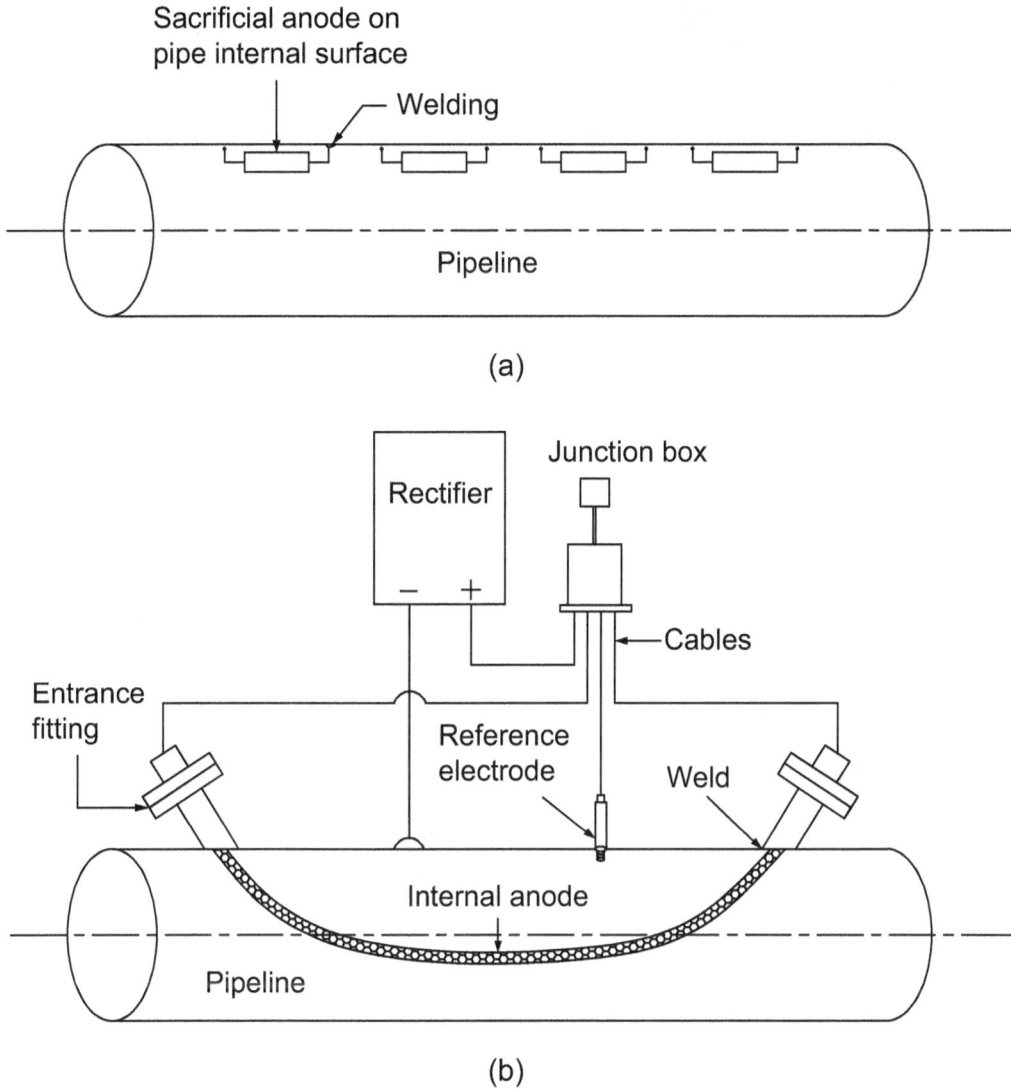

FIGURE 3.47 An internal impressed current flexible linear anodes.

Internal linear anodes can simply be laid alongside a new pipeline; cable plowed next to an existing pipeline, or installed utilizing horizontal directional drilling under an existing structure. Linear anodes require only a small trench for installation, ideal for congested areas and minimizing land-owner "right of way" issues.

The anode assembly is installed in the pipeline through entrance and exit fittings that are welded into the pipe and allow the anode segments to be suspended along the length of the pipeline. The internal linear anode is a complete system that consists of one or more MMO linear anode assemblies and fittings depending on the pipe configuration and final design. An internal flexible linear anode is shown Figure 3.47.

The principle of the auxiliary anode method is shown in Figure 3.48 and protection of the tubesheet and the end closure when titanium tubes are installed in an exchanger is shown in Figure 3.48(c) [140].

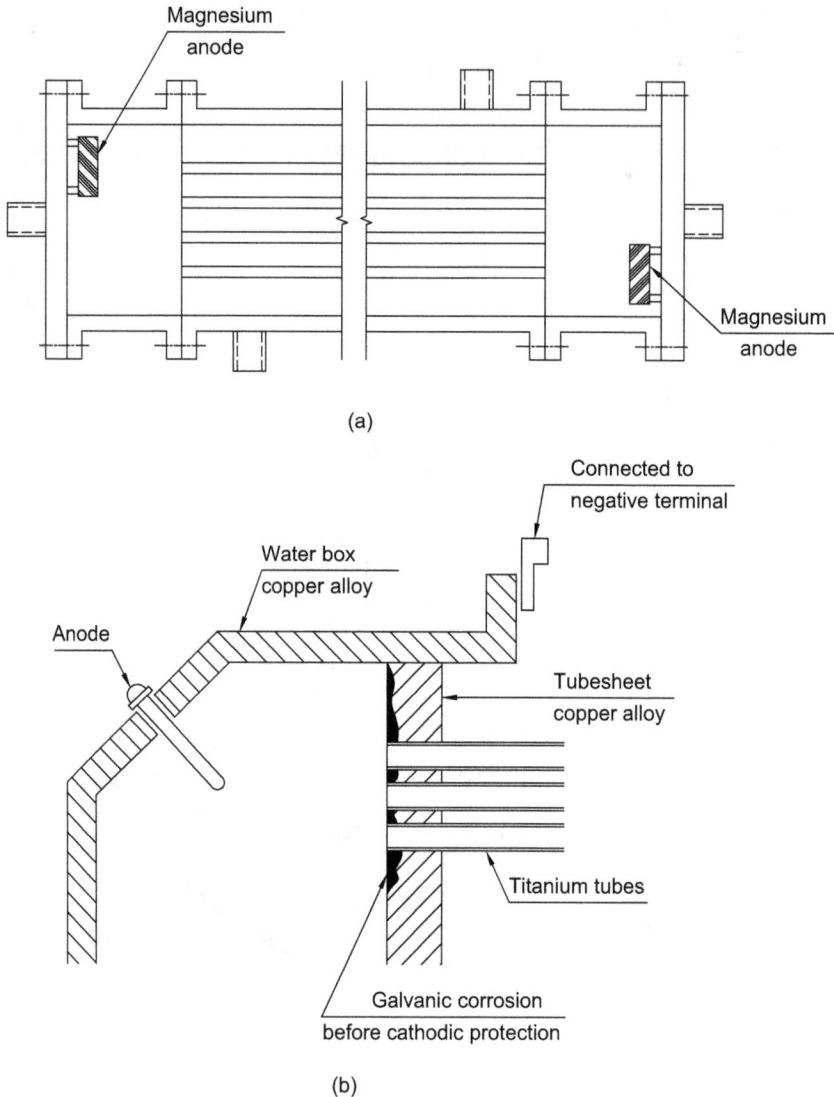

FIGURE 3.48   Cathodic protection of heat exchanger. (a) Sacrificial anode, (b) impressed current method.

Shortcomings of cathodic protection: with a cathodic protection system in seawater and in certain applications, if condenser and heat exchanger tubes happen to be cathodic, they may become covered with an undesirable thermal resistance layer. This may require a decrease in the amount of current flowing in the circuit or a change in the kind of anode, its location, or size [39]. Overprotection of a titanium tubesheet may result in the hydriding of the tubesheet.

3.7.8.3 Anodic Protection

If the potential of a metal is maintained in the range that leads to passivity, then the corrosion current density may be very low and will be stable. Consequently, the corrosion will be very low. This principle is employed in anodic protection. Anodic protection is an established, cost-effective method for corrosion control that can be applied in almost any electrically conductive solution in which a metal exhibits active–passive behavior [32,141]. Anodic coating is a technique to control the corrosion of a metal surface by making it the anode of an electrochemical cell - examples include galvanised steel and clad aluminum brazing sheet.

"Passivation" is the conditioning of a metal surface to produce a protective surface film that blocks corrosive ions from reaching the metal, thereby retarding corrosion. To change a metal from an active to a passive state, the electrode potential must be raised above the passivation potential. Passivation can be achieved by methods such as the following [142]: (1) applying an external current of sufficient strength or (2) using agents of sufficient oxidizing power to give a mixed potential above the passivation potential of the metal.

3.8 CORROSION MEASUREMENT

The measurement of corrosion in combination with actions to remedy high corrosion rates allows cost-effective plant operation together with reduced life-cycle costs associated with the operation. Corrosion measurement employs a variety of techniques to determine how corrosive the environment is and at what rate metal loss is being experienced. Corrosion measurement is the quantitative method by which the effectiveness of corrosion control and prevention techniques can be evaluated and provides the feedback to enable corrosion control and prevention methods to be optimized.

3.9 CORROSION MANAGEMENT SYSTEM

A corrosion management system (CMS) is the framework for integrating sound corrosion technology and mature management practices in order to efficiently and effectively manage corrosion to protect assets and asset systems, increase organizational return on investment (ROI), and achieve broader organizational goals, such as public service and public safety. An effective CMS system should include the following [143]:

i. Plans for desired asset life.
ii. Procedures and best practices for corrosion prevention.
iii. Inspection and monitoring of existing assets.
iv. Corporate strategies, objectives, and policies.
v. Active management of the CMS system with key performance metrics.

3.9.1 BUILDING BLOCKS OF A SUCCESSFUL CMS

Requirements of an effective CMS [143]:

i. realistic asset life plans;
ii. implementation of best practices for corrosion protection;
iii. inspection monitoring;
iv. crucial policies;
v. key performance indicators (KPIs).

According to the 2016 International Measures of Prevention, Application, and Economics of Corrosion Technologies study (IMPACT) by NACE International, CMS is defined as policies, processes, and procedures for planning, executing, and continually improving an organization's ability to manage the threat of corrosion for existing and future asset. A hierarchy of elements essential to an effective CMS include those common to all asset management systems (organizational policy, strategy, and objectives; and enablers, controls, and measures), as well as those specific to corrosion work plans, procedures, and working practices [144].

NACE has created a comprehensive guidance document that can be used as a work plan for developing and implementing an effective CMS. The document explains each aspect of the CMS, each developmental step, and each concept encompassed within the system.

3.9.2 IMPORTANCE OF CORROSION MANAGEMENT

It typically includes the following [144]:

(1) optimizing corrosion control actions and minimizing lifecycle corrosion costs, and
(2) meeting safety and environmental goals.

A CMS is most successful when it is integrated within an organization's overall management system and tools are provided to create consistent processes and operating procedures so that employees from the top down can respond holistically and effectively when faced with corrosion-related incidents.

3.9.3 ESSENTIAL ELEMENTS OF A SUCCESSFUL CORROSION MANAGEMENT PROGRAM

Essential elements of a successful corrosion management program [145, 146]:

1. Life-cycle cost analysis. Once credible corrosion threats are identified, the next step in the process is to determine the feasibility of using low-cost construction materials such as carbon steel. This life-cycle cost analysis provides the means to assess the viability of CS.
2. Corrosion control strategies. Once the analysis is completed, the next step in the process of developing a corrosion management strategy is to determine the corrosion control technologies to be used. Numerous industry standards provide guidelines for the threat assessment and control selection, such as follows:
   i. Material selection.
   ii. External coatings, like paint.
   iii. Internal linings are effective barriers for internal corrosion due to process fluid. These internal linings often require special application processes and equipment, and they are normally suitable for large tanks and vessels.
   iv. Cladding with a corrosion resistant alloy can be a solution when maintenance of other internal linings is not practical.
   v. Coating is often selected in conjunction with cathodic protection (CP), which can be used on buried or immersed metallic components and structures.
   vi. Sacrificial anode (passive) and impressed current (active) are two popular techniques of CP.
   vii. Use of corrosion inhibitors.
3. Identification of corrosion threats.
4. Life-cycle cost analysis.
5. Corrosion control strategies.
6. Corrosion monitoring and inspection.

Once corrosion threats and the respective controls are identified, the next step in management is to select proper monitoring and inspection techniques to determine the effectiveness of the deployed barriers. Some barriers are active and require constant surveillance or monitoring, while routine inspection may be adequate for other barriers.

3.10 CORROSION MONITORING

Corrosion monitoring is the practice of measuring the corrosivity of process stream conditions by the use of "probes," which are inserted into the process stream and which are continuously exposed to the process stream condition. Corrosion monitoring "probes" can be mechanical, electrical, or electrochemical devices. Corrosion monitoring has been discussed in detail by Britton et al. [147] and in Ref. [148]. The methods of corrosion monitoring include the following:

i.   Weight loss coupons.
ii. Electrical resistance.
iii. Linear polarization.
iv. Hydrogen penetration.
v.   Galvanic current.

Some corrosion measurement techniques can be used online, constantly exposed to the process stream, while others require offline measurement, such as those determined in a laboratory analysis. Some techniques give a direct measure of metal loss or corrosion rate, while others are used to inform that a corrosive environment may exist.

3.10.1 APPROACHES TO CORROSION MONITORING

In general, there are three approaches for corrosion monitoring [149]: (1) local approach; (2) component approach; and (3) systems approach. The local approach involves investigations of corrosion in terms of local conditions. The component approach involves investigating plant components and their corrosion phenomena that arise due to the complex environmental and operational conditions. The systems approach considers the plant system in its totality. This approach deals with interrelations of phenomena occurring in different components of the system.

3.10.2 CORROSION MONITORING TECHNIQUES

Several techniques are used in corrosion monitoring, and the techniques are generally classified as either online or offline. Various online and offline techniques are described next.

3.10.2.1 Online Monitoring Techniques

Online corrosion monitoring is conducted to assess the corrosivity of the process stream and for detecting changes that may occur in operation. Some of the most common monitoring techniques include online methods such as mass loss coupons, electrical resistance (ER) probes, linear polarization resistance (LPR) probes, and hydrogen probes.

Online corrosion data are obtained from probes or sensors inserted into the system at accessible points that reproduce the particular area of interest. Various online monitoring techniques include corrosion coupons, electrical resistance principle, pitting potential, linear polarization principle and Tafel plots, hydrogen test probe, galvanic measurements, pH measurements, dimensional changes through online ultrasonic testing, radiography, and acoustic emission technique [147]. Some of these techniques are discussed next. Some of the commonly used inspection techniques are visual inspection, radiographic testing (RT), ultrasound testing (UT), eddy current testing (ECT), magnetic particle testing (MT), and liquid penetrant testing (PT) [150].

Corrosion coupons: the most common online monitoring technique is with corrosion coupons. Corrosion coupons can be made in any size or shape, such that they are retrievable from process equipment without shutting down the unit. Normally, the corrosion coupons are carefully weighed before insertion and weighed after retrieval.

Weight change: when results of other techniques are in question, they are usually verified with weight loss testing. However, weight loss determination would give misleading information in the case of localized attack such as pitting.

Hydrogen diffusion: atomic hydrogen can diffuse into steel and form molecules. If hydrogen diffusion is detected, it shows imminent danger due to hydrogen damage/hydrogen attack. Hydrogen diffusion can be measured using either a hydrogen probe (pressure measurement) or a hydrogen monitoring system (electrochemical).

3.10.2.2 Electrochemical Techniques

Electrochemical techniques are particularly useful in determining the corrosion rate that is actually happening in a metal at any given time. The three techniques most often used involve (1) electrical resistance principles using zero-resistance ammeters; (2) polarization curves; and (3) linear polarization curves.

Other methods include Tafel extrapolation (where the linear portion of the anodic or cathodic polarization curve is extrapolated back to the corrosion potential to locate the current density associated with corrosion rate at that potential) and current measurements at constant potential (where corrosion rate is monitored for a given oxidizing condition) [151].

Monitoring of pitting potential: the potential at which pit initiation occurs is called the pitting potential. Pitting potential is determined by electrochemical techniques, which consists of measuring current and potential potentiostatically, either stepwise or by applying a constant-potential sweep rate in a standard chloride-containing solution.

3.10.2.3 Online Monitoring of Water Purity in Thermal Power Stations

Corrosion inhibition in a cooling-water system involves monitoring calcium hardness, alkalinity, total solids, pH, dissolved gases (oxygen and hydrogen), total dissolved salts, etc. Automatic analyzers continuously monitor water and steam purities of thermal power stations. For more details on online monitoring of cooling water refer to Refs. [152, 153].

Offline monitoring techniques: offline monitoring techniques involve various nondestructive examination methods to determine the thickness and integrity of the heat exchanger and pressure vessel components. Various NDT methods employed are as follows:

1. Visual examination with or without optical aids such as borescopes.
2. Eddy current testing.
3. Magnetic particle examination.
4. Liquid penetrant test.
5. Ultrasonic examination.
6. Radiography.
7. Thermography.

3.10.2.4 Corrosion Monitoring of Condensers by Systematic Examination of the State of the Tubes

This procedure involves extracting representative tubes and examining them in the laboratory with modern analytical equipment [153]. In a tube investigation procedure, one considers the three zones (inlet, center, and outlet) separately, because they are exposed to different conditions. On all samples, the following four criteria are checked:

1. Microscopic examination of the condition of the tube surface.
2. Residual wall thickness.
3. Weight of the surface layer.
4. Composition of the surface layer.

3.10.2.5 Limitations of Corrosion Monitoring

It is important to be aware of the limitations of corrosion monitoring data. Some of the important limitations of corrosion monitoring are [147] the following:

1. The data are only a qualitative guide to the actual behavior of the plant or process.
2. A confidence factor is established only through experience, and particularly through comparison with other sources of information, notably that provided by NDT.

3.10.2.6 Requirements for Success of Corrosion Monitoring Systems

Many factors contribute to the success of a monitoring program. Some of the most important considerations include the following [147]:

1. Use of correct technique.
2. Correct location of monitoring probes.
3. Reliability of the equipment and instrumentation.
4. Data obtained must be straightforward to interpret.

3.11 COOLING-WATER SYSTEM CORROSION

A huge part of the water used in industry is used for cooling a product or a process. The availability of water in most industrialized areas and its high heat capacity have made water the favored heat transfer medium in industrial and utility type applications. In recent years, the use of water for cooling has come under increasing pressure. As a result cooling-water use patterns are changing and will continue to do so. For many systems, water is used in a single-pass and returned to the water body.

3.11.1 CORROSION PROCESSES IN WATER SYSTEMS

Water, the most commonly used cooling medium, removes unwanted heat from process fluids. Corrosion of metals in contact with water is electromechanical in nature. A generalized anodic oxidation reaction for ferrous metal has already been presented and copper and aluminum materials are represented by [74, 154] the following:

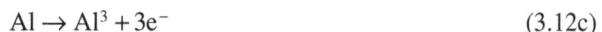

$$M \rightarrow M^{n+} + ne^- \tag{3.12a}$$

$$Cu \rightarrow Cu^+ + e^- \tag{3.12b}$$

$$Al \rightarrow Al^3 + 3e^- \tag{3.12c}$$

where M represents the metal that has been oxidized to its ionic form having a valence of n^+ and the release of n electrons. The reactions at the cathodic site on the metal surface are given by Ref. [74] as in the following:

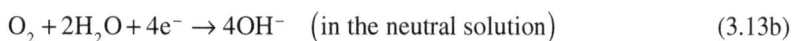

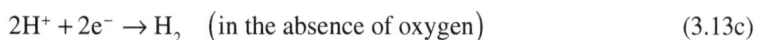

$$O_2 + 4H^+ + 4e^- \rightarrow 2H_2O \quad (\text{in the presence of oxygen}) \tag{3.13a}$$

$$O_2 + 2H_2O + 4e^- \rightarrow 4OH^- \quad (\text{in the neutral solution}) \tag{3.13b}$$

$$2H^+ + 2e^- \rightarrow H_2 \quad (\text{in the absence of oxygen}) \tag{3.13c}$$

3.11.2 FACTORS INFLUENCING COOLING-WATER CORROSION

Factors influencing cooling-water corrosion such as chemical, physical, and biological are given below [154].

3.11.2.1 Chemical Factors

The main chemical factors, which influence the tendency of cooling water to corrode, are mineral salt content, pH value, and dissolved gases, notably oxygen and carbon dioxide. Other conditions that may modify the local environment of the metal surface are deposits of corrosion products, scale, and biological slimes.

3.11.2.2 Physical Factors

The main physical factors contributing to corrosion are temperature, dissimilar metals contact, and suspended solids.

3.11.2.3 Biological Factors

Sulfate reducing bacteria together with sulfates, are present in most waters and if oxygen is depleted locally, e.g., beneath corrosion deposits are able to reduce sulfates to sulfides. Corrosion from this source may be of the pitting type, the pits being filled with soft, foul-smelling sulfides.

3.11.3 CAUSES OF CORROSION IN COOLING-WATER SYSTEMS

Depending on the sources of water—river, lake, ocean, or sea—cooling-water corrosion that attacks heat exchangers is due to one or more of the following impurities [11,74,123,133]:

1. Dissolved solids and water hardness.
2. Chloride.
3. Sulfate.
4. Silica.
5. Oil.
6. Iron and manganese.
7. Suspended mater such as turbidity, dirt, clay, silt, sand, etc.
8. Dry residue.
9. Dissolved gases such as O_2, carbon dioxide, hydrogen sulfide, ammonia.
10. Dissolved organic matter.
11. Microbiological organisms.

Apart from these factors, water temperature, pH, and flow velocity also influence corrosion. Effects of these factors on the cooling-water corrosion are discussed next.

3.11.3.1 Dissolved Solids and Water Hardness

All waters containing calcium and/or magnesium salts such as calcium carbonate ($CaCO_3$), calcium bicarbonate [$Ca(HCO_3)_2$], magnesium carbonate ($MgCO_3$), and magnesium bicarbonate [$Mg(HCO_3)_2$] in considerable amounts are called "hard." The following classifications for hardness are often applied to freshwater [155]:

Soft: 0–50 ppm.
Medium hard: 50–100 ppm.
Hard: >100 ppm.

The measure total dissolved solids (TDS) is the concentration of dissolved solids that cannot be removed by filtration. Seawater is many times higher in dissolved solids than freshwater.

Total hardness is the amount of calcium and magnesium salts, which may be present as bicarbonates (temporary hardness) or as sulfates, chlorides, nitrates, or carbonates (permanent hardness).

Scale deposition: water-formed deposits are commonly referred to as scale. This scale increases the resistance for heat transfer and increases pressure drop, and promotes localized hot corrosion spots. Some hardness in the water may be helpful if the pH and alkalinity can be controlled to permit formation of a thin protective scale (termed stabilization of the water). Unfortunately, the buildup of a thick scale on the tube surface seriously interferes with heat transfer. The normal scale-forming compounds commonly found in cooling-water systems are calcium carbonate, calcium sulfate, calcium phosphate, and silicates. Of these compounds, calcium carbonate has very low solubility and perhaps is the principal scale-forming material in cooling waters. Bicarbonate ions react with hydroxyl ions generated at the cathode to produce carbonate, which precipitates with calcium from the water as calcium carbonate. The equation for calcium carbonate scaling is given by

$$Ca\left(HCO_3\right)_2 + OH^- \rightarrow CaCO_3 + HCO_3 + H_2O \qquad (3.14)$$

Prevention of calcium carbonate scale: scale-forming tendencies of cooling water can be predicted by determining the Langelier saturation index (LSI), also known as the calcium carbonate saturation index, of the water. Some prefer the Ryznar stability index. The LSI is defined as the difference between the actual pH value of the water and the value of pH that the water would have if it was in equilibrium with $CaCO_3$, also known as saturation pHs. Accordingly, the LSI is given by

$$LSI = pH - pH_s \qquad (3.15)$$

The LSI is not applicable to seawater because of the high salt content [155].

Determination of LSI: the LSI is calculated from (1) the alkalinity; (2) the calcium hardness; (3) the ionic strength (total solids); (4) the pH value; and (5) the temperature. The index can also be found out by direct determination of the pH of water as such and after saturating it with calcium carbonate. A condition of carbonate equilibrium with the absence of scale can be accomplished by adjusting any of the variables of the LSI. The common practice is to lower the alkalinity and pH by adding sulfuric acid to the cooling water [75]. A chart for calculating the saturation index is shown in Figure 3.49.

Scaling control: if the LSI is positive, scale will be deposited and will probably stifle corrosion of the metal. If it is negative, the water is unsaturated with $CaCO_3$, and it is likely to be corrosive and may dissolve existing hardness salts or may not be scale forming. In other words, for LSI > 0, water is super saturated and tends to precipitate a scale layer of $CaCO_3$, and for LSI = 0, water is saturated (in equilibrium) with $CaCO_3$. A scale layer of $CaCO_3$ is neither precipitated nor dissolved and, for LSI < 0, water is under saturated and tends to dissolve solid $CaCO_3$. It is also worth noting that the LSI is temperature sensitive. The LSI becomes more positive as the water temperature increases. Conversely, systems that reduce water temperature will have less scaling. LSI and the probable nature of water is given next [35]:

For LSI = −2.0, water is corrosive and nonscale forming.
For LSI = −0.5, water is mildly corrosive and nonscale forming.
For LSI = 0.0, water is mildly corrosive.
For LSI = +0.5, water is mildly corrosive and scale forming.
For LSI = + 2.0, water is not corrosive but definitely scale forming.

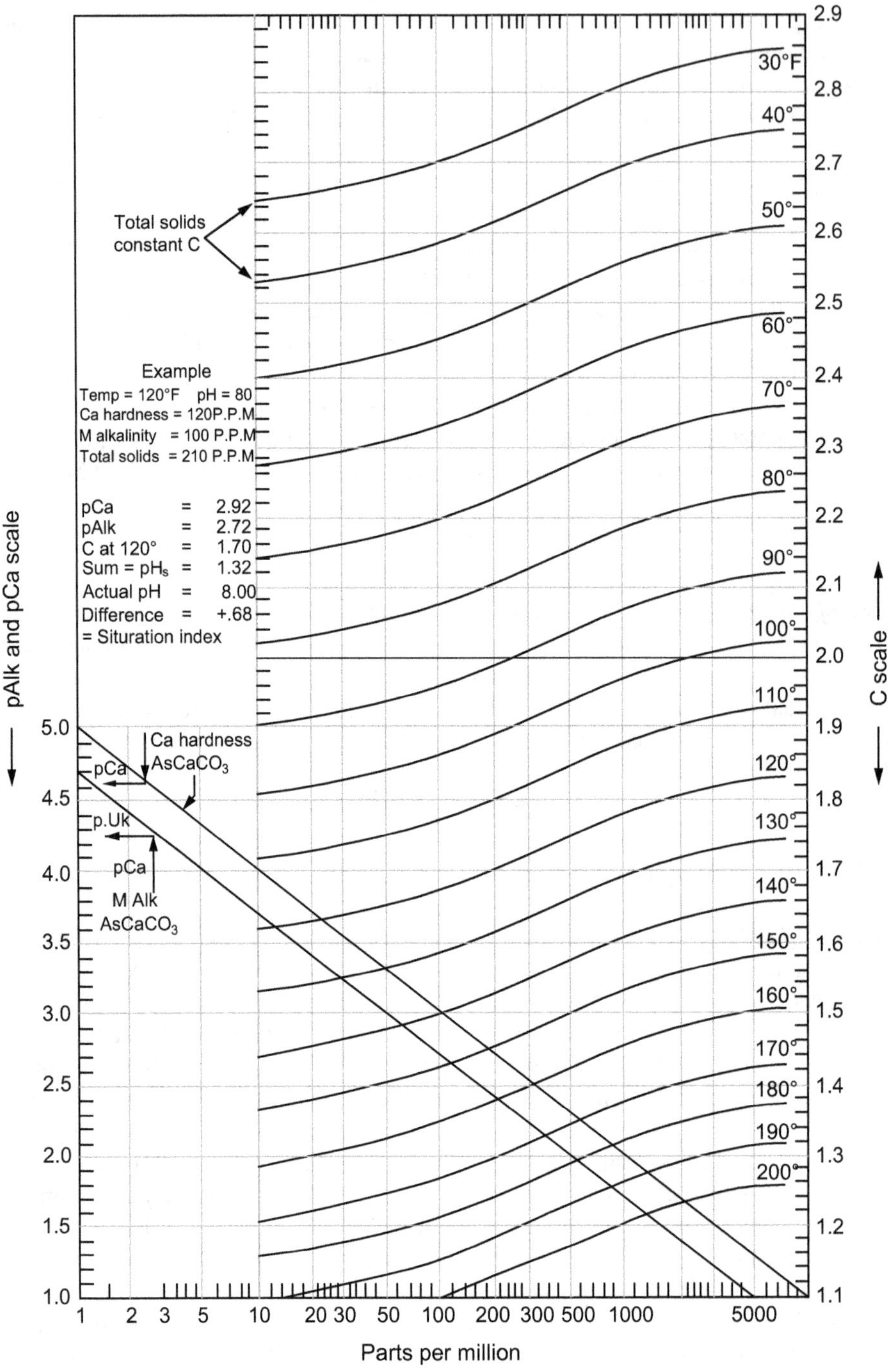
Total solids
constant C
Example
Temp = 120°F pH = 80
Ca hardness = 120P.P.M.
M alkalinity = 100 P.P.M.
Total solids = 210 P.P.M.
pCa = 2.92
pAlk = 2.72
C at 120° = 1.70
Sum = pHs = 1.32
Actual pH = 8.00
Difference = +.68
= Situration index
Ca hardness
AsCaCO3
pCa
p.Uk
pCa
M Alk
AsCaCO3
pAlk and pCa scale
C scale
Parts per million
30°F
40°
50°
60°
70°
80°
90°
100°
110°
120°
130°
140°
150°
160°
170°
180°
190°
200°

FIGURE 3.49 Chart for calculating saturation index. (Ca and alkalinity expressed as ppm $CaCO_3$; temperature in degrees F.)

Source: From Powell, S. et al., *Ind. Eng. Chem.*, 37, 842, 1945.

From these, the main choice appears to be between keeping the LSI low and controlling scale formation, or having a high index and controlling the corrosion. Methods to control scale formation are discussed in Chapter 2.

3.11.3.2 Chloride

Chlorides are the main salts in seawater, around 19,000 ppm [155]. High-chloride waters are usually corrosive. High chloride ion concentrations easily destroy the protective surface films including passive films on a large number of metals [90], or the chloride forms complex ions with dissolved iron, thereby preventing or interfering with the formation of protective corrosion product scales [20]. In the case of austenitic stainless steels, chloride ions will easily destroy the passive film and result in pitting corrosion or when the metal is subjected to simultaneous tensile stress can cause chloride SCC. Aluminum is also easily attacked by chloride.

3.11.3.3 Sulfates

Sulfates are not as corrosive as chlorides [155]. However, concrete cooling-water basins are endangered as soon as the water contains more than 250 ppm of sulfate ions [133].

3.11.3.4 Silica

Silica will react with magnesium or calcium to form deposits of insoluble magnesium or calcium silicates. Also it will form siliceous glassy scales. To avoid such deposits, the silica concentration must be limited.

3.11.3.5 Oil

Oil is one of the most undesirable contaminants of industrial cooling-water systems. Oils usually leak into the system as a result of poor maintenance. They interfere with the action of corrosion inhibitors; by forming a thin film on metal surfaces, oils retard heat transfer, and they become nutrients for biological organisms [11].

3.11.3.6 Iron and Manganese

Iron oxide is produced either by internal corrosion of steel heat exchanger surfaces and transmission lines or from precipitation of soluble iron brought into the cooling system from the earth by the makeup water. Less than 0.2 ppm of iron and manganese is desirable. With higher contents, sludge-type hydroxides are precipitated in the presence of oxygen [133]. Iron oxide scale, which forms as easily dislodged "plate-like" layers, becomes lodged in the tubes and leads to localized impingement attack. Iron oxide scale is prevented by applying cathodic protection or by protective coatings [37].

3.11.3.7 Suspended Matter

Suspended matter such as clays, silt, and corrosion products present in once-through or open cooling-water system is capable of depositing on metal surfaces in low-velocity areas (<1 m/s), forming a physical barrier. This buildup will contribute to the formation of differential aeration cells and will promote localized corrosion [74].

3.11.3.8 Dry Residue

Dry residue after evaporation includes all the dissolved substances in the water. If dry residue expressed as salt content is greater than 500 ppm, the conductivity of water when it comes into contact with the dry residue will increase, which in turn will enhance corrosion [133].

3.11.3.9 Dissolved Gases

The most common dissolved gases are oxygen, carbon dioxide, hydrogen sulfide, and ammonia. Their role in cooling-water corrosion is discussed next.

Oxygen: cooling water saturated with oxygen will lead to increased corrosion. Oxygen can behave as a depolarizer and can increase the rate of corrosion by speeding up the cathodic reaction. It can also act as a passivator because it promotes the formation of a stable passive film. Dissolved oxygen is a major factor contributing to the natural corrosion of steel. Oxygen in traces promotes SCC of brass in the presence of ammonia; oxygen when coexisting with ammonia leads to the highest rate of sulfide attack; and oxygen also promotes condensate corrosion. Nevertheless, a minimum quantity of oxygen is needed to form the carbonate-rust protective film. This should amount to about 5–6 ppm at normal ambient temperature [35].

Removal of oxygen: control of oxygen corrosion is critical to the reliability of a steam generator system. Mechanical deaeration and oxygen scavenging by chemical means effectively reduce oxygen levels in boiler feedwater systems [156]. Some of the successful devices to reduce the influence of oxygen on cooling-water corrosion are mentioned in Ref. [35]. One of these methods is the mechanical method, which consists of raising the water temperature and lowering the pressure (vacuum) so that the gas is liberated (all gases are less soluble in water with increasing temperature and decreasing pressure). This method is applicable for removing other gases also from water. Another mechanical method consists of passing water over scrap iron, which removes the dissolved oxygen and is applicable to closed systems. There is also the chemical method that consists of adding a slight excess of sodium sulfite. The sodium sulfite reacts rapidly with the dissolved oxygen to form sodium sulfate:

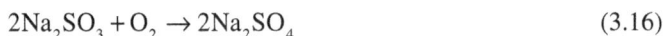

$$2Na_2SO_3 + O_2 \rightarrow 2Na_2SO_4 \tag{3.16}$$

This method has been economically used for years for the removal of the last traces of oxygen from boiler water.

Carbon dioxide: natural cooling water always contains free carbon dioxide, in larger or smaller quantities. Dissolved free carbon dioxide makes water acidic. The free carbon dioxide is defined as the quantity that is required to retain the alkaline earth carbonate in accordance with the equation:

$$CaCO_3 + CO_2 + H_2O \rightleftharpoons Ca^{++} + 2HCO_3^- \tag{3.17}$$

Cooling waters without excess free carbon dioxide lead to the formation of a protective carbonate-rust scale, which represents one of the most important protective surface layers [90]. However, excessive carbon dioxide will cause corrosion. Excess carbon dioxide is removed from cooling water by spraying water over a mass of coke or other inert material exposed to the air. Sometimes by this process the carbon dioxide content can be reduced to 5–10 ppm. The second method involves treating the water with lime [35].

Hydrogen sulfide: hydrogen sulfide in cooling waters generally results from the action of bacteria on organic matter, particularly sewage. The presence of even few parts per million may change the corrosivity of water drastically. The presence of hydrogen sulfide in a water has the tendency to reduce inlet-end or impingement corrosion, but increases the probability of pitting corrosion occurring along the entire length of a tube [155]. Additionally, the salts created by this attack (copper sulfide, iron sulfide, etc.) are insoluble and precipitate on the metal surface as loose, noninhibitive deposits that promote galvanic couple and thus promote corrosion [11].

Ammonia: ammonia is originally derived from the thermal degradation of various nitrogen-containing compounds added to the boiler feedwater to reduce corrosion. Ammonia does not cause excess corrosion on ferrous metals. However, traces of ammonia will cause both rapid general thinning and SCC in brasses. Pure copper and copper–nickel alloys are not susceptible to ammonia SCC, but a higher rate of uniform corrosion will take place with higher ammonia contents [133].

Dissolved organic matter: dissolved organic matter in excess quantity tends to produce sludge-type deposits and fouling. Dissolved organic matter is detected indirectly by determining the potassium permanganate ($KMnO_4$) consumption. If the $KMnO_4$ consumption exceeds about 25 mg/L water, then the water is "dirty," which tends to produce sludge-type deposits and fouling [133].

Microbiological organisms: corrosion due to microbiological organisms has been discussed in the section on MIC.

pH value: the pH of the water can have several specific effects. High pH levels make the formation of heavy calcium carbonate and calcium phosphates scales likely. From Equation 3.6, it is evident that where a greater number of hydrogen ions (low pH) are available, the corrosion reaction should occur more rapidly. Natural waters produce more or less neutral solution (pH value about 7). Clean seawater has a pH of about 8. In general, the neutral and weakly alkaline waters usually cause no or only very slight corrosion on metallic materials, but weakly to strongly acidic water is always corrosive. Type of water also influences corrosion. The corrosiveness of a natural freshwater with a pH of 7–8 usually is much less than that of seawater with a similar pH. The difference is due to the presence of more mineral salts in seawater. Nevertheless, pH is not the only determining factor. Water hardness, particularly the carbonate hardness, is of decided importance [133]. An additional factor that decides the water corrosion is the dissolved mineral salts.

Temperature: in general, an increasing temperature will increase corrosion rates. Temperature plays a dual role with respect to oxygen corrosion. For an open system, beyond 175°F (80°C), increasing the temperature will reduce oxygen solubility, whereas for a closed system in which the oxygen cannot escape, corrosion continues to increase linearly with temperature [69]. Another effect of increased temperature is the increase in fermentation of organic materials with formation of sulfides, which promote corrosion [155]. In certain waters, the corrosion rate drops off with an increase in temperature because the water deposits a thin scale. Calcium carbonate forms as a result of the decomposition of calcium bicarbonate at elevated temperatures [35]:

$$Ca\left(HCO_3\right)_2 \leftrightarrows CaCO_3 + H_2O + CO_2 \qquad (3.18)$$

Hot wall effect: severe local corrosion or pitting can rapidly perforate condenser and heat exchanger tubes on the cooling-water side where bubbles of air or foreign material separate from the water (known as blanketing of base metal) and hence localized heating of heat transfer surface (Figure 3.50). Here the metal is hotter, and the corrosion proceeds at a correspondingly higher rate. This is known as the hot wall effect.

Flow velocities: corrosion is favored by too low or too high cooling-water velocities. High velocity helps to prevent the accumulation and deposition of corrosion products, which might create anodic sites to initiate corrosion and helps to maintain clean surfaces free from fouling deposits, stagnant areas, and hot wall effect. On the other hand, too high a velocity can destroy the protective surface film and result in erosion–corrosion or impingement corrosion, especially on metals such as copper and aluminum [133]. The velocity, in general, should not be less than about 1 m/s.

The following values apply in respect to the maximum admissible flow velocities for water, V_w:

Pure aluminum	1.2–1.8 m/s
Pure copper	1.8 m/s
Copper-containing arsenic	2.1 m/s
Aluminum brass	2.0–2.4 m/s
Naval brasses	2.0–2.4 m/s
CuNiFe 90/10	2.5–3.0 m/s
CuNiFe 70/30	3.0–4.5 m/s
Monel 400	4.0 m/s
Carbon steel	3 m/s
Austenitic stainless steel	4.0–5.0 m/s
Ni-Fe-Cr alloys	4.0–5.0 m/s
AL-6XN®	7.0–8.0 m/s
SEA-CURE®	9.0–10 m/s
Titanium	Up to 30.0 m/s

For other liquids, allowable velocity V_1 is given by

$$V_1 = V_w \left[\rho_w / \rho_1 \right]^{0.5} \qquad (3.19)$$

where
  ρ_w is density of water,
  ρ_1 is density of liquid,

For gases and dry vapors, allowable velocity for steel tubing is given by

$$V_g = V_w \frac{1800}{\sqrt{p_a \times M_w}} \text{ ft/s} \qquad (3.20)$$

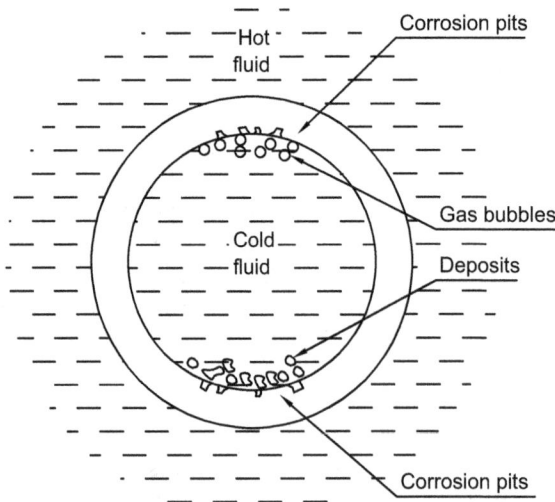

FIGURE 3.50 Hot wall effect.

Note: assume cold fluid as cooling water.

where

    V_g is gas/vapor velocity,

    p_a is absolute pressure of gas/vapor, psia,

    M_w is the molecular weight of gas/vapor.

Allowable velocities for other metals may be taken to be in the same ratio as for water.

3.11.4 Cooling-Water Systems

An understanding of the relationship between cooling water and the buildup of deposits and corrosion of heat transfer surfaces requires an awareness of cooling-system characteristics. There are basically three types of cooling systems:

    a. Once-through cooling systems (Refer Figure 2.30a).
    b. Open recirculating or closed cycles with cooling towers (Refer Figure 2.30b).
    c. Closed recirculating systems (Refer Figure 2.30c).

A choice among these depends on the quantity and quality of water available, water temperature, type of process plant, heat exchanger size, and means for water disposal.

3.11.4.1 Once-Through Cooling-Water Systems

In a once-through cooling-water system, the water is pumped from a water source and supplied through the equipment and discharged. This is used where cheap supplies of water in large quantity are available (e.g., river, lake, sea). The possibilities of water treatment of the large quantity of water handled for corrosion control are practically nil due to economic reasons and environmental regulation on discharges [133]. Only upstream filtration or inoculation with dispersing agents or ferrous sulfate dosing can be considered. Additionally, cathodic protection control may be instituted.

Discharge water temperature must be compatible with state and federal regulations. Permits for once-through systems, except for coastal (seawater) applications, can be difficult to obtain. Various antifouling materials including polyacrylates, naturalorganic materials, and other organic polymers are used to control deposition of solids. The Langelier Saturation Index will provide some indication as to whether the water will tend to be scale-forming, corrosive, or neither.

Calcium carbonate, the most common scale in once-through systems, is usually inhibited by applying one or a combination of several deposit control materials, such as polyphosphates, phosphonates, and polyacrylates. Shock treatment with chlorine may also be used to control biological fouling. As with recirculating water systems, chlorine residuals in once-through cooling-water system discharges must comply with environmental regulations.

3.11.4.2 Open Recirculating Systems

In an open recirculating cooling-water system, warm water from the coolers, condensers, etc., is routed to a cooling tower where cooling of the water is effected primarily by evaporation of a portion of the circulating water. This concentrates the dissolved solids in the water; the level of concentration of dissolved solids is controlled by blowdown of some of the circulating water. At equilibrium, the rate of dissolved-solids loss with blowdown and windage or drift (droplets of water leaving the tower) equals that gained with the water added as makeup to replace water lost by evaporation, drift, and blowdown. Water treatment is employed to prevent or minimize the below:

    i.   scaling or heat exchanger surfaces by salts of hardness ions, silica, or silicates;
    ii. corrosion of the system by oxygen or low pH;

   iii. attack of cooling tower wood by algae, bacteria, or fungi; and

   iv. fouling of heat exchanger surfaces by suspended solids and marine organisms.

Open recirculating cooling systems are inherently subject to more treatment-related problems than once-through systems [157]:

   i. cooling by evaporation increases the dissolved solids' concentration in the water, raising corrosion, and deposition tendencies;

   ii. the relatively higher temperatures significantly increase corrosion potential;

   iii. the longer retention time and warmer water in an open recirculating system increase the tendency for biological growth;

   iv. airborne gases such as sulfur dioxide, ammonia, or hydrogen sulfide can be absorbed from the air, causing higher corrosion rates;

   v. microorganisms, nutrients, and potential foulants can also be absorbed into the water across the tower.

The treatment of cooling water of this system calls for the following measures [133]:

   a. Monitoring and control of pH.
   b. Adding a corrosion inhibitor.
   c. Adding an algae-destroying agent.
   d. An additive (stabilizer) to prevent calcium carbonate precipitations.
   e. Possible makeup water treatment.
   f. Partial filtration.

Cooling Towers

Cooling towers are the most common method used to dissipate heat in open recirculating cooling systems. They are designed to provide intimate air/water contact. Heat rejection is primarily by evaporation of part of the cooling water. Some sensible heat loss (direct cooling of the water by the air) also occurs, but it is only a minor portion of the total heat rejection.

Scale Control

Scale is an adherent deposit laid down during operation, causing impaired heat transfer and restricting flow in the cooling-water system. The scale-forming tendency of a water increases with increasing cycles of concentration of dissolved solids in a recirculating system.

    Calcium carbonate, the principal scale found in cooling-water systems, forms when calcium and alkalinity levels become too high. Because it has an inverse temperature solubility, calcium carbonate tends to deposit on warmer heat transfer surfaces. Several scaling indices have been developed and are often used as guides to predict the tendency of cooling water to form scale. These include the Langelier Saturation Index (LSI) and the Ryznar Stability Index (RSI).

    The main objective in using the Langelier is to adjust the cooling water to a nonscaling, noncorrosive condition. A positive LSI indicates a tendency to deposit calcium carbonate scale; a negative LSI indicates an unsaturated condition with respect to calcium carbonate and a tendency to dissolve any existing calcium carbonate and to be corrosive.

Corrosion Control

Factors contributing to the corrosiveness of cooling water are as follows: dissolved oxygen concentration, pH, calcium and alkalinity concentrations, dissolved solids concentration, water temperature, and circulating-system metallurgy.

    Successful application of inhibitors requires a detailed understanding of the various corrosion processes that are possible in a given system. Typical inhibitors are chromates and dichromates, silicates, nitrates, ferrocyanides, and molybdates. Inhibitors specific to control of copper and copper alloy corrosion are sometimes used, particularly when the circulating water pH tends to be low. Polyphosphates were used at one time as corrosion inhibitors in concentrations of 10 to 15 ppmw.

Biological Fouling

A cooling tower is an ideal environment for the development and growth of microorganisms. Algae can develop in the tower where sunlight is present while slime can develop in almost any part of the system. These deposits can cause local corrosion as well as plugging and reduction of heat transfer. Growth of the microorganisms can be controlled by using chlorine, chlorinated phenols, organometallic salts, quaternary ammonium compounds, and various other biocides.

Chlorination is probably the most widely used control method. Chlorination programs may be either continuous or intermittent. Chlorine can cause deterioration of cooling tower wood, so prolonged exposure to concentrations over 1 ppmw should be avoided. Limitations on chlorine concentrations in effluent waters have necessitated closer control of chlorine dosage.

3.11.4.3 Closed Recirculating Systems

Closed recirculating systems continuously recirculate the same water. Closed recirculating cooling-water systems are well suited to the cooling of gas engines and compressors. Diesel engines in stationary and locomotive service normally use radiator systems similar to the familiar automobile cooling system. This system has little water loss and is free from algal growth. Because there are no evaporation losses, the requirement for makeup water is minimal, and the mineral content remains essentially constant. Makeup water is needed only when leakage has occurred at pump packings or when water has been drained to allow system repair. However, corrosion by-products can foul the heat exchanger. Hence, fouling deposits should be removed periodically by suitable offline cleaning methods. The completely closed cycle represents the ideal possibility for a clean and effective water treatment. Corrosion control by inhibitors is the most important control measure.

3.11.5 CORROSION CONTROL METHODS FOR COOLING-WATER SYSTEMS

In principle, damage to cooling systems can be checked in many ways:

1. Proper material selection.
2. Cooling-water system design.
3. Continuous water treatment system.
4. Use of inhibitors.
5. Ferrous sulfate dosing.
6. Protective coatings.
7. Cathodic protection.
8. Passivation.
9. Biological control.
10. Scale control.
11. Systematic cleaning.

Corrosion control measures such as protective coating and cathodic protection, biological control, and various scale control measures have been discussed earlier in this chapter. Additional measures to control scaling as a foulant were discussed in Chapter 2, Fouling.

3.11.5.1 Material Selection

Corrosion control by means of cooling-water design should take into account two possibilities: (1) the water treatment program and (2) selection of corrosion-resistant material. Proper material selection involves selecting a material that can be exposed to the cooling water without the danger of corrosion. The question of whether corrosion in cooling systems can be checked by means of suitable material selection at the design stage or by instituting a permanent water treatment program when the unit starts functioning must be considered on the basis of corrosion damage and economic considerations [90]. On the material side, cupronickel and titanium are costly compared to carbon steel, aluminum and copper, and certain copper alloys such as brasses. On the water treatment side, continuous online treatment including dosing of chemicals and additives involves money, especially when the quantity of water handled is large, as in a once-through cooling-water system. The objectives of such a strategy are to ensure that the operating costs are minimized throughout the life of the plant. This can be established by studying the following technical and economic considerations [37]:

Estimated life of candidate tube materials from trials, manufacturer's literature, and experience.
Predicted number of outages and associated costs.
Number of retubes required.
Capital costs of materials and retubing costs.
Heat transfer efficiency and thermal performance effects.
Compatibility with other cooling-water system materials.
Cooling-water system design.

In planning a heat exchanger cooled by natural water, the first step is to obtain the following information about the water [36]:

Analysis of water.
Water temperature upon entry.
Solids' content.
Content of organic matter including H_2S.
Certain guiding principles to be considered at the design stage are as follows [133, 136]:

Empty the cooling system completely when in the standstill condition. Many more cases of corrosion have taken place when the cooling system was at a standstill than in operation.
    When welding on the cooling-water side, care should be taken to avoid protruding weldments or crevices as a result of unsatisfactory weld penetration.
    Design sealing in such a way that seals cannot lift on the waterside.
    Narrow gaps should be avoided in the design stage. If this is not possible, the gap width should preferably be large (>0.5 mm).
    The flow velocities should be neither too low nor too high. Except for coppers and brasses, maintain water velocities as high as practical. Velocities of 3–8 ft/s are the minimum desired.
    Beware of galvanic coupling.
    Always maintain vents on top tubesheets.
    Locate water inlet and outlet nozzles on the shellside as near the tubesheets as possible to reduce "dead pockets."
    Place water on the tubeside, and the process liquid on the shellside.
Additional guidelines were given by Forchhammer [36] as follows:

In the construction phase, check the integrity of system components for tightness with low-chloride water free from solids.
To form the first oxide layer in the tubes, the unit should run several weeks without interruption.

3.11.5.2 Water Treatment

By water treatment, we either remove the aggressive components to a great extent or add specific chemicals to the water. In this way, both corrosion and fouling are avoided. Water treatments can be generally subdivided into three main groups:

1. Chlorination/settling/filtration to remove the turbidity and microorganisms.
2. Water softening to remove water hardness.
3. Partial or full demineralization for the removal of hardness and all dissolved salts.

Removal of hardness is effected by (1) the cold lime process; (2) the sodium cation exchange process; (3) demineralization; and (4) the acid process.

3.11.5.3 Corrosion Inhibitors

The use of corrosion inhibitors is one of the foremost methods of controlling corrosion in a cooling-water system. The main effect that corrosion inhibitors have in aqueous ferrous systems is to reduce the initial corrosion rate sufficiently to allow the gamma-iron oxide passive film to form and, in some cases, to take part directly in film formation. Corrosion control by inhibitors has been discussed in the section on corrosion prevention and control.

3.11.5.4 Ferrous Sulfate Dosing

It is difficult to form protective films on copper alloys such as aluminum brass while subjected to brackish water and seawater. The addition of ferrous sulfate offers a possibility of building up a protective film and diminishing the danger of erosion–corrosion [158]. Experience has shown that the best results are obtained under the following conditions [36,140]:

Ferrous sulfate dosing, tube cleaning by sponge balls or brushes, and deaeration must start working from the very beginning, as startup after the first corrosion attack has occurred cannot stop corrosion.
Cleanliness of tube surfaces is especially important while commissioning.
A water temperature above 8°C–10°C.
Association with continuous tube cleaning.
Solution injected at water box inlet.
Chlorination stopped during injection, since investigation shows that a synergistic effect between iron and chlorine can lead to rapid attack and early failure.

The optimum dose to form the protective film depends on the temperature and content of organic matter in seawater. Typical injection values range between 0.5 and 1 ppm of Fe^{2+} 1 h/day or continuous injection of 0.02–0.05 ppm [159].

The concept of cooling-water treatment for corrosion, deposit, and biofouling is discussed in Refs. [160a, 160b].

3.11.5.5 Cooling water treatment for corrosion, deposition and biofouling control

The concept of cooling water treatment for corrosion, deposit and biofouling control is discussed in Refs.[160a, 160b] and shown in Figure 3.51[160a]

3.11.5.6 Passivation

Donohue et al. [142] discussed passivation of clean metal surfaces in cooling-water system through pretreatment of surfaces by application of a polyphosphate–surfactant combination and prefilming

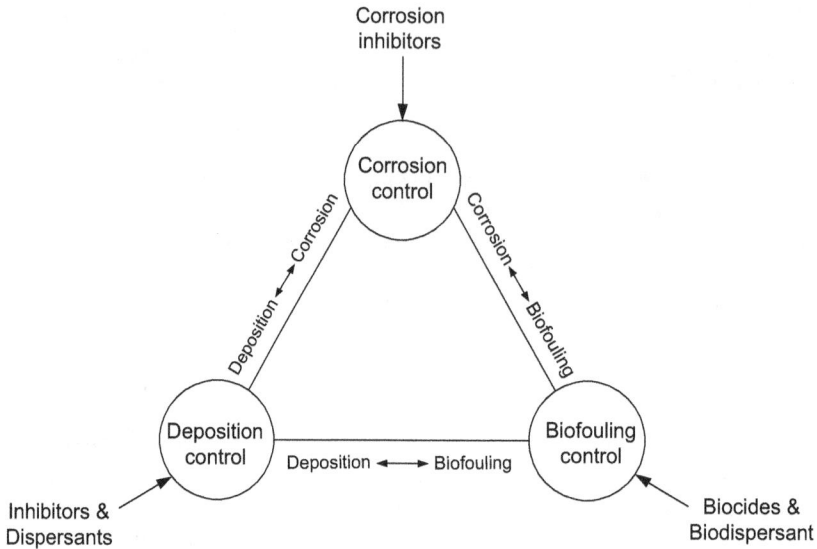

FIGURE 3.51 The concept of cooling water treatment for corrosion control, deposition control and biofouling control.

with chromate or nonchromate zinc polyphosphate or polymer polyphosphates. Chromate, nitrite, and orthophosphate (in the presence of oxygen) can promote passivity on clean iron surfaces.

3.11.6 INFLUENCE OF COOLING-WATER TYPES ON CORROSION

Cooling waters may vary from the purest of distilled or demineralized waters to full-strength seawater. High-conductivity chloride-bearing water, saturated with oxygen, leads to concentration cell corrosion or deposit attack on almost all alloys, ferrous or nonferrous, regardless of their inherent resistance to water.

3.11.6.1 Fresh Water

In general, if fresh water is involved, no serious corrosion problems are likely to arise. Fresh waters are normally handled with tubes of arsenical copper, inhibited admiralty brass, inhibited aluminum brass, aluminum bronze, red brass, or copper–nickels [155].

3.11.6.2 Seawater Corrosion

Seawater environments are highly corrosive since they contain high concentration of salts (mainly sodium chloride), dissolved oxygen, carbon dioxide and biological marine life. Under stagnant or polluted conditions, additional corrosive species such as ammonia or sulfide compounds, and/or sulfate reducing bacteria (SRB) may affect some material's performance. For seawater applications use ferritic, super ferritic, duplex, or super austenitic stainless steels [161].

In seawater applications, arsenical copper and red brass 85:15 would not be used for tubes in power plant condensers cooled with seawater. Inhibited aluminum brass tubes are most widely used. Aluminum bronze (5%) would be suitable for acid-polluted seawater. Copper–nickels will give better service than other tubes in polluted waters [155].

3.11.6.3 Brackish Waters

Brackish waters are a mixture of seawater and fresh water. They are generally less corrosive than seawater except where contaminated with industrial wastes and sewage. In general, when contaminants are present, aluminum brass and aluminum bronze prove more satisfactory than other alloys. In the absence of contaminants, arsenical admiralty and the cupronickel alloys are satisfactory. The choice between these two depends upon water velocity [35]. Among ferrous alloys, the ferritic stainless steels, such as type 410 (12% chrome) and type 430 (18% chrome), are unsatisfactory in brackish cooling water due to insufficient alloy content to repair the passive films that are broken down by the deposit attack. Hence, the use of straight chrome stainless steel is to be avoided in brackish cooling water [136].

3.11.6.4 Boiler Feedwaters

Generally, boiler feedwaters are only slightly corrosive toward copper and its alloys due to deaeration and water treatment. The selection of the proper alloy is generally based on the temperature and pressure requirements. Copper and arsenical admiralty brass are most commonly used for the low-pressure and low-temperature feedwater heaters, whereas cupronickels are used for the higher-pressure and higher-temperature heaters because of their higher creep strength [35].

3.11.7 Corrosion of Individual Metals in Cooling-Water Systems

Aluminum: for aluminum, pitting is the most common form of corrosion. Pitting is usually produced by the presence of halide ions, of which chloride (Cl^-) is most frequently encountered in liquid cooling loops. Waters containing small amounts of the heavy metal ions such as iron, copper, and mercury will also cause pitting corrosion. Waters that are essentially neutral offer no problem. Pitting of aluminum may be experienced beneath deposits. Use of the Alclad products often makes aluminum usable in cooling-water services like automobile radiators and adds an inhibitor when using water with aluminum to maintain a clean heat transfer surface.

Copper and its alloys: copper and its alloys are very important heat exchanger materials because of their generally high thermal conductivity and resistance to corrosion in waters. When copper corrodes, it is more often degraded by general corrosion than by pitting. General corrosion will often result when copper is exposed to ammonia, oxygen, or fluids with high sulfur content. Another source of corrosion affecting copper is dissolved salts in the fluid, such as chlorides, sulfates, and bicarbonates. Copper alloys too can suffer corrosion under deposits, erosion–corrosion, dezincification of the brasses, and sulfide attack.

Steel: the corrosion of steel in cooling water is governed mainly by the oxygen availability and water composition [136]. Corrosion attack is overcome by addition of inhibitors and cathodic protection.

Austenitic stainless steel: the corrosion resistance of chromium–nickel (18-8) stainless steels depends upon a passive surface film. This film is readily penetrated by the chloride ion. For this reason, the stainless steels are subject to pitting and SCC from the chlorides. The pitting rate of the 18-8 alloys is usually quite low in closed-cycle cooling waters and in inhibited water associated with periodic tube cleaning.

Superferritics and superaustenitics: these steels have been primarily developed for seawater applications.

Titanium: titanium is an ideal material for seawater applications in the absence of fouling deposits.

3.11.8 Forms of Corrosion in Cooling Water

A wide spectrum of materials is used on the cooling-water side. These materials are affected by most forms of corrosion discussed earlier. The corrosion prevention and control measures discussed therein are equally applicable to combat cooling-water corrosion also. Various forms of corrosion, susceptible alloys or susceptible components, and remedial measures are mentioned very briefly in the following paragraphs.

3.11.8.1   Uniform Corrosion

All alloys, especially copper and its alloys, suffer uniform corrosion, which is controlled by pH adjustment, inhibition, and proper material selection.

3.11.8.2   Galvanic Corrosion

Less noble metals experience galvanic corrosion. Susceptible components are tubesheets, water boxes, bolts, and flanges. Alloys close to one another in the galvanic series should be used. Use insulation to separate the active metal from the noble metal.

3.11.8.3   Pitting Corrosion

Susceptible materials include aluminum, stainless steels, and copper alloys. Pitting corrosion takes place on the tubesheet and tube wall in the form of pits after local destruction of the protective film by deleterious chemical species such as chlorine or mercury ions. Additionally, at water speeds below about 3 ft/s, suspended solids tend to settle on the tube and may initiate pitting corrosion under the deposits due to differential cell action [104]. Most of these potential causes of pitting corrosion can be avoided by tube cleaning or removing chloride ions and others involving the water chemistry; use steel containing Mo.

3.11.8.4   Crevice Corrosion

Crevice corrosion takes place in the area between the tube and the tubesheet, at gasketed joints, beneath deposits, and at bolt holes. Proper tube expanding will prevent the circumferential tube crack in tube-to-tubesheet joints. While pitting and crevice corrosion normally occur fairly uniformly throughout the heat exchanger, the attack may predominate on the bottom of the internal surfaces of the tubes if mud or sediment has collected there and promotes crevice corrosion [30].

3.11.8.5   Stress Corrosion Cracking

Most common metals and alloys are susceptible to SCC. This form of corrosion takes place at the tube-to-tubesheet expanded joints and at U-tube bends.

3.11.8.6   Corrosion Fatigue and Fretting Wear

All alloys, at the tube and baffle interface, and tubes at midspan due to collision between adjacent tubes are susceptible to corrosion fatigue and fretting wear.

3.11.8.7   Erosion of Tube Inlet

When the cooling water contains large quantities of suspended abrasive solid particles, any protective film formed is rapidly removed and the tube is liable for thinning [158].

The problem of inlet-end tube erosion (erosion–corrosion) in high-pressure feedheaters with carbon steel tubes and condensers with copper alloy tubes is discussed here. In general, the factors governing erosion–corrosion in feedwater heaters are [161] the following:

Feedwater velocity (turbulence).
pH value of feedwater (protective layer of Fe_3O_4).
O_2 content of feedwater (corrosion).
Feedwater temperature (protective layer).

It has been suggested that at the usual pH values (9–9.5) and O_2, contents (0.002–0.007 ppm), and with water chambers and tube inlets designed to promote uniform water flow, erosion at the tubes and tube inlets can be eliminated, even for high water velocity [161].

3.11.8.8 Dezincification
Tubes and tubesheets made of brasses are susceptible to dezincification.

3.11.8.9 Microbiologically Induced Corrosion
All but the higher nickel–chromium alloys and titanium have been found to be subject to microbiological corrosion. This form of failure was discussed earlier.

3.12 PREVENTING CORROSION IN AUTOMOTIVE COOLING SYSTEMS

Water and water-and-glycol solutions are common heat transfer fluids used in automotive cooling systems. Quality of water needs to be considered to prevent corrosion. Chloride is corrosive, and use of tap water should be avoided if it contains more than 100 ppm of chloride. Hardness of water also needs to be considered because it forms scale on the metal surfaces. Deionized water or demineralized water is highly recommended in order to avoid chloride and scale buildup. A suitable corrosion inhibitor must be added to the deionized or demineralized water.

REFERENCES

1. Munger, G. C., Corrosion as related to coatings, in *Corrosion Prevention by Protective Coatings*, National Association of Corrosion Engineers, Houston, TX, 1984, pp. 19–46.
2. Uhlig, H. H., Environmental effects on properties of materials, *Mater. Protect. Perform.*, July, 23–27 (1972).
3. *Resistance to Corrosion*, 4th edn., Inco Alloys International, Huntington, WV, 1985.
4. Jarman, R. A. and Shreir, L. L., The basic principles of corrosion, *Weld. Metal Fabr.*, October, 370–379 (1987).
5. Baboian, R. ed., *Corrosion Tests and Standards—Application and Interpretation*, ASTM, New York, 1995.
6. Brosilow, R., Can you weld to avoid corrosion, *Weld. Design Fabr.*, September, 67–74 (1982).
7. Roser, W. R. and Rizzo, F. E., The corrosion engineer's look at passive alloys, *Mater. Protect. Perform.*, April, 51–54 (1973).
8. *Corrosion Basics—An Introduction, Second Edition*, Pierre R. Roberge, ed. (Houston, TX: NACE International, 2006), pp. 13–14.
9. www.twi-global.com/what-we-do/research-and-technology/technologies/materials-and-corrosion-management/corrosion-testing/high-temperature-corrosion
10. A.S. Khanna, in Handbook of Environmental Degradation of Materials (Second Edition), 2012, 5.23.3 Coal-Based Power Plants. www.sciencedirect.com/topics/chemistry/high-temperature-corrosion
11. Silverstein, M. R., Cooling water, *Client. Eng.*, August, 84–94 (1971).
12. Fontana, M. G. and Greene, N. D., *Corrosion Engineering*, 2nd edn., McGraw-Hill, New York, 1978.
13. Uhlig, H. H., *Corrosion and Corrosion Control*, John Wiley & Son, Inc., New York, 1963.
14. L. S. Van Delinder, ed., Atmospheric corrosion, in *Corrosion Basics—An Introduction*, NACE, Houston, TX, 1984, pp. 221–222.
15. Merrick, R. D., Design of pressure vessels and tanks to minimize corrosion, *Mater. Perform.*, January, 29–37 (1987).
16. *AL-6XN^R Alloy*, Allegheny Ludlum Steel Corporation, Pittsburgh, PA.
17. Kobrin, G., Materials selection, in *Metals Handbook*, 9th edn., Vol. 13, Corrosion, American Society for Metals, Metals Park, OH, 1987, pp. 321–337.
18. www.corrosionclinic.com/Corrosion-Rate-Units-Converter_uA-per-cm2_mdd_gmd_mpy_um-per-y.htm

19. Elliott, P., Design details to minimize corrosion, in *Metals Handbook*, 9th edn., Vol. 13, Corrosion, American Society for Metals, Metals Park, OH, 1987, pp. 338–343.

20. Freedman, A. J. and Boies, D. B., Causes and cures for cooling water corrosion problems, *Petroleum Refiner*, April, 157–162 (1961).

21. Stanfield, B. M., Polarization cell: A common cure for corrosion, *Mater. Protect. Perform.*, March, 32–35 (1973).

22. Steven, P. L., General corrosion, in *Metals Handbook*, 9th edn., Vol. 13, Corrosion, American Society for Metals, Metals Park, OH, 1987, pp. 80–103.

23. L. S. Van Delinder, ed., Open pitting corrosion, in *Corrosion Basics—An Introduction*, NACE, Houston, TX, 1984, p. 95.

24. Speidel, M. O. and Atrens, A. eds., *Brown Bowri Symposium on Corrosion in Power Generating Equipment*, Plenum Press, New York, 1983.

25. Lichtenstein, J., Fundamentals of corrosion causes and mitigation, *Mater. Perform.*, March, 29–31 (1978).

26. Wilten, H. M., Microscope speeds diagnosis of refinery corrosion, *Hydrocarb. Process. Petrol. Refiner*, *40*, 151–280 (1961).

27. Asphahani, I. A., Corrosion resistance of high performance alloys, *Mater. Perform.*, *19*, 33–43 (1980).

28. Uhlig, H. H., Discussion—An essay on pitting, crevice corrosion, and related potentials, *Mater. Perform.*, August, 35–36 (1983).

29. Dexter, C. S., Localized corrosion, in *Metals Handbook*, 9th edn., Vol. 13, Corrosion, American Society for Metals, Metals Park, OH, 1987, pp. 104–122.

30. Syrett, B. C. and Coit, R. L., Causes and prevention of power plant condenser tube failures, *Mater. Perform.*, February, 44–49 (1983).

31. Jarman, R. A. and Shreir, L. L., Corrosion of jointed structures—Part II, *Weld. Metal Fabr.*, 445–449 (1987).

32. Henthorne, M., Understanding corrosion, *Chem. Eng.*, Desk Book Issue, December 4, 19–33 (1972).

33. Guide to selecting engineered materials, metals, polymers, ceramics, composites, *Adv. Mater. Process.*, *137*(6) (1990).

34. Stengerwald, R., Metallurgically influenced corrosion, in *Metals Handbook*, 9th edn., Vol. 13, Corrosion, American Society for Metals, Metals Park, OH, 1987, pp. 123–135.

35. *Condenser Tube Manual*, Bridgeport Brass Company, CT, USA, 1964.

36. Forchhammer, P., Results of damage research on corrosion failures in heat exchangers, *Heat Transfer Eng.*, 5, 19–32 (1984).

37. Woodward, A. R., Howard, D. L., and Andrews, E. F. C., Condensers, pumps and cooling water plant, in *Modern Power Station Practice, Vol. C, Turbines, Generators, and Associated Plant* (D. J. Littler, E. J. Davies, H. E. Johnson, F. Kirkby, P. B. Myerscouh, and W. Wright, eds.), Pergamon Press, New York.

38. Crook, P., Practical guide to wear for corrosion engineers, *Mater. Perform.*, February, 64–66 (1991).

39. Van Delinder, L. S. ed., Erosion-corrosion, in *Corrosion Basics—An Introduction*, NACE, Houston, TX, 1984.

40. Yokell, S., *A Working Guide to Shell and Tube Heat Exchangers*, McGraw-Hill, New York, 1990.

41. TEMA, *Standards of Tubular Exchanger Manufacturers Association*, 10th edn., Tubular Exchanger Manufacturers Association, Tarrytown, NY, 2019.

42. Van Delinder, L. S. ed., Cavitation, in *Corrosion Basics—An Introduction*, NACE, Houston, TX, 1984.

43. Dawson, J. L., Shih, C. C., Gearey, D., and Miller, R. G., Flow effects on erosion-corrosion, *Mater. Perform.*, April, 57–60 (1991).

44. Dawson, J. L. and Shih, C. C., Corrosion under flowing conditions—An overview and model, *CORROSION/90*, Paper No. 21, Houston, TX, 1990.

45. 90/10 Copper-nickel vs Sea-Cure stainless steel—A functional comparison of two condenser tube alloys, FineweldR tube, Technical Letter, Olin Brass.

46. Syrett, B. C., Corrosion in fossil fuel power plants, in *Metals Handbook*, 9th edn., Vol. 13, Corrosion, American Society for Metals, Metals Park, OH, 1987, pp. 985–1010.

47. Theus, G. J. and Daniel, P. L., Corrosion in steam generating systems, in *Brown Bowri Symposium on Corrosion in Power Generating Equipment* (M. O. Speidel and A. Atrens, eds.), Plenum Press, New York, 1983, pp. 185–231.

48. Irving, B., Stress corrosion cracking: Welding's no. 1 nemesis, *Welding J.*, December, 37–40 (1992).

49. Kedzie, D. P. and Rizzo, F. E., The corrosion engineer's look at stress corrosion, *Mater. Protect. Perform.*, October, 53–56 (1972).

50. Logan, H. L., *The Stress Corrosion Cracking of Metals*, John Wiley & Sons, New York, 1966.

51. McIntyre, D. R. and Dillon, C. P., *Guidelines for Preventing Stress Corrosion Cracking in the Chemical Process Industries*, Publication 15, Materials Technology Institute, Columbus, OH, 1985, pp. 8–14.

52. Berry, W. E., Stress corrosion cracking—A nontechnical introduction to the problem, DMIC Report 144, Defence Metals Information Centre, Battelle Memorial Institute, Columbus, Ohio, January 6, 1961.

53. Pement, F. W., Wilson, I. L. W., Reynolds, S. D., and Fletcher, W. D., Microanalytical characterization of stress corrosion cracking in power plant heat exchanger tubing, *Mater. Perform.*, 26–39 (1977).

54. Craig, B., Environmentally induced cracking, in *Metals Handbook*, 9th edn., Vol. 13, Corrosion, American Society for Metals, Metals Park, OH, 1987, pp. 145–149.

55. Brown, B. F., *Stress Corrosion Cracking in High Strength Steels and in Titanium and Aluminum Alloys*, Naval Research Laboratory, Washington, DC, 1972.

56. Parkins, R. N., An overview: Prevention and control of stress corrosion cracking, *Materials Performance*, Vol. 24, No. 8, NACE, Houston, TX, August 1985, pp. 9–24.

57. McIntyre, D. R., How to prevent stress corrosion cracking in stainless steels—Part I, in *The Chemical Engineering Guide to Corrosion Control in the Process Industries* (R. W. Greene, ed.), McGraw-Hill Publications Co., New York, 1986, pp. 4–7.

58. Partridge, E. P., Hydrogen damage in power boilers, *Trans. ASME, J. Eng. Power*, July, 311–320 (1964).

59. Glaser, W. and Wright, I. G., Forms of corrosion—Mechanically assisted degradation, in *Metals Handbook*, 9th edn., Vol. 13, Corrosion, American Society for Metals, Metals Park, OH, 1987, pp. 136–144.

60. L. S. Van Delinder, ed., Corrosion fatigue, in *Corrosion Basics—An Introduction*, NACE, Houston, TX, 1984.

61. Tatnall, R. E., MIC: A corrosion engineer's enigma? *Mater. Perform.*, August, 56 (1988).

62. Pollock, W. I., What is MIC? *Mater. Perform.*, August, 56 (1988).

63. Scott, P. J. B., Goldie, J., and Davies, M., Ranking alloys for susceptibility to MIC—A preliminary report on high-Mo alloys, *Mater. Perform.*, January, 55–57 (1991).

64. Kobrin, G., Corrosion by microbiological organisms in natural waters, *Mater. Perform.*, July, 38–42 (1976).

65. Pope, D. H., Duquette, D. J., Arland, J. H., and Wayner, P. C., Microbiologically influenced corrosion of industrial alloys, *Mater. Perform.*, April, 14–18 (1984).

66. Pope, D. H., Soracco, R. J., and Wilde, E. W., Studies on biologically induced corrosion in heat exchanger system at the Savannah River Plant, Aiken, SC., *Mater. Perform.*, July, 43–50 (1982).

67. Borenstein, S. W., Microbiologically influenced corrosion of austenitic stainless steel weldments, *Mater. Perform.*, January, 52–54 (1991).

68. www.gti.energy/wp-content/uploads/2019/02/MIC_brochure2018_v4HiRes.pdf

69. Borenstein, S. W., Microbiologically influenced corrosion failures of austenitic stainless steel welds, *Mater. Perform.*, August, 62–66 (1988).

70. Puckorius, P. R., Massive utility condenser failure caused by sulfide producing bacteria, *Mater. Perform.*, December, 19–22 (1983).

71. Powell, C. A., Preventing biofouling with copper alloys, Copper Development Association, Publication No. 113, U.K., 1995.

72. Tuthill, A. H., Guidelines for the use of copper alloys in seawater, *Mater. Perform.*, September, 12–22 (1987).

73. Schutz, R. W., A case for titanium's resistance to microbiologically influenced corrosion, *Mater. Perform.*, January, 58–61 (1991).

74. Boffardi, P., Control of environmental variables in water-recirculating systems, *Metals Handbook*, 9th edn., Vol. 13, Corrosion, American Society for Metals, Metals Park, OH, 1987, pp. 487–497.

75. Poole, J. S. and Bacon, H. E., Monitoring cooling water systems, *Mater. Protect. Perform.*, March, 16–25 (1972).

76. Borenstien, S. W. and Lindsay, P. B., Microbiologically influenced corrosion failure analyses, *Mater. Perform.*, March, 51–54 (1988).

77. Gooch, T. G., Stress corrosion cracking of welded austenitic stainless steel, *Weld. World*, *22*, 64–76 (1984).

78. Corrosion of weldments, in *Metals Handbook*, 9th edn., Vol. 13, Corrosion, American Society for Metals, Metals Park, OH, 1987, pp. 344–368.

79. Warren, D., Hydrogen effects on steel, *Mater. Perform.*, January, 38–46 (1987).

80. Sauvage, M., Requirements for the manufacture of pressure vessels for wet H_2S service, *WRC Bulletin*, *374*, 72–81.

81. Terasaki, F., Ohtani, H., Ikeda, A., and Nakanishi, M., Steel plates for pressure vessels in sour environment applications, *Proc. Inst. Mech. Eng.*, *200*, 141–158.

82. Mirabal, E., Bhattacharjee, S., and Pazos, N., Carbonate-type cracking in an FCC wet gas compressor station, *Mater. Perform.*, July, 41–45 (1991).

83. Kane, R. D. and Cayard, M. S., Improve corrosion control in refining processes, *Hydrocarb. Process.*, November, 129–142 (1995).

84. *Methods and Controls to Prevent In-Service Cracking of Carbon Steel (P-l) Welds in Corrosive Petroleum Refinery Environments*, NACE RP-04–72 (1976 Rev.), National Association of Corrosion Engineers, Houston, TX, 1976.

85. Evaluation of pipeline and pressure vessel steels for resistance to hydrogen induced cracking, NACE Standard TM-02-84, NACE International, Houston, TX, 2011.

86. Tech. Services Bull. No. 778: A516 Steels Including HIC-Tested A516 Steels, Rev. November 1992, Lukens Steel Company, Coatesville, PA, 1992.

87. www.ndt.net/apcndt2001/papers/1154/1154.htm

88. www.mistrasgroup.com/how-we-help/special-emphasis/htha/

89. www.twi-global.com/technical-knowledge/faqs/what-is-high-temperature-hydrogen-attack-htha-hot-hydrogen-attack

90. Inspection for High Temperature Hydrogen Attack (Draft4 for comment ballot, March 2021) API Recommended Practice 586, Section 2, First Edition, TBD 2022

91. Britton, J. N. (2002), Corrosion at pipe supports: Causes and solutions (I-Rod® brand pipe supports).

92. https://stoprust.com/technical-library-items/06-pipe-supports/

93. www.corrosionpedia.com/definition/1702/touch-point-corrosion-tpc Last updated: August 25, 2017

94. www.nationalboard.org/Index.aspx?pageID=184

95. www.corrosionpedia.com/definition/1219/corrosion-under-insulation-cui

96. Yin, K., Yang, Y., and Frank, C. Permeability of coal tar enamel coating to cathodic protection current on pipelines. *Constr Buil Mate*, *192*, 20–27 (2018).

97. https://rti.rockwool.com/learnings/corrosion-under-insulation/

98. Della Anggabrata, Reviewed by Raghvendra Gopal, Understanding the Causes and Cures for Corrosion Under Insulation, Published: April 13, 2021, Industrial Insulation, Stress Corrosion Cracking of Austenitic Stainless Steel, TECHNICAL BULLETIN, IND—TB009 02/21/2020 (Replaces 01/19/18), 2020 Johns Manville, Denver, CO.

99. https://corrosion-doctors.org/Forms-crevice/CUI.htm

100. https://inspectioneering.com/tag/corrosion+under+insulation

101. www.jmsse.in/files/122corrosion

102. https://solutions.borderstates.com/corrosion-protection-for-pipelines/

103. www.zavenir.com/insight-category/general-articles/pipeline-corrosion-prevention-with-rust-inhibition-technology…

104. Smith, P. *Process Piping Design Handbook, Volume One The Fundamentals of Piping Design Drafting and Design Methods for Process Applications*, Gulf Publishing Company Houston, Texas, 2007.

105. Baker Jr, M. and Fessler, R. R. Pipeline Corrosion Final Reprot Submitted to U.S. Department of Transportation Pipeline and Hazardous Materials Safety Administration, Office of Pipeline Safety Integrity Management Program Under Delivery Order DTRS56-02-D-70036, November 2008, pp. 1–72.

106. CRTS, Inc., Specialist in Pipeline Internal Field Joint Coatings, 1807 N. 170th E. Ave. Tulsa, OK 74116 U.S.A., www.coatingrobotics.com

107. www.mannesmann-linepipe.com/en/supply-range/oil-gas-line-pipe/coatings-linings/fusion-bonded-epoxy.html

108. www.thermacor.com/wp-content/themes/thrma/flat-theme/pdf/specs/FBE%20Complete.pdf

109. www.thomasnet.com/articles/chemicals/epoxy-fusion-bonded

110. Siddiqui, E. and Prasad Mishra, D., *Electrical Market Division (EMD-Infra)* 3M India Limited, Bangalore

111. Kehr, J. A. and Enos, D. G. *FBE, a Foundation for Pipeline Corrosion Coatings*, 3M Company, Austin, TX, pp. 1–20.

112. https://coloradogeologicalsurvey.org/hazards/corrosive-soil/

113. www.giatecscientific.com/education/understanding-concrete-corrosion/

114. www.edswaterproofing.com/eds_wp_new_site_here/wp-content/uploads/2015/11/Corrosion-in-Rei nforced-Concrete.pdf

115. www.cement.org/Learn/concrete-technology/durability/corrosion-resistance-of-concrete

116. Swindeman, R. W. and Gold, M., Developments in ferrous alloy technology for high-temperature service, *Trans. ASME, J. Pressure Vessel Technol.*, *113*, 133–140 (1991).

117. Collins, J. A., Effect of design, fabrication and installation in the performance of stainless steel equipment, *Corrosion, 11*(1), pp. 27–34 (1955).

118. Landrum, J. R., Designing equipment for corrosion resistance, in *Materials Engineering, Part I: Selecting Materials for Process Equipment* (K. J. McNaughton, ed.), McGraw-Hill, New York, 1980, pp. 157–164.

119. Shabin, H. I., Corrosion inhibition in a cooling water system, *Corros. Prevent. Control*, February, 9–12 (1982).

120. Van Delinder, L. S. ed., Inhibitors, in *Corrosion Basics—An Introduction*, NACE, Houston, TX, 1984, p. 127.

121. Freedman, A. J., Cooling water technology in the 1980s, *Mater. Perform.*, November, 9–16 (1984).

122. Kaplan, R. I. and Ekis, E. W., Jr., The on-line removal of iron deposits from cooling water systems, *Mater. Perform.*, November, 40–44 (1984).

123. Boffardi, B. P., Corrosion control of industrial cooling water systems, *Mater. Perform.*, November, 17–24 (1984).

124. Beynon, E., Cooper, N. R., and Haningan, H. J., Cooling system corrosion in relation to design and materials, in *Engine Coolant Testing: State of the Art, ASTM STP 705* (W. H. Ailor, ed.), American Society for Testing and Materials, Philadelphia, PA, 1980, pp. 310–326.

125. Hinchcliffe, D. and Town, J., Experience with nonchromate cooling water treatment (case histories), *Mater. Perform.*, September, 36–38 (1977).

126. Lula, R. A., Compiler, *Source Book on the Ferritic Stainless Steels*, American Society for Metals, Metals Park, OH.

127. Debold, T. A., Select the right stainless steel, *Chem. Eng. Prog.*, November, 38–43 (1991).

128. Saunders, E. A. D., *Heat Exchangers: Selection, Design and Construction*, Addison Wesley Longman, Reading, MA, 1989.

129. Llewellyn, D. T., *Steels, Metallurgy and Applications*, Butterworth-Heineman, Oxford, U.K., 1992.

130. Munger, C. G., Corrosion prevention by protective coatings, *Mater. Perform.*, April, 57–58 (1987).

131. Van Delinder, L. S. ed., Coatings and linings, in *Corrosion Basics—An Introduction*, NACE, Houston, TX, 1984, pp. 47–48.

132. www.laserax.com/blog/thermal-spray-coating

133. Weber, J., Corrosion and deposits in cooling systems—Their causes and prevention, *Sulzer Tech. Rev.*, *3*, 219–232 (1972).

134. McGill, W. A. and Weinbaum, M. J., Aluminum vapor diffused steels resist refinery corrosion, *Mater. Protect. Perform.*, July, 28–32 (1972).

135 www.tcenergy.com/siteassets/pdfs/commitment/safety/pipelines-and-operations/tc-cathodic-protect ion.pdf

136. Asbaugh, W. G., Corrosion in heat exchangers using brackish water, *Chem. Eng.*, July 5, 146–152 (1965).

137. https://inspectioneering.com/tag/cathodic+protection

138. Shailesh, J., Impressed current cathodic protection system for water pipeline internal surfaces, Paper No. ICP23, NIGIS, CORCON 2017, 17–20 September, Mumbai, India.

139. Flexible Anode Cathodic Protection Technology. News—Shaanxi Elade New Material Technology Co., Ltd (eladeanode.com)

140. Paren, J. and Pouzenc, C., Design of power station condensers, *GEC Alsthom Tech. Rev.*, 6, 19–38 (1991).

141. Singbell, D. and Garner, A., Anodic protection to prevent the stress corrosion cracking of pressure vessel steels in alkaline sulfide solutions, *Mater. Perform.*, April, 31–36 (1987).

142. Donohue, J. M. and James, E. W., Passivation and its application in cooling water systems, *Mater. Perform.*, October, 34–36 (1977).

143. www.advancedfrpsystems.com/the-complete-guide-to-corrosion-management-system/

144. https://corrosionillinois.com/solutions/corrosion-prevention-management-systems/

145. BenDu Bose on 5/4/2020 1:41 PM, Essential Elements of a Successful Corrosion Management program, Materials Selection & Design and Identification of Corrosion Threats

146. www.materialsperformance.com/articles/material-selection-design/2016/07/essential-elements-of-a-successful-corrosion-management-program

147. Britton, C. F. and Tofield, B. C., Effective corrosion monitoring, *Mater. Perform.*, April, 41–44 (1988).

148. Schweitzer, P. A., ed., *Corrosion and Corrosion Protection Handbook*, 2nd edn., Marcel Dekker, New York, 1983.

149. Somm, E., Schlachter, W., and Schwarzenbach, A., Power generating equipment, in *Brown Bowri Symposium on Corrosion in Power Generating Equipment* (M. O. Speidel and A. Atrens, eds.), Plenum Press, New York, 1983, pp. 3–25.

150. www.materialsperformance.com/articles/material-selection-design/2016/07/essential-elements-of-a-successful-corrosion-management-program

151. Chance, R. L., Walker, M. S., and Rowe, L. C., Evaluation of engine coolants by electrochemical methods, in *Engine Coolant Testing: Second Symposium, ASTM STP 887* (R. E. Beal, ed.), American Society for Testing and Materials, Philadelphia, PA, 1986, pp. 99–122.

152. Cramer, S. D. and Covino, Jr., B. S. eds., *Corrosion: Fundamentals, Testing, and Protection*, Volume 13A, 10th edn., ASM International, Metals Park, OH, 2003.

153. Garverick, L., ed., *Corrosion in Petrochemical Industries*, 1st edn., ASM International, Metals Park, OH, 1994.

154. Sustainable Water Use In Chemical, Paper, Textile And Food Industries, Summary Report (Generic + Industrial Data) On Scaling, Fouling And Corrosion parameters UCM, HOL, PTS, VITO, 24th of May of 2010, pp. 1–109.

155. Tracy, A. W., Nole, V. F., and Duffy, E. J., Water analyses and their functions in selecting condenser tube alloys, *Trans. ASME, J. Eng. Power*, July, 329–332 (1965).

156. Cotton, I. J., Oxygen scavengers—The chemistry of sulfide under hydrothermal conditions, *Mater. Perform.*, March, 41–47 (1987).

157. www.suezwatertechnologies.com/handbook/chapter-31-open-recirculating-cooling-systems

158. Lockhart, A. M., Reducing condenser tube corrosion at Kincardine Generating Station with ferrous sulphate, *Proc. Inst. Mech. Eng.*, 495–512 (1964–1965).

159. Sato, S., Nagata, K., and Yamauchi, S., Evaluation of various preventive measures against corrosion of copper alloy condenser tubes by sea water, *NACE National Conference*, Unpublished Paper No. 195, Toronto, Ontario, Canada, 1981.

160a. www.gmc.biz.vn/blogs/san-pham/cooling-water-treatment

160b. www.power-eng.com/om/advanced-cooling-water-treatment-concepts-part-1/#gref

161. Sonnenmoser, A., Design and operation of large high pressure feedheaters, *Brown Boveri Rev.*, 7/8–73, 352–359.

BIBLIOGRAPHY

Bostwick, T. W., Reducing corrosion of power plant condenser tubing, *Corrosion*, Vol. 17, pp. 12–19, August (1961).

Brittalan, O. J., Declerks, D. H., and Urhis, F. H., Linings, *Chem. Eng.*, Deskhook Issue, *12* October, 127–135 (1972).

Coburn, S. K., ed., *Corrosion Source Book*, American Society for Metals, Metals Park, and National Association of Corrosion Engineers, Houston, TX, 1984.

Craig, B., Forms of corrosion: Introduction, in *Metals Handbook*, 9th edn., Vol. 13, Corrosion, American Society for Metals, Metals Park, OH, 1987, p. 79.

Greene, R. W., ed., *The Chemical Engineering Guide to Corrosion*, McGraw-Hill, New York, 1986.

McDonald, M. M., Corrosion of brazed joints, in *Metals Handbook*, 9th edn., Vol. 13, Corrosion, American Society for Metals, Metals Park, OH, 1987, pp. 876–886.

McNaughton, K. J., ed., *Materials Engineering, Part I: Selecting Materials for Process Equipment; Part II: Controlling Corrosion in Process Equipment*, McGraw-Hill, New York, 1980.

Morrow, S. J., *Materials Selection For Seawater Pumps*, ITT Corporation, Industrial Process, Lancaster, Pennsylvania, pp. 73–80.

Pludek, R. V., *Design and Corrosion Control*, Macmillan, London, U.K., 1977.

Sankara Narayanan, T. S. N., Surface pretreatment by phosphate conversion coatings, A Review, *Rev. Adv. Mater. Sci.* 9 (2005), Advanced Study Center Co. Ltd., pp. 130–177

Schreiber, C. F. and Coley, F. H., *CORROSION/75*, Paper No. 36, NACE, Houston, TX, 1975.

4 Boiler, Thermal Power Plant, and Heat Exchangers of Cooling and Feedwater Systems

4.1 INTRODUCTION

According to American Boiler Manufacturers Association, "Boiler is a pressure vessel that provides a heat transfer surface (generally a set of tubes) between the combustion products and the water. A boiler is usually integrated into a system with many components."

Boiler is a pressure vessel into which water is fed, and by the application of heat, it is evaporated into steam. A boiler, which includes drums, tubes, flues, ducts, auxiliary equipment, etc., is subject to continual stress variation resulting from expansion and contraction and from elevated temperatures when it is in service. Boilers must be made adequately strong and with suitable materials to withstand these forces and temperatures.

The basic concept of a boiler involves a heat source (furnace) and a feed water heater, which allows water to be heated above its boiling point. The type of heat source and the method of heat exchange are what primarily define different types of boilers. Combustion boilers are widely used to generate steam for industrial applications and power generation. Major parts and assemblies of a boiler are given below:

a. Air-fuel supply arrangement.
b. Burner: the burner initiates the combustion reaction of air-fuel mixture inside the boiler. A nozzle in the burner that turns the fuel pumped from the fuel source atomizes and ignites it to create and sustain the combustion.
c. Combustion chamber: it burns the fuel and generates heat, which is transferred to the heat exchanger.
d. Economizer: an important heat exchanger is the economizer to raise the feedwater temperature and hence increase the thermal efficiency of a boiler.

Steam drum and mud drum: the steam drum collects the steam while the mud drum is placed beneath the steam drum and collects the solid, which is removed periodically.

Boiler types can be grouped into two broad categories: water-tube boilers and fire-tube boilers. In the water-tube boilers, tubes containing water are heated by combustion gases that flow outside the tubes, while in the fire-tube boilers hot combustion gases flow inside the tubes and water flows outside. Key design parameters to determine the boiler size and power are the output steam mass flow rate, pressure, and temperature. Since the first boilers were used in the 18th century, the design of boilers has evolved so as to increase efficiency as well as steam pressure and temperature. In

DOI: 10.1201/9781003352068-4

the industrialized countries, more than 50% of the industrial boilers are fired by natural gas. The basic elements of a boiler include the fuel supply arrangement, burner combustion chamber, exhaust stack, and controls. Important references on boiler and steam generation and use are Refs. [1–4].

4.1.1 Steam Generator

The boiler is that part of steam generator where phase changes (or boiling) occurs from liquid (water) to vapor (steam), essentially at constant temperature and pressure. The steam generator can be divided into two general sections, the furnace and the convection pass. The furnace provides a large volume with an opening and water-cooled enclosure walls inside of which combustion takes place. The convection pass contains tube bundles, which compose the superheater, reheater, boiler bank, and economizer. The air heater usually follows the convection pass.

4.1.1.1 Definitions Related to Steam

Saturated Water

When water just begins to boil, it is called saturated water. As more heat is added (at constant pressure), the fluid temperature will remain at the saturation temperature until all of the water is converted to steam. The heat input or enthalpy necessary to convert saturated water to saturated steam is called the heat of vaporization. The conversion of water to steam requires much more energy beyond that required to reach the boiling point.

Boiling point. The term boiling point is used to identify conditions at atmospheric pressure (29.92 inches of mercury). For instance, the boiling point of water at atmospheric pressure is 100°C (212°F). The boiling point is actually a function of pressure and increases as pressure increases. At higher pressures, more heat energy is required to raise the fluid temperature to the boiling point.

Enthalpy. The amount of heat energy contained in the fluid is termed enthalpy and is measured in BTUs/lb.

Heat of vaporization. The points at which all of the water has been converted to steam are indicated by the saturated steam line. The heat input or enthalpy necessary to convert saturated water to saturated steam is termed the heat of vaporization and is indicated for a given temperature by the horizontal constant pressure lines.

Steam quality. Quality is the percent by weight of vapor in a steam/water mixture. As more water is converted to steam, quality increases. Water on the saturated water line has a quality of 0%. Superheated and saturated steams have a quality of 100%. Water that has been heated to saturation and has sufficient additional heat added to convert half of it to steam has a quality of 50%.

Superheated Steam

Superheated steam is steam heated to a temperature higher than its boiling point corresponding to the operating pressure. At normal atmospheric pressure, superheated steam has a temperature above 100°C (212°F). Use of superheated steam permits more efficient operation of devices that convert heat into mechanical work.

4.1.2 Boiler Mountings

The boiler mountings are fittings that are mounted on the boiler for its proper functioning. They are essential for safety, economics, and convenience. These mountings include water level gauges, safety or relief valves, drain and blowdown valves, vent valves, water and steam sample connections, blow off cock, stop-check valves, fusible plug, etc.

4.1.3 BOILER ACCESSORIES

The boiler accessories are devices, which form an integral part of a boiler but are not mounted on it. They are installed to increase the efficiency of steam power plant. They improve the working conditions and help in maximum utilization of heat energy from the flue gases. Here is a list of major accessories of a boiler: economizer, air preheater, reheater, superheater, soot blower, steam trap, steam separator, and boiler feed water pump, etc.

4.1.3.1 Air Preheater

The heat carried out with the flue gases coming out of the economizer are further utilized for preheating the air before supplying to the combustion chamber. It is a necessary equipment for supply of hot air for drying the coal in pulverized fuel systems to facilitate grinding and satisfactory combustion of fuel in the furnace.

4.1.3.2 Economizer

The purpose of the economizer is to preheat the boiler feed water before it is introduced into the steam drum by recovering heat from the flue gases leaving the boiler. The economizer is located in the boiler rear gas pass below the rear horizontal superheater. The economizer is continuous unpinned loop type and water flows in upward direction and gas in the downward direction. Advantages of the economizer include the following.

   i.   Fuel economy—used to save fuel and increase overall efficiency of boiler plant.
   ii. Reducing size of boiler—as the feed water is preheated in the economizer and enter boiler tube at elevated temperature. The heat transfer area required for evaporation reduced considerably.

4.1.3.3 Superheater

A superheater heats steam to high temepratures. It consists of a group of tubes made of special alloys such as chromium–molybdenum. The wet steam from the boiler drum enters into the superheater and gets heated up to a temperature of 535°C and pressure of 170kg/sq cm by gaining heat from the flue gases leaving the furnace. The steam entering the superheater and then the secondary super-heater to the tertiary superheater.

4.1.3.4 Reheater

Power plant furnaces may have a reheater section containing tubes heated by hot flue gases outside the tubes. The function of the reheater is to reheat the steam coming out from the high-pressure tur-bine to a temperature of 540°C. The reheater is composed of two sections, the front pendant section and the rear pendant section. The rear pendant section is located above the furnace arc and the rear water wall and front pendant section is located between the rear water hanger tubes and the super-heater platen section.

4.1.4 SOOT BLOWERS

A soot blower is a device, which is designed to blast soot and ash away from the walls of a furnace or similar piece of equipment. Soot blowers operate at set intervals, with a cleaning cycle that can vary in length, depending on the device and the size of the equipment that needs to be cleaned. The basic principle of the soot blower is the cleaning of heating surfaces by multiple impacts of high-pressure air, steam, or water from opposing nozzle orifices at the end of a translating-rotating tube.

A traveling lance with nozzle jets penetrates the narrow openings in the boiler tube banks to blast the tubes clean. The tubes must be kept clean to allow optimum boiler output and efficiency.

4.1.5 BOILER CODES AND STANDARDS

There are a number of codes and standards, acts and regulations covering boilers and related equipment that should be considered when designing a system. Regulatory requirements are dictated by a variety of sources and are all focused primarily on safety. Boiler codes and design standards provide the basic guidelines for boiler application, design, construction, and operations.

4.1.5.1 ASME Boiler and Pressure Vessel Codes

Section I Power Boilers

Provides requirements for all methods of construction of power, electric, and miniature boilers; high-temperature water boilers, heat recovery steam generators, solar receiver steam generators, certain fired pressure vessels, and liquid phase thermal fluid heaters to be used in stationary service; and power boilers used in locomotive, portable, and traction service.

Section II Materials

Part A covers ferrous material;
Part B covers nonferrous material;
Part C covers welding rods, electrodes, and filler metals; and
Part D covers material properties in both customary and metric units of measure.

Together, these four parts of Section II comprise a "Service Code" to other BPVC Sections, providing material specifications adequate for safety in the field of pressure equipment.

i. Section IV Heating Boilers
    Provides requirements for design, fabrication, installation, and inspection of steam heating, hot water heating, hot water supply boilers, and potable water heaters intended for low-pressure service that are directly fired by oil, gas, electricity, coal, or other solid or liquid fuels.
ii. Section VI Care and Operation of Heating Boilers
    Covers operation guidelines applicable to steel and cast-iron boilers limited to the operating ranges of Section IV Heating Boilers.
iii. Section VII Care of Power Boilers
    Provides guidelines to assist those directly responsible for operating, maintaining, and examining power boilers.
iv. Section VIII Pressure Vessels, Div 1, 2 and 3.

4.1.6 FUEL

Coal, fuel oil, and natural gas are the main types of fuel, but coal plays a large role in power generation because of the enormous reserves of coal in many countries like China, USA, India, etc., around the world. These fuels can release a substantial quantity of heat when burnt. These fuels consist of a large number of complex compounds comprised of five principal elements: carbon (C), hydrogen (H), oxygen (O), sulfur (S), and nitrogen (N).

4.1.6.1 Coal

Coal use involves unloading, storage, and handling facilities; preparation before firing using crushers and pulverizers. There is a wide variation in the properties of coal and its ash. As a result, while the design of the steam-producing unit must provide optimum performance when firing the intended coals, it must also accommodate reasonable alternate coals if necessary.

4.1.7 BOILER WATER CHEMISTRY

The chemistry of boiler water should meet certain minimum requirements laid out by the boilermakers. The boiler water if contains high TDS the carryover of solids will be more. Since the efficiency of the separator is fixed, the only way to bring carryover is to bring down the boiler water TDS. In boiler water, the presence of free NaOH must be eliminated by practicing coordinated phosphate control. High alkalinity and the presence of suspended impurities increase carryover [5].

When the suspended impurities are more, there is a blanket formation over the water-steam inter-face and this prevents the easy passage of steam bubbles from the boiler water. The steam bubbles out of the skin formed by the impurities throwing the suspended impurities to the steam driers. The silica in boiler water is to be controlled more stringently based on pressures. Silica carryover is more at pressures above 28 kg/cm². Boiler water is blown down to control the amount of total dissolved solids (TDS) in the boiler. This water is pressurized, hot and dirty, creating large volumes of flash steam and possible disposal problems. A heat recovery system can reclaim large amounts of energy during this essential process.

4.1.8 FEEDWATER CHEMISTRY

The chemistry of feedwater should meet certain minimum requirements laid out by the boilermakers. The feedwater if contains oil or organic matter, the same leads to foaming. The carryover of solids is very high. If the feedwater has dissolved iron, which may be either from condensate or from the makeup water or generated within the boiler, will again lead to foaming. If the feedwater contains suspended impurities then also carryover will be high. Dissolved oxygen in feedwater leads to corrosion of economizer and this leads to generation of iron oxide, which again leads to increased suspended iron in the boiler water [2].

4.1.9 BOILER SELECTION CONSIDERATIONS

The following criteria should be considered when selecting a boiler to meet the application needs. The criteria are as follows:

1. Pressure vessel codes and standards' requirements.
2. Steam or hot water requirement.
3. Boiler load.
4. Performance considerations.
5. Safety.
6. Life cycle cost.
7. Emissions.

4.1.10 INSTANTANEOUS DEMAND

Instantaneous demand is a sudden peak load change that is usually of short duration. The instantaneous load demand is important to consider when selecting a boiler to ensure that these load variations are taken into account. If the instantaneous demand is not included in the system load calculations, the boiler may be undersized.

TABLE 4.1
Classification of Boiler

Features	Types
Areas of application	Utility, Industrial boiler
Steam-water circulation method	(a) Natural; (b) forced circulation; (c) once-through; (d) once-through with superimposed circulation
Fuels' firing method	Stoker, cyclone furnace, fluidized bed, fixed or moving grate.
Steam condition	Saturated, superheated and supercritical steam
Operating temperature and pressure	Subcritical, supercritical, ultrasupercritical

4.1.11 BOILER CLASSIFICATION

Boilers are differentiated by their configuration, size, and the quality of the steam or hot water produced. Table 4.1 shows a generic method of classification of boilers.

Boiler size is most often measured by the fuel input in million Btu per hour (MMBtu/hr). Size may also be measured by output in pounds of steam per hour (pph). Output may also be measured in boiler horsepower.

The boiler horsepower (BHP) is the amount of energy required to produce *34.5 pounds* of steam per hour at a pressure and temperature of *0 psig and 212 °F*, with feedwater at *0 psig and 212 °F*. One boiler horsepower is equal to 33,475 Btu/hr (9811W or 8436 kcal/hr) evaporation capacity.

Fire-tube units are typically the smallest boilers, with most units less than 10 MMBtu/hr in capacity. Water-tube boilers can be separated into two classes by size. Most existing water-tube boilers have a capacity of less than 10 MMBtu/hr. A smaller number of water-tube boilers are between 10 and 10,000 MMBtu/hr, including almost all large industrial and power generation boilers. Although fewer in number than the fire-tube and small water-tube boilers, these large water-tube boilers account for most steam production. The large water-tube boilers are also the boilers most likely to use solid fuel and there are several different solid fuel combustion technologies in use at these larger size ranges, including stoker, fluidized bed, and PC boilers. In PC boilers, the coal is pulverized to a dust-like consistency and blown with air into the boiler, where it burns in suspension in the furnace. PC boilers are characterized by the burner configuration (tangential, wall, cyclone) and whether the bottom ash exits the boiler in solid or molten form (wet bottom vs dry bottom) [6].

4.1.11.1 Stoker Fired Boiler

This type of boiler was capable of burning a wide range of coals, from bituminous to lignite, as well as by-products of waste fuels. In stoker-fired boilers coal is pushed, dropped, or thrown onto a grate to form a fuel-bed. Stokers are divided into two general classes: overfeed, in which fuel is fed from above, and underfeed, wherein fuel is fed from below. However, over the years, this type of boiler has become less popular due to more efficient technological advancement.

4.1.11.2 Natural Circulation Boiler

Boilers in which the motion of the working fluid in the evaporator is caused by the thermosiphon effect on heating the tubes are called "natural circulation boilers." Circulation of water in natural circulation boilers depends on the difference between the density of a descending body of relatively cool and steam-free water and an ascending mixture of hot water and steam. All natural circulation boilers are drum-type boilers.

4.1.11.3 Forced Circulation Boiler

The density difference between the saturated liquid and saturated vapor starts diminishing at 18 MPa or higher fluid pressure, thus it is difficult to maintain natural circulation of fluid flow in boiler tubes. In such cases, fluid flow is ensured with the help of forced/assisted circulation using pumps. The forced/assisted circulation principle applies equally in both supercritical and subcritical ranges.

4.1.11.4 Fire-Tube Boiler

A fire-tube boiler is a type of boiler in which hot gases pass from a fire through one or more tubes running through a sealed container of water. The heat of the gases is transferred through the walls of the tubes by thermal conduction, heating the water and ultimately creating steam. The tank may be installed either horizontally or vertically. Today, they are used extensively in the stationary engineering field, typically for low-pressure steam use, such as for heating a building. This type of boiler was used on virtually all steam locomotives in the horizontal "locomotive" form. This has a cylindrical barrel containing the fire tubes, but also has an extension at one end to house the "firebox." The horizontal fire-tube boiler is also typical of marine applications, using the Scotch boiler; thus, these boilers are commonly referred to as "scotch-marine" or "marine"-type boilers. A fire-tube boiler is shown in Figure 4.1 with other types of boilers.

4.1.11.5 Water-Tube Boiler

In a water-tube boiler water circulates in tubes heated externally by the hot flue gas. Fuel is burned inside the furnace, creating hot gas that heats up the water in the steam generating tubes. The water-tube boiler can be built with boiler shell, burner, mud drum or mud ring, furnace, safety valve, strainer, sight glass, feed check valve, steam stop valve, etc. The types of water-tube boilers include simple vertical boiler, stirling boiler, and Babcock and Wilcox boilers. While fire-tube boilers are less costly and tend to be easy to operate, they have relatively small steam capacities. Water-tube boilers are capable of producing high-pressure steam and are often used in applications that require higher capacities. A water-tube boiler is shown in Figure 4.1 with other types of boilers.

4.1.11.6 Fluidized-Bed Boiler

A bed of solid particles is said to be **fluidized** when the **pressurized fluid** (liquid or gas) is passed through the medium and causes the solid particles to behave like a fluid under certain conditions [7]. Fluidized-bed combustors (FBCs) are developed for solid fuel combustion. FBCs burn solid fuel suspended in a bed of inert material at the base of the furnace. Combustion air is injected from the bottom of the combustor to keep the bed in a floating or "fluidized" state. Mixing in the fluidized bed provides efficient heat transfer that allows a more compact design than in conventional water-tube designs. The efficient mixing allows improved combustion at a lower temperature, which reduces the formation of nitrogen oxides (NOx). The excellent mixing of the bed makes FBCs well suited to burn solid refuse, wood waste, waste coals, and other nonstandard fuels. Use of limestone as the bed material helps remove sulfur dioxide (SO_2) from the flue gas.

4.1.11.7 Package Boilers

At the lower range of industrial boilers, 16–45 t/h, package boiler is the normal type of water-tube boiler supplied. These have the distinct advantages that they are cheaper than the conventional field-erected boiler, can be pretested, checked and all controls can be set before being dispatched; coupling-up time at site if well arranged can be 1 or 2 weeks, thus achieving earlier commissioning. It comes as a complete package. Once delivered to the site, it requires only the steam, water pipe work, fuel supply and electrical connections to be made for it to become operational. Packaged boilers are generally of shell type with fire-tube design so as to achieve high heat transfer rates by both radiation and convection.

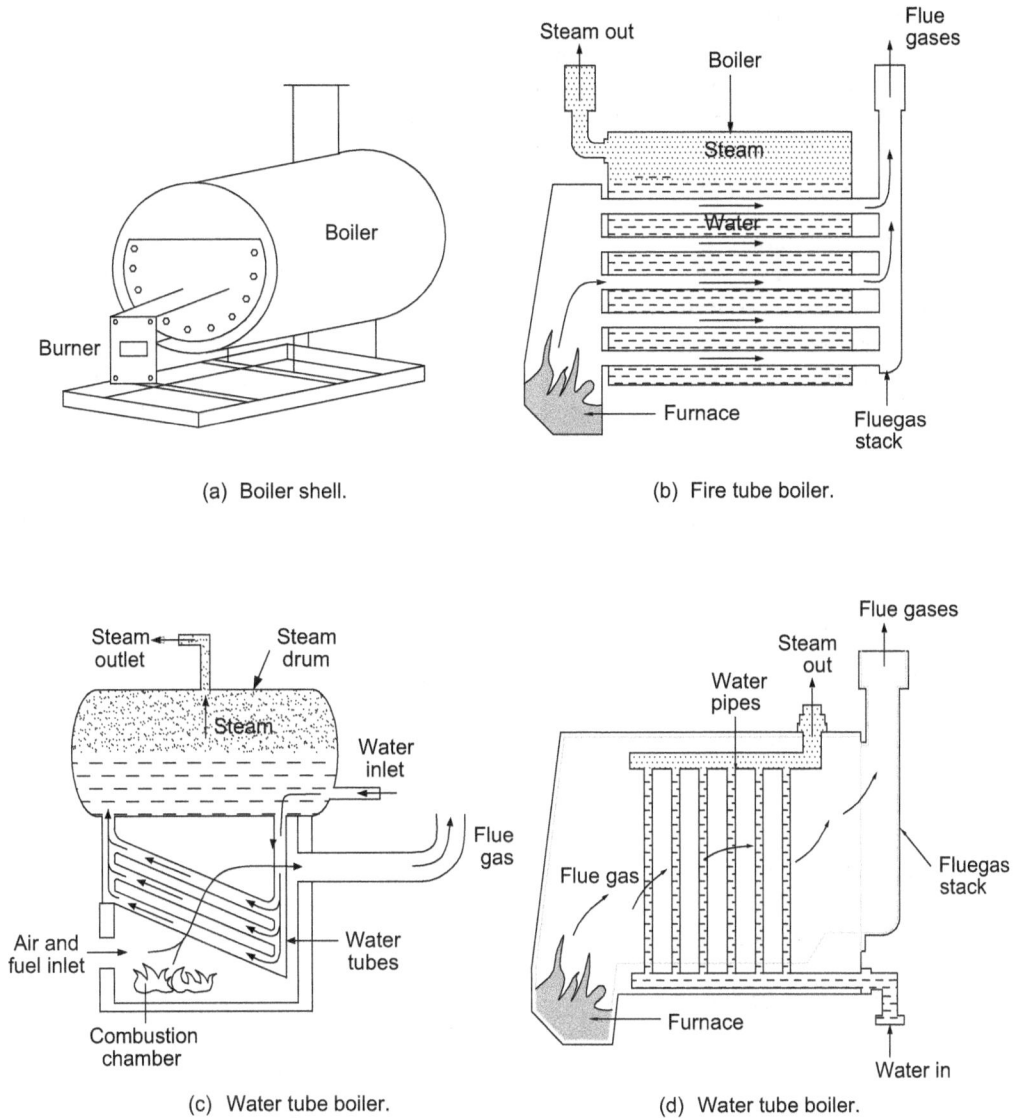

(a) Boiler shell.

(b) Fire tube boiler.

(c) Water tube boiler.

(d) Water tube boiler.

FIGURE 4.1 Boiler (a) General-Shell boiler, (b) Fire tube boiler and (c and d) Water tube boiler.

4.1.11.8 Waste Heat Recovery Boiler

Waste heat recovery boiler (WHRG) is a system that recovers various kinds of waste heat generated from the production process of steel, nonferrous metal, chemical, cement, etc., and those equipment of industrial furnaces, refuse incinerators, industrial waste incinerators, and convert such recovered heat into useful and effective thermal energy [8]. WHRB is contributing to industries in terms of improvement of thermal efficiency, energy saving, environmental protection, etc. Conditions of the origin of waste heat depend upon each kind of facility, which discharges waste heat. Conditions like gas temperature, pressure, corrosiveness, dust content, etc., vary depending on the conditions of each case. It is also common for these boilers to receive additional heat by supplementary firing of other by-products or by burning gas or light oil to raise the boiler inlet gas temperature to a level commensurate with the desired steam conditions.

4.1.11.9   Cyclone-Fired Boilers

Cyclone-fired boilers, for coals that produce low melting ash, use crushed rather than pulverized coal. Combustion occurs in horizontal cyclone furnaces attached to the boiler firebox. Because these furnaces are small, they have high heat release rates and resulting high peak flame temperatures, which melt coal ash to form slag. This slag collects in a slag tank beneath the furnace. High temperatures also result in high NOx emissions. Cyclones are wet-bottom boilers, so called because molten ash is collected at the bottom of the furnace. A small number of pulverized coal-fired boilers also have wet-bottom furnaces.

4.1.11.10   Oil- and Gas-Fired Boilers

Oil- and gas-fired boilers are similar in design to pulverized coal boilers, but are smaller because these fuels burn more easily than coal. Most oil- and gas-burning boilers are wall-fired or tangentially fired, although a small number of cyclone boilers have been converted from coal service.

4.1.11.11   Downshot Boiler

For combustion of anthracite coals with low volatile content, a downshot boiler is used, with burners firing down into the furnace in order to maximize residence time and furnace temperature to increase burnout [9].

4.1.11.12   Circulating Fluidized-Bed Boiler

Circulating fluidized-bed (CFB) boilers are an eco-friendly power plant design where the air and fuel are injected simultaneously and combustion occurs during a recirculating cycle, significantly reducing emissions such as nitrogen and sulfur oxides. This technology is unique in allowing complete combustion of low-grade and anthracitic coals [9].

4.1.11.13   OxyFuel Boiler

OxyFuel combustion is a technology to maximize the recovery of carbon dioxide by burning coal in oxygen instead of air. Comparing with other CO_2 recovery technologies, it is the only technology to capture total CO_2 in the flue gas by physically condensing the steam [9].

4.1.11.14   Industrial Boilers

Industrial boilers are utilized in many different industries for a wide variety of purposes and the main product is process steam. Industrial boiler operation can vary significantly between seasons, daily, and even hourly depending on the steam demand. Industrial boilers can be designed for the above fuels as well as coarsely crushed coal for stoker firing and a wide range of biomass or by-product fuels. Industrial boilers have preferably multifuel burning capacity. In the past, industrial boilers were designed to burn just a single fuel. With multifuel capability wider flexibility with fuel properties is achieved.

4.1.11.15   Utility Boiler

A utility boiler generates steam for the sole purpose of powering turbines to produce electricity. Utility boilers are water-tube boilers; combustion takes place in an enclosed furnace and heat is transferred from the furnace to water in tubes. In the furnace itself, heat is transferred by radiation from the combustion gases to tubes lining the walls. As gases cool and leave the furnace, the primary heat transfer mechanism becomes convection. Utility boilers can be designed for subcritical or supercritical pressure operation. At subcritical pressures, the flow circuits must be designed to accommodate the two-phase steam-water flow and boiling phenomena. At supercritical pressures, the water acts as a single-phase fluid with a continuous increase in temperature as it passes through the boiler.

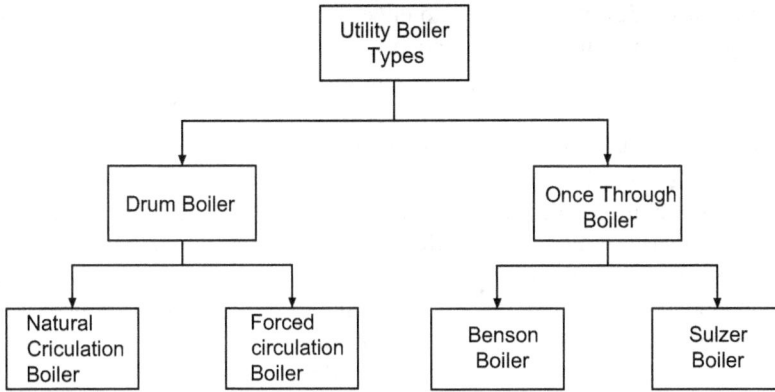

FIGURE 4.2 Classification of utility boilers.

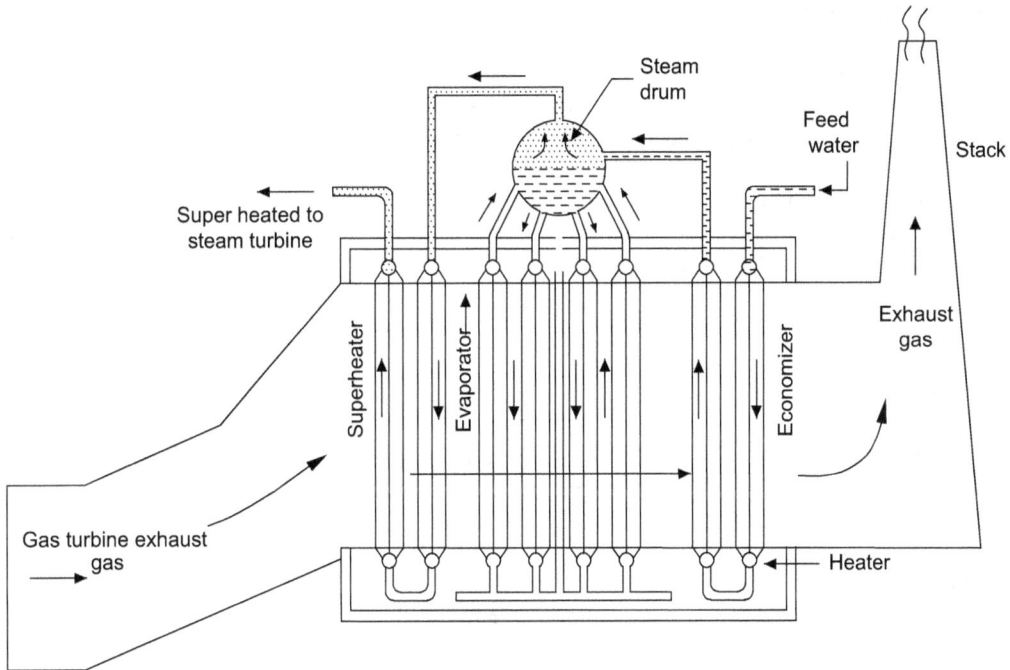

FIGURE 4.3 A Heat recovery steam generator (HRSG) (schematic)

Types of Utility Boiler

Generally there are two types of utility boilers—drum boilers and once through boilers.

a. Drum boiler

In a drum boiler the drum acts as a reservoir for the working fluid. The drum is connected to cold downcomers and hot riser tubes through which circulation of water takes place. A drum-type boiler can be either the natural circulation type or forced/assisted circulation type. Drum-type boilers are essentially subcritical boilers; they operate below the critical pressure of the working fluid. The economic design pressure limit of fluid in a drum-type boiler is around 18 MPa. In the steam drum, saturated steam is separated from a recalculating steam/water mixture. The recalculation flow is from the steam drum via down comer tubes to either the mud drum or the water wall header, and from there through riser tubes back to the steam drum.

b. Once-through boiler

Once-through (OT) boiler does not have a steam drum. Simply put, a once-through boiler is merely a length of tube through which water is pumped, heat is applied, and the water is converted into steam. In actual practice, the single tube is replaced by numerous small tubes arranged to provide effective heat transfer surface. Feedwater in this type of boiler enters the bottom of each tube and discharges as steam from the top of the tube. The working fluid passes through each tube only once and water is continuously converted to steam. As a result there is no distinct boundary between the economizing, evaporating, and superheating zones. An important reference source on OT boiler is Benzon boiler. Types of once-through boiler includes Benzon boiler and Sulzer boiler. Figure 4.2 shows classification of utility boilers.

4.1.11.16   HRSG

Heat recovery steam generators (HRSGs) are used in power generation to recover heat from hot flue gases (500–600°C), usually originating from a gas turbine or diesel engine. A HRSG used in a combined cycle power plant (CCPP), where hot exhaust gas from a gas turbine is fed to the HRSG to generate steam is shown in Figure 4.3.The HRSG consists of the same heat transfer surfaces as other boilers, except for the furnace. Since no fuel is combusted in a HRSG, the HRSG have convention-based evaporator surfaces, where water evaporates into steam. A HRSG can have a horizontal or vertical layout, depending on the available space. Essentially, the HRSG is composed of several heat exchangers making it a large heat exchanger. The heat exchanger tubes are set in **different modules or sections** known as follows [10]:

  ii. Economizer.
  ii. Evaporator.
  iii. Superheater.

The HSRG system is integrated by the above three units whose functions are economizer, where cold water exits as saturated liquid; evaporator, where saturated liquid is converted to steam, and superheater, where saturated steam is dried by overheating it beyond its saturation point.

Working of HRSG
The water is first passed through the economizer from where it passes through evaporator and then to superheater. The hot gases first come into contact with the superheater and then with evaporator and then with economizer. The temperature of the superheater section is highest in the system because it is the closest to the input of hot gases. The temperature of the economizer is the lowest among the three major components because it is farthest from the hot gases' input.

4.2   COAL-BASED THERMAL POWER PLANT

4.2.1   Steam Generators of Power Plants

Steam generators, or boilers as they are often called, form an essential part of any power plant or cogeneration plant. The steam-based Rankine cycle has been adopted for power generation for centuries. Though steam parameters, such as pressure and temperature, have been steadily increasing from subcritical (SC) to supercritical, ultrasupercritical (USC) and presently advanced ultrasupercritical (A-USC) during the last several decades, the function of the boiler remains the same, namely, to generate steam at the desired conditions efficiently and with low operating costs.

4.2.2 Definition of Subcritical, Supercritical, and Ultrasupercritical States

Subcritical steam boilers feature a drum, which is a reservoir providing working fluid circulation that allows the separation of water from steam. Drum-type boilers operate only at subcritical steam pressures and the pressure of the boiler is based on the firing rate and steam flow. In a subcritical unit, it is easy to separate water from steam. Above the supercritical pressure, steam and water cannot be separated. This requires a different design and operating procedures than subcritical boilers.

The term "supercritical" refers to conditions above the critical point (22.1MPa and 374.1°C) where distinct liquid and gas phases do not exist as it becomes a single working fluid. Another name for supercritical units normally used is "once-through" since there is no need for a steam drum. PC plants can have different operating conditions depending on the design and materials used for the boiler. Higher temperature and pressure operating conditions can increase plant efficiency and reduce emissions (on a per energy unit basis). The pressure and temperature ranges for subcritical and supercritical and ultrasupercritical steam generation systems are given below [11]:

i. Subcritical—operating conditions are generally around 2400 psi and 1000°F. Plant efficiencies are in the range of 37% (LHV).
ii. Supercritical—operating conditions are generally around 3550 psi and 1050°F. Plant efficiencies are in the range of 40% (LHV).
iii. Ultrasupercritical—operating conditions are generally around 4350 psi and 1100°F. Plant efficiencies are in the range of 42% (LHV).

Up to an operating pressure of around 193.74 Kg/cm^2 in the evaporator part of the boiler, the cycle is subcritical. In this case a drum-type boiler is used because the steam needs to be separated from water in the drum of the boiler before it is superheated and led into the turbine. Above an operating pressure of 224.337 Kg/cm^2 in the evaporator part of the boiler, the cycle is supercritical.

Definition of supercritical and ultrasupercritical boiler pressure and temperature profiles differs from one country to another. Thus the usage of the term ultrasupercritical with regards to its pressure and temperature range varies somewhat. Compared to subcritical unit, supercritical and ultrasupercritical technologies offer improved efficiency, proven and reliable baseload operations, and flexibility.

4.2.3 Supercritical Technology

Supercritical is a thermodynamic expression describing the state of the substance where there is no clear distinction between the critical and gaseous phase (i.e., they are a homogenous fluid). Water reaches this phase at a pressure above around 220 bar (225.56 Kg/cm^2) and temperature 374.15 K. In addition, there is no surface tension in a supercritical fluid as there is no boundary between liquid and gas phase. By changing the pressure and temperature of the fluid, the properties can be "tuned" to be more liquid- or more gas-like. Once-through boilers are therefore used in supercritical cycles. Once-through boilers are better suited to frequent load variations than drum-type boilers, since the drum is a component with a high wall thickness, requiring controlled heating [2].

4.2.4 Steam Requirements

Modern turbines are designed for nominal conditions of 2400 psi/1000°F/1000°F (16.5 MPa/538°C/538°C), i.e., main steam pressure, main steam temperature and reheat temperature) for subcritical systems and 3500 psi/1050°F/1100°F (24.1 MPa/566°C/593°C) for supercritical systems.

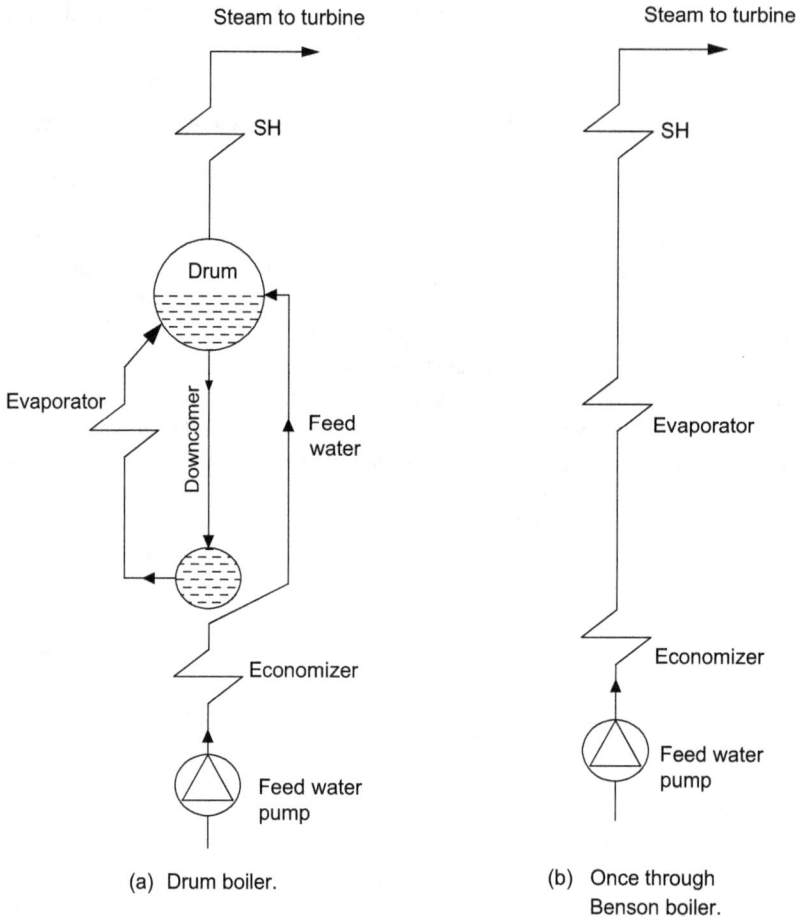

(a) Drum boiler.   (b) Once through
                                              Benson boiler.

FIGURE 4.4 Steam generation principle (a) Drum boiler and (b) Benson once through boiler.

Note: SH is superheater.

4.2.5 UTILITY BOILER

The utility boilers are large-capacity steam generators used purely for steam generation. Generally there are two types of utility boilers—drum boilers and once-through boilers.

4.2.5.1 Drum Boiler

In a drum boiler, The evaporation zones are comprised of the boiler drum, downcomers, and riser tubes or simply risers. The feedwater from the economizer enters the boiler drum, flows down through the downcomers, passes through the pipes to the water-wall bottom header and rises through the riser tubes. In the riser tubes, the feedwater absorbs the heat, part of which is converted to steam and reenters the boiler drum as a steam-water mixture. From this mixture, steam is separated in the boiler drum and purified.

Dry saturated steam from the boiler drum is lead out through the saturated steam pipes to the superheaters. The process of separation and purification of steam in the boiler drum is accomplished by drum internals, e.g., cyclones, baffles, etc., chemical and feedwater admission piping, blowdown lines, etc. The process includes three steps: separation, steam washing, and scrubbing [2].

The steam generation principle in a drum boiler and Benson once-through boilers are shown in Figure 4.4 and in Figure 4.5.

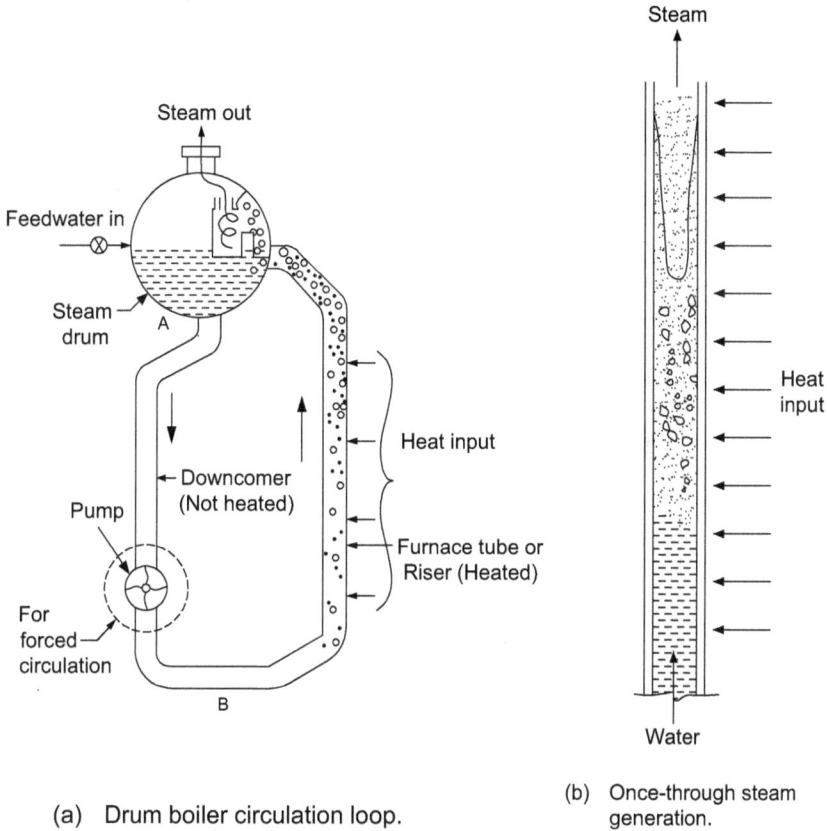

(a) Drum boiler circulation loop.

(b) Once-through steam generation.

FIGURE 4.5 Basics of steam generation in a boiler (a) Drum boiler and (b) Benson once through boiler.

Note: ST-steam turbine, SCR-selective catalytic reactor, ESP-elctrostatic precipitator, FGD- flue gas desulfuriser.

4.2.5.2 Once-Through Boiler

a. BENSON Once-Through Boiler

Benson once-through boiler has no drum. This means that during the once-through operation the fluid passes through the economizer, evaporator, and superheater without any recirculation. The feedwater, pumped into the unit as a subcooled liquid, passes sequentially through all the pressure parts heating surfaces where it is converted to superheated steam as it absorbs heat; it leaves as steam at the desired temperature. The working fluid passes through each tube only once and water is continuously converted to steam. As a result there is no distinct boundary between the economizing, evaporating, and superheating zones (SH) as shown in Figures 4.4 and 4.5. There is no recirculation of water within the unit and, for this reason, a conventional drum is not required to separate water from steam. These boilers can be operated either at subcritical or at supercritical pressures.

b. Sulzer Boiler

In contrast to the Benson boiler, the Sulzer Boiler operates by means of the separator vessel during low load operation as well as during the once-through operation. Due to the separator vessel, the evaporation endpoint is fixed locally. The water level in the separator vessel represents the balance between the feedwater and the fuel mass flow. Depending on the water level of the separator vessel, the feedwater mass flow has to be changed.

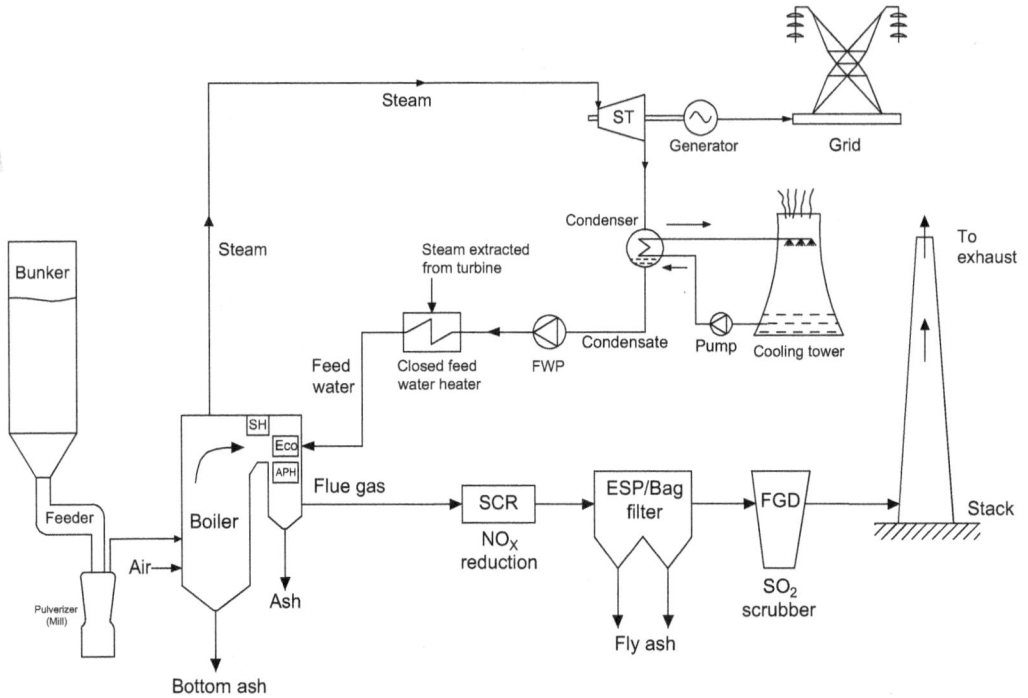

FIGURE 4.6 General arrangement of a pulveried coal fired boiler based power plant.

Note: ST-steam turbine, SCR-selective catalytic reactor, ESP-elctrostatic precipitator, FGD- flue gas desulfuriser.

4.2.6 PRINCIPLE OF POWER GENERATION AT A COAL-BASED POWER PLANT

In a fossil fueled power plant, heat, from the burning of coal, oil, or natural gas, converts water into steam. The main steam from the boiler is expanded through a steam turbine to generate electricity. After expansion through the high-pressure turbine stage, steam is typically sent back to the boiler to be reheated before expanding through the intermediate- and low-pressure turbine stages. Reheating, single or double, increases the cycle efficiency by raising the mean temperature of heat addition to the cycle. The exhaust steam from the turbine is then condensed in the condenser and the condensate is there after being pumped to the boiler where it again receives heat and the cycle is repeated.

The basic theoretical working cycle is of a steam power plant is "the Rankine cycle." The modern steam power plant uses " the modified Rankine cycle," which includes reheating, superheating, and regenerative feedwater heating. Figure 4.6 shows general arrangement of a pulverized coal (PC) fired boiler-based power plant and Figure 4.7 shows a coal-based thermal power plant, including the feedwater system. References [11–15] detail the working of steam-based thermal power plants.

4.2.7 THERMAL POWER PLANT

In a conventional steam power plant, a boiler consists of a furnace in which fuel is burned, surfaces to transmit heat from the combustion products to the water, and a space where steam can form and collect. Most coal fired power station boilers use pulverized coal. In a typical coal fired thermal power plant, coal is conveyed from an external stack and ground to a very fine powder in the pulverized coal mill. There it is mixed with preheated air driven by the forced draught fan. The hot air-fuel mixture is forced at high pressure into the boiler where it rapidly ignites. The coal is pulverized to a fine

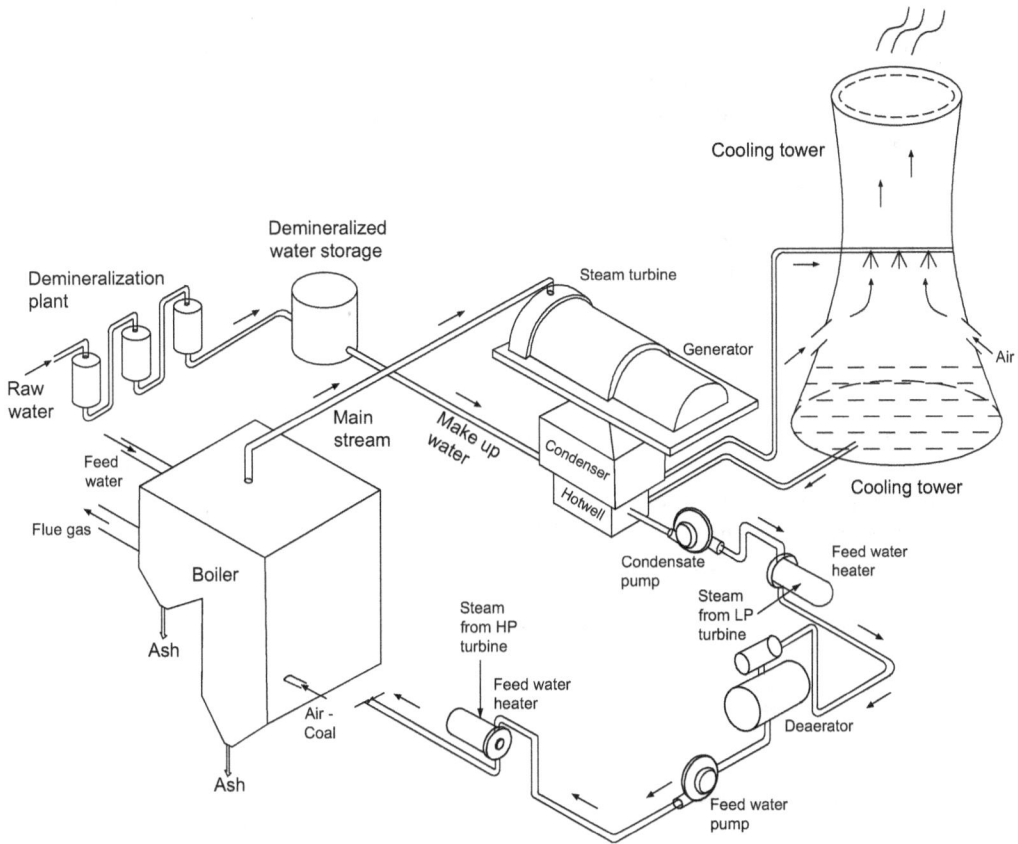

FIGURE 4.7   Coal based thermal power plant with feedwater system layout.

powder, so that less than 2% is +300 μm and 70–75 % is below 75 microns, for a bituminous coal. This system has many advantages such as the ability to fire varying quality of coal, quick responses to changes in load, use of high preheat air temperatures, etc. One of the most popular systems for firing pulverized coal is the tangential.

4.2.7.1 Fuels Flow

Coal Bunker
These are in process storage silos used for storing crushed coal from the coal handling system. Generally, these are made up of welded steel plates. These are located on top of the mills so as to aid in gravity feeding of coal.

Coal Feeders
Each mill is provided with a drag link chain/rotary/gravimetric feeder to transport raw coal from the bunker to the inlet chute, leading to mill at a desired rate.

Mills
Pulverized fuel firing is a method whereby the crushed coal, generally reduced to fineness such that 70–80% passes through a 200 mesh sieve, is carried forward by air through pipes directly to burners or storage bins from where it is passed to burners. When discharged into a combustion chamber, the mixture of air and coal ignites and burns in suspension.

Furnace

A boiler furnace is that space under or adjacent to a boiler in which fuel is burned and from which the combustion products pass into the boiler proper. It provides a chamber in which the combustion reaction can be isolated and confined so that the reaction remains a controlled force. In addition it provides support or enclosure for the firing equipment.

Burners

There are a total of 24 pulverized coal burners for corner fired C.E. type boilers and 12 oil burners provided each in between two pulverized fuel burner. The pulverized coal burners are arranged in such a way that six mills supply the coal to the burners at four corners, of the furnace. All the nozzles of the burners are interlinked and can be tilted as a single unit from to a convenient position. The oil burners are fed with heavy fuel oil until boiler load reaches about 25%.

4.2.7.2   Waste Heat Recovery from Flue Gas

In a modern steam generator, various components are arranged in the flow path of products of combustion so that they efficiently absorb heat from the products of combustion and provide steam at the rated temperature, pressure, and capacity. These components include the boiler surface/furnace water wall, superheater, reheater, economizer, and air heater.

4.2.7.3   Working of a Coal Fired Power Station

The basic working of generating steam from coal in a thermal power station is described below. The illustration given in Figure 4.6 shows the general arrangement of a boiler burning pulverized coal.

4.2.7.4   System Arrangement and Key Components

The major subsystems of a modern coal fired power generating facility include the following:

    i.   fuel (coal) receiving and preparation,
    ii.   pulverizing of coal,
    iii. fuel-air supply to the furnace and combustion,
    iii. steam generation,
    iv. superheating and reheating of steam,
    iv. flue gas passage through environmental protection equipments also known as air quality control system (AQCS),
    v.   steam flow to steam turbines and control,
    vi. electric generator,
    vii. condenser cooling-water system including cooling tower,
    viii. condensate return system including deareator, feed water heaters, etc.

The Heat Transfer Process

The heat transfer process encompasses the following circuits or systems:

    a.   Water and steam circuit.
    b.   Fuel-air supply and furnace.
    c.   Fuel-burning system.
    d.   Draft system.
    e.   Heat recovery system.

Heat produced due to burning of coal is utilized in converting water contained in boiler drum or OT boiler into steam at suitable pressure and temperature. The steam generated is passed through the superheater. Superheated steam then flows through the turbine. The steam is exhausted from the high-pressure turbine, and the steam with reduced pressure and temperature is returned to the boiler reheater. The reheated steam is then passed to the intermediate pressure turbine, and then to the low-pressure turbine set. The outlet of the low-pressure turbine (LPT) is condensed in a surface condenser. This condensed water is collected in the hot well and is again sent to the boiler in a closed cycle. The rotational energy imparted to the turbine by high-pressure steam is converted to electrical energy by the generator.

4.2.7.5 Fuel Flow

Raw coal is discharged from the feeders to the pulverizers. Pulverized coal is transported by the primary air to the burners through a system of pressurized fuel and air piping. The burners are located on the furnace walls with opposed firing (burners on the front and rear walls).

4.2.7.6 Coal Pulverising Mill

In a typical coal-fired thermal power plant, coal is conveyed from an external stack and ground to a very fine powder in the pulverized coal mill. The mill usually consists of a round metal table on which large steel rollers or balls are positioned. The table revolves, forcing the coal under the rollers or balls, which crush it. There it is mixed with preheated air driven by the forced draught fan. The hot air-fuel mixture is forced at high pressure into the boiler where it rapidly ignites.

4.2.7.7 Feedwater Flow

Water of a high purity flows vertically up the tube-lined walls of the boiler, where it turns into steam, and is passed to the boiler drum, where steam is separated from any remaining water.

4.2.7.8 Furnace

In the furnace of a steam generator combustion of fuel takes place with atmospheric air to release heat. Hence, it is essential to ensure complete combustion of fuel under all operating conditions to harness the maximum heat potential of the fuel under combustion.

Air taken from the atmosphere is first passed through the air preheater, where it is heated by flue gases. The hot air then passes through the furnace. The flue gases after passing over boiler and superheater tubes, flow through economizer, air preheater, the dust collector and other air quality control devices and finally they are exhausted to the atmosphere through the chimney.

4.2.7.9 Draft system

For sustaining a healthy combustion process in a furnace it needs to receive a steady flow of air and at the same time remove products of combustion from the furnace without any interruption. When only a chimney is used for the release of products of combustion, the system is called a natural draft system. When the chimney is augmented with forced draft (FD) fans and/or induced draft (ID) fans the system is called a mechanical draft system. Small steam generators use natural draft, while large steam generators need mechanical draft to move large volumes of air and gas against flow resistance.

4.2.7.10 Fuel-Burning System

The primary function of a fuel-burning system is to provide controlled and efficient conversion of chemical energy of the fossil fuel into heat energy. The heat energy thus generated is transferred to the heat-absorbing surfaces of the steam generator. The main equipment for firing fossil fuel in a furnace is the burner. Generally in modern boilers a tangential firing system is used, i.e., the coal nozzles/guns form tangent to a circle. The temperature in fire ball is of the order of 1300°C. The boiler is a water-tube boiler hanging from the top. Water is converted to steam in the boiler and steam is separated from water in the boiler drum.

4.2.7.11   Superheaters and Reheaters

Steam produced from a boiler without a superheater will either be dry saturated or more likely, wet. In works where steam is transmitted over long distances, the inevitable heat loss from pipe surfaces causes the steam to become even wetter at the point of use unless a superheater is fitted to the boiler plant. The function of the superheaters and reheaters is to raise the total heat content of the steam (superheat the steam). The steam superheater is used to produce superheated steam or to convert the wet steam to dry steam, generated by a boiler. Radiative superheaters are used as the first heat transfer surfaces after the furnace. Often a platen arrangement is preferred to create a fouling-resistant construction. Superheaters are the hottest heat transfer surfaces in the boiler, so they are often built with temperature-resistant alloys. As these superheaters operate at flue gas temperature ranges of 800–1200°C, they often need to be built with high-temperature corrosion resistant materials. Superheater material should be capable of withstanding high temperature and corrosion. Thermodynamically there is no limit on the maximum temperature for superheating. Thus the maximum temperature depends on the metallurgy of the tubes.

Types of Superheater
On the basis of the mode of heat transfer, the types are (i) radiant superheater and (ii) convective superheater.

Reheater is a part of the boiler, which reheats steam output from the first level of the steam turbine. Reheated steam will again absorb the heat energy from the boiler to be used in the next level steam turbine. Reheater is one way to improve the thermal efficiency of the Rankine cycle

Superheaters and reheaters are specially designed to increase the temperature of the saturated steam and to help control the steam outlet temperature. They are simple single-phase heat exchangers with steam flowing inside and the flue gas passing outside, generally in the cross flow. Superheaters and reheaters are often divided into multiple sections to help control steam temperature and optimize heat recovery, and their heating surfaces can be arranged either horizontally or vertically.

There are two fundamental advantages of providing superheaters and reheaters: first, gain in efficiency of the thermodynamic cycle and second, enhancement of life in the last stage of the turbine by reduction of moisture content of the flowing steam.

Superheater (SH) and reheater (RH) tubing are required to have high mean creep strength (> 100 MPa or 14.5 ksi for 100 000 h), good thermal fatigue strength, good weldability, resistance to fire-side corrosion and erosion, and resistance to steam-side oxidation [16].

4.2.7.12   Water and Steam Flow

The condensed water from the hot well of the condenser is passed by a feed pump through a deaerator, and prewarmed, first in a feed heater powered by steam drawn from the high-pressure set, and then in the economizer, before being returned to the boiler drum. The cooling water from the condenser is cooled due to evaporation in a cooling tower, creating a highly visible plume of water vapor, before being pumped back to the condenser in cooling-water cycle. A basic feedwater–steam flowchart is given below:

feedwater from economizer → boiler drum → low-temperature superheater → radiant superheater → final superheater → high-pressure turbine → reheater → intermediate pressure turbine → low-pressure turbine → steam condenses in condenser-condensate-feedwater heater-economizer.

4.2.7.13   Flue Gas System

The flue gases after passing over superheater tubes, reheater, economizer, air preheater, flow through the dust collector and finally they are exhausted to the atmosphere through the chimney. The electrostatic precipitator consists of metal plates, which are electrically charged. Dust (fly ash) contained in the flue gases is attracted on to these plates, so that they do not pass up the stack to pollute the

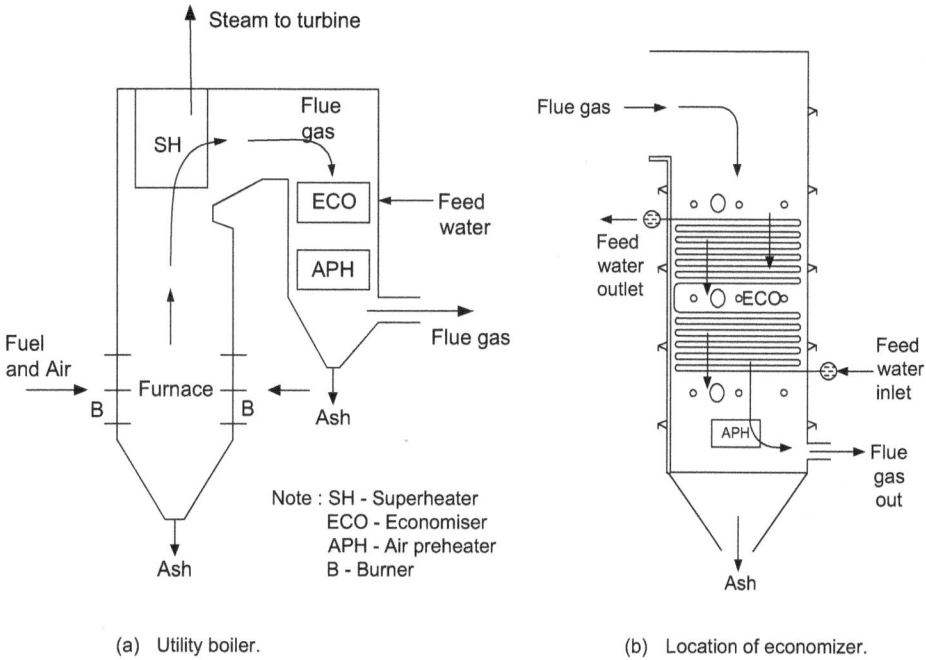

(a) Utility boiler.   (b) Location of economizer.

FIGURE 4.8   Flue gas system path with APH and economizer (a). Utility boiler and (b) Location of economizer.

atmosphere. Finally the flue gases are drawn by the induced draught fan into the main flue and to the stack. Flue gas system path is shown in Figure 4.8.

4.2.7.14   Flues and ducts

Flues are used to interconnect boiler outlets, economizers, air heaters, and stack. Ducts are used to interconnect forced-draft fans, air heaters, and wind boxes or combustion air plenums. Flues and ducts are usually made of steel. Expansion joints are provided to allow for expansion and contraction. All flues or ducts carrying heated air or gases should be insulated to minimize radiation losses.

4.2.8   AIR HEATERS

Air heater or air preheaters (APHs) are basically heat exchangers installed in the exit flue gas duct of the boiler. The purpose of the air preheater is to recover the heat from the boiler flue gas to heat the incoming air for combustion as it increases the thermal efficiency of the boiler by reducing the fuel consumption. Figure 4.9 shows an air preheter installed in flue gas exhaust path of a boiler. There are two types of air preheaters for use in steam generators in thermal power stations: one is a tubular type built into the boiler flue gas ducting, and the other is a regenerative air preheater (refer to Chapter 6, Heat Exchangers: Classification, Selection, and Thermal Design).

4.2.8.1   Construction Details of Air Heater

The air heater is made as multipass design in order to recover the heat with less heat transfer surface [17].

   i.   Tube size

The tube size is usually 63.5 mm OD × 2.01 mm thick. Corten steel tubes are used for improving the tube life against cold end corrosion. Tube size smaller than 63.5 mm can pose problems, such as choking with fuel ash. Smaller tubes also increase the pressure drop across the air heater on the tube side.

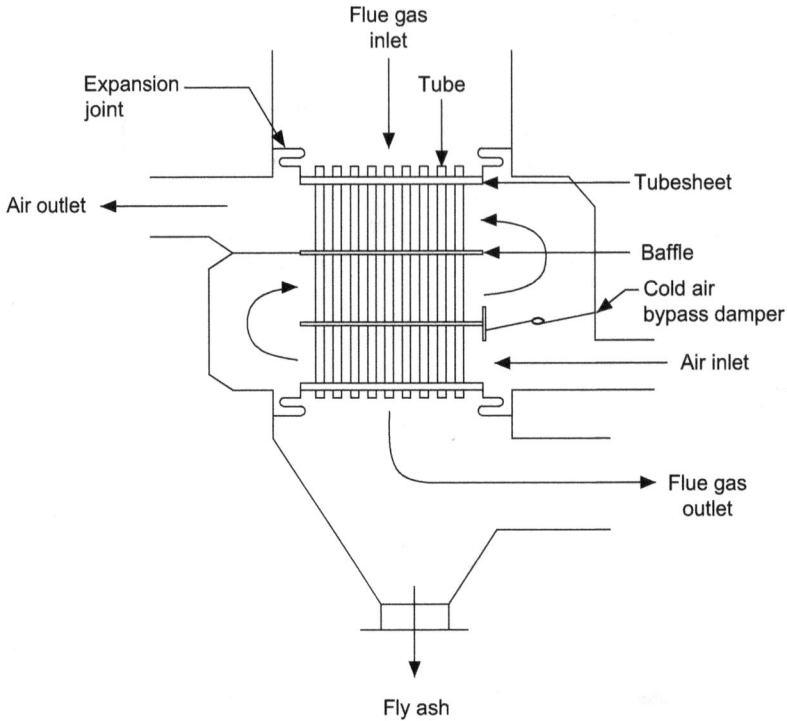

FIGURE 4.9   An air preheter installed in flue gas exhaust path of a boiler.

   ii.   Tube length

The tube length is usually limited depending on the access available below or above the air heater for tube replacement. Long tubes are certainly vulnerable for vibration. Also the longer tube lengths need expansion joint at air heater casing to account for differential expansion between the casing and the tubes.

   iii.   Tube pitch

Tube pitch is so selected to have a minimum ligament of 15 mm.

   iv.   Tube length

The tube length per pass is usually limited to 3 m. Proper deflectors or guide vanes have to be provided to ensure the flow across the air heater is properly distributed.

   v.   Tube layout

Tube can be arranged in line or staggered. However when the gas is dust laden and when the passes outside the tubes, the tubes have to be in line.

   vi.   Tube fitup

The tubes are expanded into the tubesheet. Some tubes are welded in addition to expansion for the structural stability of the air heater. The expansion of the tubes also facilitates easy removal of tubes by collapsing the tube ends.

vii. Air-side velocity

The air-side velocity is usually around 5 to 6 m/s in order to have an optimum draft loss.

viii. Gas-side velocity

The gas-side velocity is limited to 18 m/sec. Higher gas-side velocities could lead to the erosion of tubes.

4.2.8.2 Cold Corners in Air Preheaters

Of all the equipment in a heater, the ones most susceptible to acid dew-point corrosion are air preheaters (APHs). These heat exchangers transfer the heat from flue gases to the combustion air before it enters the burner. When fuels containing sulfurous compounds combust, the resulting flue gases can condense into sulfuric acid (H_2SO_4), sulfurous acid (H_2SO_3), and other aggressive substances. These condensed acids will corrode the tubes and other extended surfaces in convection sections, flue ducts, stacks, air preheater, etc. The amount of sulfur-bearing compounds, mostly in the form of hydrogen sulfide (H_2S) in the fuel, is directly correlated with the concentration of the acid droplets and, therefore, with the degree of corrosion. The concept of cold end corrosion in an air preheater is shown in Figure 4.10.

4.2.8.3 Control of Cold End Corrosion

Basically there are few approaches used by engineers to combat the problem of cold end corrosion [17]:

i. By-passing a portion or all of the cold air to increase the metal temperature.
iii. Recirculate a portion of the hot air back to the cold end, so that the air inlet temperature is more.
iv. Use a parallel flow air heater to improve the metal temperature.
v. Use a steam coil air preheater to preheat the air before admitting into the air heater.
vi. Use tubes made of corten steel (i.e., copper bearing) to minimize corrosion.

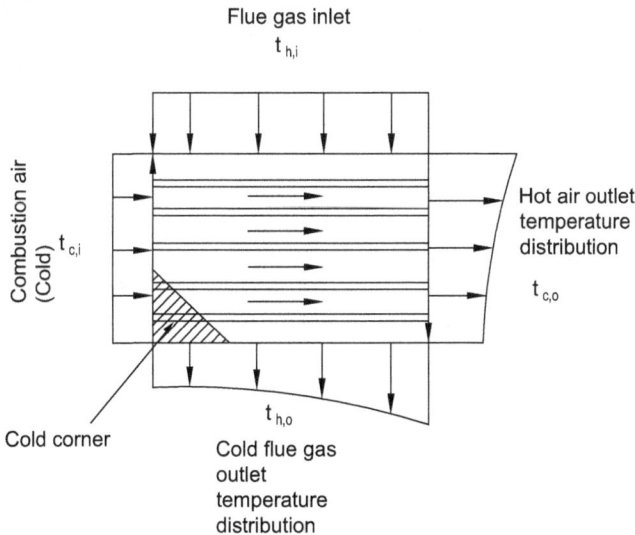

FIGURE 4.10 Concept of cold end corrosion in an air preheater.

4.2.8.4 Dew Point Corrosion

The combustion of most fossil fuels, natural gas being one exception, produces flue gases that contain sulfur dioxide, sulfur trioxide, and water vapor. At some temperature, these gases condense to form sulfurous and sulfuric acids. The exact dew point depends on the concentration of these gaseous species, but it is around 300°F. Thus surfaces cooler than this temperature are likely locations for dew point corrosion. Any point along the flue gas path, from combustion in the furnace to the top of the chimney, is a possible site. Any flue gas leak can also cause this type of corrosion. Dew point corrosion is exacerbated in coal fired boilers by the presence of fly ash [18].

4.2.9 ECONOMIZERS

Economizers are basically tubular heat exchangers used to preheat boiler feedwater before it enters the steam drum (recirculating units) or furnace surfaces (once-through units) [15,16]. Figure 4.8 shows an economizer location on a coal fired boiler flue gas path.

Economizers reduce the potential of thermal shock and strong water temperature fluctuations as the feedwater enters the drum or waterwalls. A few types of economizers are shown in Figure 4.11.

4.2.9.1 Economizer Surface Types Based on Ref. [19].

   a. Bare Tube

The most common and reliable economizer design is the bare tube, in-line, crossflow type. When coal is fired, the fly ash creates a high fouling and erosive environment. The bare tube, in-line arrangement minimizes the likelihood of erosion and trapping of ash as compared to a staggered arrangement shown in Figure 4.12. It is also the easiest geometry to be kept clean by soot blowers. However, these benefits must be evaluated against the possible larger weight, volume, and cost of this arrangement.

   b. Economizer with Finned Tube

To improve heat transfer, economizers with a variety of fin types are manufactured. Fins can reduce the overall size and cost of an economizer. However, successful application is very sensitive to the flue gas environment. Surface cleanability is a key concern. The basic design of an economizer with finned tubes involves a bundle of tubes that are arranged in parallel and connected by headers. The hot flue gas from the boiler passes over the outside of the finned tubes, while the colder feedwater passes through the inside of the tubes.

   c. Helical Fins

Helically finned tubes have been successfully applied to some coal, oil, and gas fired units. The fins can be tightly spaced in the case of gas firing due to the absence of coal fly ash or oil ash. Smaller fin spacings promote plugging with oil ash, while greater spacings reduce the amount of heating surface per unit length. An in-line arrangement also facilitates cleaning and provides a lower gas-side resistance.

   4. Baffles

The tube ends should be fully baffled to minimize flue gas bypass around finned bundles. Such bypass flow can reduce heat transfer, produce excessive casing temperatures, and with coal firing can lead to tube bend erosion because of very high gas velocities. Baffling is also used with bare

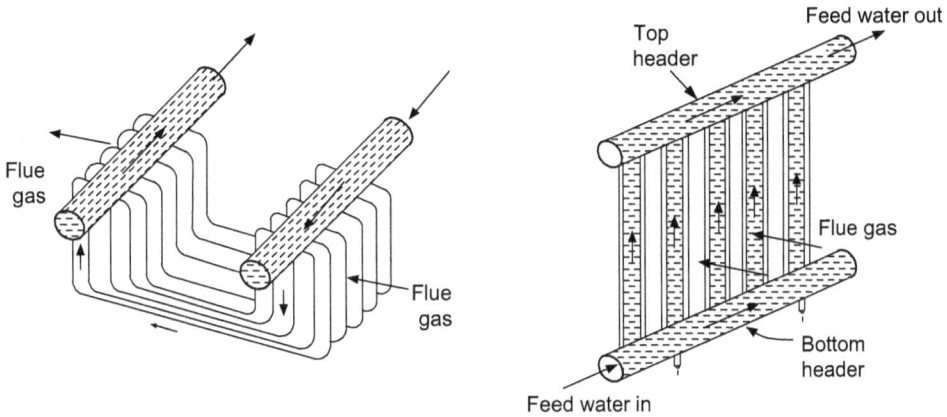

(a) Horizontal tube economizer.

(b) Vertical tube economizer.

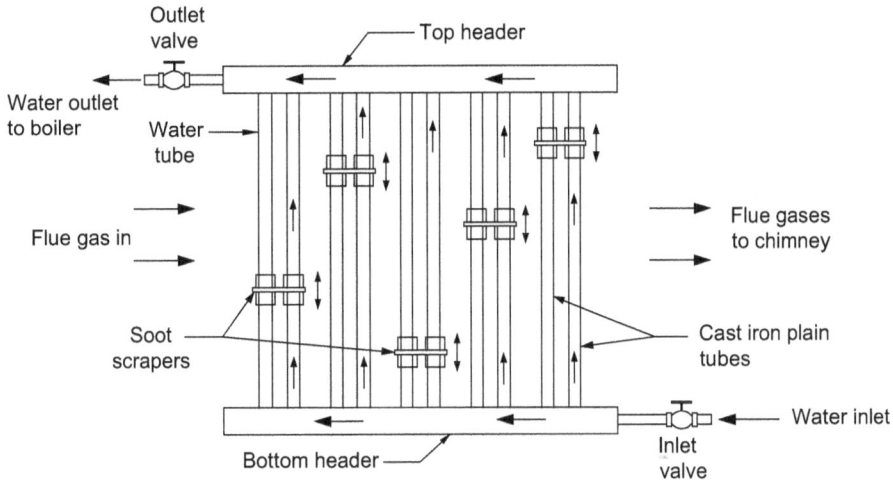

(c) Vertical tube economizer with
soot scrapers.

FIGURE 4.11 Few types of economizers.

tube bundles but is not as important as for finned tube bundles. Tube bend erosion can be alleviated by shielding the bends.

5. Velocity Limits

The ultimate goal of economizer design is to achieve the necessary heat transfer at minimum cost. A key design criterion for economizers is the maximum allowable flue gas velocity (defined at the minimum cross-sectional free flow area in the tube bundle). Higher velocities provide better heat transfer and reduce capital cost. For clean burning fuels, such as gas and low ash oil, velocities are typically set by the maximum economical pressure loss. For high ash oil and coal, gas-side velocities are limited by the erosion potential of the fly ash. This erosion potential is primarily determined

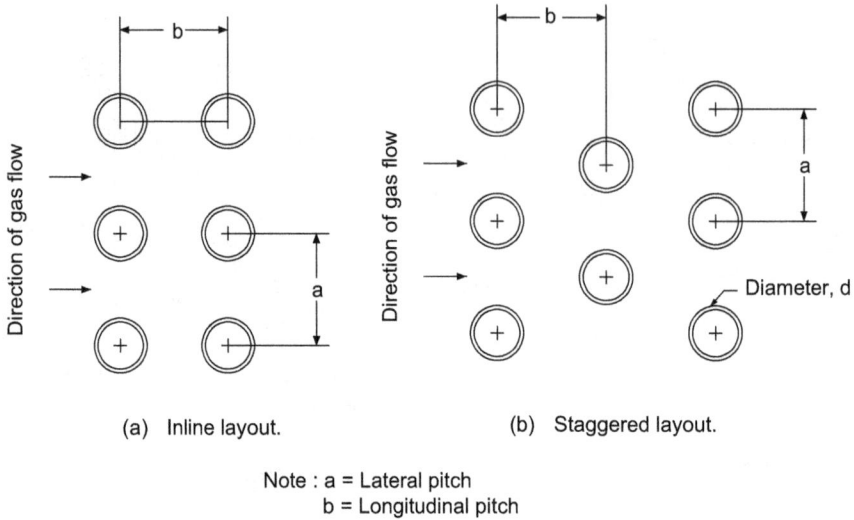

(a) Inline layout.   (b) Staggered layout.

Note : a = Lateral pitch
b = Longitudinal pitch

FIGURE 4.12   Tube layout patterens(a) Inline arrangement and (b) Staggered arrangement.

by the percentage of Al_2O_3, and SiO_2 in the ash, the total ash in the fuel and the gas maximum velocity. Based on experience, the acceptable flue gas velocities are fixed [19].

4.2.10 Ash Handling Systems

Furnace ash falls into water at the base of the boiler and then is mechanically removed from the water for transport by conveyor to one of the old opencast sites. There is a hopper for 1 day of ash production in case the conveyor fails. Rail is available in the case of longer periods of conveyor failure.

4.2.11 Waste Disposal

a. The solid waste streams from a PC boiler are furnace bottom ash, fly ash from bag filter or electrostatic precipitator, gypsum or other waste products resulting from the sulfur capture.
b. Liquid waste: liquid waste is primarily from boiler blowdown in drum-type subcritical boilers, and from cooling tower blowdown. In addition, the wet FGD process may generate a bleed waste stream.
c. Effluent from water treatment processes.
   The above wastes are to be handled and disposed of in an environmentally friendly manner.

4.2.12 Environmental Considerations

After the boiler, a flue gas desulfurization unit cleans the flue gases by injection of lime slurry, and a bag filtration system collects the ash and spent absorbent. Primary NOx emissions are kept low by the use of a combination of low-NOx burners and over-fire secondary air.

The boiler is designed with low NOx burners, an enlarged furnace, staged combustion (with overfire air or NOx ports) to minimize NOx emissions, and a selective catalytic reduction (SCR) system for further NOx control. Boilers emit varying levels of NOx, SO_2, particulate, and other compounds into the air and they discharge ash. Careful review and economic evaluation of all local regulations and permitting requirements are needed to achieve the required environmental

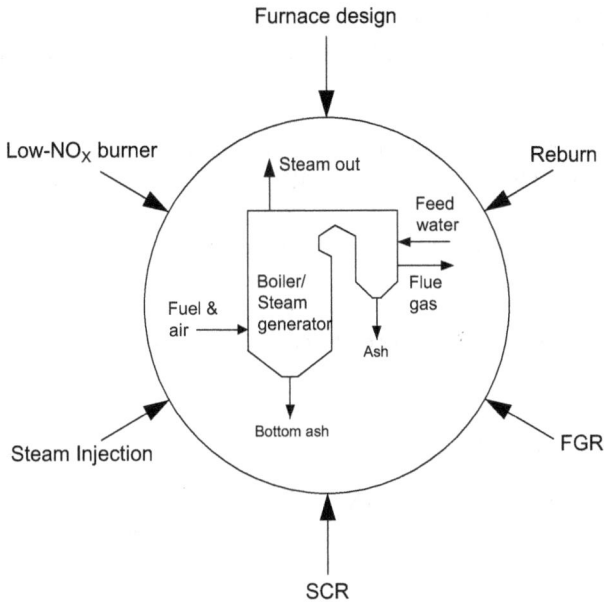

FIGURE 4.13 Basic environmental protection methods and emissions control of coal based power plant.

protection in the most cost-effective manner. Basic principle of environmental protection methods and emissions control is shown in Figure 4.13

4.2.12.1   Environmental Protection
Environmental protection is an integral part of the boiler and power system design process. The emission characteristics of industrial boilers are largely dependent on the type of fuel(s) combusted and the use of selected air pollution control systems and technologies. Air pollution control with emphasis on particulate, NOx, SO_2, and mercury emissions is perhaps the most significant environmental concern for fired systems. However, minimizing aqueous discharges and safely disposing of solid by-products are also key issues for modern power systems [20].

4.2.12.2   Emission Control
A pulverized coal fired boiler is equipped with low NOx burners to reduce NOx emissions, an electrostatic precipitator (ESP) to control particulate emissions, and a wet flue gas desulfurization (FGD) system to control emissions of SO_2. The most common methods employed to control NOx emissions from industrial boilers involve combustion modifications using low NOx burner designs, staged combustion with overfire air ports, and flue gas recirculation techniques.

4.2.12.3   Particulate Matter Control
The post-combustion control of PM emissions from coal fired combustion sources can be accomplished by using one or more or the following particulate control devices:

  i. Electrostatic precipitator (ESP).
 ii. Fabric filter (or baghouse).
iii. Wet scrubber.
 iv. Cyclone or multiclone collector.
  v. Side stream separator.

4.2.12.4 Post-combustion Emission Control

Post-combustion emission control is done by the use of one or more of the following systems, i.e., emissions vs methods of control:

(1) particulate—electrostatic precipitators (ESP) or fabric filters (baghouses);
(2) NOx—selective catalytic reduction (SCR) systems or selective noncatalytic reduction (SNCR) systems; and
(3) SO_2—wet or dry flue gas desulfurization (FGD) scrubbing systems or other forms of sorbent injection.

An electrostatic precipitator or fabric filter collects particulate, while either a wet flue gas desulfurization system (wet scrubber) or a spray dryer absorber (dry scrubber) removes most of the SO_2 .

4.2.13 Plant Performance Indicators

Understanding key performance indicators (KPIs) can serve as a guide for a plant performance and heat rate improvement program. Examples of important key performance indicators (KPIs) for a coal thermal power plant is shown in Box 4.1 [11].

BOX 4.1   KEY PERFORMANCE INDICATORS (KPIS) FOR A COAL-BASED THERMAL POWER PLANT [11]

1. Condenser back pressure
2. Auxiliary power consumption
3. Turbine efficiency (LP, IP, HP)
4. Cycle losses due to leaky vents, drains or internal turbine leakage
5. De-superheating spray flows
6. Coal quality
7. Feedwater flow
8. Steam flow
9. Make-up water
10. Unburned carbon in bottom ash and fly Ash
11. Flue gas temperatures
12. Main steam pressure
13. Main steam temperatures (SH and RH)
14. Feedwater heater performance
15. Mill horsepower per ton consumption
16. Coal flow
17. Combustion airflow (all); stoichiometry

18. Excess oxygen/excess air
19. Deviation between combustion airflow and theoretical or excess airflow
20. Total air-fuel ratio
21. System air in-leakage
22. Soot blower effectiveness
23. Tube metal thermocouples—variations and acceptable ranges
24. Ambient conditions (i.e., air and water inlet temperatures)
25. Mill inlet primary airflow temperature
26. Mill outlet primary airflow temperature
27. Air and gas system draft pressure measurements
28. Air preheater heat exchange efficiency and leakage.
29. Air preheater exit gas temperature (corrected to no-leakage)

4.3 CORROSION IN BOILER

Corrosion is one of the main causes of reduced reliability in steam generating systems. Corrosion has several causes, but many relate to the chemistry of the water. Acidity and dissolved oxygen and solids can contribute to corrosion in a boiler. Many corrosion problems occur in the hottest areas of the boiler—the water wall, screen, and superheater tubes. Other common problem areas include deaerators, feedwater heaters, and economizers.

Corrosion, erosion, scaling, clogging, etc., can significantly shorten the functional life of the boiler and its accessories. A boiler feedwater treatment system can help to protect the equipment by removing dissolved oxygen, carbon dioxide, calcium, magnesium, silica, and other problematic substances. By evaporation of the boiler water, the amount of total dissolved solids in the boiler water will increase, resulting in corrosion, sedimentation, and scale build-up. Scaling of hardness salts in the boiler will cause decreased heat transfer and under-deposit corrosion. Due to scaling, the boiler tubes can burst due to overheating. The most common causes of corrosion are dissolved gases (primarily oxygen and carbon dioxide), under-deposit attack, low pH, and attack of areas weakened by mechanical stress, leading to stress and fatigue cracking. Methods of corrosion control vary depending upon the type of corrosion encountered. The most common causes of corrosion are dissolved gases (primarily oxygen and carbon dioxide), under-deposit attack, low pH, and attack of areas weakened by mechanical stress, leading to stress and fatigue cracking [21].

4.3.1 Major Causes of Corrosion in Boilers

Corrosion in the boiler proper generally occurs when the boiler water alkalinity is low or when the metal is exposed to oxygen bearing water either during operation or idle periods. Protection of steel in a boiler system depends on temperature, pH, and oxygen content. Mechanical and operation factors such as velocities, metal stresses, and severity of service can strongly influence corrosion rates. In the steam and condensate system, corrosion is generally the result of contamination with carbon dioxide and oxygen. Specific contaminants such as ammonia or sulfur bearing gases may increase attack on copper alloys in the system. Systems vary in corrosion tendencies and should be evaluated individually. Corrosion can occur by a high amount of chlorides in the boiler, formed by a too high concentration factor. A much found problem is oxygen corrosion (pitting). Corrosion and scaling in steam boiler systems can greatly reduce boiler efficiency as well as shorten component life. A majority of corrosion is directly related to improper water treatment and maintenance of boiler water quality [21].

4.3.2 Design Conditions Affecting Corrosion

Many corrosion problems are the result of mechanical and operational problems. The following practices help to minimize these corrosion problems [21]:

i.   selection of heat and corrosion-resistant metals;
ii. reduction of mechanical stress where possible (e.g., use of proper welding procedures and stress-relieving welds);
iii. minimization of thermal and mechanical stresses during operation;
iv. operation within design load specifications, without over-firing, along with proper start-up and shutdown procedures;
v.   maintenance of clean systems, including the use of high-purity feedwater, effective and closely controlled chemical treatment, and acid cleaning when required.

4.3.3 Water Chemistry

Maintaining optimal boiler tube reliability, requires strict adherence to the water chemistry limits. Quality water chemistry is the primary and first building block for reliable power generating facility operation. Water chemistry impacts all equipment coming into contact with water or steam. A power generating facility cannot operate reliably without a quality water chemistry program. Oxygenated treatment is recommended as the best water chemistry program to provide reliable long-term operation of supercritical units. Good water chemistry is important for minimizing corrosion and the formation of scale in boilers. Steam side cleanliness should be maintained in water-tube as well as fire-tube boilers. For drum boilers, maintain proper boiler water chemistry according to ABMA or ASME guidelines by using proper continuous blowdown rates on a regular basis.

4.3.4 Deposits

Deposition is a major problem in the operation of steam generating equipment. The accumulation of material on boiler surfaces can cause overheating and/or corrosion. Both of these conditions frequently result in unscheduled downtime. Common feedwater contaminants that can form boiler deposits include calcium, magnesium, iron, copper, aluminum, silica, and (to a lesser extent) silt and oil. Most deposits can be classified as one of two types [22]:

  i.   scale that crystallized directly onto tube surfaces;
  ii. sludge is the accumulation of solids that precipitate in the bulk boiler water or enter the boiler as suspended solids.

4.3.5 Types of Corrosion

Corrosion control techniques vary according to the type of corrosion encountered. Major methods of corrosion control include maintenance of the proper pH, control of oxygen, control of deposits, and reduction of stresses through design and operational practices [21, 23].

4.3.5.1 Galvanic Corrosion

The most common type of galvanic corrosion in a boiler system is caused by the contact of dissimilar metals, such as iron and copper. These differential cells can also be formed when deposits are present.

4.3.5.2 Pitting

Pitting of boiler tube banks has been encountered due to metallic copper deposits. Such deposits may form during acid cleaning procedures if the procedures do not completely compensate for the amount of copper oxides in the deposits or if a copper removal step is not included.

4.3.5.3 Caustic Corrosion

Concentration of caustic (NaOH) can occur either as a result of steam blanketing (which allows salts to concentrate on boiler metal surfaces) or by localized boiling beneath porous deposits on tube surfaces.

Caustic corrosion (gouging) occurs when caustic is concentrated and dissolves the protective magnetite (Fe_3O_4) layer.

Steam blanketing is a condition that occurs when a steam layer forms between the boiler water and the tube wall. Under this condition, insufficient water reaches the tube surface for efficient heat transfer. The water that does reach the overheated boiler wall is rapidly vaporized, leaving behind a concentrated caustic solution, which is corrosive.

Boiler feedwater systems using demineralized or evaporated makeup or pure condensate may be protected from caustic attack through coordinated phosphate/pH control.

4.3.5.4 Acidic Corrosion

Low makeup or feedwater pH can cause serious acid attack on metal surfaces in the preboiler and boiler system. Acid corrosion can also be caused by chemical cleaning operations. Failure to neutralize acid solvents completely before start-up has also caused problems.

4.3.5.5 Hydrogen Embrittlement

Hydrogen embrittlement is rarely encountered in industrial boilers. The problem usually occurs only in units operating at or above 1500 psi.

Hydrogen embrittlement of mild steel boiler tubing occurs in high-pressure boilers when atomic hydrogen forms at the boiler tube surface as a result of corrosion. Hydrogen permeates the tube metal, where it can react with iron carbides to form methane gas, or with other hydrogen atoms to form hydrogen gas.

Coordinated phosphate/pH control can be used to minimize the decrease in boiler water pH that results from condenser leakage.

4.3.5.6 Oxygen Attack

Oxygen is highly corrosive when present in hot water. Because pits can penetrate deep into the metal, oxygen corrosion can result in rapid failure of feedwater lines, economizers, boiler tubes, and condensate lines.

If oxygen is not removed from the waterside, the tubes can suffer localized pitting and corrosion, especially underneath deposits. The majority of oxygen-pitting corrosion can occur during layup periods.

Major sources of oxygen in an operating system include poor deaerator operation, in-leakage of air on the suction side of pumps, the breathing action of receiving tanks, and leakage of undeaerated water used for pump seals.

Without proper mechanical and chemical deaeration, oxygen in the feedwater will enter the boiler.

For complete protection from oxygen corrosion, a chemical scavenger is required following mechanical deaeration.

4.3.5.7 Oxygen Control

Inadequate control of oxygen content is a major source of corrosion in boiler water tubes and feed lines. Boiler feedwater, which includes both makeup water and condensate return is deaerated to remove most of the dissolved oxygen. Levels should be reduced to the parts per billion range. The remaining dissolved oxygen should then be scavenged by chemical treatment.

4.3.5.8 Carbon Dioxide Attack

Carbon dioxide exists in aqueous solutions as free carbon dioxide and the combined forms of carbonate and bicarbonate ions. Corrosion is the principal effect of dissolved carbon dioxide. The gas will dissolve in water, producing corrosive carbonic acid.

Carbon dioxide corrosion is frequently encountered in condensate systems and less commonly in water distribution systems.

4.3.5.9 Internal Oxygen Pitting

Pitting is a form of corrosion that occurs only while the unit is offline. Oxygen pitting damage is caused by exposure of the internal surfaces of waterwall, superheater, or reheater tubes and headers to residual moisture after unit shut down. It is typically caused by oxygen-saturated stagnant water. Left unchecked, stagnant water and dissolved oxygen will lead to damage of the protective magnetite layer found on boiler tube surfaces. The root causes of this damage are due to the poor lay-up practices, poor shutdown practices, forced cooling and/or improper draining and venting procedures, improper chemical cleaning or deposition of chemicals on susceptible surfaces. There are two basic corrosion categories in boiler tubes [21]:

i. Internal corrosion such as hydrogen damage, acid phosphate corrosion, caustic gouging, and pitting.
ii. External corrosion: water wall fireside corrosion, superheater (SH)/reheater (RH) fireside corrosion, and ash dew-point corrosion.

4.3.6   Guidelines for Corrosion Control

The following conditions may be controlled through the following procedures [21]:

    i.   maintenance of proper pH and alkalinity levels;
    ii. control of oxygen and boiler feedwater contamination;
    iii. reduction of mechanical stresses;
    iv. operation within design specifications, especially for temperature and pressure;
    vi. proper precautions during start-up and shutdown;
    vii. effective monitoring and control of parameters.

4.3.7   Basic Corrosion Prevention Methods

To minimize the effect of corrosion before they happen, *use a boiler logbook*. Regularly tracking the normal operation of boiler room equipment makes it easy to spot when something critical changes. pH changes could indicate problems with water treatment or process contamination [24]. The common methods for prevention of corrosion include the following [25]:

1. Filtration of solid suspended impurities and particles from water.
2. Removing dissolved oxygen from the boiler feedwater.
3. Maintaining alkaline conditions in the boiler water.
4. Keeping the boiler internal surfaces clean and free from deposits.
5. Protecting boilers during out of service periods.
6. Using a chemical treatment program to counteract corrosive gases in steam and condensate systems.

The selection and control of chemicals for preventing corrosion requires an understanding of the causes and corrective measures.

4.3.8   Methods of Corrosion Control

Corrosion control techniques vary according to the type of corrosion encountered. Major methods include maintenance of the proper pH, control of oxygen, control of deposits, and reduction of stresses through design and operational practices.

Deaeration is used to avoid corrosion by removing the dissolved gasses (mainly O_2 and CO_2).

Corrosion control techniques vary according to the type of corrosion encountered. Corrosion control techniques include the following [21, 26]:

    i.   Major methods include maintenance of the proper pH, control of oxygen, control of deposits, and reduction of stresses through design and operational practices.
    ii. Deaeration is used to avoid corrosion removing the dissolved gasses (mainly O_2 and CO_2).
    iii. Protection of steel in a boiler system depends on temperature, pH, and oxygen content. Generally, higher temperatures, high or low pH levels and higher oxygen concentrations increase steel corrosion rates.
    iv. Mechanical and operation factors such as velocities, metal stresses, and severity of service can strongly influence corrosion rates.

4.3.9   Operational Problems in Boilers

Boilers operate under a wide variety of pressures, steaming rates, service cycles, and applications. Designs vary widely. All these combine to place requirements on operating procedures, water quality, and chemical treatment. There are numerous problems that commonly occur in boiler systems. These can be classified as follows [26]:

  i. preboiler corrosion (boiler make-up and feedwater system);
  ii. boiler system corrosion;
  iii. boiler system deposition;
  iv. carryover of steam purity problems;
  v. corrosion due to condensate.

4.3.9.1 Condenser in Leakage

It is generally agreed that the condenser is the greatest source of contamination. Improper maintenance of the condenser will result in unnecessary contamination of the cycle resulting in other major equipment failures (i.e., steam generator tube leaks, turbine blade failure, etc.). Leaking condenser tubes allow cooling water to enter the condensate and feedwater supply. Any contaminants introduced to the condensate and feedwater supply can become active components of numerous corrosion mechanisms. The three most harmful contaminants introduced by condenser leaks are sodium, chlorides, and sulfates.

4.3.10 INDUSTRIAL BOILER WATER TREATMENT METHODS

4.3.10.1 Water Chemistry

Water chemistry impacts all equipment coming in contact with water or steam. A power generating facility cannot operate reliably without a quality water chemistry program. Hardness of boiler feedwater needs to be treated by chemical means. The hardness and chemicals will react to form sludge (precipitate). Sludge deposits can form when precipitated material in the boiler water adhere to surfaces. To avoid the formation of these deposits, organic compounds are added.

4.3.10.2 Water Hardness

Water containing high amounts of calcium and magnesium salts is a hard water. Calcium and magnesium compounds are relatively insoluble in water and tend to precipitate out. This causes scale and deposit problems. Such water must be treated to make it suitable for steam generation.

4.3.10.3 Boiler Water Treatment

Untreated water contains dissolved minerals, gases, and particulates. Water quality is primarily an issue with steam boilers that use a lot of make-up water. Primary indicators of boiler water treatment are pH, total dissolved solids (TDS), total suspended solids (TSS), and hardness. Boiler system water treatment is classified as either external or internal [27].

Boiler water treatment consists of mechanical and chemical water treatment as, pretreatment of make-up water (softening, degassing) and the dosage of chemicals for boiler water treatment. The boiler water treatment products take care of scavenging dissolved oxygen, prevention from scaling, dispersion of solids, prevention from carryover, corrosion inhibition, and neutralization of volatile acidic components.

4.3.10.4 External Water Treatment

Solid removal methods include settling, clarification, and filtration. Selection of the appropriate suspended solids' removal system depends upon the nature and concentration of suspended solids and the degree of solids' removal required. Such external treatment may include the following [28]:

  i. clarification;
  ii. filtration;

iii. softening and demineralization;
iv. dealkalization;
v. deaeration and heating.

4.3.10.5 Internal Water Treatment

Internal boiler water treatment is applied to minimize the potential problems and to avoid any catastrophic failure, regardless of external treatment malfunction:

i. addition of chemicals (pH control, oxygen removal, other);
ii. blowdown (removes accumulated solids from boiler water).

4.3.11 FEED WATER CHEMISTRY

4.3.11.1 pH Control

Maintenance of proper pH throughout the boiler feedwater, boiler, and condensate systems is essential for corrosion control. The corrosion rate of carbon steel at feedwater temperatures approaches a minimum value in the pH range of 9.2–9.6.

The best pH for protection of copper alloys is somewhat lower than the optimum level for carbon steel. For systems that contain both metals, the condensate and feedwater pH is often maintained between 8.8 and 9.2 for corrosion protection of both metals. The optimum pH varies from system to system and depends on many factors, including the alloy used.

To elevate pH, neutralizing amines should be used instead of ammonia, which (especially in the presence of oxygen) accelerates copper alloy corrosion rates.

4.3.11.2 Oxygen Scavenging

Some oxygen remains in the boiler feedwater after deaeration and this residual oxygen must be removed to control boiler system corrosion. Chemical oxygen scavengers such as sodium sulfite and hydrazine are commonly used. Oxygen scavengers are usually introduced into the boiler system immediately downstream of deaeration, frequently into the deaerator storage tank, in order to minimize corrosion of as much of the system as possible.

4.3.11.3 Oxygen Scavengers

List of oxygen scavangers are given below [29]:

a. Mechanical Oxygen Removal

There are several ways to minimize corrosion in a boiler system and protect the metal surfaces throughout. The first step to reducing dissolved oxygen should be through mechanical means. The most effective way to do this is with a deaerator.

b. Sulfite

Sulfite, in the form of either sodium sulfite, sodium bisulfite, or sodium metabisulfite, is the most commonly used oxygen scavenger in low-pressure boilers, primarily due to cost. Sulfite reacts with dissolved oxygen to form sulfate.

c. Hydrazine

Hydrazine is a simple, volatile, oxygen scavenger that does not contribute to dissolved solids in its reaction with dissolved oxygen.

d. Neutralizing Amines

Carbon steel must be in contact with water in an alkaline pH range in order to minimize corrosion. The absolute pH values, and ranges of values, differs based on the water purity and whether the system also contains copper alloy (e.g., feedwater heaters or condensers).

4.3.12 IMPURITIES IN BOILER FEEDWATER

Feedwater is mainly condensate, which gets collected after the steam gets condensed in a condenser and addition of makeup water if required. Both condensate and treated makeup water are mixed in feed tank from where they are fed to the boiler. The different impurities in the feedwater can be grossly classified into three classes [30]:

i. dissolved gases such as oxygen and carbon dioxide;
ii. dissolved solids; and
iii. suspended solids.

Feedwater is to be treated for each of these types of impurities. Feedwater must be pretreated to remove impurities to control deposition, carryover, and corrosion in the boiler system. Poor quality water gives poor quality steam. The first step in any treatment is filtration of suspended solids. On the basis of proven satisfactory performance, cost, and other considerations, cartridge filters are a practical solution to most problems of water clean-up.

4.3.12.1 Dissolved Gases

Carbon dioxide mixes with water and forms carbonic acid. Oxygen results in corrosion and pitting of boiler tubes. Deareator head ensures that condensate, makeup water and flash steam are mixed rigorously and oxygen and other dissolved gases get separated. Apart from this, solubility of oxygen in water goes on decreasing as the temperature of feedwater increases. A dissolved oxygen level of 2 ppm in boiler feedwater is acceptable. Oxygen scavengers can also be added to reduce the dissolved oxygen levels in the feedwater. Carbonic acid can be removed from feedwater by neutralization with alkalis.

4.3.12.2 Dissolved Solids

Dissolved solids form scales, which can be extremely difficult to remove. Hence, it is quite essential that they are restricted from going inside the boiler. Dissolved solids include impurities like hardness, sulfates carbonates, etc. Hard water can be converted into soft water by using a softener, which removes these ions from the water by the process of ion exchange. Along with this, water also has chlorides and sulfates, which can be removed by the process of deionization.

4.3.12.3 Suspended Solids

Suspended solids are sediments or organic matter, which comes along with the makeup water.

4.3.12.4 Methods to Remove Water Impurities

Small amounts of impurities can be effectively treated inside the boiler to keep them in solution or allow them to be discharged via blowdown. Ion exchange processes include softening, demineralization, and dealkalization. Precipitation softening usually involves the addition of lime or a combination of lime and soda ash to remove various constituents from water by precipitation; this is followed by clarification and usually filtration.

4.3.13 ABMA AND ASME STANDARDS FOR BOILER WATER SOLIDS IN INDUSTRIAL BOILERS

The American Boiler Manufacturers' Association (ABMA) and the American Society of Mechanical Engineers (ASME) have developed suggested limits for boiler water composition, which depend upon the type of boiler and the boiler operating pressure. These control limits for boiler water solids are based upon one or more of the following factors [28]:

Sludge and total suspended solids—these result from the precipitation in the boiler of feedwater hardness constituents due to heat and to interaction of treatment chemicals, and from corrosion products in the feedwater. They can contribute to boiler tube deposits and enhance foaming characteristics, leading to increased carryover.

Total dissolved solids—these consist of all salts naturally present in the feedwater, of soluble silica, and of any chemical treatment added. Dissolved solids do not normally contribute to scale formation but excessively high concentrations can cause foaming and carryover or can enhance "under deposit" boiler tube corrosion. Methods like clarification, filtration, and chemical treatment are used to limit suspended solids from the feedwater.

Silica—this may be the blowdown controlling factor in softened waters containing high silica. High boiler water silica content can result in silica vaporization with the steam, and under certain circumstances, siliceous scale.

Iron—occasionally in high-pressure boilers where the iron content is high in relation to total solids, blowdown may be based upon controlling iron concentrations. High concentrations of suspended iron in boiler water can produce serious boiler deposit problems and are often indications of potentially serious corrosion in the steam/steam condensate systems.

4.3.13.1 Chemical Treatment

In boiler operations, it is important to following established guidelines for reliable operation. In the area of boiler operations, some guidelines that are available include the following [28]:

i. ASME Consensus on operating practices for the control of feedwater and boiler water chemistry in modem industrial boilers—Book No. 100367.
ii. ASME Consensus for the lay-up of steam generating systems practical guide to avoiding steam purity problems in the industrial plant (ASME CRTD, Vol. 35).
iii. ASME Boiler and Pressure Vessel Code—Recommended rules for care of power boilers.
iv. Electric Power Research Institute (EPRI), Project 2712: Consensus Guidelines on Fossil Plant Cycle Chemistry.

4.3.13.2 Boiler Water Treatment Philosophies

Understanding boiler system design is very important to effectively chemically treat the systems. The boiler water treatment philosophies are as follows [28]:

i. Solubilizing type chemical treatment programs should not be used in a fire-tube boiler due to design and control issues.
ii. HRSG's systems must always be treated using demineralized or equivalent feedwater and an equilibrium PO_4-type chemical treatment.
iii. Proper lay-up, start-up, and shutdown procedures must be in place.
iv. All applicable regulations must be followed.
v. Systems should be monitored as per established guidelines and inspected.

4.3.13.3 Preboiler Corrosion

Most of the oxygen present in feedwater is removed in the deaerator by "mechanical means." The scrubbing section of deaerators heat incoming water by mixing it with steam. Because oxygen can be harmful to feedwater systems even at low concentration, chemicals are generally fed to the storage section of the deaerator to react with the remaining molecular oxygen. Dissolved oxygen in the feedwater can attack the feedwater line, heaters, and the economizer. Oxygen-induced corrosion of low-carbon steel results in localized pitting of the metal surface. The rate of reaction and severity of this type of corrosion will depend upon the level of dissolved oxygen, temperature, and pH of the feedwater [21].

4.3.13.4 Boiler System Corrosion and Deposition

Many solids, soluble in boiler feedwater, precipitate in the boiler water. Calcium, magnesium, iron, copper, aluminum, silica, and to a lesser extent silt and oil, are common contaminants in boiler feedwater that can form deposits including scale. The basic approaches for selecting industrial internal boiler water treatment for the control of deposition include the following [28]:

 i.   solubilizing programs;
 ii, precipitation programs; and
 iii. for high-purity water systems, coordinated phosphate-pH programs.

4.3.13.5 Oxygen Corrosion

Most of the oxygen present in feedwater is removed in the deaerator by so-called "mechanical means." The scrubbing section of deaerators heat incoming water by mixing it with steam. The solubility of oxygen and other dissolved gases is greatly reduced at elevated temperatures and most of the gases are vented to the atmosphere. Oxygen attack on steel can be decreased by maintaining an alkaline pH and limiting the oxygen concentration.

4.3.13.6 Caustic Corrosion

To avoid caustic corrosion, maintain caustic-free bulk water in addition to keeping the system free of deposits. This can be accomplished with a coordinated phosphate-pH program in which sodium phosphates act as buffering agents.

4.3.13.7 Steam Purity

Maintenance of effective boiler water level control, minimizing exceeding maximum load rating and rapid load changes, use of good boiler feedwater and attemperation water quality and use of effective steam purification equipment will minimize steam purity problems.

4.3.13.8 Condensate System Corrosion

In a condensate system, the water contains varying amounts of contaminants that promote the corrosion reaction, the most common being dissolved oxygen and carbon dioxide. In the presence of oxygen, the corrosion of iron proceeds rapidly. With proper mechanical deaeration and chemical oxygen scavenging, virtually all oxygen can be eliminated from boiler feedwater. Good system design is required to minimize air contact with the condensate and subsequent oxygen absorption [28].

4.3.13.9 Monitoring Water Quality

Water quality monitoring varies from weekly litmus test strips to continuous electronic instrumentation and automated chemical treatment. The size of the boiler, the importance of water quality, and the skills of the boiler operators are all factors in deciding how best to monitor boiler water quality.

4.3.14 Carryover

Carryover is a general term used to describe moisture and entrained boiler water solids, which pass over with the steam from the boiler to turbines and process systems. The presence of salts can accelerate corrosion of boiler steel. Carryover can occur in two ways, mechanical (leaving small droplets of water with steam due to poor separation equipment and high drum levels) and vapor carryover. Vapor carryover is very difficult to control; the best method is to reduce unwanted solids in the feedwater. Foaming and priming in the boiler steam drum are causes of carryover. Foaming is the formation of small, stable bubbles on the steam release surface. This may be due to excessive levels of dissolved and suspended solids in the boiler water, high alkalinities and oil, process or sugar and other organic contamination. Foaming is adverse to the boiler system as it results in wet steam with boiler water that contains a high level of dissolved and suspended solids. It causes contamination in heat exchangers, control valves, and steam traps. The concentration of impurities in the steam boiler is termed total dissolved solids (TDS). Priming is the mechanical lifting or surging of boiler water into the steam outlet.

Carryover of impurities with steam is a major concern in boilers having superheaters and also if steam is used in a steam turbine. Carryover results from both ineffective mechanical separation methods and vaporous carryover of certain salts. The factors responsible for carryover are as follows [31]:

i. amount of dissolved solids in boiler water;
ii. chemical nature of dissolve solids;
iii. suspended solids in boiler water;
iv. boiler design;
v. boiler operating condition.

Carryover from a steam drum causes significant issues downstream such as corrosion of high-temperature superheater/reheater circuits. Also, erosion, fouling, and stress-corrosion cracking issues in HP and LP turbines can occur.

4.3.14.1 Preventing carryover

The basic preventive measure is to maintain the concentration of solids in the boiler water not exceeding the recommended levels. High water levels, excessive boiler loads and sudden load changes are to be avoided. Very often contaminated condensate returned to the boiler system causes carryover. The return condensate should be filtered to remove suspended solids before being fed back to the boiler. Efforts should be made to trace the source of any excessive contamination and the problem rectified. The use of chemical antifoams is effective in controlling carryover due to concentration of impurities in the boiler water. Steam-separating equipment must be inspected for proper installation.

4.3.14.2 Methods of Carryover Prevention

Methods used to prevent carryover include the following [32]:

i. Control of boiler water level and load swings.
ii. Control of steaming load within rated capacities.
iii. Proper firing distribution.
iv. Installation and maintenance of steam separation equipment.
v. Control of boiler water dissolved and suspended solids by blowdown.
vi. Control of boiler water chemistry.
vii. Elimination of oil, process, or other organic contamination.
viii. Addition of chemical antifoams.

4.3.14.3 Foam Control

Foaming can be caused by high levels of dissolved solids, suspended solids, alkalinity, or by the introduction of foaming-promoting materials into the boiler, for example, by the use of oil-contaminated steam condensate. Although effective antifoam agents are available to suppress foam formation, it is usually more economical to reduce or eliminate the problem by adjusting boiler water treatment (external and/or internal), increasing boiler foam-promoting contaminants from recycled steam condensate, etc. Foaming can be controlled by monitoring its causes, such as the high amount of suspended solids, contamination by fats and oils, and high alkalinity. Even with proper monitoring of these factors, the process plant needs to ensure proper control of the high level of total dissolved solids (TDS) in the boiler water to prevent foaming and carryover. TDS is expressed in ppm units and can be ensured by the conductivity or density method [33].

4.3.15 UTILITY BOILERS

4.3.15.1 Boiler Feedwater Chemistry

Boiler feedwater is the water supplied to the boiler. Often, steam is condensed and returned to the boiler as part of the feedwater. The water needed to supplement the returned condensate is termed make-up water. Make-up water is usually filtered and treated before use. Feedwater composition therefore depends on the quality of the make-up water and the amount of condensate returned.

Feedwater chemistry is critical to the overall reliability of supercritical generating facilities. Safe, reliable operation of large power generating plants depends upon the establishment of chemical conditions throughout the steam-water circuit that minimize the corrosion of construction materials and suppress the formation of deposits. The feedwater (make-up water) from outside needs to be treated for the reduction or removal of impurities by first filtration, and then followed by softening, evaporation, deariation, ion exchange etc. Internal treatment is also required for the conditioning of impurities within the boiler system, to control corrosion, as reactions occur in the boiler itself and the steam pipelines [11].

4.3.15.2 Water Chemistry Balance

Water chemistry impacts all equipment coming into contact with water or steam. Good water chemistry is important for minimizing corrosion and the formation of scale in boilers. Steam side cleanliness should be maintained in water-tube as well as fire-tube boilers. Oxygenated treatment is recommended as the best water chemistry program to provide reliable long-term operation of supercritical units.

Cation Conductivity

To provide reliable long life for equipment, feedwater cation conductivity should be maintained at the lowest feasible level. A well-maintained condensate polishing unit can provide a cation conductivity of 0.06 µmho. Therefore, it is reasonable to establish that anytime the feedwater cation conductivity exceeds 0.1 µmho, an abnormal condition exists and the cause should be investigated. To maximize the reliable life of this equipment, a best practice is to set the abnormal cation conductivity limits at 0.1 µmho.

Dissolved Oxygen

Dissolved oxygen must be maintained between 30 and 150 ppb in both the condensate and feedwater systems. Air or oxygen injection may be required to accomplish this and deaerator vents should normally be closed, although cycling of the vents may be required to avoid exceeding 150 ppb in the feedwater cycle.

pH

A recommend range is 8.0–9.8. This range would allow operation of the unit with a pH as low as 8.0 potentially requiring less ammonia and extending condensate polisher life. However, a caveat to this recommendation is given if two-phase flow-accelerated corrosion (FAC) is experienced, suggesting that it may be necessary to raise pH to the upper limit of 9.8 to stem this phenomenon.

4.3.16   Corrosion in Utility Boilers—Corrosion Mechanisms in the Water/Steam Cycle

4.3.16.1   Corrosion in the Water-Steam Circuits of Power Plants

The purpose of good cycle chemistry is to reduce corrosion and deposits in the water-steam circuit of power plants. In conventional fossil and HRSG plants, several types of corrosion can occur in the water-steam cycle. These include the following [34]:

i.   Under deposit corrosion which can lead to hydrogen damage and caustic gouging failures in drum units typically operating above ~8 MPa (~1100 psi).
ii.   Flow-accelerated corrosion (FAC), due to the accelerated dissolution of the protective oxide (magnetite) on the surface of carbon steel components caused by flow.
iii.   Corrosion fatigue (CF), due to repetitive applied stress causing damage to the internal protective oxide layer (magnetite).
iv.   Pitting corrosion due to inadequate shutdown procedures throughout the cycle, and
v.   Stress corrosion cracking (SCC) of sensitive steel components in the superheater, reheater, and steam turbine due to the presence of impurities, such as sodium hydroxide and chloride.

The feedwater system is the major source of corrosion products in the fossil fuel boiler or HRSG evaporator and then deposit on the heat transfer surfaces of the water/steam cycle path.

Leaks in water-cooled condensers are a common source of impurities, such as chloride and sulfate, entering the water/steam circuit,

4.3.16.2   Corrosion Tendencies of Boiler System Components

Proper treatment of boiler feedwater effectively protects against corrosion of feedwater heaters, economizers, and deaerators. The ASME Consensus for Industrial boiler specifies maximum levels of contaminants for corrosion and deposition control in boiler systems. The consensus is that feedwater oxygen, iron, and copper content should be considerable and that pH should be maintained between 8.5 and 9.5 for system corrosion protection.

4.3.16.3   Boiler System Major Components

In order to minimize boiler system corrosion, an understanding of the operational requirements for all critical system components is necessary [21].

a.   Feedwater Heaters

The primary problems are corrosion due to oxygen and improper pH, and erosion from the tubeside or the shellside.

Erosion is common in the shellside, due to high-velocity steam impingement on tubes and baffles.

Corrosion can be minimized through proper design (to minimize erosion), periodic cleaning, control of oxygen, proper pH control, and the use of high-quality feedwater.

b.   Deaerators

Corrosion fatigue at or near welds is a major problem in deaerators. Operational problems such as water/steam hammer can also be a factor. Other forms of corrosive attack in deaerators include stress corrosion cracking of the stainless steel tray chamber, inlet spray valve spring cracking, corrosion of vent condensers due to oxygen pitting, and erosion of the impingement baffles near the steam inlet connection.

The mechanism of most deaerator cracking has been identified as environmentally assisted fatigue cracking. Steps taken to minimize the potential for cracking includes, for example, stress relieving of welds and minimization of thermal and mechanical stress during operation. In addition, water chemistry should be designed to minimize corrosion.

c. Economizers

Economizer heat transfer surfaces are subject to corrosion product buildup and deposition of incoming metal oxides. These deposits can slough off during operational load and chemical changes. Corrosion can also occur on the gas-side of the economizer due to contaminants in the flue gas, forming low-pH compounds.

Water-tube economizers are often subject to the serious damage of oxygen pitting. The most severe damage occurs at the economizer inlet and, when present, at the tube weld seams. Other common causes of economizer failure include fatigue cracking at the rolled tube ends and fireside corrosion caused by the condensation of acid from the boiler flue gas.

d. Superheaters

One major problem is the oxidation of superheater metal due to high gas temperatures, usually occurring during transition periods, such as start-up and shutdown. Deposits due to carryover can contribute to the problem. Oxygen pitting, particularly in the pendant loop area, is another major corrosion problem in superheaters.

4.3.16.4 Condensate Piping Corrosion and Boiler Deposits

Carbon dioxide, CO_2, is formed in boilers by the decomposition of carbonates that naturally occur in water. Carbon dioxide escapes the boiler with the steam and condenses in the condensate system forming carbonic acid which is corrosive. CO_2 corrosion is indicated by thinning of condensate lines. Carbon dioxide lowers the pH of the condensate.

4.3.16.5 Blowdown

In drum boilers, in the process of evaporating water to form steam, scale and sludge deposits form on the heated surfaces of a boiler tube. Insoluble substances settle on the tube surfaces, forming a scale and leading to an increase in tube wall temperatures. The purpose of boiler blowdown is to control solids in boiler water. The dissolved solids' concentration of boiler water is intermittently or continuously reduced by blowing down some of the boiler water and replacing it with feedwater. Blowdown rate is generally expressed as a percent relative to the steam flow rate from the drum. Blowdown is accomplished through a pressure letdown valve and flash tank. Heat loss is often minimized by use of a regenerative heat exchanger. Excess blowdown wastes energy, water, and chemical treatment. However, insufficient blowdown can result in scaling and sludge even when chemical treatment is applied. To save energy, water, and chemical treatment, the boiler should be operated near the maximum levels of total dissolved solids or silica, depending on which is the controlling parameter. Blowdown rates typically range from 4% to 8% of boiler feedwater flow rate, but can be as high as 10% when makeup water has a high solids content. The concept of blowdown is shown in Figure 4.14a.

Automatic Blowdown Control Systems. These systems optimize surface blowdown by regulating water volume discharged in relation to amount of dissolved solids present. Conductivity, TDS, silica or chlorides concentrations, and/or alkalinity are reliable indicators of salts and other contaminants dissolved in boiler water. A probe provides feedback to a controller driving a modulating blowdown valve. An alternative is proportional control – with the blowdown rate set proportional to the makeup water flow.

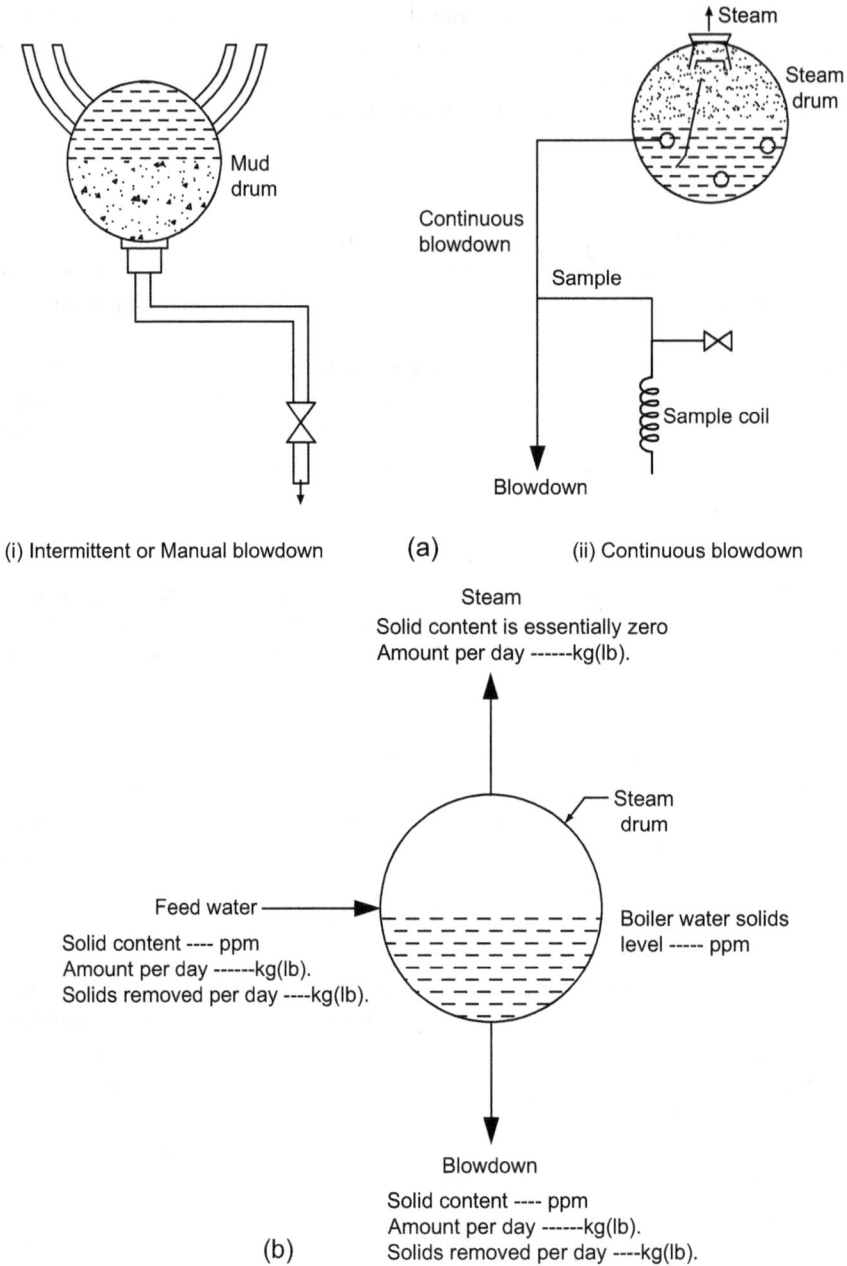

FIGURE 4.14 (a) The concept of blowdown and (b)The concept of calculaton of CoC.

Cycles of Concentration (CoC) "Cycles of concentration" refers to the accumulation of impurities in the boiler water. If the boiler water contains 10 times the level of impurities in the makeup water, it is said to have 10 cycles of concentration. The calculation CoC is shown in Figure 4.14b.

4.3.16.6 Deposition in Drum Boiler

Sludge or easily removable deposits accumulate at the bottom of the tubes in the mud drum and should be removed by intermittent blowdown, generally once per shift. Based on conductivity readings, the frequency may be increased or decreased. Continuous blowdown is usually taken from the steam drum a few inches from above the waterline, where the concentration of solids is the highest.

Deposition. The accumulation of material on boiler surfaces can cause overheating and/or corrosion. Both of these conditions frequently result in unscheduled downtime. Common feedwater contaminants that can form boiler deposits include calcium, magnesium, iron, copper, aluminum, silica, and (to a lesser extent) silt and oil. Most deposits can be classified as one of two types [22]:

iii. scale that crystallized directly onto tube surfaces;
iv. sludge is the accumulation of solids that precipitate in the bulk boiler water or enter the boiler as suspended solids.

4.3.16.7 Scale and Sludge

In boilers, water evaporates continuously and the concentration of the dissolved salts increases progressively. When their concentrations reach saturation point, they are thrown out of water in the form of precipitates on the inner walls of the boiler. If the precipitation takes place in the form of loose and slimy precipitate, it is called sludge. On the other hand, if the precipitated matter forms a hard, adhering crust/coating on the inner walls of the boiler, it is called scale [35].

Scale and sludge reduce boiler efficiency. The insulating properties of scale will reduce heat transfer capacity in the affected area. The best method for preventing scale formation is to remove hardness from make-up water. External treatment such as a water softener is designed to remove minerals from water before they enter the boiler system [22].

4.3.16.8 Scaling

Most scale-forming substances have a decreasing solubility in water with an increase in temperature. In any boiler water treatment program, generally the objective is to add chemicals to prevent scale formation caused by feedwater hardness constituents, such as calcium and magnesium compounds and to provide pH control in the boiler to enhance maintenance of a protective oxide film on boiler water surfaces. There are methods such a phosphate-hydroxide, coordinated phosphate, chelant treatment, and polymer treatment methods. In medium- and low-pressure boilers, all these methods have been used.

4.3.16.9 Dissolved Solids

Boiler water that recirculates in drum and steam generation circuits has a relatively high concentration of dissolved solids that have been left behind by water evaporation. Water chemistry must be carefully controlled to assure that this concentrate does not precipitate solids or cause corrosion within the boiler circuitry. Boiler water chemistry must also be controlled to prevent excessive carryover of impurities or chemicals with the steam.

4.3.17 CONDENSATE POLISHER SYSTEMS

Condensate polishing normally applies to the treatment of condensed steam from turbines before it is returned to the boiler. The ultimate goal of condensate polishing is to remove all soluble impurities and protect the high-pressure boilers. These contaminants can come from many sources [36]:

iii. Cooling water from condenser in-leakage.
ii. Make-up water.
iii. The conditioning chemicals used to protect the circuit, e.g., ammonia, morpholine, and hydrazine can be regarded as sources of contamination if they are to be removed by the condensate purification plant.
iv. Insoluble impurities will usually be the corrosion products from the materials in the system with copper and iron are the predominant species.

Condensate polishing is a proven method of producing high purity demineralized water from boiler condensate. The condensate can be returned to an ultrapure state by simple polishing rather than using fresh make-up process water. For details on condensate polishing refer to Refs. [37–39].

Condensate polishers are resin-based ion exchange systems that are commonly used in power plant condensate systems to remove dissolved contaminants (chlorides and silica, etc.) and suspended contaminants (iron or copper oxide particulates). The removal of minerals also maintains the pH balance of the water by removing acidic ions. The condensate polishing system is installed between the condensate pumps and low-pressure feedwater heaters. The cost-effective alternative to boiler blow-down is to "polish" the untreated condensate for reuse, using a sodium cycle condensate polisher. More condensate will be used and requires less make-up water, which reduces pretreatment chemical consumption and fuel costs [40,41].

4.3.17.1   All-Volatile Treatment, AVT

As a part of commissioning activities prior to keeping the boiler in operation, acid cleaning is performed in the boilers. During this process, an intentionally black layer of magnetite (Fe_3O_4) is made to form on the boiler tube surfaces, which prevents further corrosion of the parent material. Oxygen that enters the condensate system will oxidize the protective layer of magnetite of boiler tube to brownish red color ferric oxide (Fe_2O_3).

Feedwater chemistry is critical to the overall reliability of fossil and HRSG plants. Corrosion takes place in the feedwater system, and the associated corrosion products flow into the boiler/HRSG where they may deposit in various areas. The choice of feedwater chemistry depends primarily on the materials of construction and secondly on the feasibility of maintaining purity around the water-steam cycle.

A volatile alkalizing agent, usually ammonia is added to the feedwater to increase the pH. In some specific cases, an organic amine can be added instead. There are three variations of volatile conditioning that can be applied to the feedwater [11, 41]:

i.   Reducing all-volatile treatment, AVT(R).
ii. Oxidizing all-volatile treatment, AVT(O).
iii. Oxygenated treatment (OT).

For all-volatile treatment AVT(R), AVT(O), or OT, no solid chemicals are added to the boiler or preboiler cycle. Boiler water chemistry control is by boiler feedwater treatment only. Feedwater pH is controlled with ammonia or an alternative amine. For traditional all-volatile treatment, as opposed to oxygen treatment, hydrazine or a suitable alternative is added to scavenge residual oxygen. Because all volatile treatment adds no solids to the boiler water, solids' carryover is generally minimized. All-volatile treatment can be, but rarely is, used below 6.9 MPa (1000 psig). Supercritical boilers use all-volatile treatments, generally consisting of ammonia and hydrazine. Because of the extreme potential for deposit formation and steam contamination, no solids can be tolerated in supercritical once-through boiler water, including treatment solids.

4.3.17.2   AVT(R)—All-Volatile Treatment (Reducing)

This treatment involves the addition of ammonia (or an amine or blend of amines of lower volatility than ammonia) and a reducing agent (usually hydrazine or one of the acceptable substitutes) to the condensate or feedwater of the plant. AVT(R) provides protection to copper-based alloys in mixed-metallurgy feedwater systems.

4.3.17.3 AVT(O) —All-volatile Treatment (Oxidizing)

This all-volatile treatment has emerged as the much preferred treatment for feedwater systems, which only contain all-ferrous materials (copper alloys can nevertheless be present in the condenser). In these cases, a reducing agent should not be used during any operating or shutdown period. Ammonia (or an amine or blend of amines of lower volatility than ammonia) is added at the polisher outlet (if a polisher is included within the cycle). This is the treatment of choice for multipressure combined cycle/HRSG plants, which have no copper alloys in the feedwater.

4.3.17.4 Oxygenated Treatment, OT

For conventional fossil plants, optimized OT involves one oxygen injection location at the condensate polisher outlet, operating with the vents on the feedwater heaters and deaerator closed, and with knowledge of the total iron levels at the economizer inlet and in the feedwater heater cascading drain lines. Ammonia is added at the condensate polisher outlet. There is often a minimum level of oxygen, which is required to provide full passivation of the single-phase flow locations in the main feedwater line and the drain lines, and to maintain this protection. With oxygen treatment, the feedwater pH can be reduced, e.g., down to 8.0 to 8.5. An advantage of oxygen treatment is decreased chemical cleaning frequencies for the boiler. In addition, when oxygen treatment is used in combination with lower pH, the condensate polisher regeneration frequency is reduced.

Too low or total absence of oxygen is no longer considered the best corrosion control for BFW and condensate. Oxygenated treatments that deliberately inject oxygen into the condensate and BFW system or the use of oxygen scavenger at reduced concentrations may be necessary to maintain oxygen levels within the desired range to mitigate various failures including flow-accelerated corrosion (FAC).

4.3.17.5 Coordinated Phosphate Treatment

Coordinated phosphate pH treatment controls boiler water alkalinity with mixtures of disodium and trisodium phosphate added to the drum through a chemical feed pipe. The objective of this treatment is largely to keep the pH of boiler water and underdeposit boiler water concentrates within an acceptable range.

4.3.18 STEAM AND WATER ANALYSIS SYSTEM (SWAS)

Corrosion and erosion are major concerns in thermal power plants operating on steam. The steam reaching the turbines need to be ultrapure and hence needs to be monitored for its quality. A major goal of plant water chemical control is preventing solids' buildup and corrosion in the plant. A steam and water analysis system (SWAS) is a system dedicated to the analysis of steam or water. **SWAS** is a system dedicated to the analysis of steam or water. SWAS effectively monitors the parameters such as pH, conductivity, silica, sodium, phosphates, dissolved oxygen, etc., and thus helps in maintaining healthy operation of a power plant as given below [42, 43]:

i. *Silica.* Silica volatizes in steam and it is carried into the turbine. Although the silica deposits are not corrosive, but they severely reduce the efficiency.

ii. *Hydrazine.* It works as an oxygen scavenger, destroying traces of dissolved oxygen. Hydrazine is also a passivating agent and produces a protective oxide coating on metal tube wall.

iii. *pH.* pH analysis of pH helps detect changes that impact the effectiveness of closing and potential corrosion of the boiler tubes.

    iv. *Cation conductivity.* Provides an indication of total dissolved solids and susceptibility to scaling.

    v. *Dissolved oxygen.* Monitoring of dissolved oxygen level helps to keep the corrosion in check as, dissolved oxygen analysis help to monitor the efficiency of the deaerator.

4.3.19 BOILER CORROSION MONITORING TECHNIQUES

Boiler corrosion monitoring techniques include the following [44, 45]:

    a. Temperature excursion monitoring. Monitoring of temperature excursion of SH and RH area and trending.

    b. Five core chemical parameters (pH, Na, DO, NH_3, and PO_4) are required to be monitored daily. These are water and steam pH, sodium (dissolved oxygen) in saturated steam, dissolved oxygen in dearator and condenser, ammonia in water and phosphate in the drum.

    c. Dissolved oxygen in condenser/dearator. Level of dissolved oxygen in dearator and condenser is to be regularly checked and actions are to be taken to bring down within limits. Reducing oxygen level in condensate water reduces corrosive action. It has been experienced that attachment failures and weld joint defect failures are reduced by maintaining low DO.

    d. ASLD (acoustic steam leak detection) installation. In order to have early detection of BTL and stoppage of unit at an early stage to reduce secondary damages particularly in RH, SH and div SH area, It has been experienced that in the units where ASLD is not installed secondary damages cause heavy loss of generation due to increased time of repair.

4.4 BOILER DEGRADATION MECHANISMS

4.4.1 BOILER PRESSURE COMPONENT FAILURE

One of the most complex, critical, and vulnerable systems in fossil power generation plants is the boiler pressure components. In fossil power generation plants, the boiler pressure components include economizer, superheater, reheater, water wall panel, mud drum, steam drum, deaerator tank, blow down tank, in bed assembly, etc., are referred to as the parts of the boiler that are exposed to the high interior pressure of water or steam. Failures have been related to poor design, fabrication practices, fuel changes, operation, maintenance, and water cycle chemistry. Boiler tube failures continue to be the major cause of forced outages. Boiler pressure component failures have historically contributed to the highest percentage of loss of availability. Failures have been related to poor design, fabrication practices, fuel changes, operation, maintenance, feed water chemistry, etc. Boiler tube failures continue to be the major cause of forced outages [45, 46].

4.4.2 MATERIALS USED FOR PRESSURE PARTS

4.4.2.1 Materials for Boiler Tubes [47]

    a. Carbon steels

Carbon steels show mild corrosion resistance and fair strength up to 1000°F. However, their usage above 800°F must take into account the susceptibility to graphitization. Graphitization has not been a significant problem in thicknesses encountered in boiler tubing, e.g., SA 178

b. Carbon-moly steels

Carbon-moly steels exhibit higher creep strengths than plain carbon steels, and are widely used in high-temperature boiler service. These steels nominally contain 0.5% molybdenum, e.g., SA 209.

c. Intermediate chrome alloys

SA213-T2, SA213-T12, SA213-T11, SA213-T22, SA213-T9.

d. Austenitic stainless steels

Austenitic stainless steels SA213-T304, SA213-T316, SA213-T321, and T347.

e. Pressure parts

Radiant superheater—SA-213 T22, SA-213 T9, SA-213 T91, SA-213 TP304H (chrome-moly, austenitic stainless).

Low-temperature reheater—SA-213 T12, SA-213 T11 (carbon-moly alloy, crome-moly alloy).

Water wall—SA-178 A, SA-178 C, SA-192, SA-210A1, SA- 210C (low carbon, medium carbon).

4.4.3 Discussion of Boiler Degradation Mechanisms

Boiler degradation mechanisms are generally categorized as fireside and waterside. List of boiler failures due to inadequacies in design, operation, and maintenance are shown in Box 4.2 and a list of corrosion oriented failures in a boiler are given in Box 4.3 (compiled from Refs. [11, 46, and 48]). Various mechanisms of boiler degradation based on Refs. [45, 48–50] are discussed below.

BOX 4.2 LIST OF BOILER FAILURES DUE TO INADEQUACIES IN DESIGN, OPERATION, AND MAINTENANCE

1. Low-temperature creep
2. Chemical cleaning damage
3. Fatigue in water-cooled circuits
4. Pitting in water-cooled tubes
5. Coal particle erosion
6. Falling slag damage
7. Acid dew-point corrosion
8. Steam-touched tubes
9. Long-term overheating/creep
10. Fireside corrosion in coal-fired units
11. Fireside corrosion in oil-fired units
12. Dissimilar metal welds
13. Short-term overheating
14. Chemical cleaning damage
15. Fatigue in water-cooled circuits
16. Flow-accelerated corrosion
17. Pitting in water-cooled tubes
18. Coal particle erosion
19. Falling slag damage
20. Acid phosphate corrosion
21. Steam-touched tubes
22. Long-term overheating/creep
23. Fireside corrosion in coal-fired units
24. Fireside corrosion in oil-fired units
25. Dissimilar metal welds
26. Short-term overheating
27. Caustic gouging
28. Hydrogen damage
29. Corrosion fatigue
30. Erosion-corrosion
31. Soot blower erosion

BOX 4.3   LIST OF CORROSION ORIENTED DAMAGES IN A BOILER

    i.   galvanic corrosion
   ii.   pitting corrosion
  iii.   stress corrosion cracking
   iv.   caustic corrosion
    v.   caustic embrittlement
   vi.   caustic gouging
  vii.   acidic corrosion
 viii.   hydrogen embrittlement
   ix.   oxygen attack
    x.   carbon dioxide attack
   xi.   corrosion fatigue
  xii.   waterwall corrosion
 xiii.   low-temperature corrosion
  xiv.   flow-accelerated corrosion

4.4.3.1   Corrosion

Corrosion occurs inside and outside the tubes, pipes, drums, and headers of these lower-temperature components. Internal corrosion is usually associated with the boiler water, contaminants in the water, and improper chemical cleaning or poor storage procedures. External corrosion can be caused by corrosive combustion products, a reducing atmosphere in the furnace, moisture between insulation and a component, and acid formed on components in the colder flue gas zones when the temperature reaches the acid dew point.

4.4.3.2   Erosion

Erosion is a metal removal process caused by fluid laden particles striking the metal's surface. Erosion usually occurs due to excessive velocities. Various mechanisms, such as fly ash erosion, soot blowing erosion, falling slag erosion, and coal particle erosion can cause erosion on the boiler tubes. Fly ash erosion, is a significant boiler tube failure concern, occurs in the regions with high local flue gas velocities, with high ash loading, and with abrasive particles such as quartz.

Erosion of boiler components is a function of the percent ash in the fuel, ash composition, and local gas velocity or soot blower activity. Changing fuels to a high-ash fuel may lead to more erosion, slagging, and fouling problems.

Where two-phase flow (steam and water) exists, failures due to erosion are caused by the impact of the fluid against a surface. Equipment vulnerable to erosion includes turbine blades, low-pressure steam piping, and heat exchangers that are subjected to wet steam.

Erosion is common near soot blowers; on the leading edges of economizers, superheaters, and reheaters; and where there are vortices or around eddies in the flue gas at changes in gas velocity or direction.

4.4.3.3   DNB

Departure from nucleate boiling (DNB) is caused by the localized overheating of heat transfer surfaces under the deposition. Departure from nucleate boiling (DNB) occurs when nucleate boiling is converted to film boiling above a certain heat flux, known as the critical heat flux (CHF).

4.4.3.4   Fatigue Cracking

Fatigue cracking (due to repeated cyclic stress) can lead to metal failure. The metal failure occurs at the point of the highest concentration of cyclic stress. Examples of this type of failure include cracks in boiler components at support brackets or rolled in tubes when a boiler undergoes thermal fatigue due to repeated start-ups and shutdowns.

4.4.3.5 Soot Blower Erosion

Soot blower erosion causes accelerated OD tube wastage by direct material removal and removal of the fireside oxide that increases the oxidation rate. When the soot blower is operated, supply air or steam entrains and accelerates abrasive ash particles. This causes local tube OD erosion on impacted tube surfaces that result in thinned tube walls. Root causes of soot blower erosion can be improper soot blower operation and/or maintenance, and improper soot blower alignment.

4.4.3.6 Oxygen Attack

Formation of reddish brown hematite (Fe_2O_3) or "rust" deposits or tubercles accompanied by hemispherical pitting is the most familiar form of oxygen attack. Dissolved oxygen is also a significant component in ammonia corrosion of copper alloys, stress-corrosion cracking of austenitic stainless steels and chelant corrosion [48].

4.4.3.7 Thermal Fatigue

This damage results from cyclic and excessive thermal fluctuations accompanied by mechanical constraint. Excessive temperature gradients can add to internal strain to initiate or enhance the cracking process. Thermal fatigue can occur in water walls or other areas subjected to DNB or rapidly fluctuating flows.

4.4.3.8 Overheating

Overheating failures are often classified as either short-term or long-term.

Short-term overheating frequently exhibits a thin-lipped longitudinal rupture, accompanied by noticeable tube bulging, which creates the large fish-mouth appearance.

Tube operating temperatures above 850°F(454°C), or slightly above the oxidation limits of the tube steels, can lead to blistering, tube bulging or thick lipped creep rupture failures [48]. Long-term overheating usually occurs in superheaters, reheaters, and water walls as a result of gradual accumulation of deposits or scale, partially restricted steam or water flow, excessive heat input from burners or undesired channeling of fireside gases. Tubes subjected to steam blanketing are also prone to long-term overheating failures.

4.4.3.9 Long-Term Overheating Failures by Creep Damage

The root-causes of long-term overheating/creep fatigue are overheating of tube metal by a variety of causes, increased tube stress level caused by wall thinning and inadequate original design and/ or material choices. High-temperature, stress-rupture strength and oxidation resistance is increased by alloying additives of chromium and molybdenum. Stress rupture failures are dependent on time, temperature, and stress [48]. This relationship can be used to estimate stress rupture life expectancy by the Larsen–Miller method defined as

$$P = T \left(\text{Log } t_r + 20 \right)$$

where
P is the Larsen-Miller parameter,
T is absolute metal temperature, degrees Rankine (°F + 460),
t_r is time for rupture, hour.

4.4.3.10 High-Temperature Pressure Component Degradation

The high-temperature pressure components are defined here as those that operate at or above 482°C (900°F) metal temperature. The components normally at this higher temperature are the superheaters, reheaters, and attemperators with their connecting tubing and piping. These components operate in a temperature range high enough where they can experience degradation from oxidation, fuel ash corrosion, carburization, graphitization, and creep. These are the high-temperature failure modes. High-temperature creep rupture and creep fatigue failure are the two primary aging mechanisms for the high-temperature components.

4.4.3.11   Low-Temperature Creep

Low-temperature creep is creep cracking driven dominantly by residual stress created during manufacture. The root-causes of failure are unanticipated sources of high residual stress and service stress and/or high tube bend differential hardness in relatively low-temperature tubing. It can be found in either water or steam circuits in that it occurs between 298°C and 421°C. Cracking is initiated at the OD. Areas of interest are in the low-temperature sections of the reheater and primary superheater or high-temperature regions of the economizer.

4.4.3.12   Corrosion Fatigue

Corrosion fatigue occurs through the combination of cyclic thermal stress, mechanical strain, and corrosive waterside environment. The corrosion component during corrosion fatigue consists of the following:

    i.   low ph;
   ii.   excessive oxygenated water;
  iii.   contaminants in the form of chlorides and sulfides.

Maintaining optimal boiler tube reliability, requires strict adherence to the water chemistry limits. Corrosion may also be accelerated by the thermal fatigue stresses associated with startup and shutdown cycles. Corrosion fatigue can occur in the steam drum around rolled tube joints. The residual stresses from the tube rolling process are additive to the welding and operating pressure stresses. Corrosion from chemical cleaning and water chemistry upsets acts on this highly stressed area to produce cracking around the seal weld or the tube hole.

Chemical Control of Feedwater
Proper alloy selection, either in the initial construction or as equipment is replaced, should be carefully considered. Once the decision is made, the water chemistry program must follow to minimize corrosion of the feedwater equipment and deposits in the boiler and turbine.

Feedwater pH Control. The pH limits recommended on all ferrous-alloy condensate and feedwater piping are now a minimum of 9.2 with an upper limit of 9.8 or even 10.0 in systems with an air-cooled condenser.

4.4.3.13   Flow-Accelerated Corrosion (FAC)

FAC can be considered as the accelerated dissolution of the normally protective iron oxide film into a flowing stream of water or a water-steam mixture. Under certain water chemistry, fluid velocity, and operating conditions, FAC can cause internal wall thinning of condensate and feedwater piping, heater drip and drain lines, and other carbon steel piping. Chemical dissolution of ID oxide is most aggressive at water temperatures of 130°C–140°C with reducing water conditions and flow-accelerated removal at higher water temperatures up to approximately 280°C. This means that most instances of damage occur in feedwater or economizer sections.

    The risk of FAC depends on the combination of the factors such as water chemistry (promoting the dissolution of the protective film of oxide magnetite), pH, materials of construction, temperature (ranging from 200°F to 400°F), turbulence, quality of fluid (single or two phase) and materials of construction, etc. In HRSGs, economizers, evaporators, IP and LP drums are susceptible to FAC occurring. Material selection plays an important role in single-phase as well as two-phase FAC. Material upgrade to at least 1-1/4 Cr steel is recommended in chronic FAC cases. For carbon steel, higher pH values are better for the production and stability of magnetite. Operating with low pH values in the feedwater and condensate destabilizes magnetite and increases the rate of FAC on carbon steel in the feedwater system [50].

Turbulence. The laminar flow in smooth tubing/piping does not cause FAC. It needs a turbulent flow (reducers, bends, deaerator shells, valves, and the like) for mass transportation of dissolved oxides. Magnetite dissolution will be significantly affected by the pH of the feedwater; it is the second-most-important factor for FAC. Magnetite dissolution is less with higher pH. However, there is a limitation on pH in drum boilers containing mixed metallurgy in the feedwater train because copper tends to corrode at elevated pH. HRSG units may need to be controlled with different pH for individual pressure stages (if possible) to mitigate FAC issues [50].

Flow-assisted corrosion is most often found in systems with all ferrous metallurgy. The incorporation of as little as 0.1% chromium into low carbon steels has been shown to have benefits in reducing FAC. FAC is often observed in HRSG pipes in power plants. In HRSGs, economizers, evaporators, IP and LP drums are susceptible to FAC occurrence [50].

Prevention/Mitigation of FAC

The pH, temperature, and oxygen concentration are the main parameters that can affect the potential for FAC. Boiler feed water pH from 9.2 to 9.6 is often recommended. Upgrading the material to Cr-Mo steel usually solves the problem [11].

4.4.3.14 Hydrogen Damage

Hydrogen damage is the principle mode of waterside under deposit corrosion. Hydrogen damage occurs in boilers operating usually above 1000 psig (6.9 MPa) and under heavy deposits or other areas where corrosion releases atomic hydrogen. Hydrogen is being generated electrochemically and diffused into the unprotected tube surface. Diffused hydrogen reacts with iron carbide to form methane gas that is captive, pressurized, and essentially explodes causing micro-fissures to form, the material to carburize and become brittle.

The root causes of hydrogen damage are excessive deposits initiated by flow disruption and a source of high heat flux (flame impingement, burner misalignment) that causes departure from nucleate boiling or DNB, wick boiling or steam blanketing. Hydrogen damage is caused by the presence of three conditions simultaneously: flow disruption with high boiler deposits, acidic contamination of the boiler water, and high heat flux. Hydrogen damage is caused by the reaction of iron carbides in boiler tube steel, located in high heat-flux zones, with hydrogen produced as a result of corrosion reactions, particularly those taking place in low pH water.

4.4.3.15 Carburization

Incomplete combustion leads to increased amounts of CO and unburned fuel contacting the superheater tubes. Metal tubes require an oxidizing atmosphere to form a protective oxide film on the surface of the tube. When localized reducing conditions exist around the tube at an elevated temperature, the tube cannot protect itself from wastage from the sulfur and other corrosive species.

4.4.3.16 Graphitization

Graphitization is micro-structural material degradation caused by the decomposition of iron carbide into ferrite and graphite after prolonged exposure to elevated temperatures. Graphitization is not as prevalent in the power industry because of improved materials but remains a concern with local graphitization in repair welds (weld flaw).

4.4.3.17   Dissimilar Metal Weld Failure

Dissimilar metal weld failure is caused by welding austenitic stainless steel (SA-213 321H, SA-213 347H, and SA-213 304 H) materials to ferritic alloy (SA-213 T22) materials with high service temperatures and/or mechanical stresses using nickel-based weld filler material. Dissimilar metal welds experience creep and mechanical fatigue at the ferritic/nickel-based weld joint, which leads to stress accumulation and eventual weld failure.

4.4.3.18   Oxygen Pitting

Localized corrosion occurs in the form of sever pitting especially at the location where the water is not easily drained during shut down, which comes into contact with air.

4.4.3.19   ID Oxygen Pitting Corrosion

Oxygen pitting damage is caused by exposure of the internal surfaces of waterwall, superheater, or reheater tubes and headers to residual moisture after unit shut down. The highly oxygenated and/or low pH condensate damages the protective oxide layer and initiates pitting. The root causes of this damage are the influences of poor lay-up practices, poor shutdown practices, forced cooling and/or improper draining and venting procedures, improper chemical cleaning or deposition of chemicals on susceptible surfaces. Another form of pitting is initiated by the deposition of contaminant salts onto boiler tube surfaces. This occurs in sections of the boiler/steam generator where during normal operation only dry steam flow occurs, such as reheaters [11].

4.4.3.20   Fretting Wear

Fretting damage results from adjacent superheater or reheater tubes or attachments coming into direct contact through vibration, impact, rubbing, etc., with one another causing material wear and accelerated oxidation of the surfaces by continuous removal of the protective oxide layer.

4.4.3.21   Chemical Cleaning Damage

The root-causes are one or more improper procedural operations during the cleaning process.

4.4.3.22   Coal Particle Abrasion

Coal particle erosion causes waterwall tube OD thinning from direct contact with the preignition fuel stream. This occurs when tube protection fails or there is fuel stream and/or tubing misalignment.

4.4.3.23   Stress Corrosion Cracking, SCC

The risk for SCC can be reduced by eliminating the water chemistry related initiating steps of pit creation. As such, it is extremely important to perform layup and reheat drying procedures to control pitting corrosion. The best practice is to control boiler water pH and trip the unit when it drops below 7.0.

4.4.3.24   Metallic Oxides in Boiler Systems

Iron and copper surfaces are subject to corrosion, resulting in the formation of metal oxides. This condition can be controlled through careful selection of metals and maintenance of proper operating conditions. Maintenance of the proper pH, elimination of oxygen, and application of metal-conditioning agents can minimize the amount of copper alloy corrosion.

4.4.3.25   Steam Side Burning

Steam side burning is a chemical reaction between steam and the tube metal. It is caused by excessive heat input or poor circulation, resulting in insufficient flow to cool the tubes. Under such conditions, an insulating superheated steam film develops. Once the tube metal temperature has reached 750°F in boiler tubes or 950–1000°F in superheater tubes (assuming low alloy steel construction), the rate of oxidation increases dramatically; this oxidation occurs repeatedly and consumes the base metal. The problem is most frequently encountered in superheaters and in horizontal generating tubes heated from the top [28].

4.4.3.26   Caustic Attack

It is a form of under deposit corrosion(UDC). Caustic concentrates within deposits by evaporation of water leaving behind the hydroxide. The wastage occurs between the thick deposit and the steel. A second form is caustic gouging. Under the right flow and pH conditions, caustic may concentrate at the edges of steam bubbles formed. An upset of the boiler chemistry with too much sodium hydroxide (NaOH) or localized concentration of hydroxide increases the tube wastage dramatically. In this case, no deposit exists and the resultant morphology is one of a smooth, rippled, clean ID surface [50].

Caustic Gouging (Chemically Induced Failure)
Sodium hydroxide (NaOH) is used extensively in boiler water treatment to maintain the optimum hydroxyl ion concentration range to form a protective magnetite film on steel surfaces. However, excessive sodium hydroxide can destroy the protective film and corrode the base metal. NaOH can concentrate during departure from nucleate boiling (DNB), film boiling or steam blanketing conditions. Concentration also occurs when normal boiler water evaporates beneath deposits leaving behind the caustic at the metal surface.

4.4.3.27   Caustic Corrosion/Gouging in Boiler

Caustic gouging is caused by high heat transfer and poor circulation of boiler water at the high heat transfer location in the boiler. Any condition that promotes the formation of high concentrations of caustic levels can result in caustic gouging. One type of caustic gouging that can occur in low-pressure boilers is caused by departure from nucleate boiling or DNB.

Caustic corrosion occurs when caustic is concentrated and dissolves the protective magnetite (Fe_3O_4) layer, causing a loss of base metal and eventual failure [51]. Concentration of caustic (NaOH) can occur as a result of steam blanketing. Stem blanketing is a condition that occurs when a steam layer forms between the boiler water and the tube wall. Under this condition, insufficient water reaches the tube surface for efficient heat transfer. The water that reaches the overheated boiler wall is rapidly vaporized, leaving behind a concentrated caustic solution, which is corrosive.

Boiler feedwater systems using demineralized or evaporated make-up or pure condensate may be protected from caustic attack through coordinated phosphate/pH control. Phosphate buffers the boiler water, reducing the chance of large pH changes due to the development of high caustic concentrations. This results in the prevention of caustic buildup beneath deposits or within a crevice where leakage is occurring.

4.4.3.28   Caustic Embrittlement

Caustic embrittlement is the phenomenon in which the material of a boiler becomes brittle due to the accumulation of caustic substances. As water evaporates in the boiler, the concentration of sodium carbonate increases in the boiler. In high-pressure boilers, sodium carbonate is used in softening of water by lime soda process, due to this some sodium carbonate may be left behind in the water. As the concentration of sodium carbonate increases, it undergoes hydrolysis to form sodium hydroxide [52]:

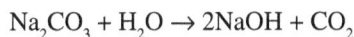

$$Na_2CO_3 + H_2O \rightarrow 2NaOH + CO_2$$

There are many causes of caustic embrittlement, including the combined action of the following three components [53]:

- A susceptible material.
- A given chemical species.
- Tensile stress.

The presence of alkali in the crevices, found around the rivet heads and other hot spots, combined with fabrication stress around rivet holes, causes cracks in the steel boiler shells and tube plates. This corrosion at high pH levels produces hydrogen, which attacks the crystal structure of iron, making it hard and brittle. This is highly dangerous because the tube can then fail at the boiler's normal operating temperature [54–56].

4.4.3.29   Fly Ash Erosion
Fly ash erosion accelerates tube wastage by direct material removal and by removal of protective fireside oxide of the base material. This increases the oxidation rate. After coal is combusted and the boiler's effective area is reduced, fly ash with high local velocity removes the tube materials protective oxide scale resulting in accelerated wall thinning.

4.4.3.30   Fire Side Corrosion
The root-cause of fireside corrosion are local "reducing" environment and the occurrence and deposition of pyro-sulfates, burning coal with unusually aggressive ash/slag (in cyclone units), direct impingement of carbonaceous particles removing the tubes protective oxide, coal particle erosion-corrosion, nonuniform mixing of fuel and air in the flame, etc. Release of sulfur and chloride compound by unburnt coal particles due to incomplete combustion, can eat away the tube surface leading to thinning or puncture under hot condition.

4.4.3.31   Waterside Corrosion
Deposition, dissolved oxygen, low pH, and the scale depositions are the main reason of waterside corrosion. Maintaining the water chemistry with respect to operating pressure and boiler type is a key to avoid the water corrosion.

4.4.3.32   Dissolved Oxygen (O_2)
Pitting corrosion is the leading cause of premature failure of a steam boiler. In water, O_2 is inversely soluble to the heat applied to it. This means that as a boiler is heated, any and all O_2 is released within the confines of the boiler itself. As this O_2 is released, it causes extreme and deleterious pitting corrosion,

4.4.3.33   Corrosion on the Fireside of Boiler Components
All fossil fuels, with the possible exception of natural gas, contain constituents that will promote corrosion on the fireside of boiler components. Those constituents include compounds of sulfur, vanadium, and sodium; but in the case of municipal-refuse boilers, chlorine is an increasing concern. There are three temperature regimes where fireside corrosion occurs [57]:

1. Less than 300°F. Dew point corrosion occurs when sulfuric acid condenses.
2. 500°–750°F. Waterwall-tube corrosion occurs in coal-fired units by the formation of pyrosulfates of sodium and potassium. In refuse-fired boilers, mixtures of chlorides of zinc, lead, iron, and sodium are the likely causes of corrosion.
3. Temperatures greater than 1000°F. The cause of superheater and reheater ash corrosion depends on the fuel. The corrosive species are different for coal and oil-fired boilers. Mixtures of vanadium pentoxide and sodium oxide or vanadium pentoxide and sodium sulfate are the principal offensive compounds in oil ash. For coal-fired boilers, sodium and potassium iron trisulfates are the liquid species blamed for high-temperature corrosion.

4.4.3.34 Acid Phosphate Corrosion

Acid phosphate corrosion is the second mode of waterside UDC. Phosphate corrosion occurs where thick deposits are present with the right combination of feedwater chemicals and localized flow disruption or a localized concentrating mechanism. Excessive addition of mono- or di-sodium phosphates in the water treatment chemicals leads to phosphate corrosion. Nowadays, acid phosphate corrosion failures are rare due to improved chemistry practices [50].

4.4.3.35 Guidelines for Prevention/Control of Waterside Degradation Mechanisms

The majority of waterside issues are related to feedwater chemistry, contamination, layup procedures, load cycling, and ramp rates. Waterside issues can be minimized with proper metallurgy selection, a suitable water treatment program, periodic sampling, and chemical cleaning, along with proper maintenance, operation, and layup procedures. The following guidelines will help in mitigating or reducing waterside issues [50]:

i. Clean boiler tubes periodically. A clean boiler can eliminate potential under deposit corrosion (UDC) and phosphate hideout issues.

ii. Avoid the combination of oxygen and moisture when the unit is idle (short-term or long-term).

iii. Conduct periodic tube sampling from representative areas. This can identify serious waterside issues before they become catastrophic.

iv. Use rifled tubing in the high-heat flux areas to reduce caustic gouging and UDC.

v. Avoid feedwater chemistry upsets, coolant flow reductions in tubes, excessive overfiring/blowdowns, and load swings, which all contribute to caustic gouging. Partial flow restrictions can also promote caustic gouging.

vi. Prevent the source of contamination in the case of thick deposits. Shutdown the unit immediately in the case of contamination ingress.

vii. Reduce flow disruptions in tubes.

viii. Control caustic levels in feedwater chemistry.

ix. Keep the amount of particulate iron oxide as low as possible through blowdown to prevent build-up of thick scales.

x. Monitor high-heat flux zones along with high-stressed areas for UDC and corrosion fatigue, respectively.

xi. Perform nondestructive examination to identify UDC and hydrogen damage, specifically on the HP evaporator tubes, slope tubes, roof tubes, and high-heat flux zones. All hydrogen-damaged tubes should be replaced; chemical cleaning cannot remove these extremely thick and tenacious deposits.

xii. Select a water treatment scheme consistent with FWH tube metallurgy and change the feedwater chemistry following FWH metallurgy upgrades. Systems with all-ferrous metallurgy (although condensers may contain copper-based materials) should be operated with either OT or AVT(O) with an optimum pH range of 9.2 to 9.6. Air in-leakage should be prevented. Units with mixed metallurgy, that is, FWHs containing copper-based materials, require a reducing environment to prevent copper corrosion. They should be operated at a pH range of 9.0 to 9.3. However, be aware that interconnecting carbon steel components may be susceptible to FAC.

xiii. Upgrading from carbon steel to low-alloy steels containing higher chromium in areas susceptible to FAC.

xiv. Control single-phase FAC with oxidizing potential. Two-phase FAC cannot be reduced with oxidizing potential, however, as oxidizing power is associated with the gas phase. The majority of the time material upgrade to low-alloy steel could be the only option to reduce two-phase FAC.

xv. Change chemistry from reducing to oxidizing potential after upgrading FWH metallurgy from copper to stainless steel, otherwise, serious FAC damage can occur in carbon steel components.

xvi. Maintain proper pH levels. Low pH conditions play an important role in pitting corrosion, corrosion fatigue, UDC, and single- and two-phase FAC.

xvii. Keep iron level in the feedwater less than 2 ppb for all-ferrous metallurgy—note that systems with OT produce iron concentration less than 1 ppb. In mixed metallurgy systems, copper level should be less than 2 ppb.

xviii. Implement a comprehensive FAC program to manage FAC damage. Inspections along with nondestructive techniques should be used to assess FAC damage.

xix. Operate multipressure HRSG systems in an oxidizing environment, such as AVT(O) or OT, without addition of oxygen scavenging chemicals. HRSGs should be designed with all-ferrous metallurgy or an upgrade should be in order to prevent FAC and UDC issues.

xx. Monitor iron concentration in the feedwater and LP drum. Results will provide an indication of whether FAC is active or not. Total iron concentration should be less than 2 ppb in the feedwater and less than 5 ppb in the drum to eliminate the possibility of active FAC.

xxi. Inspect tubes via nondestructive examination, such as linear phased array or radiography, for corrosion fatigue cracking in the areas susceptible to corrosion fatigue, and replace tubes damaged by corrosion fatigue. Follow Electric Power Research Institute guidelines to locate areas susceptible to corrosion fatigue. Pad welding should be avoided on corrosion fatigue-damaged tubes because it exacerbates corrosion fatigue due to higher residual stresses.

xxii. Reduce ramp rates because transient stresses, along with feedwater chemistry, play a significant role in developing corrosion fatigue. Corrosion fatigue occurs with the synergistic effect of stress large enough to fracture the magnetite (Fe_3O_4) scale and feedwater with high oxygen concentration or low pH.

4.4.4 CAUSE WISE BOILER DAMAGE MECHANISMS

Cause wise boiler damage mechanisms are given below:

4.4.4.1 Damage Mechanisms Due to Manufacturing Defects

QC/QA program during manufacturing influenced damage mechanisms include the following [11]:

1. weld and repair defects;
2. ID oxygen pitting corrosion;
3. OD rubbing and fretting;
4. chemical cleaning damage;
5. coal particle erosion;
6. low-temperature creep.

4.4.4.2 Water Quality Influenced Damage Mechanisms

Water quality damage mechanisms include the following [11]:

1. steamside stress corrosion cracking;
2. flow-accelerated corrosion;
3. corrosion fatigue;
4. supercritical waterwall cracking;
5. hydrogen damage.

4.4.5 Damage Mechanisms Due to Coal Quality

Coal quality damage mechanisms that attack the gas-side of tubing and headers from impurities in the fuel and fuel combustion products include the following [11]:

1. fly ash erosion;
2. soot blower erosion;
3. fireside corrosion;
4. acid dew-point corrosion;
5. stress corrosion cracking (gas-side).

4.4.5.1 Fly Ash Erosion

Fly ash erosion accelerates tube wastage by direct material removal and by removal of protective fireside oxide of the base material. This increases the oxidation rate. After coal is combusted and the boiler's effective area is reduced, fly ash with high local velocity removes the tube materials protective oxide scale resulting in accelerated wall thinning.

4.4.5.2 Soot Blower Erosion

Soot blower erosion causes accelerated OD tube wastage by direct material removal and removal of the fireside oxide that increases the oxidation rate. When the soot blower is operated, supply air or steam entrains and accelerates abrasive ash particles and possibly moisture or condensate from wet air or low enthalpy steam. Root causes of soot blower erosion can be improper soot blower operation and/or maintenance, insufficient superheat in soot blower supply steam or moisture removal in air supply and improper soot blower alignment.

4.4.5.3 Fireside Corrosion

Fireside corrosion in the superheaters and reheaters of coal-fired units is known as coal ash corrosion. Fireside corrosion in boilers is commonly divided into two categories by location, namely furnace wall corrosion in the lower furnace and fuel ash corrosion of superheaters in the upper furnace. The root-causes of fireside corrosion are local "reducing" environment and the occurrence and deposition of pyro-sulfates, burning coal with unusually aggressive ash/slag (cyclone units), direct impingement of carbonaceous particles removing the tubes protective oxide and nonuniform mixing of fuel and air in the flame. Different methods have been proposed for controlling fireside corrosion, including reducing sulfur content, installing shields, using additives and applying corrosion-resistant weld overlays. To combat fireside corrosion, ultrasonic thickness measurement (UT) for wall thinning and trending data to quantify wastage rates is very popular.

4.4.5.4 Acid Dew-Point Corrosion

Acid dew-point corrosion refers to corrosion resulting from condensation of acid gases such as sulfur dioxide, sulfur trioxide, hydrogen chloride, and others within the combustion products. At temperatures below flue gas dew points, these gases and the water vapor in the flue gas will condense to form sulfurous acid, sulfuric acid and hydrochloric acid, which can lead to severe corrosion of carbon steels, low alloy steels, stainless steels, and other alloys. Dew-point corrosion typically results in excessive metal loss, leading to perforation or rupture.

Structures affected by dew-point corrosion. All fired process heaters and boilers that burn fuels containing sulfur have the potential for sulfuric acid dew point corrosion in the economizer sections and in the stacks.

4.4.5.5 Stress Corrosion Cracking

The root-cause of stress corrosion cracking requires three aspects together: interaction with chlorine gasses, excessive tensile stress and a susceptible, sensitized material. However, an initiating defect such as a scratch, pit, or weakness in the protective oxide allows for initial corrosive attack.

4.4.6 BOILER TUBE FAILURES

The boiler tubes are under high-pressure and/or high-temperature conditions. They are subject to potential degradation by a variety of mechanical and thermal stresses and environmental attack on both the fluid and fireside. Mechanical components can fail due to creep, fatigue, erosion, and corrosion. The following discussion on boiler tube failure mechanisms is based on Refs. [46,50, 58–63].

4.4.6.1 Categories of Boiler Tube Failure Mechanisms

Identifying and correcting the root-cause of tube failures is essential to help lessen the chance of future problems. A comprehensive assessment is the most effective method of determining the root-cause of a failure. A tube failure is usually a symptom of other problems. To fully understand the cause of the failure, you must investigate all aspects of boiler operation leading to the failure in addition to evaluating the failure itself. The following boiler tube failure mechanisms are some of the most common that occur on modern operating boilers and they are organized into three sections as given below [61]:

    i.   Waterside failure mechanisms.
    ii. Fireside failure mechanisms.
    iii. General failure mechanisms.

Some of the types of boiler tube failure mechanisms of fireside failure mechanisms, waterside failure mechanisms and general failure mechanisms are listed in Table 4.2.

4.4.6.2 Discussion on Boiler Tube Failure Mechanisms

Damage mechanisms and their causes for boiler tube failures, after Ref. [59, 60].

    a. Caustic gauging

Localized corrosion due to the concentration of caustic or alkaline salts that usually occurs under evaporative or high heat transfer conditions. It can be mitigated by water chemistry control and design.

TABLE 4.2
Classification of Boiler Tube Failure Mechanisms

Waterside failure mechanisms	Fireside failure mechanisms	General failure mechanisms
Caustic attack/caustic gouging	Fuel ash corrosion	Short-term overheat
Oxygen pitting	High-temperature oxidation	Long-term overheat
Hydrogen damage	Fireside corrosion fatigue	Graphitization
Acid attack	Erosion	Dissimilar metal weld (DMW)
Pitting corrosion, Stress corrosion cracking	Mechanical fatigue	failure
Waterside corrosion fatigue	Waterwall corrosion	Exfoliation

Source: Edited from Ref. [61].

---

b. Long-term creep failures

Creep failures occurs due to high-temperature exposure of the boiler tube that slowly and continuously deformed under load below yield stress. The temperature at which creep begins depends on the alloy composition. Several factors like basic metallurgical condition, internal scale deposition, and operating factors can contribute to creep failure.

c. Creep

Creep is a time-dependent deformation that takes place at elevated temperature under constant load. Such failure results in overheating or overstressing the tube material beyond its capabilities for either a short-term or a long-term period. The two types of creep failures due to overheating are as follows:

d. Short-term overheat failure

Short-term overheat failure results in a ductile rupture of the tube metal. It is characterized by "fish mouth" opening in the tube where the fracture surface is a thin edge. Short-term overheat failures are most common during boiler start up.

e. Long-term overheat

Tube metal often has heavy external scale build-up and secondary cracking. Superheater and reheat superheater tubes commonly fail after many years of service, as a result of creep. Furnace water wall tubes also can fail from long-term overheat.

f. Erosion-corrosion

Erosion is metal removal caused by particles striking the metal's surface. Various mechanisms, such as fly ash erosion, soot blowing erosion, falling slag erosion, and coal particle erosion can cause erosion on the boiler tubes.

g. Flow-accelerated corrosion (FAC)

FAC involves the removal of the protective oxide layer. It can occur in carbon or low alloy steel piping systems under flowing, water at elevated temperature. Flow rate, pH, oxygen content, temperature, and geometry can affect the mechanism.

h. Fatigue

Fatigue is a phenomenon of damage mechanism caused by cyclic or fluctuating stresses, which are caused by mechanical loads, flow induced vibration.

i. Fireside corrosion

One of the primary challenges of reliably burning coal is managing the corrosion experienced by the furnace heat transfer surfaces. Fireside corrosion remains a leading cause of failure in superheater and reheater tubes. Fireside corrosion in the superheaters and reheaters of coal-fired units is known as coal ash corrosion.

j. High-temperature hydrogen attack (HTHA)

High-temperature hydrogen attack (HTHA), also called hot hydrogen attack, is a problem which concerns steels operating at elevated temperatures (typically above 400°F or 204°C) in hydrogen environments, in refinery, petrochemical and other chemical facilities, and, possibly, high-pressure steam boilers. It is not to be confused with hydrogen embrittlement or other forms of low-temperature hydrogen damage. HTHA is the result of hydrogen dissociating and dissolving in the steel, and then reacting with the carbon in solution in the steel to form methane [64].

k. Hydrogen damage

Hydrogen damage is most commonly associated with excessive deposition on ID tube surfaces, coupled with a boiler water low pH excursion. Under deposit releases atomic hydrogen, which migrates into the tube wall metal, reacts with carbon in the steel (decarburization) and causes intergranular separation.

l. Dissimilar metal weld (DMW) failure

DMW describes the butt weld where an austenitic (stainless steel) material joins a ferritic alloy, such as SA213 T22, material. Material fails at the ferritic side of the weld, along the weld fusion line. These failures are attributed to several factors: high stresses at the austenitic to ferritic interface due to differences in thermal expansion properties of the two materials, excessive external loading stresses and thermal cycling, and creep of the ferritic material.

m. Stress-assisted corrosion (SAC)

Water touched surfaces of industrial boiler tubes are susceptible to blunt discontinuous cracking generally referred to as stress-assisted corrosion (SAC). SAC is generally observed on the waterside of carbon steel boiler tubes at locations exhibiting the combination of a substantial external attachment weld, at tube bends, or at other areas with high stress. SAC under the weld attachment are generally associated with a significant internal oxide accumulation compared to a thin protective magnetite (Fe_3O_4) film found on the internal surface of boiler tubes. Nondestructive evaluations were not effective for detecting SAC, but metallographic evaluations revealed wall penetrations up to 30% of the tube thickness that were transgranular with bulbous features, rounded tips, and filled with a relatively dense oxide [65,66].

n. Water corrosion

Dissolved oxygen, low pH, and the scale depositions are the main reason for waterside corrosion. Maintaining the water chemistry with respect to operating pressure and boiler type is key to avoid water corrosion. High temperatures and stresses in the boiler metal tend to accelerate the corrosive mechanisms.

o. Oxygen pitting

Localized corrosion occurs in form of sever pitting especially at the location where the water is not easily drained during shut down, which comes into contact with air.

p. Stress corrosion cracking

Stress corrosion may result in either intercrystalline or transgranular cracking of carbon steel. It is caused by a combination of metal stress and the presence of a corrosive. A metallurgical examination of the failed area is required to confirm the specific type of cracking. Once this is determined, proper corrective action can be taken.

q. Stress rupture

Boiler tubes operating at high temperatures under significant pressure are vulnerable to stress rupture failures. The conditions and mechanisms that either lead to or are associated with stress rupture, include overheating, high-temperature creep, graphitization, and dissimilar metal welds

r. Thermal fatigue

Thermal fatigue can occur due to cyclic stresses caused by temperature fluctuations. Damage is in the form of cracking that may occur due to relative movement expansion and this is constrained, under repeated thermal cycling. Typically occurs at areas such as header ligaments, welded attachments, tube stub welds, circumferential external surface cracking of water wall tubes in supercritical units, and fabrication notches.

s. Waterside corrosion fatigue

In boilers, corrosion fatigue cracking can result from continued breakdown of the protective magnetite film due to cyclic stress. Corrosion fatigue is influenced by boiler design, water chemistry, boiler water oxygen content, and boiler operation. The breakdown of the protective magnetite on the ID surface of the boiler tube exposes tube to corrosion. The locations of attachments and external weldments, such as buckstay attachments, seal plates, and scallop bars, are most susceptible. The problem is most likely to progress during boiler startup cycles.

4.4.7 FACTORS INFLUENCING BOILER TUBE FAILURES

Factors influencing boiler tube failures, after Ref. [45]:

i. Water chemistry

With increases in operating pressure, feedwater quality becomes even more critical.

ii. Coal quality

Using a different type of coal for emission or economic reasons will adversely affect the capability, operability, and reliability of boiler and boiler auxiliaries.

iii. Cycling operation

Many base-load designed boilers have been placed into cycling duty, which has a major impact on the boiler reliability as indicated by occurrences of serious corrosion fatigue in water-touched circuitries, economizer inlet header shocking, thick-wall header damages, and others.

iv. NOx emission control

Deep staging combustion for NOx reduction has produced serious waterwall fire corrosion for high sulfur coal firing, especially in supercritical units.

v. Aging

A large percentage of existing fossil-fired units are exceeding "design life" without plans for retirement. These vintage units are carrying major loads in power.

vi. Material flaws

Material flaws include defects introduced into the tube during its manufacture, fabrication, storage, and/or installation. Such defects might include forging laps, inclusions, or laminations in the metal, lack of fusion of the welded seams, deep tool marks, or scores from tube piercing, extrusion, or rolling operations, gouges, punctures, corrosion, impact dents, etc.

Failures tend to be more predominant in high-temperature sections because of the interaction of flaws with the higher stress in these locations.

Lap defects are crevices in steel that are closed but not metallurgically bonded. They may occur in seamless tubes as a consequence of the presence of internal voids or cracks in the ingot from which the tube was formed. Lap defects can also be caused by faulty methods of steel rolling in the steel mill.

4.4.8 FLOW CHART TO DETERMINE BOILER FAILURE MECHANISMS

Ref. [67] presents examples of common failure mechanisms in a variety of boilers including power boilers, recovery boilers, and heat recovery steam generators (HRSGs). Visual examination may help the equipment operator decide whether further metallurgical examination and root-cause analysis is warranted. The only sure way to determine a failure mechanisms and root-cause is a full metallurgical examination. Ref. [67] divides these mechanisms into three classifications: cracks, holes/pits/gouges, and deformation.

4.4.9 PROMINENT REASONS FOR BOILER TUBE FAILURES [68]

i.   Poor water quality. With increases in operating pressure, feedwater quality becomes even more critical.
ii. Coal quality. Using a different type of coal for emission or economic reasons has adversely affected the capability, operability, and reliability of boiler and boiler auxiliaries. Coal quality damage mechanisms that attack the gas-side of tubing and headers from impurities in the fuel and fuel combustion include the following [11]:
fly ash erosion
soot blower erosion
fireside corrosion
acid dew-point corrosion
stress corrosion cracking (gas-side)
iii. Cycling operation. Many base-load-designed boilers have been placed into cycling duty, which has a major impact on the boiler reliability as indicated by occurrences of serious corrosion fatigue in water-touched circuitries, economizer inlet header shocking, thick-wall header damages, and others.
iv. NOx emission. Deep staging combustion for NOx reduction has produced serious waterwall fire corrosion for high sulfur-coal firing, especially in supercritical units.
v.   Aging. A large percentage of existing fossil-fired units are exceeding "design life" without plans for retirement. These vintage units are carrying major loads in power generation.

4.4.9.1 Economizer Effects

Finned economizer tube is such a location due to the double difficulties of poor/impossible access for thickness measurement and shielding. Economizer tube bends close to casing walls are another such location.

4.4.10 Erosion Failures

Erosion is associated with solid fuel-fired boiler tubes. The cause can be defective design, defective erection, improper operation, and maintenance as given below [69].

4.4.10.1 Causes Attributed to Design

Solid fuels depending on the amount of ash and type of ash constituents can cause erosion. There can be numerable causes under this heading. To name a few,

1. Design with high gas velocities solid fuels such as coal or rice husk can erode the boiler tubes due to their ash constituents.
2. Design without considering normal dust flow pattern expected within the tube bank.
3. Design without considering the preferential gas flow upstream/downstream of the tube bank.
4. Design without provision for controlling the preferential flow.
5. Design with narrow clearance between tubes. Adopting very closely pitched tubes is not a desired design feature for solid fuel fired boilers. Narrow clearances (20 to 30 mm) are not possible to maintain in actual case, though in drawing it is possible. Clearances come down during erection/operation at the site for various reasons such as below:
   i. Bow in tube length as received from the tube manufacturer or caused by the transport/ erection procedure.
   ii. Fuel ash deposition brings down the tube to tube clearances.
   iii. Improper support system (welded supports).
   iv. Thermal expansion brings down the clearances.
   v. Staggered pitch with narrow clearance makes the tube erode faster than the inline tube arrangement.
   vi. It is always preferable to use a minimum of 40 mm clearance for fouling fuels.
6. Design without proper lateral spacers to maintain the longitudinal/transverse pitch of tubes
7. Design with possibilities for impingement erosion.
8. Failure to provide the sacrificial tube shields near soot blowers.
9. Improper design of flow dividers.
10. Failure to provide proper seal box at places where the tubes enter inside the gas path. In other words, seal boxes are required wherever the tubes enter the flue path.

4.4.10.2 Causes Attributed to Erection

1. Improper erection methods resulting in irregular pitch of tube banks.
2. Improper/incomplete erection of protective shields/gas baffle. Failure to erect the gas baffles is a common problem due to the stringent erection schedule imposed in the project stage. Failure to overlap the shields is a common problem seen.
3. Incomplete erection of seal box. The drawing may specify that the seal box is to be erected with seal welding. Many engineers fail to read the drawing/fail to ensure the detail is carried out at site.

4.4.10.3 Causes Attributed to Operation

1. Operation of the boiler beyond the design parameters.
2. Operation of the boiler without understanding the fuel characteristics/operation of the boiler with fuels not designed for.

4.4.10.4 Causes Attributed to Maintenance

1. Failure to ensure the design pitch is maintained during tube replacement.
2. Failure to observe the pattern of erosion and to take remedial advice from the manufacturer.
3. Failure to fit the gas baffles and tube shields/sealing arrangement after the tube replacement.
4. Decision to retain the distorted/plugged coils within the flue path. Many times it is seen that the distorted coils offer sites for ash accumulations and cause the tube erosion.

4.4.11 RENEWAL OF BOILER TUBES

For any boiler retubing job, it is absolutely essential to use tubes that conform to the tube specification of the particular boiler. Much of the required information on sizes, thickness, and number of tubes per boiler is given in the manufacturer's technical manual. Boiler tubes should be replaced when they cannot be made leak tight, or when they are warped, or otherwise seriously damaged. The steps involved in a boiler tube repair involves the following [69]:

removing defective tube
cleaning of new tubes
preparing tubesheet holes
repairing tubesheets
preparing tube ends
fitting tubes
expanding and rolling/or welding of tubes
conducting leak test.

4.4.12 CLEANING INDUSTRIAL BOILER FIRESIDES AND WATERSIDES

Boiler heat transfer surfaces must be kept clean to provide for safe and economical boiler operation. The methods and procedures involved in fireside and waterside cleaning is discussed below [70].

4.4.12.1 Cleaning Firesides

Excessive fireside deposits of soot, scale, and slag cause the following conditions: reduced boiler efficiency, corrosion failure of tubes and parts, reduced heat transfer rates and boiler capacity, blocking of gas passages with high draft loss and excessive fan power consumption, and fire hazards. Methods for cleaning boiler firesides include wire brush and scraper cleaning, hot-water washing, wet-steam lancing, and sweating.

i.   wire brush and scraper cleaning.
ii. hot-water washing.
iii. wet-steam lancing.
iv. sweating.

Fireside slag can be removed from the convection superheaters by forming a sweat on the outside of the tubes. Cold water is circulated through the tubes, and moisture from the air condenses on the tubes to produce sweat. The hard slag is changed into mud by the sweat, and the mud can be blown off by an air or a steam lance.

4.4.12.2 Cleaning Watersides

Except for oil, these deposits are not usually soluble enough to be removed by washing or boiling out the boiler.

  i. Mechanical Cleaning

There are several types available, but perhaps one of the most common is the pneumatic turbine-driven tube cleaner. After washing, thoroughly dry out the boiler watersides.

  ii. Chemical Cleaning

In most cases mechanical cleaning is the preferred method for cleaning watersides. Chemical (acid) cleaning requires special authorization since it requires elaborate and costly equipment and rather extensive safety precautions. Inhibited acid cleaning is used to remove Mill scale from the watersides of new or recently serviced boilers.

  iii. Acids for Cleaning

The following acids are used to clean boilers: hydrocholoric acid, phosphoric acid, sulfamic acid, citric acid, and sulfuric acid.

  iv. Inhibitors

Without inhibitors, acid solutions attack the boiler metal as readily as they attack the deposits. With the addition of suitable inhibitors, the reaction with the boiler metal is greatly reduced.

4.4.12.3 Acid Cleaning Procedures

Boiler units can be acid cleaned by either the "circulation" or "fill and soak" method. The fill and soak method is used for cleaning units with natural circulation. The boiler unit is filled with the inhibited acid solution at the correct temperature and allowed to soak for the estimated time.

4.4.12.4 Flushing and Neutralizing

After acid cleaning, drain and then flush the unit with clean, warm water until the flushing water effluent is free of acid and soluble iron salts.

4.4.12.5 Boiling Out

New boilers, or boilers that have been fouled with grease or scale, should be boiled out with a solution of boiler compound. New boilers must be washed out thoroughly.

4.4.13 NDT Techniques for Crack Detection

The techniques under this category are the following [71]:

   (i)   Liquid penetrant testing (LPT)
   (ii)   Magnetic particle testing (MPT)
   (iii)   Radiography (RT)
   (iv)   Ultrasonic Testing (UT)
   (v)   Eddy current testing (ECT)
   (vi)   Magnetic flux leakage (MFL) technique
   (vii) Potential drop technique or electric perturbation technique

(viii)   Magnetic Barkhausen noise (MBN) analysis technique
(viii)   Magnetic Barkhausen noise (MBN) analysis technique
(ix)   Visual testing and equipment aided visual testing
(x)   *In situ* metallography and replica metallography
(xi)   Thermography
(xii)   Acoustic emission testing (AET)
(xiii)   Vibration monitoring. The main issues here are the detection and sizing of cracks so that a correct decision between replacement of the component or repair of the cracked region can be taken.
(xiv)   Remote field electromagnetic technique (RFET) can be used to detect, ID, or OD flaws on the hot side half of the boiler water wall tubes.
(xv)   Low-frequency electromagnetic technique (LFET) is used for tube scanning.

4.4.13.1   NDT Methods to Detect Weld Integrity

With the boiler out of service, apply the following nondestructive testing methods to detect the flaws to assure weld integrity [11]:

i.   Cracking—UT, PT, MT, VT
ii.   Inadequate joint preparation—RT, UT
iii. Incomplete fusion—UT
iv.   Laminations—UT, PT, MT, VT
v.   Overlap—PT, MT
vi.   Porosity—RT, PT, VT
vii. Slag inclusions—RT, UT
viii. Undercut—RT, VT
ix.   Butt joints—RT, UT, PT, MT, VT
x.   Corner joints—UT, PT, MT, VT
xi.   Tee joints—UT, PT, MT, VT
xii. Lap joints—PT, MT, VT

(Note: VT-visual testing or examination, RT-radiographic testing, UT-ultrasonic testing, PT-dye penetrant testing, MI-magnetic particle testing.

4.4.14   FACTORS INFLUENCING BOILER TUBE FAILURES

Factors influencing boiler tube failures, after Ref. [45]:

i. Water chemistry

With increases in operating pressure, feedwater quality becomes even more critical.

ii. Coal quality

Using a different type of coal for emission or economic reasons will adversely affect the capability, operability, and reliability of boiler and boiler auxiliaries.

iii. Cycling operation

Many base-load designed boilers have been placed into cycling duty, which has a major impact on the boiler reliability as indicated by occurrences of serious corrosion fatigue in water-touched circuitries, economizer inlet header shocking, thick-wall header damages, and others.

iv. NOx emission control

Deep staging combustion for NOx reduction has produced serious waterwall fire corrosion for high sulfur coal firing, especially in supercritical units.

v. Aging

A large percentage of existing fossil-fired units are exceeding "design life" without plans for retirement. These vintage units are carrying major loads in power.

vi. Material flaws

Material flaws include defects introduced into the tube during its manufacture, fabrication, storage, and/or installation. Such defects might include forging laps, inclusions, or laminations in the metal, lack of fusion of the welded seams, deep tool marks or scores from tube piercing, extrusion, or rolling operations, gouges, punctures, corrosion, impact dents, etc.

Failures tend to be more predominant in high-temperature sections because of the interaction of flaws with the higher stress in these locations.

Lap defects are crevices in steel that are closed but not metallurgically bonded. They may occur in seamless tubes as a consequence of the presence of internal voids or cracks in the ingot from which the tube was formed. Lap defects can also be caused by faulty methods of steel rolling in the steel mill.

4.4.15 Approach for the Control of Boiler Tube Leaks (BTL)

The boiler tubes are under high-pressure and/or high-temperature conditions. They are subject to potential degradation by a variety of mechanical and thermal stresses and environmental attack on both the fluidside and fireside. Mechanical components can fail due to creep, fatigue, erosion, and corrosion.

4.4.15.1 Preventing Boiler Tube Leakage

SG tube failure is the leading cause of unavailability in power generating facilities. It is therefore imperative to be attentive to the root-causes of tube failures. Four causes of tube failure are as follows [11]:

i.   ID pitting,
ii.   stress corrosion cracking,
iii. corrosion fatigue, and
iv.   waterwall cracking.

It is generally agreed that the condenser is the greatest source of contamination. Improper maintenance of the condenser will result in unnecessary contamination of the cycle resulting in other major equipment failures (i.e., steam generator tube leaks, turbine blade failure, etc.). The three most harmful contaminants introduced by condenser leaks are sodium, chlorides, and sulfates. An effective program of inspection and repair must be developed to ensure reliable condenser operation.

4.4.15.2 Tube Pitting

Causative factors associated with steam generator tube pitting: sludge, tube scale, oxidizing species, acid chloride, and steaming conditions. Pitting reduces the local tube wall thickness and thus the margin to tube leakage. In addition to the consequences of a leak in service, pitting may require the following measures:

i.   plugging of tubes and thus removal of tubes from service,
ii. frequent shutdowns for eddy current inspections,
iii. lengthy shutdown for repairs by sleeving, and
iv. additional evaluation of steam generator integrity of pitted tubes.

4.5 DEAERATOR, FEEDWATER HEATER, CONDENSER, AND FEEDWATER SYSTEM

4.5.1 DEAERATOR

Deaerators are used in industrial applications such as power plants and chemical process industries to reduce corrosion and extend the life of a steam-generating boiler. A deaerator achieves this task by reducing the level of dissolved oxygen and carbon dioxide in the feedwater, which in turn reduces the amount of corrosive compounds within the steam system over time. When the deaerated water exits the deaerator, carbon dioxide concentrations will be zero and dissolved oxygen concentrations will be less than or equal to 7 ppb (parts per billion) (this can be tested through an extraction point). Deaerators also reduce the need for water treatment chemicals, such as an oxygen scavenger. In addition to reducing dissolved gasses in the feedwater to reduce corrosion damage, a deaerator also raises the temperature of the feedwater before it enters the boiler, thus requiring less fuel for the boiler to heat the water hot enough to produce steam. As a result, the boiler system performs better, is more efficient overall, and helps maintain operating costs. The concept of deaeration and mixing of condensate is shown in Figure 4.15. For details on deaerator, refer to Refs. [72–75]. Also refer to Standards for Tray Type Deaerators, 9th Edition, 2011, Heat Exchange Institute (HEI), Cleveland, Ohio.

4.5.1.1 The Principle of Deaeration

Deaerators operate on Henry's Law of partial pressure that says that the quantity of dissolved gas in a liquid is directly proportional to the partial pressure above that liquid. When deaeration takes place, the partial pressure above the boiler feedwater is decreased as the water's temperature rises. Simply stated, the solubility of the gases in the water is decreased as the temperature of the water increases.

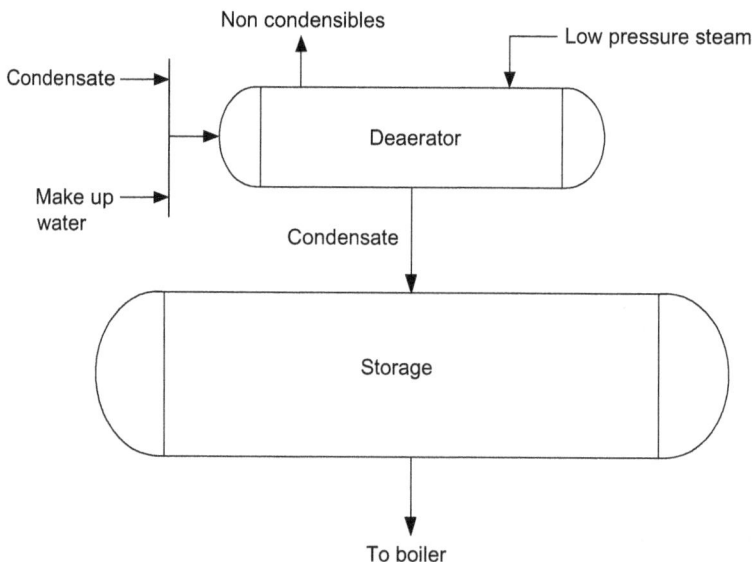

FIGURE 4.15 The concept of deaeration and mixing of condensate.

4.5.1.2 Types of Deaerators

In general, there are two main types of deaerators used today [72–74]:

i. tray type deaerators and
ii. spray type deaerator

In the tray type Figure 4.16(a), the water is distributed over trays and steam is injected to strip the dissolved gases from the water as it cascades down from tray to tray. The spray type Figure 4.16(b) uses spray nozzles to "atomize" the water into droplets. While deaerators can reduce oxygen to very low concentration levels, chemical treatment may still be needed to remove the last traces.

4.5.1.3 Working of a Deaerator

The deaerator is part of the feedwater heating system and receives extraction steam from the turbine. The condensate to be heated and the extraction steam are intimately mixed in the deaerator by a system of spray nozzles and cascading trays between which the steam percolates as shown in Figure 4.16 the condensate is heated to saturated conditions and the steam condensed in the process. Any dissolved gases in the condensate are released in this process and removed from the deaerator by venting to the atmosphere or to the main condenser. This ensures removal of oxygen from the system particularly during startup of the turbine and minimizes the risk of corrosion within the rest of the system.

4.5.2 CLOSED FEEDWATER HEATERS

A feedwater heater is a shell and tube heat exchanger designed to preheat feedwater by means of condensing steam extracted or bleed from steam turbines as shown in Figure 4.17. Feedwater heaters are used in a regenerative water-system cycle to improve the thermodynamics efficiency, resulting in a reduction of fuel consumption and thermal pollution. The boiler feedwater is heated up by steam extracted from suitable turbine stages. The heater is classified as closed, since the tubeside fluid remains in a closed circuit and does not mix with the shellside condensate, as is the case with open feedwater heaters. They are unfired since the heat transfer within the vessel does not occur by means of combustion, but by convection and by condensation.

The feedwater heating system uses extraction steam to preheat the feedwater in order to optimize thermodynamic efficiency and to raise the temperature at the desired value for admission in the steam generators. The basic massflow diagram for a FWH is shown Figure 4.18 [76].

A feedwater heater is an unfired heat exchanger designed to preheat feedwater by means of condensing steam extracted or bled from a steam turbine and the objective is to improve the thermodynamic efficiency of the cycle. The heater is classified as closed, since the tubeside fluid remains in a closed circuit and does not mix with the shellside condensate, as is the case with open feedwater heaters. They are unfired since the heat transfer within the vessel does not occur by means of combustion, but by convection and condensation. Most feedwater heaters are of a standard shell and tube configuration, although some are of header type. The majority uses U-tubes, which are relatively tolerant to the thermal expansion during operation. They are designed as per HEI Standard for closed feedwater heaters [77]. Based on operating pressure it is as classified as low-pressure feedwater heaters and low-pressure feedwater heaters.

The feedwater heaters placed between the main condensate pumps and the boiler feedwater pumps are named LP feedwater heaters. They are U-tube bundle heat exchangers. The recommended tube material is stainless steel. The advantage is the resistance to corrosion and erosion. The tubes are longitudinally welded, which allows small tube wall thickness to be realized. The LP feedwater heaters are arranged horizontally.

Leading feedwater heaters manufacturers include Alstrom, Holtec International, Balcke-Durr, an SPX Company, Godrej Americas (Yuba® Feedwater Heater and Ecolaire® Steam Surface Condenser

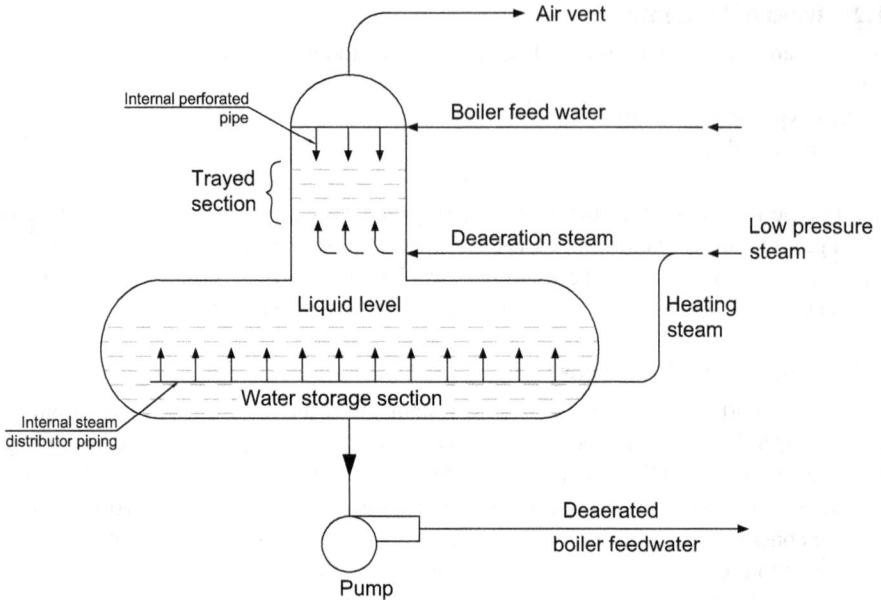

Air vent

Internal perforated
pipe

Boiler feed water

Trayed
section

Deaeration steam

Low pressure
steam

Liquid level

Heating
steam

Water storage section

Internal steam
distributor piping

Deaerated
boiler feedwater

Pump

(a)   Tray type.

Boiler feed water

Air vent

Steam

A

B   B   C

D

E   F   F

Steam
sparger

Deaerated
boiler feedwater

Pump

Note : A = Spray nozzle.
        B = Sparay nozzel shroud.
        C = Baffle.
        D = Steam supply pipe.
        E = Preheating section.
        F = Deaeration section.

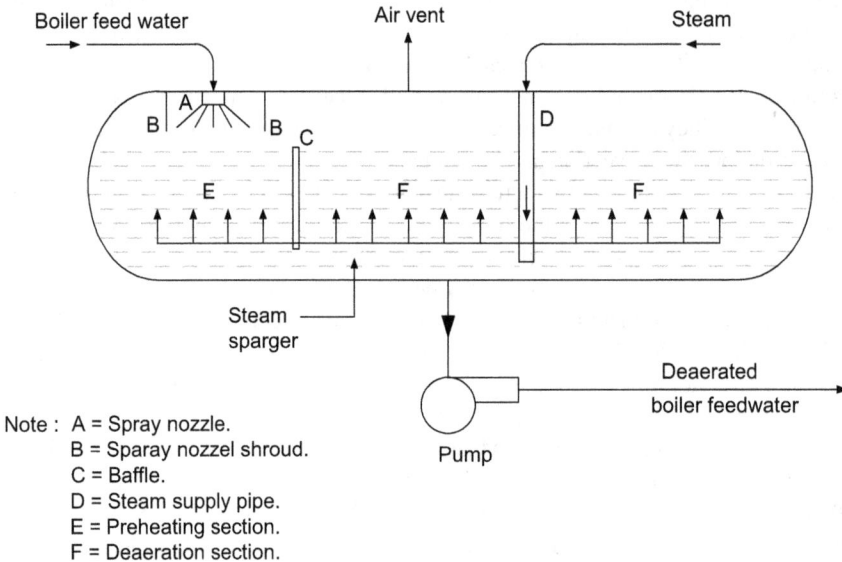

(b)   Spray type dearator.

FIGURE 4.16   Deareatortype (a) Tray type deaerator and (b) Spray type deaerator.

brands) and Westinghouse among others. Most feedwater heaters are of a standard shell and tube configuration, although some are of header type. The majority uses U-tubes, which are relatively tolerant to the thermal expansion during operation. They are designed as per HEI Standard for closed feedwater heaters or other national standards.

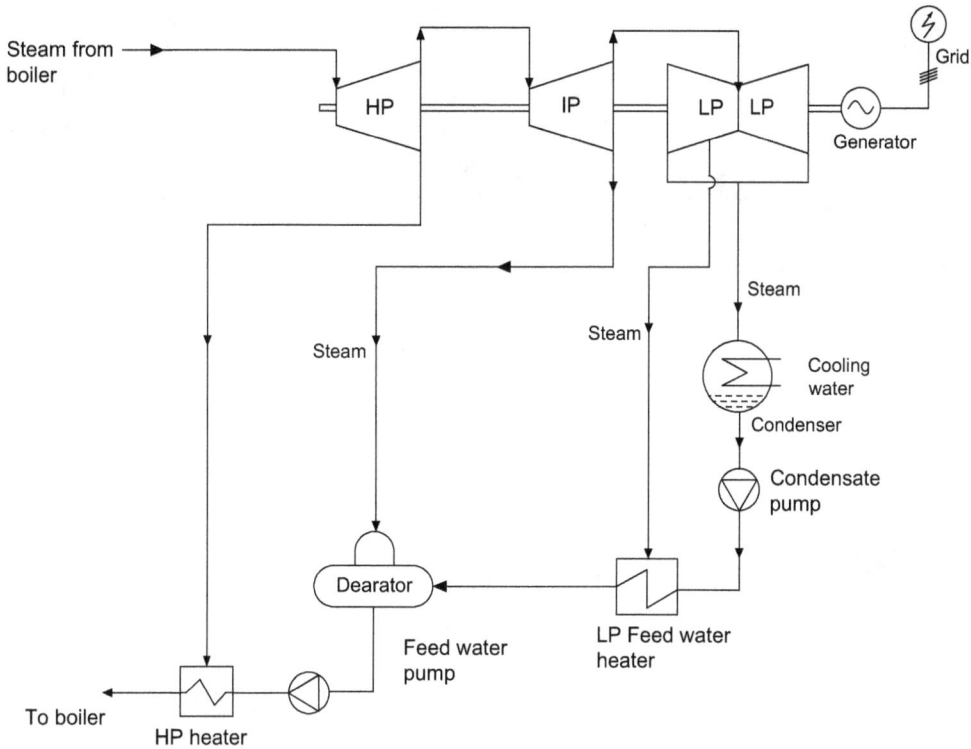

FIGURE 4.17 Steam extraction from steam turbines for feed water heating(an example).

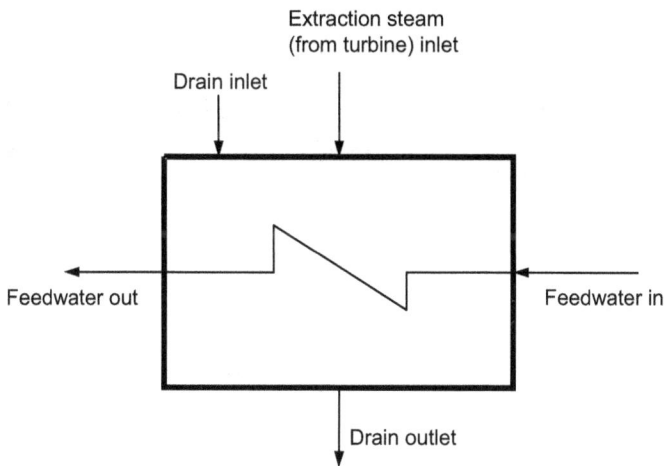

FIGURE 4.18 The massflow diagram for a FWH.

4.5.2.1 High-Pressure Feedwater Heaters

The feedwater heaters placed between boiler feedwater pump and boiler are named high-pressure feedwater heaters. The tube material is mostly specified by the consumer. The standard is using seamless low-alloy carbon steel tubes or others. The advantages are the good thermal conductivity and its good mechanical properties at high temperatures. Figure 4.19 shows a low-pressure and a high-pressure feedwater heater.

(a)

(b)

FIGURE 4.19 Feedwater heater. (a) LP feedwater heater and (b) HP feedwater heater.(Courtesy of Holtec International, Marlton, NJ.)

4.5.2.2 Feedwater Heater Design Based on Number of Zones

There are three basic types:

i.   Single zone type—condensing zone only.
ii. Two zone type—condensing and drain cooling zones common to lower pressure and lower temperature nuclear steam cycles.
iii. Three zones type—de-super heating, condensing and drain cooling zones—mostly found in higher pressure and higher temperature fossil steam cycles.

Feedwater heater types based on shellside zones are shown in Figure 4.20.

(a) 1 Zone single nozzle.

(b) 2 Zones single nozzle.

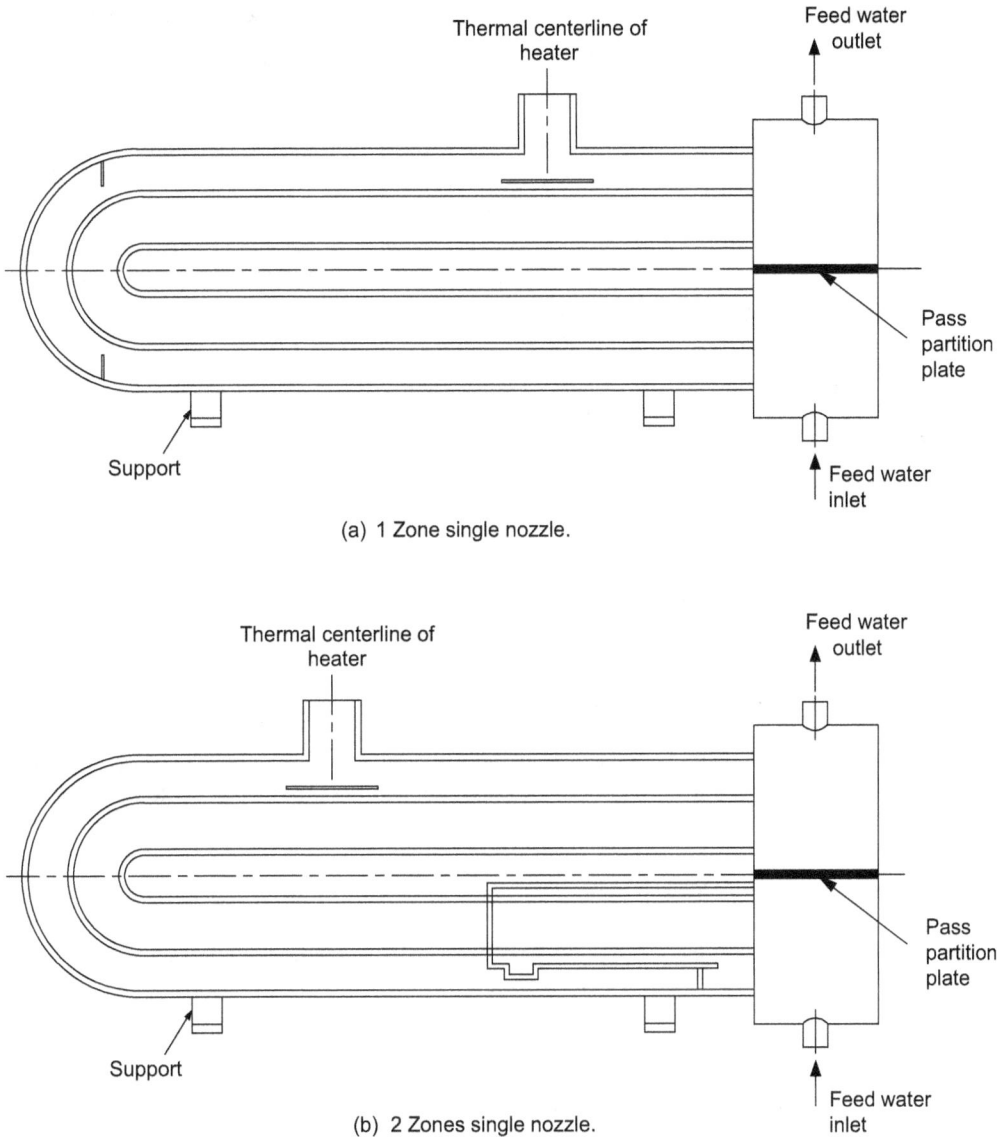

FIGURE 4.20 Feedwater heater types based on shellside zones (a) Single zone type and (b) Two zones type.

4.5.2.3 Zones within a FWH

There are three separate zones within the shell in a feedwater heater and their details are given below:

(i) Condensing zone: all of the steam is condensed in this area, and any remaining noncondensable gases must be removed. A large percentage of the energy added by the heater occurs here. All feedwaters have this zone.

(ii) The condensed steam subcooling zone (optional) at the saturation temperature and is cooled by convective heat transfer from the incoming feedwater.

(iii) Desuperheating zone (optional): the incoming steam enters this zone, giving up most of its superheat to the feedwater.

The desuperheating zone removes a portion of the sensible heat of the superheated extraction steam to elevate the temperature of the feedwater.

4.5.2.4 Steam Flow

The bled steam first enters the desuperheating zone enclosure and is cooled while raising the temperature of the feedwater leaving the heater to a level approaching or equal to the steam saturation temperature. The condensing zone is the largest heat transfer region within the heater shell. The major portion of heat transfer takes place here as the steam condenses and gives up its latent heat. The subcooling zone, which is enclosed in a separate shrouded area within the shell, further cools the condensed steam while heating the incoming feedwater.

4.5.2.5 Basic Terms

The basic terms related to FWH as per Standards for Closed Feedwater Heaters, 9th Edition, 2015 [77] are below:

   a. Heater Duty

Feedwater heater duty consists of the net heat transferred to the feedwater and is expressed in Btu/hr. The heat duty obtained from feedwater flow and enthalpy parameters shall be calculated for each applicable zone. The sum of individual zone heat duties shall be the overall "heater duty."

   b. Terminal Temperature Difference (TTD)

Terminal temperature difference is the difference between saturation temperature corresponding to the entering extraction steam and the outlet feedwater temperature. This value could be either positive or negative.

   c. Drain Subcooler Approach (DCA)

The drain subcooler approach is the temperature difference between the drains leaving the shell side of the heater and the entering feedwater on the tube side.

   d. Logarithmic Mean Temperature Difference (LMTD)

Logarithmic mean temperature difference is the ratio of the difference between the greater temperature difference and the lesser temperature difference to the Napierian Logarithm (Natural Log) of the ratio of the greater temperature difference to the lesser temperature difference.

   e. Desuperheating Zone (DSH)

The desuperheating zone removes a portion of the sensible heat of the superheated extraction steam to elevate the temperature of the feedwater.

   f. Condensing Zone

The condensing zone heats the feedwater by condensing steam.

   g. Drain Subcooling Zone (DC)

The drain subcooling zone reduces the temperature of the drains leaving the condensing zone below the saturation temperature by transferring heat to the entering feedwater.

h. Drains

Entering drains into a feedwater heater are defined as any liquids which enter the heater from higher pressure stages or sources and combine with the shellside condensate.

4.5.2.6 Standards and Codes

Standards provide practical information on nomenclature, dimensions, testing, and performance. Use of the standards will ensure a minimum of misunderstanding between Manufacturer and Purchaser, and will assist in the proper selection of equipment best suited to the requirements of the application.

a. Standards of Heat Exchange Institute (HEI), Cleveland, Ohio

i. Standards for Closed Feedwater Heaters, 9th Edition, 2015
ii. Standards for Shell and Tube Heat Exchangers, 5th Edition, 2013
iii. Standards for Tray Type Deaerators, 9th Edition, 2011
iv. Performance Standards for Liquid Ring, Vacuum Pumps, 4th Edition, 2010
v. Standards for Direct Contact Barometric and Low Level Condensers, 8th Edition, 2010
vi. Standards for Steam Jet Vacuum Systems, 7th Edition, 2012
vii. Standards for Steam Surface Condensers, 11th Edition, 2012.

b. HEI Closed Feedwater Heater Tech Sheets

i. TECH SHEET #106: Specification of Tube Hole Sizes and Tolerances for Support and Baffle Plates
ii. TECH SHEET #127: Basics of Closed Feedwater Heaters
iii. TECH SHEET #128: Typical Feedwater Heater Cold Start-Up Closed Feedwater Heater Specification Sheets—English Units and SI Units
iv. TECH SHEET #132 Comparison of OD and ID Eddy-Current Inspection of Tubing
v. TECH SHEET #134 Heat Treatment of Tubes for Condenser, Feedwater Heater, and Shell & Tube Heat Exchangers
vi. TECH SHEET #135 A Comparison of Welded & Seamless Heat Exchanger Tubing
vii. TECH SHEET #124 Relief Valves vs. Rupture Discs

c. Construction Code

ASME BPVC Sec VIII, Div 1 or any other national Code.

4.5.2.7 Orientation

A wide variety of configurations are offered such as horizontal, vertical, channel up, channel down, one zone, two zones, three zones, duplex, or combinations that are best suited to meet needs. High-pressure heater channel closures can be cup forged with bolted covers, pressure seals, breech locks, or hemi-head/manways. Figure 4.21 shows horizontal orientation of a FWH.

4.5.2.8 Construction

Feedwater heater construction basic details are given below [78]:

1. Heater body
   The heater body proper is of welded rolled steel plates. If the heating temperature is high, 0.5% Mo steel plates are used. In the low-pressure type, the water chamber is fastened to the heater with flanges or welded to the heater body; in the high-pressure and high-temperature type, they are welded together for complete air tightness.

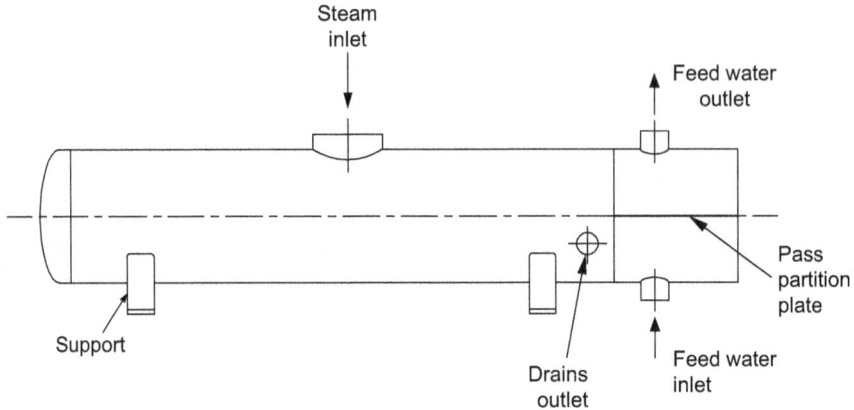

FIGURE 4.21   Horizontal orientation of a FWH.

2. Tubesheets. Tubesheets of the low-pressure type are made of steel plates while those of the high-pressure type are made of steel forgings formed integrally with the water chamber. They are drilled with special care to receive the heating tubes.
3. Water chambers. In the high-pressure type heater, the water chamber is specially constructed to withstand high-pressure feedwater.
4. Tubes. Tubes are U-shaped seamless tubes. For low-pressure heaters, aluminum brass tubes, stainless steel tubes, or carbon steel tubes are used. For high-pressure heaters, copper nickel tubes, 70–30 nickel copper (Monel metal) tubes, or carbon steel tubes are used.

4.5.2.9   Working

The low-pressure feedwater heaters receive extraction steam from the low-pressure turbine for heating the feed water. Condensate extraction pumps, pump condensate from the condenser hotwell through the low-pressure heaters to the deaerator.

The high-pressure feedwater heaters receive extraction steam from the high-pressure turbine. The condensed steam collects in the bottom of the shell and is drained away. Feedwater pumps, pump feedwater from the deaerator storage tank through the high-pressure heaters to the steam generator. The conventional design for a feedwater heater is that of a horizontal cylindrical shell inside which is a bank of U-tubes connected to a divided header at one end, as shown in Figure 4.22. Generally, the drains from the high-pressure heaters are cascaded through lower-pressure heaters to the deaerator, and those from the low-pressure heaters likewise to the condenser. Like the reheaters and deaerator, the feedwater heaters are self-regulating and draw only as much extraction steam from the turbine as they require to heat the feedwater.

4.5.2.10   Temperature Profiles for a High-Pressure Feedwater Heater

The steam condenses almost isothermally, and the condensate is subcooled below the saturation temperature. In the subcooling zone heater surface is assigned to extract heat from the condensate (drains) from the condensing zone. Figure 4.23 depicts the temperature profiles for a high-pressure feedwater heater, which receives superheated steam extracted from a high-pressure turbine [79]. A heater may have neither a desuperheating zone nor a drain cooling zone. In the condensing zone, the tubes are supported by plates or grids of rods. The desuperheating and drain-cooling zones are contained within the shell by a shroud or wrapper, and are usually well baffled to both support the tubes and promote a satisfactorily high shellside heat transfer coefficient. Sometimes other types of a baffle support, based on some form of grid or array of rods, are used to minimize the risk of tube flow induced vibration. Feedwater heaters can be located

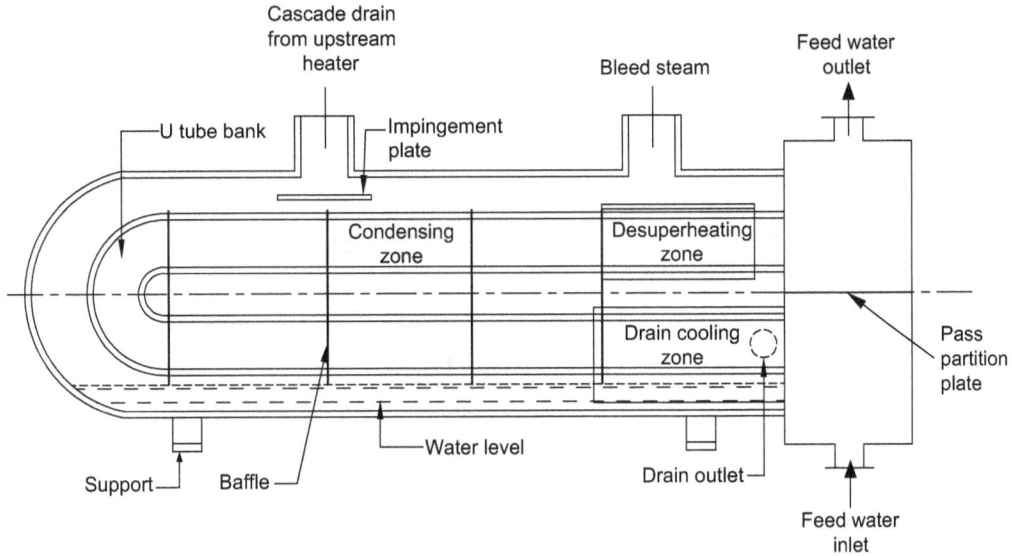

FIGURE 4.22 Internal construction details of a horizontal, high-pressure feedwater heater.

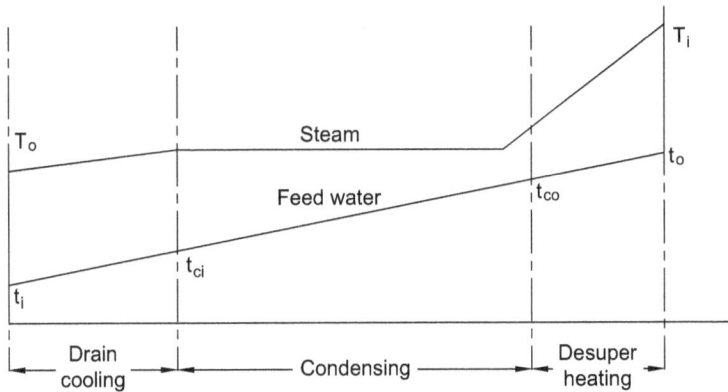

FIGURE 4.23 Temperature profiles for a high pressure feedwater heater.

either horizontally or vertically. The horizontal orientation is more common, but vertical heaters are sometimes preferred. A feedwater heater must be equipped with a vent to allow removal of noncondensing gases.

4.5.2.11 Thermal Design Considerations

Thermal design of a feedwater heater requires an economic optimization of many factors, including material and operating costs. Special attention must be paid to avoidance of (a) wet-wall conditions in the desuperheating section in order to avoid erosion-corrosion problems and (b) excessive pressure drop in the drain cooler, which could cause flashing, and consequent tube damage.

4.5.2.12 Performance Measures

Feedwater heater performance is quantified in three ways: the feedwater temperature rise (the difference between the outlet and inlet feedwater temperatures), the terminal temperature difference (the saturation temperature of the heating steam and the feedwater outlet temperature), and the drain cooler approach (the temperature difference between the drain cooler outlet and the feedwater inlet temperature) as given below:

    i.   feedwater (FW) temperature rise
    ii. terminal temperature difference (TTD)
   iii. drain cooler approach (DCA)

The above three important measurements can be taken that will provide information about the operating efficiency of a feedwater heater as follows [76]:

Feedwater temperature rise is the difference between the feedwater outlet temperature and the feedwater inlet temperature. A properly performing heater should meet the manufacturer's design specifications, provided the level controls are up to the task.

Terminal temperature difference (TTD) is defined as the saturation temperature of the extraction steam minus the feedwater outlet temperature. An increase in TTD indicates a reduction in heat transfer while a decrease is an improvement. Typical ranges for TTD on a high-pressure heater with and without a desuperheating zone are $-3°F$ to $-5°F$ and $0°F$, respectively. The TTD for low-pressure heaters is typically around $5°F$. Steam tables and an accurate pressure reading are required to complete this calculation.

Drain cooler approach (DCA) temperature is the temperature difference between the drain cooler outlet and the feedwater inlet temperatures. The DCA infers the condensate levels present within a feedwater heater. An increasing DCA temperature difference indicates the level is decreasing, whereas a decreasing DCA indicates a rise in level. A typical value for DCA is $10°F$. Figure 4.24 shows the three performance parameters for feedwater heater.

Performance Indicators [80]:

    a. Feedwater temperature rise

       Feedwater temperature rise is the difference between the feedwater outlet temperature and the feedwater inlet temperature:

       Feedwater temperature rise $= T_{FW\ Out} - T_{FW\ In}$

A properly performing heater should meet its design specifications, provided the level controls are working properly.

    b. Terminal temperature difference (TTD)

Terminal temperature difference, or TTD, provides an excellent indication of the performance of a feedwater heater. As shown in Figure 4.24, the TTD is calculated by subtracting the tube-side feedwater outlet temperature from the feedwater heater shellside steam saturation temperature (at the operating shell pressure):

$$TTD = T_{Sat\ Shell} - T_{FW\ Out}$$

An increase in TTD indicates reduced heat transfer. TTD increases can be caused by several internal issues such as tubesheet bypass, fouling, etc. A high drain level begins to cover tubes in the condensing section and will also increase TTD.

    c. Drain cooler approach (DCA)

A drain cooler is usually an extra tube surface area in the FWH that is baffled off from the rest of the tube bundle so that all of the condensate draining from the FWH flows through the drain cooler. This

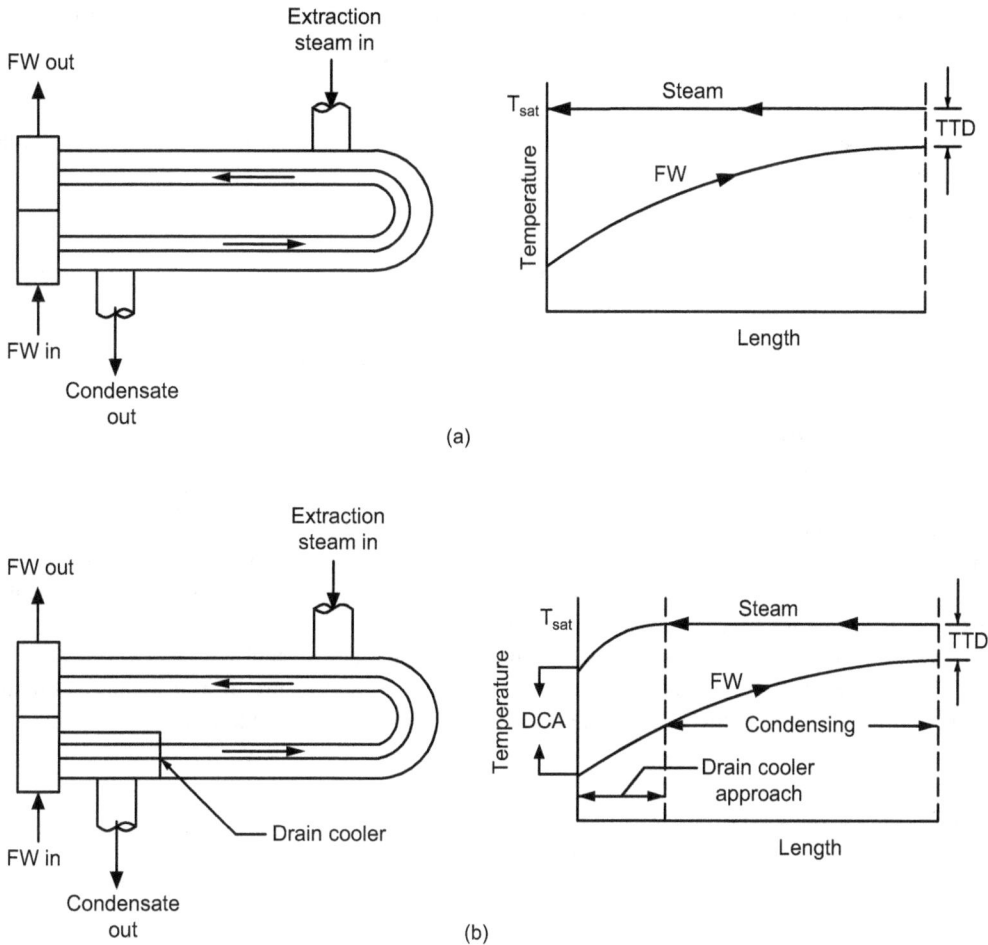

FIGURE 4.24 The three performance parameters for feedwater heater (a)TTD Illustrated in a Straight condensing feedwater heater and (b) Feedwater heater with drain cooler.

allows the condensate draining from the FWH to be used to heat the incoming feedwater. A measure of the effectiveness of the drain cooler is the DCA. As shown in Figure 4.24, DCA is calculated by subtracting the temperature of the feedwater entering the heater from the shell condensate drain temperature:

$$DCA = T_{Drain} - T_{FW\,In}$$

4.5.2.13 Feedwater Heater Design and Maintenance [11]

Minimizing cycling of heaters
It is well known that frequently cycling units from service shortens their life and reduces reliability. Heaters can also be impacted by cycling. Implementing practices to minimize cycling can extend their life.

Utilizing instrumentation to enable timely leak detection
Timely leak detection provides significant payback. Just as boiler leaks often result in collateral damage, heater leaks left unrepaired can damage adjacent tubes. These heater leaks also impact the

heater performance and unit heat rate. Available instrumentation already installed for other purposes can often be used for leak detection.

Acoustic leak detection
Acoustic leak detection can provide a means of detecting heater leaks. This has been successfully implemented by several plants.

Flow monitoring
Installing and monitoring flow instrumentation can be an inexpensive method of early leak detection by alarming the difference between flows before and after heaters. Many plants monitor boiler feed pump suction flow (before heaters) vs. feedwater flow (after heaters). Since these instruments are already installed for other purposes, there is very little cost required to implement this monitoring.

Implement safe heater repair practices
Online heater tube repair should not be attempted unless heater can be isolated with double valving.

Monitor heater performance
Installing monitoring equipment with the capability to utilize the existing instruments to monitor feedwater terminal temperature difference (TTD) and drain cooler approach (DCA) temperature enables tracking of heater performance. This facilitates detection of such problems as partition plate bypass.

In addition to the parameters displayed on this template, carbon steel HP heaters can develop scale on the internal tube surfaces that can result in feedwater pressure drop, causing additional pump head thereby impacting heat rate and capacity. Correcting this condition typically requires chemical cleaning of the feedwater heaters.

4.5.2.14 Feedwater Heater Performance Issues
Typical feedwater heater problems in a power plant are as follows [80, 81]:

i.   Poor level control.
ii. High level covers the condensing tubes and reduces heat transfer causing a high TTD.
iii. Low level uncovers tubes and may even allow steam to blow through without condensing. If the heater has an integral drain cooler, it will have a high DCA.
iv. Insufficient venting—accumulation of noncondensable gases in the heater shell reduces heat transfer.
v.   Leaking emergency drain valves TTD, DHA.
vi. Mechanical damage inside the heater.

Desuperheater or drain cooler Baffle damage in heater shell
   Waterbox partition leaks.

4.5.2.15 Controls for Auxiliary Equipment
Over recent years, power stations have tended to centralize the control and monitoring of various devices at their central control rooms. The control method is often pneumatic, and data transmitted to the central control room by pneumatic pressure and fluids of direct detectors are not introduced into indicating controllers [82].

Water level control
The water level of the condenser, feedwater heater, deaerator, and various tanks must be maintained at a certain level. The most important aspect to feedwater heater performance is a precise and reliable level control under all operating conditions. Accurate level control ensures the unit is operating

in the area of greatest efficiency (straight condensation) so that optimum heat transfer occurs while preventing undue wear and tear on the feedwater heater and other system components.

Pressure control

Equipment sections requiring pressure control (such as the heating steam system for deaerator), which transform steam extracted from the turbine into use detection mechanisms composed of pressure transmitters and pressure indicating controllers, and the pneumatic pressure transmitted by indicating controllers operates the regulator valves.

Temperature control

When using superheated steam by reducing its temperature, as in the case of a temperature reducing device installed in a heating steam system for a deaerator, or a steam transmission system of a factory, the steam temperature is controlled by detecting the secondary side temperature of the temperature reducing device.

Control valves

Many control valves operated by the pneumatic pressure transmitted by the indicating controller are equipped with a diaphragm in their drive part, and they function by pneumatic pressure acting upon the diaphragm. If a large driving force is required because of the valve's unbalancing force, a piston-type driver is employed, and the piston functions by using high pneumatic or hydraulic pressure.

Alarm switch

As alarm switches for the water level, pressure, and temperature are provided.

4.5.2.16   Failures of the Feedwater Heater

OD erosion is the most common type of damage in carbon steel tubes. The locations most susceptible to OD wear are the subcooling zone and the desuperheating zone. Pitting can occur in tubes of stainless steel and copper alloys. Conventional eddy current testing is applied for nonferromagnetic materials. Remote field eddy current testing is quite effective for inspection of carbon steel tubing.

4.5.2.17   Tube Failures Due to Corrosion

Feedwater heaters in nuclear and fossil applications often experience failures and aging degradation from mechanisms such as flow vibration induced damage, flow-accelerated corrosion, or surface area loss due to tube leakage and plugging that require replacement to restore plant and operational performance. Failures have been identified in the following major physical locations: tubes, condensing, and drain cooler zones. The major corrosion mechanisms identified to affect the feedwater heaters are as follows [82]:

i.   uniform or general corrosion
ii. pitting and crevice corrosion
iii. galvanic corrosion
iv. erosion corrosion or waterside "impingement attack"—a form of localized corrosion that occurs on the waterside of the tubes in areas where the turbulence intensity at the metal surface is high enough to cause mechanical disruption of the protective oxide film.
v.   stress corrosion cracking (SCC)

4.5.3   Steam Surface Condenser

The condenser is simply a large shell and tube heat exchanger with tubes usually horizontally mounted. The shell can be a circular or rectangle in shape. The purpose of the condenser is to condense the exhaust steam from the low-pressure turbine so that it can be returned to the system for reuse as feedwater. The condenser has thousands of small tubes that are made out of admiralty brass, copper, ferritic/super austenitic stainless steel, titanium, etc. The condensate generated is usually recirculated back into the boiler and reused. Steam enters the condenser shell through the steam inlet connection usually located at the top of the condenser. It is distributed longitudinally over the tubes through the space designated as dome area. When the steam contacts the relatively cold tubes, it condenses. This condensing effect is a rapid change in state from a gas to a liquid. This change in state results in a great reduction in specific volume, and it is this reduction in volume that creates the vacuum in the condenser. The vacuum produced by condensation will be maintained as long as the condenser is kept free of air. The steam side (or shell) of condensers typically has a vacuum corresponding to an absolute pressure of 1 or 2 inches of mercury, while the cooling water (or tube) side is at a positive pressure. For maximum thermal efficiency, corresponding to a minimum back pressure, a vacuum is maintained in the condenser. However, this vacuum encourages air in-leakage. A vacuum venting system is utilized to support the condenser vacuum by continually removing any air entering the system.

The condensate is continually removed from the hotwell by condensate pump(s) and is discharged into the condensate system. The air in the system, generally due to leakage in piping, around shaft seals, valves, etc., enters the condenser and mixes with the steam. The saturated air is removed from the condenser by the vacuum venting equipment such as steam jet air ejectors, liquid ring vacuum pumps, or a combination of both. It is necessary to continuously remove air from the system in order to maintain the desired vacuum. An increasing amount of air in the condenser would reduce its capacity and cause the pressure to rise. For standards on steam surface condensers refer to Ref. [83]. At the bottom, there is a condensate extraction pump that helps extract the condensed water from the

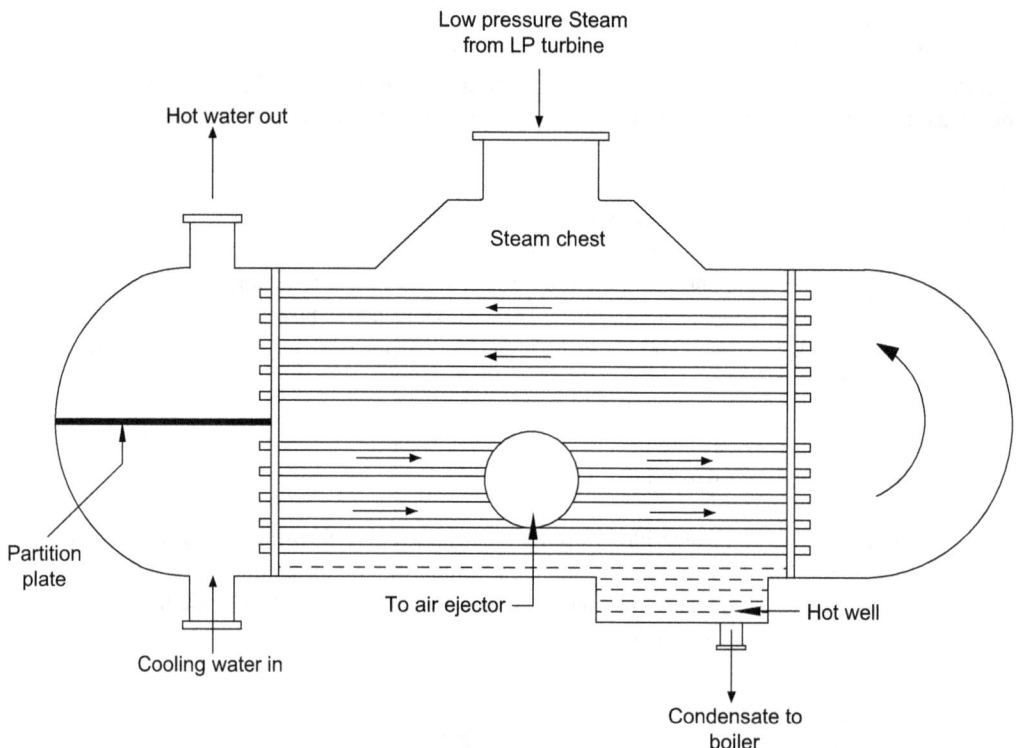

FIGURE 4.25   Steam surface condenser (schematic).

FIGURE 4.26   Steam surface condenser. (Courtesy of GEA Iberica S.A., Barrio de San Juan, Vizcaya, Spain.)

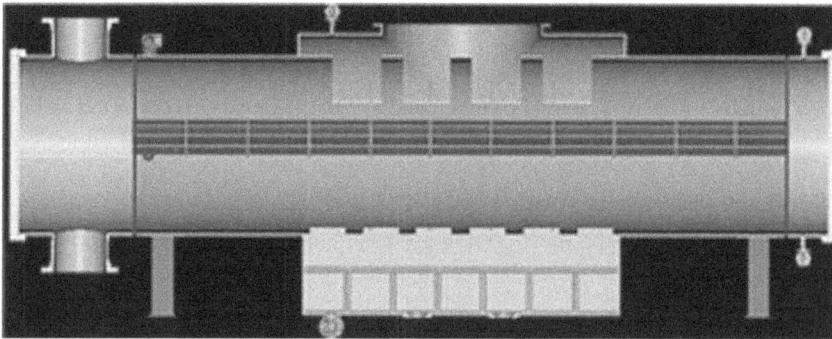

FIGURE 4.27   Steam surface condenser showing dome for steam entry at the top, air removal section in the middle of the core, and condensate collection well at the bottom. (Courtesy of GEA Iberica S.A., Barrio de San Juan, Vizcaya, Spain.)

condenser. The condenser hotwell serves as the entry point for demineralized make-up water, which is required to compensate for leakage losses. Level sensors indicate when there is a need for additional water. The condensate is continually removed from the hotwell by condensate pump(s), and is discharged into the condensate system. A typical steam surface condenser is shown in Figure 4.25 (schematic) and Figure 4.26. Figure 4.27 shows the cross section of a surface condenser with an air removing section.

4.5.3.1   Condenser Types

a. There are two primary types of condensers that can be used in a power plant:

   (i) direct contact condenser
   (i) surface contact or steam surface condenser

Direct contact condensers condense the turbine exhaust steam by mixing it directly with cooling water. The older type barometric and jet-type condensers operate on similar principles. Hellar system uses direct contact condenser.

b. *In yet another classification, there are two types of surface condensers:*

    i.   water-cooled surface condenser
   ii.   air-cooled surface condenser

An air-cooled condenser can be used in thermal power plants, where cooling water is in short supply. However, an air-cooled condenser is significantly more expensive and cannot achieve as low a steam turbine exhaust pressure (and temperature) as a water-cooled surface condenser.

4.5.3.2 Types of Surface Condenser

Surface condensers can be classified into various types depending upon the position of condensate extraction pump, flow of condensate and arrangement of tubes [84].

    a. Down Flow Surface Condenser

In these down flow condensers the exhaust steam travels downwards with cooling water flowing through the tubes. Down flow surface condenser is shown in Figure 4.28(a) [84].

    b. Central Flow Surface Condenser

In this central flow surface condenser type steam enters around the shell's circumference. The condensate flows radially towards the center of tube bundle. The condensate is collected at the condenser bottom and pumped to the hotwell. Central flow surface condenser is shown in Figure 4.28(b) [84].

4.5.3.3 Surface Condenser of Power Plants

Steam surface condensers are the most commonly used condensers in modern power plants. The exhaust steam from the turbine flows on the shellside (under vacuum) of the condenser, while

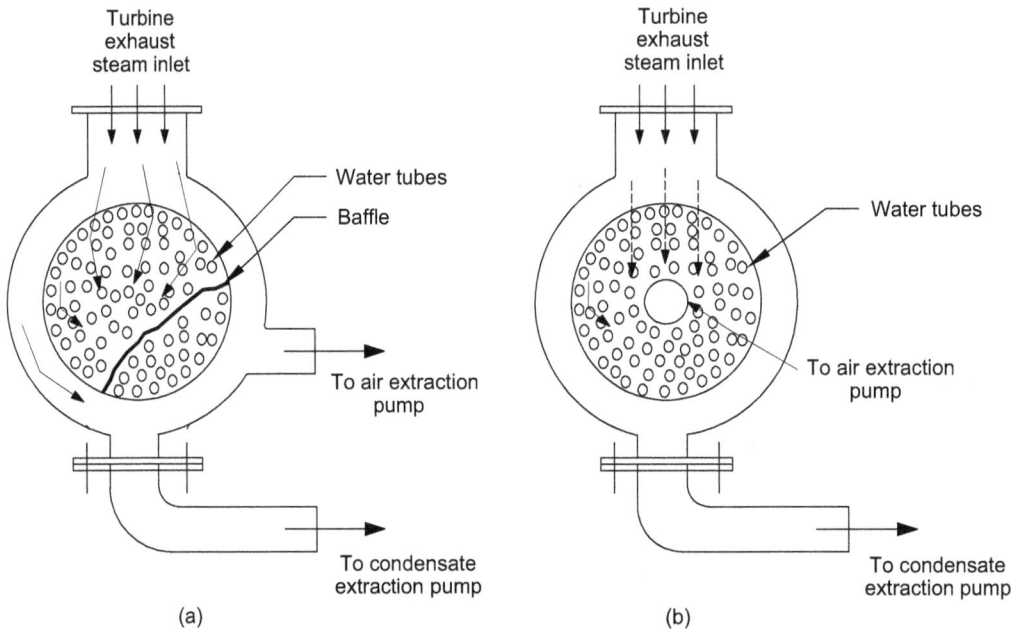

FIGURE 4.28 Surface condense with a cylindrical shell (a) Down flow type and (b) Ccentral flow type.

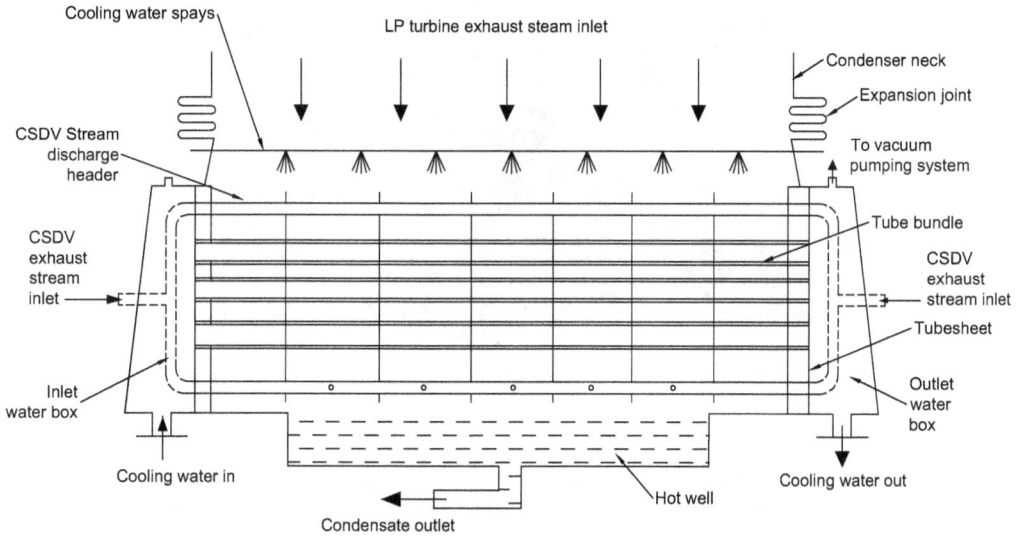

Note : CSDV - Condenser steam discharge valves.

(a)

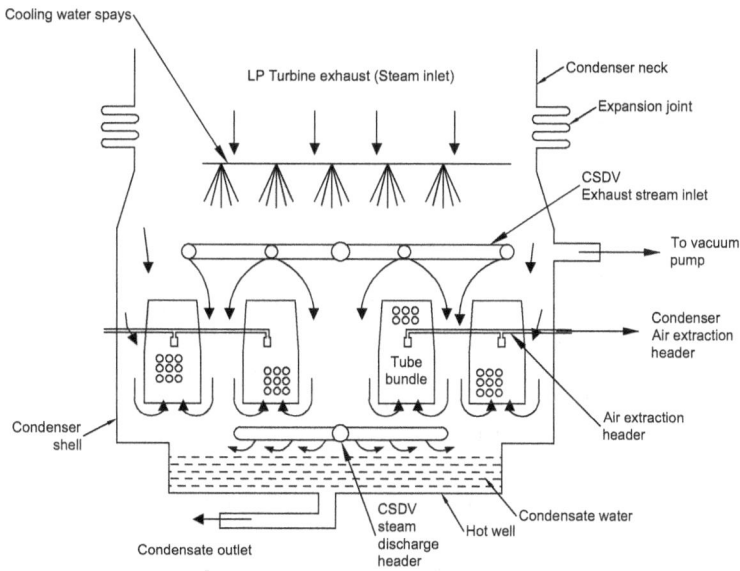

Note : CSDV - Condenser steam discharge valves.

(b)

FIGURE 4.29 Typical power plant condenser (longitudinal section). Adopted and modified from R.A. Chaplin, Ref. (72).

the plant's circulating cooling water flows in the tube side. The source of the circulating water can be either a closed-loop (i.e., cooling tower) or once through (i.e., from a lake, ocean, or river). The condensed steam from the turbine, called condensate, is collected in the bottom of the condenser, which is called a hotwell. The condensate is then pumped back to the steam generator to repeat the cycle. Figure 4.29 [72] depicts a typical water-cooled surface condenser as used in power stations

FIGURE 4.30 Circular shell surface condenser under fabrication. (Courtesy of CaldemonIberica, S.A., a unit of GEA Heat Exchangers, Cantabria, Spain.)

to condense the exhaust steam from a steam turbine driving an electrical generator as well in other applications. Figure 4.30 shows a circular shell surface condenser under fabrication.

4.5.4 PARTS OF CONDENSER

(1) Steam inlet; (2) exhaust connection for turbine; (3) impingement protection—a plate (perforated or solid), dummy tubes, or solid rods used to protect the tubes against high entrance impingement velocity; (4) condenser shell—cylindrical or rectangular body; (5) baffle plate; (6) shell expansion joint; (7) tubes; (8) dome area—an open area above the tubes that permits the steam to easily distribute throughout the length of the bundle without stagnant or overloaded zones; (9) shell flange; (10) air removal section; (11) air ejector; (12) hotwell-storage area with volume sufficient to contain all the condensate produced in the condenser in a given time period. Normally 1 min retention time is specified under design operating conditions. Bath tub or cylindrical types may be used, depending upon the volume and deaeration requirements; (13) condensate outlet(s); (14) support saddles; (15) tubesheets; (16) waterbox—commonly referred to as inlet waterbox, outlet waterbox, return waterbox, return bonnet: provides a directional pathway for circulating water through the tube bundle; (17) waterbox cover flat plate bolted to the ends of channel type waterboxes; (18) waterbox flanges; (19) pass partitions ribs [85].

4.5.5 STRUCTURAL FEATURES OF THE CONDENSER

The condenser is round or rectangular, a shape which can efficiently use the space of the turbine pedestal. Structures and engineering materials of various condenser parts [78, 85,86]:

1. Shell or condenser body: the condenser body is of welded steel plates whose construction possesses adequate mechanical strength to withstand external pressure. When required by

the selected design, intermediate plates are installed to serve as baffle plates that provide the desired flow path of the condensing steam. Large condenser bodies are conveyed in sections and assembled at their installation sites, while the condenser body is usually shop-erected and minutely inspected before being shipped. After assembly, it is filled with water containing fluorescent dye and tested for leakage.

2. Baffle plate: baffle plates, made of rolled steel plates, are carefully drilled for fitting is cooling tubes. The number and position of the support plates are chosen so that they can prevent the cooling tubes from vibrating during operation. They also serve to reinforce the condenser shell so that the shell strength is adequate to withstand external pressure.

4. Water boxes: water boxes are made of cast iron or steel plates applied with an anticorrosive lining. A large one is equipped with a manhole and a small one with an inspection hole. A water box is subjected to a hydraulic test by using a testing pressure 0.4 kg/cm^2 higher than its design pressure.

5. Cooling tubes
   Many of the cooling tubes are aluminum brass series tubes; however, cupro nickel or titanium tubes are sometimes used for the air cooler section.

6. Hotwell

At the bottom of the shell, where the condensate collects, an outlet is installed. In some designs, a sump (often referred to as the hotwell) is provided. Condensate is pumped from the outlet or the hotwell for reuse as boiler feedwater.

7. Tubes and Condenser Tube Material

The tubes are supported by properly located baffle plates to help prevent deflection and flow-induced vibration of the tubes. The tube holes in the supports are de-burred on each side to prevent damage to the tubes. The tubes in the condenser are normally expanded into the tubesheets at both ends.

Condenser tube material selection is one of the most important decisions faced by designers. Typical criteria for evaluating condenser tube materials include waterside erosion–corrosion resistance, steam side corrosion resistance and resistance against impingement attack, resistance to pitting due to chlorides and sulfate-reducing bacteria, resistance to ammonia attack and SCC, compatibility with other system materials to avoid galvanic corrosion, heat transfer capability, economics, etc. A partial list of condenser tube material is given as follows:

Tube Material	ASTM Spec.
Titanium Gr1, 2, 3	B338
304L	A249
304N	A249
316L	A249
2205	A789
904L	B674
254SMO™	B676
AL-6XN™	B676
Sea-Cure™	A268
AL29-4C™	A268
Copper (iron bearing)	B111/B543
Admiralty brass (inhibited)	B111/B543
Al brass	B111/B543
Al bronze	B111/B543
Arsenical copper	B111/B543

CuNi 70/30	B111/B543
CuNi 90/10	B111/B543
Carbon steel	A179
Carbon steel	A214

A few guidelines on tube material selection

A few guidelines on tube material selection based on Ref. [87] are given below:

a. Generally the tubes are made of stainless steel, copper alloys such as brass or bronze, cupronickel, or titanium depending on several selection criteria. Common condenser tube materials include various grades of copper brasses such as admiralty brass, 90:10 and 70:30 Cu-Ni. In copper alloy condensers, stainless steel tubes are used in the air removal section.

b. Series 300 stainless steel alloys, such as 304 and 316L, in new construction in both fresh-water and seawater applications.

c. Higher-grade duplex stainless steels, such as AL-6XN and 6 Mo stainless steel alloys, may also be used.

d. Superferritic stainless steel, such as Sea-Cure, has also been used for cooling water where high levels of chloride are present. These improved materials are required when using brackish and other alternative water sources.

e. In the air-removal section, since the tubes are exposed to an oxygenated, ammonia-rich environment, which promotes corrosion (grooving) in copper-alloy tube, the tube materials in this section is to be provided with more corrosion resistant alloy such as stainless steel.

f. The use of copper bearing alloys such as brass or cupronickel is rare in new plants, due to environmental concerns of toxic copper alloys. Also depending on the steam cycle water treatment for the boiler, it may be desirable to avoid tube materials containing copper. Titanium condenser tubes are usually the best technical choice; however the use of titanium condenser tubes has been restricted by the sharp increases in the costs for this material.

g. Often, when retubing an existing condenser, the material selection may be limited due to the lower heat transfer coefficients of stainless steel and titanium alloys compared with copper alloys.

h. When considering a turbine upgrade, it is good to examine the condenser design to ensure that the changed steam flow rate does not create areas of midspan collisions, fretting, or fatigue in the condenser tubes.

8. Tubesheet

The waterbox/tubesheet/shell joints are fastened together in three ways, depending upon the tubesheet design.

1. *Tubesheet is flanged to the shell*—The waterboxes on either end of the shell are bolted to the tubesheets and shell flanges utilizing staked studs and stud bolts.

2. *Tubesheet is welded to the shell*—The tubesheet outside diameter is larger than the shell, and it extends to form a flange. In this case, the waterbox is simply bolted to the tubesheet with through bolts. All of the through bolts must be removed in this type of design in order to remove the waterboxes.

3. *Tubesheet is welded to the shell and to the waterbox*—The waterboxes are not removable. The waterbox covers can be removed by simply removing all of the through bolts.

Rectangular double tubesheeets

In many heat exchange applications, intermingling of shellside and tubeside fluids may cause undesirable results, loss of quality of product, safety hazard. Therefore, prevention of any fluid

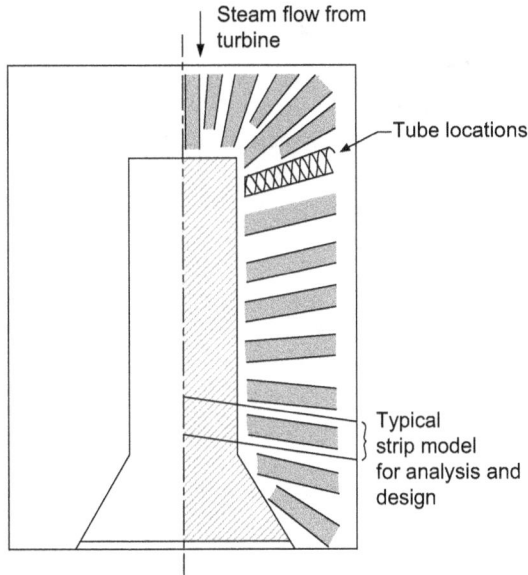

FIGURE 4.31a Rectangular tubesheet of a surface condenser(partially drilled). Adopted from Bernstein, M. D. and Soler, A. I, Ref. (88).

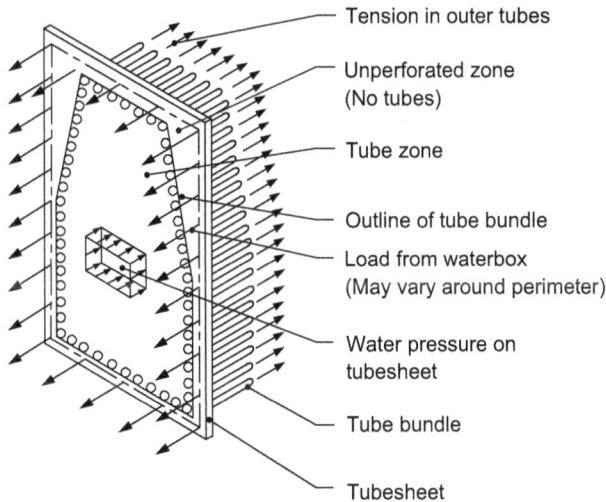

FIGURE 4.31b Idealized representation of rectangular tubesheet loading. Adopted from Bernstein, M. D. and Soler, A. I, Ref. (88).

leakage between shellside and tubeside of shell and tube heat exchangers becomes a prime design consideration. One method to inhibit mixing of component fluids is to employ double tubesheet construction. Double tubesheet construction has also been found in large power plant condensers. In the condenser application to large rectangular tubesheets, the primary concern has been prevention of contamination of treated and demineralized water due to the leakage of circulating water (raw water) into the condenser steam space. The rectangular tubesheet of a surface condenser and idealized representation of tubesheet loading are shown in Figure 4.31a and Figure 4.31b.

For rectangular tubesheet design refer to HEI Standards [83]. The basis of the rectangular tubesheet design is discussed by Bernstein et al. [88], in Singh and Soler [89] and [90].

9. Auxiliary equipments

To efficiently operate a steam turbine, it is necessary that each of its auxiliary equipment functions well, provides reliability, and harmonizes with the overall plant system. Principal auxiliary equipment consists of condensers, air ejectors, feedwater heaters, deaerators, moisture separators, evaporators, and heat exchangers for cooling equipment.

10. Air removal section

Condensers must continually vent noncondensable gases to prevent air binding and the loss of heat transfer capability. An air extraction system removes air that is in the system before start-up and that may leak into the system during normal operation. The air is extracted by vacuum pumps, which run continuously to maintain condenser vacuum. The air-removal section is located toward the bottom of, or deep within, the tube bundles where the condensate and water vapor temperature tends to be lower. This region of tubes is surrounded by a shroud (roof and side panels) to protect the tubes from being heated by descending condensate and steam.

11. Steam jet air ejectors

The air ejector is a device for discharging any air which leaks into the condenser to the atmosphere. The air ejector draws out the air and vapors that are released from the condensing steam in the condenser. If the air were not removed from the system it could cause corrosion problems in the boiler. Also, air present in the condenser would affect the condensing process and cause a back pressure in the condenser. When starting the turbine, a starting air ejector is employed to rapidly discharge a large amount of air from the condenser, and a priming air ejector is used to prime the cooling water system of the condenser. Figure 4.32 shows a steam jet air ejector.

Parts of Ejector
It generally consists of converging nozzle, diverging nozzle, diffuser throat, inlet and outlet pipes.

12. Waterboxes

The tubesheet at each end with tube ends rolled, for each end of the condenser is closed by a fabricated box cover known as a waterbox, with flanged connection to the tubesheet or condenser

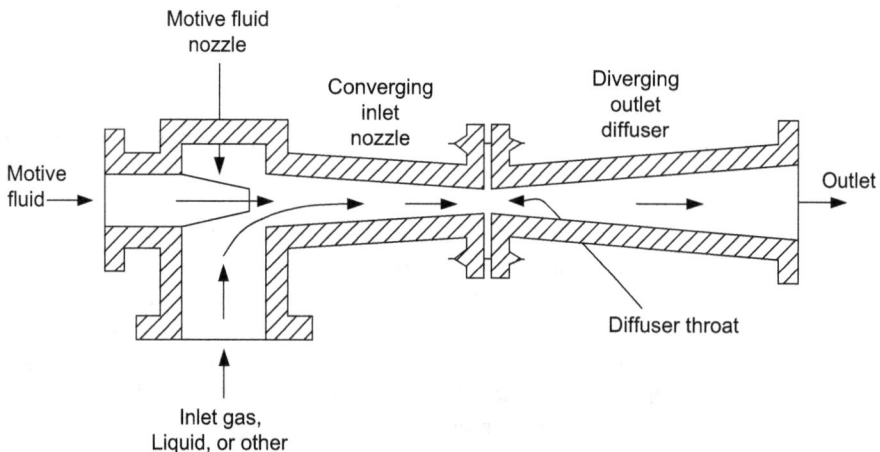

FIGURE 4.32 Steam jet air ejector.

shell. The waterbox is usually provided with man holes on hinged covers to allow inspection and cleaning. These waterboxes on inlet side will also have flanged connections for cooling water inlet butterfly valves, small vent pipe with hand valve for air venting at higher level, and hand operated drain valve at bottom to drain the waterbox for maintenance.

13. Condensate Pumps

Condensate pumps are those kinds of pumps that are used to collect and transport condensate back into a steam system for reheating and reuse, or to remove unwanted condensate. Condensate pumps have a tank in which condensate can accumulate.

14. Circulating Pumps

Condenser circulating pumps are used to pump cooing water through the condenser. The source of the cooling water can be the sea, lake, river or a cooling tower. Low speed–horizontal-double suction-volute centrifugal pumps are used for this application. This pump has a simple but rugged design that allows ready access to interior for examination and rapid dismantling if repairs are required.

4.5.6 CONDENSER PERFORMANCE

A poor performing condenser can significantly affect both the efficiency of the turbine and reduce the power generated. The factors required to optimize condenser performance include

  i. quality water treatment required to maintain the performance, and
  ii. cleaning to restore design conditions.

4.5.7 CONDENSATE SYSTEM

The condensate system takes water from the condenser hotwell and pumps it through a series of components to the suction of the feed pump. These components include plant exhaust and steam air ejector condensers, condensate polishers (if installed), low-pressure feedwater heaters, and deaerator (if installed). Some main features of the condensate system are known to contribute to steam generator corrosion [91]:

  (1) The parts of the system on the suction side of the condensate pumps are under vacuum and are a frequent source of air inleakage.
  (2) Copper alloy feedwater heater tubes have been a source of copper and copper oxide, which have aggravated corrosion attack in the steam generators.
  (3) The lack of condensate polishers has resulted in corrosion products and ionic impurities not being effectively removed from the condensate.
  (4) The lack of deaerators in the system has resulted in less than optimum deaeration of the condensate.

These problems can be addressed by use of features such as bellows or diaphragmn-sealed valves, stainless steel tubes in feedwater heaters, and installation of condensate polishers and deaerators.

4.5.7.1 Condensate Polishing System

The condensate polishing system installed downstream of the hotwell but upstream of the first stage of low-pressure feedwater heaters, processes approximately 70% of the final feedwater flow (but 100%

of condensate flow), and functions both as an ion exchanger and as a corrosion product filter. Plants with well designed and operated polisher systems generally operate with higher purity feedwater, based on the currently established control parameters, than plants without such systems [91].

4.5.7.2 Feed Water System

The feed system serves to pump condensate to a high pressure, pass it through high-pressure feedwater heaters, and introduce it into the steam generators. The main feature of feed systems, which has contributed to steam generator corrosion is the use of copper alloys for feedwater heater tubing. Copper from this and other sources (low-pressure heaters, moisture separator/reheaters, and condensers) has aggravated denting and pitting problems. Replacement of copper alloy tubing with stainless steel tubing and then chemically removing copper buildup in the feed system is recommended [91].

4.5.7.3 Drain System

Condensate and feedwater is pumped through low-pressure and high-pressure feedwater heater tube sides in order to heat the condensate/feedwater and improve cycle thermal efficiency. The heater shell sides are heated by condensing steam taken from various turbine extraction stages. The condensed steam is handled by the drain system. The drain system consists of the high-pressure and low-pressure feedwater heater drains, MSR drains, gland seal and air ejector condenser drains, drain tanks, and heater drain pumps. Typically, high-pressure drains are pumped forward into the feed system, while low-pressure drains are cascaded back to the condenser; in some cases, low-pressure drains are also pumped forward. The main contributing factor to steam generator reliability is that high-pressure drains pumped forward introduce significant amounts of corrosion products into the steam generators, including copper and copper oxides if the feedwater heaters or MSRs use copper alloy tubes. Hence, it is recommended to direct drains to the condenser and then through condensate polishers if available; as a result, the drains become deaerated and filtered and replace susceptible components with corrosion-resistant materials. In addition, copper alloys should be eliminated from MSRs, especially if their drains are pumped forward or if condensate polishers are not installed [91].

4.5.8 CONDENSER HEAT TRANSFER

U is the heat transfer coefficient for clean condenser tubes.

The condenser cleanliness factor is used to adjust the condenser heat transfer coefficient so that a condenser performance curve relating the condenser pressure to the inlet water temperature, flow rate, and heat duty can be generated [11]:

$$U = \frac{Q}{A \ \Delta T_{tm}}$$

where
U is the calculated heat transfer coefficient, Btu/hr/ft^2/°F
Q is heat transfer rate, Btu/hr
A is heat exchange area of the condenser tubes, ft^2
ΔT_{lm} is log mean temperature difference for the condenser, °F

The heat exchange area of the condenser tube is found on the heat exchange specification sheet and can be calculated directly from the number and physical dimensions of the condenser tubes. The heat transfer rate is calculated by

$$Q = \dot{m}_{cw} c_p \left(T_{hw} - T_{cw} \right) = q_{cw} \rho c_p \left(T_{hw} - T_{cw} \right)$$

where

$\dot{m}_{cw}$ is mass flow rate of circulating water, lbm/hr,

q_{cw} is volumetric flow rate of circulating water, ft³/hr,

c_p is heat capacity for water, Btu/lbm/°F,

ρ is density of water,

T_{hw} is hot water temperature at cooling tower inlet, °F, = condenser outlet temperature.

T_{cw} is cold water temperature at cooling tower outlet, °F = inlet temperature for condenser.

The log mean temperature difference for the condenser is given by

$$\Delta T_{lm} = \frac{T_{hw} - T_{cw}}{In \dfrac{T_s - T_{cw}}{T_s - T_{hw}}}$$

where T_s is saturation temperature at the measured condenser pressure, °F.

The heat transfer coefficient for clean tubes is calculated from methods contained in the heat exchanger institute (HEI) Standards for Steam surface condensers. The condenser cleanliness is calculated by cleaniliness factor, C_f given by $\dfrac{U}{U_{clean}} = C_f$

4.5.9 Condenser Heat Duty

Condenser duty is the largest component of heat rejection, and the other components, losses from auxiliary equipment and heat in the exhaust gas, are estimated based on generation type. The heat removed from the steam and transferred to cooling water by a condenser is the majority of waste heat and is defined as the condenser duty of the plant. The share of condenser duty that goes to produce evaporation depends on cooling system type as well as a number of environmental conditions, such as plant elevation, air temperature, humidity, wind speed, and ambient water temperature. Generally, cooling systems transfer the bulk of condenser duty to the atmosphere through evaporation and the rest is lost through conduction and radiation [92].

4.5.10 Reliability and Performance of Condensers

Guidelines for effective steam condensation and condenser satisfactory performance include the following:

i. Prevent/control condenser inleakage,
ii. proper steam distribution and drain,
iii. optimized steam lanes in the tube layout,
iv. efficient extraction of noncondensable gases,
v. stripping of the dripping condensate to minimize oxygen content in the hotwell,
vi. leak-tight tube-to-tubesheet joints, etc.

4.5.10.1 Condenser Tube Leakage

Condenser tube leakage has an impact on the operating cost, availability, reduction of heat transfer capability, reduction of deaeration efficiency, etc. The major impact of condenser leakage is the value of the power generation lost while the unit is at part load or off line to repair the leaks [93]. Very small condenser leaks can be detected by analysis of the water chemistry of the turbine cycle. A condenser may affect plant performance as a result of the following deficiencies [93]:

   i.   reduction of heat transfer capability,
   ii. reduction of deaeration efficiency,
   iii. loss of integrity (inleakage of air or water).

4.5.11   COMMON CONDENSER PROBLEMS

Major operational and maintenance problems of condenser include the following [68]:

   i.   high condenser pressure;
   ii.   condenser leaks;
   iii. corrosion such as galvanic, erosion-corrosion, MIC, etc.;
   iv.   low cleanliness factor;
   v.   deposition, scaling and fouling;
   vi.   ingress of gases (oxygen, carbon dioxide, ammonia, etc.);
   vii. low cleanliness factor and hence low heat transfer coefficient;
   viii. air in-leakage;
   vi.   fouling of tubesheet surfaces and blocking of tube inlets, etc.

4.5.11.1   Air inleakage

Some common points of air in-leakage into the steam path are as follows [68]:

   i.   vacuum pumps;
   ii. manways;
   iii. test probe penetrations;
   iv. rupture discs;
   v.   condensate pumps;
   vi. valve packing, etc.

Operating the condenser at the highest vacuum increases plant efficiency, thereby allowing the plant to produce more electricity. When a vacuum leak occurs in the condenser, noncondensable gases are introduced and must be vented. The gases increase the operating pressure, thereby reducing the turbine output and efficiency. The gases also decrease the heat transfer of the steam to the coolant and can cause corrosion in the generator. To prevent excessive accumulation, most power plants use steam jet air ejectors and/or liquid ring vacuum pumps to remove the noncondensable gases. When these methods cannot keep pace with the rate of air in-leakage then the leaks must be found and repaired.

4.5.12   CONDENSER TUBE FAILURE MECHANISMS

The most important tube failure mechanisms typically result from different forms of corrosion and erosion. Tube failure can be initiated on either the steam or waterside of the condenser tube by either mechanical causes or corrosion [94]:

   i.   Steamside corrosion due to volatile gasses such as ammonia which can cause grooving of some materials.
   ii. Waterside mechanical failures.

iii. Improper rolling of tubes into the tubesheet.
iv. High water velocity due to an excessive number of plugged tubes, as the condenser approaches end of life.
v. Partial blockage of tubes due to macrofouling.

4.5.12.1 Other Failure Modes

i. Mechanical vibration can result in catastrophic condenser tube failures or leaks that are very difficult to find because they open and close depending on steam conditions.
ii. When considering a turbine upgrade, it is good to examine the condenser design to ensure that the changed steam flow rate does not create areas of midspan collisions, fretting or fatigue in the condenser tubes.
iii. Blockage of the tubesheet with debris and macrobiological fouling can be problematic for units using cooling water from lakes, rivers, or seawater.
iv. Freeze protection is also important for any condenser where temperatures drop below freezing or any using cooling water that can freeze. Some waterboxes may not drain completely, particularly if the drain is partially blocked.

4.5.13 CONDENSER OPERATION AND MAINTENANCE

4.5.13.1 Water Chemistry Recommendations

Consideration should be given to operating at high pH in order to minimize erosion-corrosion in the wet-steam piping, and thus minimize sludge buildup in the steam generators. High pH operation should be undertaken only after evaluation of potential plant-specific effects, such as accelerated corrosion of copper-bearing alloys.

4.5.13.2 Make-Up Water System

Water treatment of make-up water will depend on the source of water and boiler manufacturer requirements: suspended solids removal, dissolved solids removal, softening, pH adjustment, dosing of biocides for bacterial control, dosing of anticorrosion agents. Make-up water treatment allows to optimize the boiler processes, reduce the blowdown rate and assume a high-quality stream.

4.5.14 DEPOSITION

Maintaining high condenser performance requires a high cleanliness factor. Deposition presents the most serious barrier to the transfer of heat through a surface and can be divided into two forms—fouling and scaling. Fouling, due to suspended solids in the water, is the accumulation of water suspended materials on tower fill or heat exchanger surfaces. Deposit accumulations in cooling water systems reduce the efficiency of heat transfer and the carrying capacity of the water distribution system. In addition, the deposits cause oxygen differential cells to form. These cells accelerate corrosion and lead to process equipment failure [11].

4.5.14.1 Scale

The most common type of scale found in cooling water systems is calcium carbonate. It is important to note that as temperature increases, the solubility of calcium carbonate decreases. Another common type of scale is calcium sulfate, which results from increased levels of sulfates in the cooling water. Other types of scale found in cooling water systems include silica and magnesium compounds.

4.5.14.2 Fouling

Fouling of condenser tubes generally occurs due to the factors such as microbiological growth, scale formation, deposition of debris, corrosion of condenser tube material and blockage of tube sheets, and even a thin layer of fouling formed by microbiological growth affects the heat transfer process significantly. Chemicals such as biocides for controlling the growth of microorganisms, antiscalants for the control of scale formation, dispersants for the settling of suspended particles and corrosion inhibitors for the control of corrosion in tube materials, which are generally applied in desalination and power plants.

4.5.14.3 Deposition Monitoring and Control—Scaling Indices

The formation of scale most often occurs when the water becomes oversaturated or when water temperature increases. The rate of formation of the scale will depend on temperature, alkalinity, or acidity, and the amount of scale-forming material in the water. The most common scale-forming salts that deposit on heat transfer surfaces are those that exhibit retrograde solubility with temperature. A formation prediction can be calculated with the use of indices such as the Langelier Saturation Index (LSI). It is recommended that an LSI calculation be performed once a day, preferably in the afternoon when temperatures are warmest and conditions are most favorable for deposition [11]. A formation prediction can be calculated with the use of indices such as the Langelier Saturation Index (LSI). This index is described as follows [11]:

$$LSI = pH_s - pH$$

where
pH is the measured water pH,
pH_s is the pH at saturation in calcite or calcium carbonate.

If the LSI calculation produces a positive value, the potential to form scale exists; and if a negative value is produced, no scale potential exists. At no time should an LSI value be greater than 2.2. If the values reach 2.2, changes in the treatment program need to be made immediately to decrease the potential for deposition. The Langelier Index is a tool that can be used to prevent calcium carbonate deposition. It should always be maintained below 2.2 to prevent the formation of calcium carbonate [11].

4.5.14.4 Operational Control for Scaling Control

The most direct method of inhibiting formation of scale deposits is operation at subsaturation conditions, where scale-forming salts are soluble. For some salts, it is sufficient to operate at low cycles of concentration and/or control pH. However, in most cases, high blowdown rates and low pH are required so that solubilities are not exceeded at the heat transfer surface. In addition, it is necessary to maintain precise control of pH and concentration cycles. Minor variations in water chemistry or heat load can result in scaling [94].

4.5.14.5 Microbiological Control

The biological treatment program is controlled by the amount of residual chlorine in the bulk water. Cooling tower systems can be treated with a continuous or intermittent program. In general, a system utilizing a continuous feed is to be maintained with a 0.1–0.5 ppm free chlorine residual. Nonoxidizing chemicals can also be effective in controlling microbiological use. It is possible for sessile populations to become protected from bulk water chemistry with as secreted slime that can render oxidizing chemical programs ineffective. As such, one approach to control microbiological activity is to use both oxidizing and nonoxidizing chemicals. Also, the addition of biosurfactants to oxidizing and nonoxidizing programs also improves sessile population penetrations by carrying the biocide to the microorganism [11].

4.5.15 Condenser Tube Corrosion, After Ref. [94]

General corrosion, pitting corrosion, crevice corrosion, MIC, flow-accelerated corrosion, steam water droplet impingement/erosion of SS tubing, hydrogen embrittlement, etc.

  a. Pitting

Pitting corrosion is a highly localized attack that can result in through-wall penetration in very short order. The most common initiator of stainless steel pitting is the presence of chlorides in the condensate. Several alloying elements such as chromium, molybdenum, and nitrogen promote chloride resistance in this group of alloys. The impact of each element is represented by a formula known as pitting resistance number, PRE_N, which will determine the resistance of stainless steel to chloride pitting:

$$PRE_N = \% \ Cr + 3.3 \ (\% \ Mo) + 16 \ \%N.$$

PRE_N represents the "pitting resistance equivalent" number. The higher the PRE_N, the more chloride resistance an alloy will have.

  b. Crevice Corrosion

Crevice corrosion is caused by the same galvanic driving force as pitting corrosion. Ferritic stainless steels were found to have the highest CCT for a particular PRE_N, above the duplex grade of the same PRE_N, followed by the austenitics. Each specific stainless structure provides a separate parallel linear correlation. After a typical or minimum chemistry is determined, the PRE_N can be calculated.

  c. Maximum Chloride Levels

Although the level of chlorides in the condensate is the primary driver for pitting and crevice corrosion of stainless steel, other important factors have a significant impact, including pH, temperature, presence and type of crevices, and potential for active biological species.

  d. The Effect of Material Properties

One important concern is the tube's erosion resistance. Erosion resistance is a function of the ability of the protective layer to remain attached to the substrate and the strength (hardness) of the substrate directly below the protective layer. Two types of erosion commonly cause problems in the power industry: flow-accelerated erosion-corrosion and water droplet/steam impingement erosion.

  e. Flow-Accelerated Corrosion

When the fluid velocity is so high that it will actually "scrub" the protective film from the metal surface, then flow-accelerated erosion-corrosion is possible. At the higher flow rates, stainless steels and titanium are immune to erosion and corrosion.

f. Water Droplet/Steam Impingement Erosion

Under certain conditions, it is possible to experience erosion of the tube outside surface caused by the localized impact of high-velocity water droplets. The resistance of the tubes to this form of erosion is a direct function of the hardness of the metal substrate below the protective oxide. In general, higher hardness provides higher erosion resistance. Using a water droplet impingement device developed by Avesta Sheffield, alloys can be ranked by time to failure [93].

g. MIC

One of the most common cooling waterside corrosion mechanisms in condenser tubing is microbiologically influenced corrosion. Mechanical cleaning, whether with brushes, scrapers, foam balls, or high-pressure water lancing, is the only cleaning technique that can effectively remove all biofilm in a condenser tube.

h. Steam Water Droplet Impingement/Erosion of SS Tubing

The resistance of the tubes to this form of erosion is a direct function of the hardness of the metal substrate below the protective oxide. In general, higher hardness provides higher erosion resistance. Using a water droplet impingement device developed by Avesta Sheffield, alloys can be ranked by time to failure.

i. Hydrogen Embrittlement

Titanium and superferritic stainless steels, such as S44660 and S44735, can embrittle with exposure to monotonic hydrogen. This commonly occurs in water systems that have poorly controlled cathodic protection. The problem is prevented when the system is controlled so that the voltage is maintained at a potential more positive than -750 mV.

j. Steamside Failure Mechanisms

The most common contaminates include gases like oxygen, carbon dioxide, and ammonia (from decomposing oxygen scavengers). Stainless steels and titanium are resistant to these gases, but the copper-based alloys can be attacked by ammonia; admiralty brass and aluminum brass are the most susceptible. The ammonia can cause two types of failures: ammonia grooving and stress corrosion cracking.

4.5.16 Maximum Chloride Levels in the Condensate

Although the level of chlorides in the condensate is the primary driver for pitting and crevice corrosion of stainless steel, other important factors have significant impact, which include pH, temperature, presence and type of crevices, and potential for active biological species.

4.5.17 Recommendations to Overcome Condenser Corrosion

Few recommendations for operation and maintenance and selection of condenser tube materials to overcome condenser corrosion and leak free performance over the life of the plant are below:

i.   use of titanium tubes for seawater or brackish water-cooled condensers, and for seawater applications use low-cost materials compared to titanium like duplex and superferritic stainless steel tubes;

ii. the condenser-to-turbine expansion joint has been a frequent source of air inleakage. Improved expansion joint designs with seal welding is recommended;

iii. install a cathodic protection system;

iv. minimize air inleakage and circulating water inleakage and monitor contaminants regularly;

v. proper shutdown and layup procedures to avoid particularly pitting corrosion due to cooling water lying stagnant in tubes for any significant time period;

vi. AVT.

Corrosion must be prevented through all volatile treatment (AVT) of the water. Three modes of treatment are available for accomplishing this task. These are defined as

i. AVT (R)—all volatile treatment (reducing),

ii. AVT (O)—all volatile treatment (oxidizing), and

iii. OT—oxygenated treatment. Experience has shown that OT is the preferred water treatment for supercritical units.

4.5.18 Condenser Tube Leaks

4.5.18.1 Circulating Water Inleakage

Circulating water inleakage into the condenser has been the major source of impurities introduced into the condensate and, thus, has been a major factor in boiler corrosion. There are a number of possible causes of water inleakage, including the following:

i. Use of tube materials, such as admiralty brass, that are susceptible to erosion/corrosion.

ii. Improperly rolled tube-to-tubesheet joints.

iii. Tube failures such as pitting, erosion-corrosion, failure of tubes below steam inlet nozzle due to steam impingement attack, fretting wear and collision damage due to flow induced vibration of tubes, etc.

iv. Tube manufacturing defects.

iv. Flow induced vibration damages such as collision damage and fretting of tubes.

As a result of the numerous problems experienced with condensers, and because maintaining leaktight condensers has been found to be essential to steam generator integrity, design and maintenance recommendations have been developed to provide the required integrity. These include [11]

i. use of titanium tubes for seawater- or brackish-water-cooled condensers, use of titanium or stainless steel tubes for freshwater-cooled condensers, performing periodic eddy current tests of tubes, and use of sensitive leak location equipment;

iii. the condenser-to-turbine expansion joint has been a frequent source of air inleakage. Improved expansion joint designs, such as water-sealed rubber expansion joints, are available and are recommended.

4.5.18.2 Condenser Tube Leaks and Corrosion

Improper maintenance of the condenser will result in unnecessary contamination of the cooling-water system, resulting in other major equipment failures such as steam generator tube leaks, turbine blade failure, etc. Condenser leaks are typically the greatest source of contaminants to condensate and feedwater. The three most harmful contaminants introduced by condenser leaks are as follows [11]:

sodium,
chlorides, and
sulfates.

Sodium will bond with other ions, such as hydroxides, chlorides, or sulfates, to form compounds capable of initiating corrosion mechanisms. Quality water chemistry is the primary and first building block for reliable power generating facility operation.

Leaking condenser tubes allow cooling water to enter the condensate and feedwater supply. Cooling water is not purified to the extent of boiler/steam generator feedwater and contains many contaminants. Any contaminants introduced to the condensate and feedwater supply can cause numerous corrosion types. Condenser leaks are typically the greatest source of contaminants to condensate and feedwater.

4.5.18.3 Tracking Condenser Tube Leaks

Tubesheet diagrams locating plugged condenser tubes should be regularly updated at the plant for purposes of documenting whether the failure mechanism(s) is isolated or random. Condenser tubes will typically provide an operating life of 30 to 50 years. Condensers are typically retubed when the number of failed tubes reaches 10 % of the total number of tubes [11].

  a. Cation Conductivity

Cation conductivity provides an optimal measure of water quality to monitor condenser leaks. Cation conductivity of the hotwell(s) sample can rapidly identify a condenser tube leak and all units should monitor and record cation conductivity of the individual hotwell samples as well as the combined hotwell sample [11].

  b. Sodium Analyzer

A second online instrument that indicates the presence of condenser leaks is a sodium analyzer placed at the hotwell or condensate booster pump discharge. This analyzer measures the concentration of sodium cations in the water. One possible source for sodium ingress to the unit is a condenser leak and any increases in sodium ion concentration will be measured by a sodium [11].

Best Practices Manual for Indian Supercritical

4.5.19 Eddy Current Testing of Condenser Tubes to Locate Weak Tubes

Best practices to help ensure condenser reliability include eddy current testing (ECT) of condensers tubes to help determine the remaining life of the tubes, trend tube degradation, and/or to locate tubes which may fail before the next major planned outage so they can be plugged. Timely and effective chemical cleaning, i.e., maintaining water chemistry in a supercritical unit before heat transfer can be the life saver of a steam generator. Eddy current testing of tube failures are discussed in detail in Chapter 3, Heat Exchangers: Mechanical Design, Materials Selection, Nondestructive Testing, and Manufacturing.

4.5.20 Condenser Leak Detection

The most widely accepted method to identify air inleakage is the use of helium as a tracer gas and a mass spectrometer to detect helium. Various other methods have been used with limited success including ultrasonic leak detectors. However, the latter do not work effectively in a noisy plant environment and have limited sensitivity [11]. Various leak testing methods are discussed in Chapter 3, Heat Exchangers: Mechanical Design, Materials Selection, Nondestructive Testing, and Manufacturing.

4.5.20.1 Leak Check Methods

When a unit has or is suspected to have condenser water or air leaks, there are numerous methods that can be used to locate the leaking condenser tube(s). Methods that can be utilized for identifying air and water leaks both online and offline are summarized below [11, 68, 95].

4.5.20.2 Online Methods

Identifying and repairing leaking tubes require the ability to remove a section of the condenser from service and isolate and drain the waterside.

Ultrasonic probe: detecting tube water leaks requires placing an ultrasonic probe at the face of the tubesheet to detect the sound created by the leaking tube as the condenser vacuum draws air through the leak. Very small leaks can be detected by the method. The fluid flow pattern changes from laminar to turbulent at the leaking location causing a 25–40 kHz ultrasonic frequency to be produced. This sound is detected by the acoustic sensor.

Helium/sulfur hexafluoride (SF_6): detecting tube water leaks involves injecting either of the inert gases helium or SF_6 into a portion of a condenser tubesheet. The condenser vacuum pump exhaust is then tested for the injected gas. If gas is detected, a condenser leak is present in that area and other means such as inspection or shaving cream can be used to identify the leak. (Note: environmental regulations restrict use of the chemical SF_6).

Candle: this method involves passing a lit candle over the tube sheet. The candle flame will be pulled toward the leaking tube. This is a low sensitivity technique.

Foam/shaving cream: detecting tube water leaks involves coating one end of condenser tubes and blocking the other end or coating both ends of condenser tubes with a foam agent, such as shaving cream. A leaking tube will pull the foam into the tube. Detecting air leaks involves applying the shaving cream to joints, flanges, welds, which are suspected to be leaking and observe areas where shaving cream is sucked into the condenser.

Plastic wrap: this method involves applying a plastic cling wrap over both ends of condenser tubes. A leaking tube will pull the plastic wrap into the tube. For this method to be successful a good seal between the plastic wrap and tube ends is required.

4.5.20.3 Offline Methods

Flooding

This method can only be performed while the unit is offline. The hotwell is filled with condensate that fills and runs out of any leaking condenser tube. It is possible to mix a fluorescent dye into the water to improve the visibility of any water running from a tube.

(Note: environmental regulations restrict use of the chemical SF6).

4.5.20.4 Helium Mass Leak Detection Methods

Helium leak detection can be done either by pressurizing the vessel with helium and sniffing (detector) the welds or joints from outside or by evacuating the vessel and spraying helium on the welds from outside by a tracer probe or enclosed in a vacuum mask. In each case, the helium atoms passing through a leak are connected to a mass spectrometer, where they are ionized. This can be conducted in one of three ways [96–98]:

    i.   Detector-probe technique.
    ii.   Tracer-probe technique.
    iii. Vacuum technique or hood method.

For details refer ASME Code Sec V.

(a) Helium Mass Spectrometer Test—Tracer Probe Technique

In the helium tracer probe method as shown in Figure 4.33(a), the mass spectrometer leak detector is connected to the internal volume of an evacuated vessel or heat exchanger while helium is sprayed from the probe at each joint and is moved over the external surface to detect the specific locations of leaks. Any leakage into the vacuum side is detected by the mass spectrometer leak detector. The tracer probe is a semiquantitative technique to detect and locate leakages and should not be considered quantitative.

(b) Helium Mass Spectrometer Test—Detector Probe Technique

The detector probe technique, as shown schematically in Figure 4.33(b), involves pressurizing the vessel to be leak tested with certain percentage of pure helium and sensing the leaking helium using the detector probe, which is moved at a definite speed over all the joints to be tested. Any leakage from the pressurized system is detected by the mass spectrometer leak detector.

(c) Helium Mass Spectrometer Vacuum Test—Hood Technique

The hood test is conducted by placing the component under a vacuum and connected to the mass spectrometer. A "hood" or "envelope" is then established around a portion of the component under test, such as the tubesheet of a heat exchanger. The hood, which is normally made of a plastic material or bag, is then filled with helium to test a large area at one time. If a leak is present, the helium will be drawn into the part due to the differential pressure. The mass spectrometer is monitored to verify the presence of helium leakage. The test arrangement is shown in Figure 4.33(c).

Vacuum technique or hood method: the vacuum technique or hood method involves evacuating the vessel to be tested and filling with almost 100% helium or air–helium mixture in the enclosing mask and detecting the inflow of helium to the vessel under vacuum through leaks if any, using the instrumentation as shown schematically in Figure 4.33(d).

4.5.21 PLUGGING CONDENSER TUBE LEAKS

Once a conductivity excursion has been identified as a condenser tube leak then a plan needs to be developed to eliminate the tube leak. It is recommended a tube be simultaneously plugged at both tubesheets. Tapered solid brass plugs have proven to be very reliable. However, these plugs protrude beyond the tubesheet face and make in-service leak checking difficult.

4.5.22 AIR LEAKAGE MONITORING AND CONTROL

Air leakage can be monitored through either local or remote monitoring equipment. Vendors market instrumentation capable of monitoring air flow from each condenser as well as instrumentation capable of monitoring the air removed by each air vacuum pump or steam jet air ejector (SJAE). By monitoring air flow from each condenser, the condenser with the excess air leakage can be identified.

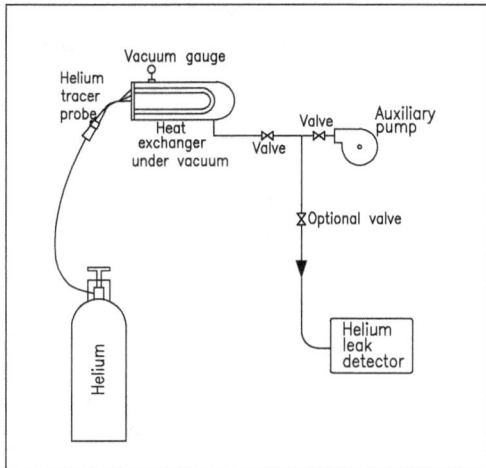

(a) Tracer probe test – Helium leak testing of evacuated vessel or system with tracer probe.

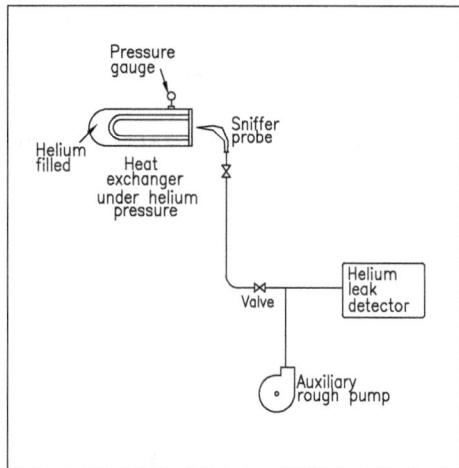

(b) Detector probe test – Helium leak testing of pressurised system with sniffing probe.

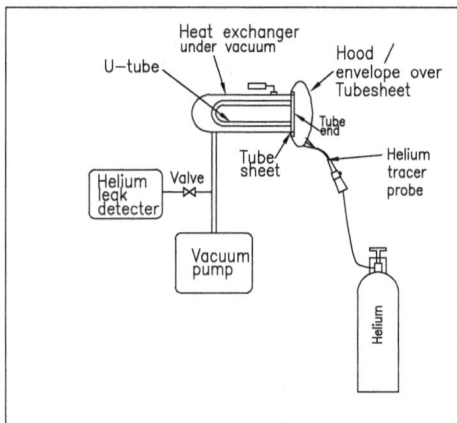

(i) Hood test – Leak testing of evacuated unit inserted into envelope sparyed with a mixture of helium on the tubeside.

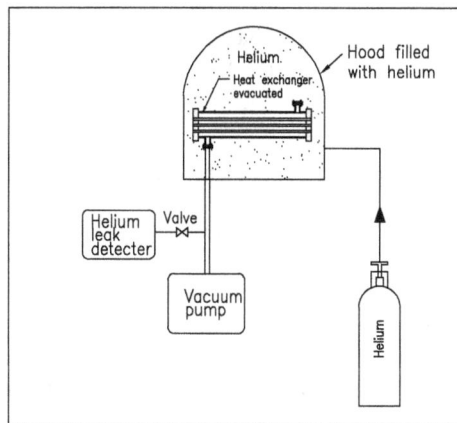

(ii) Hood test – Leak testing of evacuated unit inserted into hood with a mixture of helium on the tubeside.

(c) Hood method – Leak testing of evacuated unit .

FIGURE 4.33 Helium mass spectrometer testing. (a) Tracer probe technique, (b) detector probe technique, and (c) and (d) Vvacuum technique or hood method.

Source: Adapted from McMaster, R.C., ed., *Nondestructive Testing Handbook, Vol. 1, Leak Testing*, **2nd edn., American Society for Nondestructive Testing and American Society for Metals, Columbus, OH.**

4.5.23 CONDENSER TUBE CLEANING

Macrofouling (accumulation of debris), not only reduces the cooling-water flow rate through the tubes it can cause tube corrosion and tube erosion failures. Microfouling, deposits and scaling reduces the heat transfer coefficient and could cause under-deposit corrosion resulting in premature tube failures. Various tube cleaning options are available to reduce or eliminate the micro/

macrofouling and scaling- offline or online methods. (Sponge balls or brushes may be automatically recirculated through the condenser tubes as discussed in Chapter 2). There are many technologies available to clean condenser tubes. Three are described below [11]:

1. Scrapers provide the most effective means of cleaning condensers.
   PACE-D Technical
2. High-pressure nozzle cleaning is another method that can be used to clean the tubes.
3. Rotary brush cleaning machines are also available. These machines can remove excess loose dirt but cannot remove tenaciously attached scale.

Regular inspections of the condenser should be performed to assess the tube cleanliness and to observe, tube blockage, MIC, etc. The inlet tubesheet should be inspected for plugging during opportunity outages. Tubes at the outlet end of the waterbox should be inspected for internal surface deposits. Condensers also need to be inspected for fouling and deposition on a regular basis. Various condenser tube cleaning methods are discussed in detail in Chapter 2.

4.5.24   Condenser End-Of-Life Replacement

Condenser tubes will typically provide an operating life of 30 to 50 years. When the curve for the number of failed tubes (plotted against time) starts to become exponential, experience indicates the condenser tubes have less than 2 years of remaining life before chronic tube failures become a routine maintenance headache. Condensers are typically retubed when the number of failed tubes reaches 10% of the total number of tubes [11].

4.6   COOLING SYSTEMS

There are four main categories of cooling systems [99, 100]:

i.   once-through, where the water is used once before being returned to the water source;
ii. wet open recirculating system, where the warm cooling water is cooled and reused;
iii. dry systems, which use ambient air as the coolant. These can be further categorized into direct (standalone air-cooled condenser, ACC) and indirect (water-cooled condenser and dry tower) types; and
iv. hybrid systems, which have both dry and wet elements. Each element can be used individually or together to achieve the best features of each—wet cooling performance on the hottest days, and the water conservation ability of dry cooling when it is cooler.

4.6.1   Once-Through Cooling SYSTEM

In once-through cooling, water is withdrawn from a natural waterbody, passed through the tubes of a surface steam condenser and returned to the waterbody at a higher temperature (Refer Figure 2.20a of Chapter 3). The capital costs of once-through systems are lowest, compared with other cooling systems, but they require considerably higher water withdrawals and the large intake and subsequent discharge downstream at higher temperatures can be detrimental to aquatic life and ecosystems. Apart from corrosion, the thermal pollution is most significant when the source of the water is a river or other body with limited volume. Other drawbacks include fish impingement, especially small fish and aquatic organisms in the larvae stage and microfouling the accumulation of deposits on the inside surface of the surface condenser tube. The deposits decrease the heat transfer rate. In order to control microfouling, power plants add chemicals such as chlorine in the cooling water. The residual chemicals in the cooling water will be discharged into the source water. This could also be harmful to marine life [101].

4.6.2 Open Recirculating Cooling Systems

Recirculating wet cooling is similar to once-through cooling in that the steam is condensed in a water-cooled surface condenser; but different in that the heated cooling water is not returned to the source water body. Instead it is pumped to a cooling unit, typically a mechanical draft cooling tower and then recirculated to the condenser where cooling of the water is effected primarily by evaporation of a portion of the circulating water (Refer Figure 2.20b of Chapter 3). In cooling towers, the water is cooled by the air to near the wet-bulb temperature using the principle of evaporation. Water flows over a media called fill, which serves to increase contact time with the air and maximize heat transfer. Also, in the cooling tower a small faction (typically 1% to 2%) is evaporated in order to cool the remainder. In this system, the only water withdrawn from the environment is make-up water sufficient to replace that lost to evaporation, blowdown, and drift. This amount is typically 10 to 15 gpm per MW of steam generating capacity. Open-recirculating systems have become prevalent as water has become scarcer and environmental restrictions have been placed on bleedoff discharge. Open recirculating cooling systems save fresh water compared to the alternative method, once-through cooling. The quantity of water discharged to waste is greatly reduced in the open recirculating method, and chemical treatment is more economical. However, open recirculating cooling systems are inherently subject to more treatment-related problems than once-through systems [102]:

i. cooling by evaporation increases the dissolved solids' concentration in the water, raising corrosion and deposition tendencies;
ii. the relatively higher temperatures significantly increase corrosion potential;
iii. the longer retention time and warmer water in an open recirculating system increase the tendency for biological growth;
iv. airborne gases such as sulfur dioxide, ammonia, or hydrogen sulfide can be absorbed from the air, causing higher corrosion rates;
v. microorganisms, nutrients, and potential foulants can also be absorbed into the water across the tower.

Table 4.3 shows cooling system features for once-through, wet open recirculating system, and dry air cooled systems.

4.6.3 Closed Recirculating Cooling Systems

Closed recirculating cooling-water systems are well suited to the cooling of gas engines and compressors (Refer Figure 2.20c of Chapter 3). Diesel engines in stationary and locomotive service normally use radiator systems similar to the familiar automobile cooling system. Closed systems are also widely used in air conditioning chilled-water systems to transfer the refrigerant cooling to air washers, in which the air is chilled. In cold seasons, the same system can supply heat to air washers. Closed water-cooling systems also provide a reliable method of industrial process temperature control [103].

Advantages of Closed Systems

Closed recirculating systems have many advantages:

i. Closed systems are also less susceptible to biological fouling from slime and algae deposits than open systems.
ii. Closed systems also reduce corrosion problems drastically, because the recirculating water is not continuously saturated with oxygen, as in an open system. With the small amount of makeup water required, adequate treatment can virtually eliminate corrosion and the accumulation of corrosion products.

TABLE 4.3
Cooling System Feature

Cooling-water system type	Applications	Advantages	Disadvantages
Once-through	Coastal or natural large water sources without access restrictions	• Highest plant efficiency possible • Smallest footprint • Lowest cost	Highest regulatory burdens. High quantity of water withdrawals, impact on ecosystem. Discharge water temperature limits is stipulated to avoid thermal pollution of water bodies.
Open recirculating water system	Locations where sufficient make-up water is available	• Convenient plant site locations • save fresh water compared to the alternative method, once-through cooling. • Better performance than air-cooled units • Lower cost than air-cooled units	• Significant make-up water requirements • Large footprint • Operational issues like evaporation, blowdown, and drift. Cooling tower blowdown is a difficult stream to treat and also its disposal • subject to more treatment-related problems than once-through systems
Dry Air Cooled	Arid areas or locations where water access is prohibited or uneconomical	• Minimal water withdrawal and consumption to meet evaporation losses • Fewer water-related complications (use of air eliminates issues related to water such as corrosion, biofouling, filtration, treatment, etc.) • Fewest siting and regulatory restrictions	• Highest capital cost • Lower plant efficiency, particularly when ambient temperatures are high. • Impacted by ambient conditions • Largest footprint

4.6.4 DRY COOLING SYSTEMS

When water availability is low, a dry cooling system may be utilized. Dry cooling can be either direct or indirect. In dry cooling systems turbine exhaust steam is ducted directly to an air-cooled condenser. Heat rejection to the environment takes place in a single step in which steam is condensed in finned tube bundles, which are cooled by air blown across the exterior finned surfaces. In an indirect system, cooling water is used to condense the steam, as in a wet recirculating system. Then the cooling water flows through tube bundles that are cooled in a mechanical or natural draft cooling tower. Cooling-water make-up requirements can be nearly eliminated by use of dry cooling systems, but process and steam make-up water requirements are unaffected. Dry cooling system is further discussed in Section **4.7.4.9- 4.7.4.12**.

4.6.5 HYBRID COOLING SYSTEMS

Hybrid cooling systems employ a combination of wet and dry cooling technologies. Most systems are intended for plume abatement and are essentially all-wet systems with a small amount of dry cooling to heat the tower exhaust plume above saturation conditions during cold, high-humidity periods when the wet tower plume is likely to be visible. Hybrid systems intended for water conservation, on the other hand, are primarily dry systems with a small wet capacity to provide additional cooling during the hottest periods of the year to mitigate hot-day capacity loses associated with all-dry systems [104,105]. Hybrid cooling system is further discussed in Section **4.7.4.13**.

4.6.6 ECONOMICS OF THE CW SYSTEM

Wet recirculating systems are roughly 40% more expensive than once-through systems, while dry cooling systems are 3 to 4 times more expensive than a wet recirculating system. While most systems currently employ once-through cooling, environmental regulations and permitting requirements will likely result in developers to choose recirculating or dry cooling options in the future [106].

4.7 OPEN RECIRCULATING COOLING-WATER SYSTEMS AND COOLING TOWER

In an open recalculating cooling-water system, cooling towers are used to reject heat through the natural process of evaporation. Warm recirculating water is sent to the cooling tower where a portion of the water is evaporated into the air passing through the tower. As the water evaporates, the air absorbs heat, which lowers the temperature of the remaining water. The amount of heat that can be rejected from the water to the air is directly tied to the relative humidity of the air. Air with a lower relative humidity has a greater ability to absorb water through evaporation than air with a higher relative humidity, simply because there is less water in the air. For details of recirculating cooling-water system and cooling tower construction, performance and operation refer to Refs. [107–112].

4.7.1 COOLING TOWER

Cooling towers are most commonly used to dissipate heat in open recirculating cooling systems. They are designed to provide intimate air-water contact. Heat rejection is primarily by evaporation of part of the cooling water. Water to be cooled is distributed in the tower by spray nozzles, splash bars, or film-type fill, which exposes a very large water surface area to atmospheric air. Thermal performance of a cooling tower depends principally on the following [108]:

i. Entering and leaving water temperatures.
ii. Entering air wet-bulb temperatures.
iii. Water flow rate.

The entering air dry-bulb temperature affects the amount of water evaporated from any evaporative cooling tower. It also affects airflow though hyperbolic towers and directly establishes thermal capability in any indirect-contact cooling tower component operating in a dry mode.

4.7.2 PSYCHROMETIC ANALYSIS

A psychrometic analysis of the air passing through a cooling tower illustrates this effect (Figure 4.34 [109]). Air enters at the ambient condition Point A, absorbs heat and mass (moisture) from the water and exits at Point B in a saturated condition. The amount of heat transferred from the water to the air is proportional to the difference in enthalpy of the air between the entering and leaving conditions $(h_B—h_A)$[109].

Air heating (vector AB in Figure 4.34) may be separated into component AC, which represents the sensible portion of the heat absorbed by the air as the water is cooled, and component CB, which represents the latent portion. If the entering air condition is changed to Point D at the same wet-bulb temperature but at a higher dry-bulb temperature, the total heat transfer (vector DB) remains the

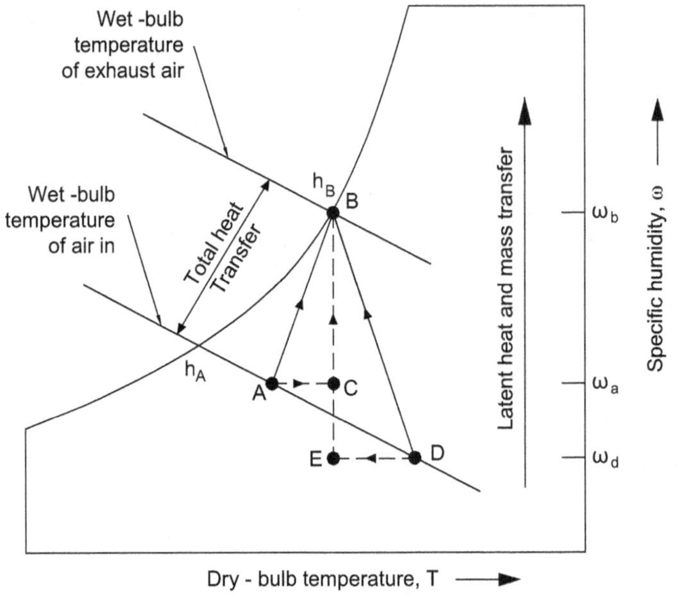

FIGURE 4.34 A psychrometic analysis of the air passing through a cooling tower.

same, but the sensible and latent components change dramatically. DE represents sensible cooling of air, while EB represents latent heating as water gives up heat and mass to the air. Thus, for the same water-cooling load, the ratio of latent to sensible heat transfer can vary significantly.

4.7.3 Cooling Tower Performance Indices

4.7.3.1 Approach

Cooling tower approach is the difference in temperature of the water entering the basin (cold) and the wet-bulb temperature. For the purpose of tower design, a tower with a smaller approach is considered superior. Modern towers commonly have approach temperatures as low as 5°F. While it is possible to have a smaller approach, it becomes cost-prohibitive since the size of the tower grows exponentially as approach lowers, which in turn requires more pumps and fans. Approach is given by

Approach = Cold Water Temperature—Wet Bulb Temperature

4.7.3.2 Cooling Tower Range

Range is the difference between the temperature of water entering the cooling tower and leaving the cooling tower. It is determined by the heat load on the tower and the water circulation rate. Range is given by

Range = Hot Water Temperature—Cold Water Temperature

Range and approach is graphically represented in Figure 4.35.

4.7.4 Classification of Cooling Towers

Cooling towers can be classified into several types based on the air draft, viz., natural draught cooling tower, and forced draft cooling tower and flow pattern—there are two main types of cooling

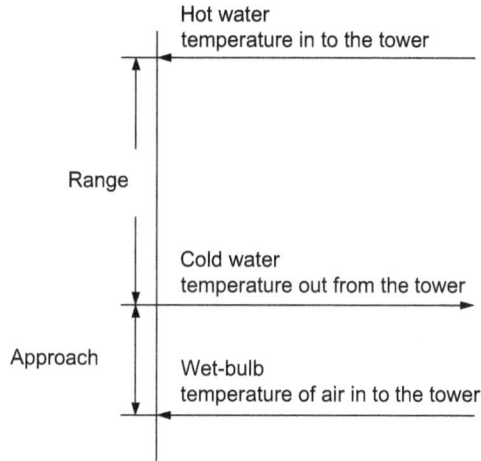

FIGURE 4.35 Graphical representations of range and approach of cooling tower.

towers that are defined by how water or air pass through them. These types include crossflow and counterflow. There are also two varieties classified solely on airflow, known as induced draft and forced draft cooling towers [113, 114].

Each type of cooling tower has its own advantages and disadvantages; thus the proper selection is needed based on the system operation. Beside type selection, the construction material selection of cooling tower is also important.

4.7.4.1 Mechanical Draft Towers

Mechanical draft towers use motor-driven fans to force or draw air through the towers. The two types are

1. Forced draft tower.
2. Induced draft tower.

Forced draft towers use a blower-type of fan at the bottom, which forces the air into the tower. Induced draft towers employ a fan at the top that pulls air up through the tower. Figure 4.36 shows mechanical draft cooling towers.

4.7.4.2 Forced Draft Cooling Towers

In such a cooling tower, the fan is used to circulate air. Forced draft means an exhaust fan placed at the base of the cooling tower, which then causes overpressure. Both axial and centrifugal fans can be used. The major components making up a forced draft cooling tower include fan stack, fan with drive, cooling fill, drift eliminators, outer shell, water basin, and water distribution system. Forced draft cooling towers consume significantly more power than induced-draft cooling towers and are relatively more prone to recirculation if not ducted.

4.7.4.3 Induced Draft Cooling Towers

Induced draft cooling towers are a type of mechanical draft tower that features one or more fans. Axial fans are always used for this type of draft. These fans will be located atop the tower, drawing air upwards against the downward flow of water. The water is passed around the wooden packing or decking. In these models, the coolest water is found at the bottom, coming into contact with the driest air since the airflow is counter to the water flow. Most induced draft cooling towers will feature with these components—spray nozzles, fill, basin, piping, casing/shell, fan and drift eliminator.

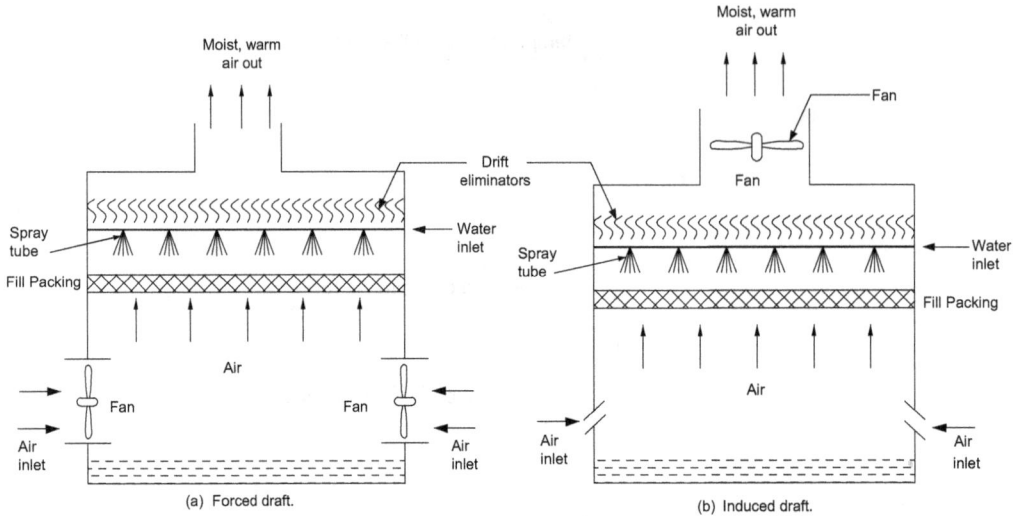

FIGURE 4.36 Mechanical draft cooling towers (a) Forced draft and (b) Induced draft

4.7.4.4 Natural Draught vs and Mechanical Draught Towers

There are two types of cooling tower based on draft:

i.   natural draught towers, which utilize the differing densities between the cooler outside (ambient) air and warm moist air within the tower. The warm air rises up the tower because of its lower density, drawing cool ambient air into the bottom portion. These towers are usually taller than mechanical draught ones; and

ii. mechanical draught towers, which use motor-driven fans to force or draw air through them. Induced draught towers employ a fan at the top to draw air up through the tower, whereas forced draught towers use a fan at the bottom to force air into the tower. Induced draught towers tend to be larger than forced draught ones.

4.7.4.5 Crossflow Cooling Towers

Crossflow cooling towers use a splash fill that allows in-flowing to move air in a horizontal path over the stream of water from the upper reservoirs. Crossflow systems are some of the more expensive equipment types, but they are also some of the easiest to maintain. However, these cooling systems are more vulnerable to frost than others.

4.7.4.6 Counterflow Cooling Towers

In a counterflow system, the in-flowing air travels in a vertical path over the splash fill as the water streams down from the reservoir above. Counterflow systems are usually smaller than their crossflow counterparts. These cooling towers are more expensive due to the fact that more energy is needed to push the air upward against the down-flowing water.

Crossflow and counterflow concept is shown in Figure 4.37.

4.7.4.7 Natural Draft Tower

In this type of cooling tower, the fan is not used to circulate air. Natural draft cooling towers utilize buoyancy via a tall chimney. The natural draft cooling tower makes use of the difference in temperature between the ambient air and the hotter air inside the tower due to their structural strength and minimum usage of material. The hyperbolic shape also aids in accelerating air flow through the

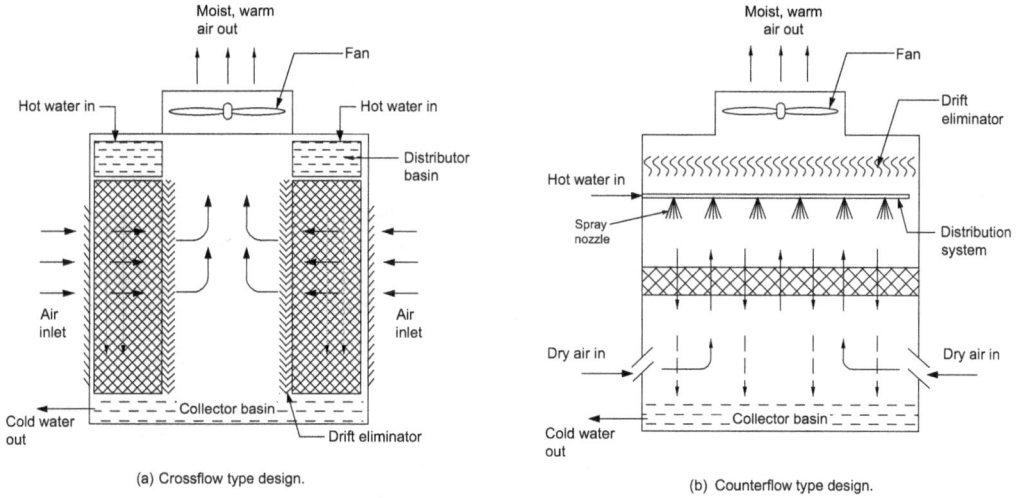

FIGURE 4.37   Cooling tower (a) Cross flow and (b) counter flow.

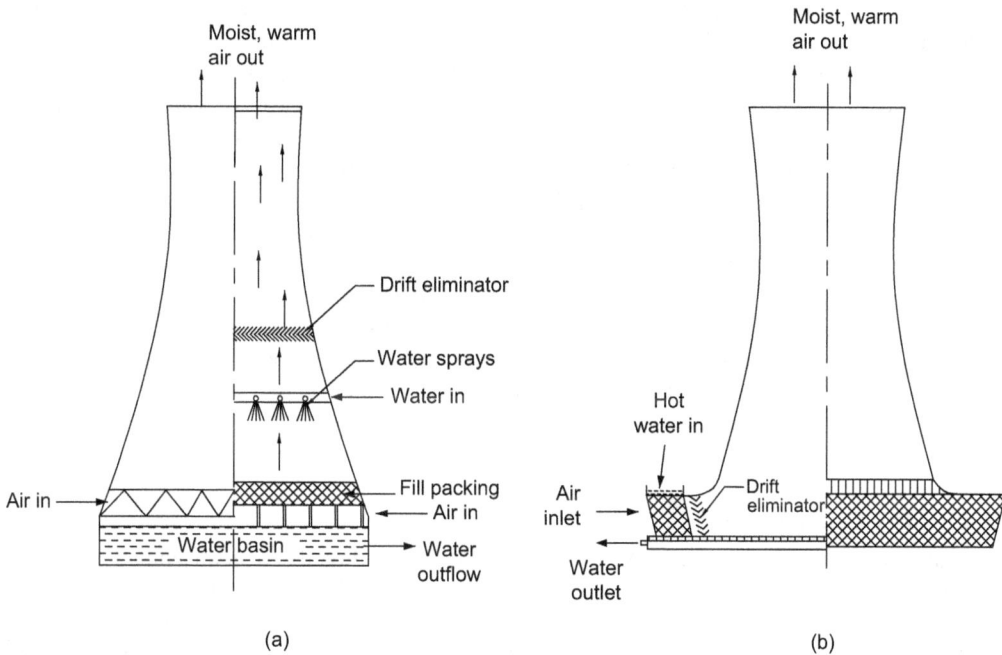

FIGURE 4.38   A natural draft cooling tower(Schematic).

tower and thus increases efficiency. As hot air moves upwards through the tower (because hot air rises), fresh cool air is drawn into the tower through an air inlet at the bottom. A natural draft tower is so called because natural flow of air occurs through the tower. Two factors are responsible for creating the natural draft; a rise in temperature and humidity of air in the column reduces its density. Due to the layout of the tower, no fan is required and there is almost no circulation of hot air that could affect the performance. Hyperbolic towers have become the design standard for all natural draft cooling towers due to their structural strength and minimum usage of material. The hyperbolic shape also aids in accelerating air flow through the tower and thus increases efficiency. Figure 4.38 shows a natural draft cooling tower.

4.7.4.8 Wet Cooling Towers

Wet cooling systems are either mechanical draft or natural draft towers. The structure of a wet cooling tower includes nozzles to spray hot water homogeneously onto the heat exchanging media (also referred to as fill or wet deck), fans which direct ambient air throughout the cooling tower, drift eliminators, which prevent water droplets from discharging with the air, and collection basin for the cooled water. The air flows from the bottom of the tower (counterflow-type tower) or perpendicular to the direction of water flow (crossflow-type tower). To achieve better performance, the heated cooling water is sprayed through fill, to increase the surface area and the time of contact between the air and water flow. Splash fill consists of material placed to interrupt the water flow causing splashing. Film fill is composed of thin sheets of material (usually PVC) upon which the water flows.

In mechanical draft tower, air is drawn through the tower by large fans and circulated water is cooled by a combination of evaporative and convective heat exchange. As ambient air is drawn past a flow of water, a small portion of the water evaporates, and the energy required to evaporate that portion of the water is taken from the remaining mass of water, thus reducing its temperature. The cooled water is collected at the bottom of the tower and pumped back to the condenser to repeat the cycle of operation. Make-up water to the cooling tower is required to replace the water that evaporates to the atmosphere and water lost to blow down to control cycle of concentration. Evaporation losses are typically the largest contributor to water consumption in a cooling tower system and can be estimated based on the cooling-water flow rate and the cooling-water temperature rise [11]. While they require less power, natural draft towers are extremely large and generally only used at facilities with high cooling-water requirements. Natural draft towers draw air through the tower by a natural chimney affect. Natural draft towers are costly to build, but have low operating and maintenance costs. Capital costs of wet cooling systems are higher than for open-loop ones, and mechanical draught towers typically have a higher parasitic load due to power consumption of the fans. Figure 4.39 shows wet cooling concept and Figures 4.36–4.38 are examples of wet cooling.

The water is lost in the cooling tower through evaporation, blowdown, and drift loss. Evaporation loss is directly related to heat load. Blowdown is used to control the concentration of the dissolved minerals in the circulating water. If the make-up water contains higher mineral content, then the blowdown volume will be higher. Cooling towers use a number of chemicals such as biocide to control biofouling, scale inhibitor to control scaling, and corrosion inhibitor to control corrosion. The blowdown may also contain these chemicals.

4.7.4.9 Dry Cooling system

Dry cooling systems in which turbine exhaust steam is ducted directly to an air-cooled condenser. Heat rejection to the environment takes place in a single step in which steam is condensed in finned tube bundles, which are cooled by air blown across the exterior finned surfaces. Although dry cooling achieves significant water savings, the capital and operating costs are much higher than they are for closed-cycle wet cooling, and the physical footprint is larger. Also, plant performance is reduced in the hotter days of the year when the steam-condensing temperature (and hence the turbine exhaust pressure) is substantially higher than it would be with wet cooling [115,116].

Direct dry cooling. Dry cooling systems are classified as either direct or indirect; direct dry cooling utilizes a or large standalone finned tube air-cooled condenser (ACC). (Figure 4.40(i) or a natural draft cooling tower(Figure 4.40(ii) Steam exiting the turbine is condensed within ACC, which are externally cooled by ambient air. The finned tubes are commonly arranged in an A-shaped frame over a forced draught fan as shown in Figure 4.40. Dry cooling offers distinct advantages for reducing water consumption, and increasing the flexibility of power plant siting. However, the capital cost of dry cooling is considerably higher than wet cooling, and the dry cooling process typically results with a penalty on power plant performance during the hottest

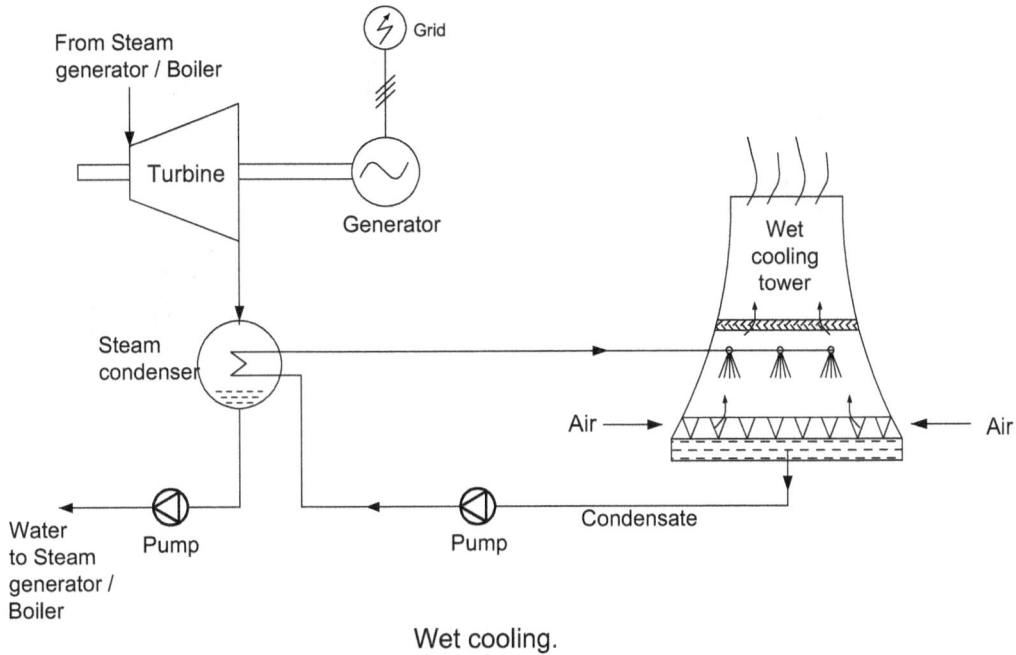

Wet cooling.

FIGURE 4.39 Wet cooling in a natural draft cooling tower.

days of the year and a larger footprint than a closed-loop wet cooling system of comparable capacity.

The condensate drains from finned tubes into condensate manifolds and then drains into a condensate tank before being pumped to the conventional condensate cycle. To reduce pressure drop in steam conveying system, ACC needs to be installed close to the turbine hall [117].

Air-Cooled Condenser

Air-cooled condenser (ACC) is a direct dry cooling system where the steam is condensed under vacuum inside air-cooled finned tubes (Figure 4.40(i)). The major components of ACC includes ducting (for transport of steam), finned tube heat exchanger, axial fans, motors, gear boxes, piping, and tanks (to collect the condensate). The ambient air flows across finned tube heat exchanger using forced draft axial fan to condense the steam. Collected condensate is recycled to utilities. The fan is driven by an electric motor with gearbox.

    a. "A-Frame" Air-Cooled Condenser

The most popular style of air-cooled condenser (ACC) is the modularized A-frame design, used in power plants of all sizes. Airflow is delivered by forced draft axial fans, driven by electric motors and gearboxes, installed below the heat exchangers (Figure 4.40).

    b. Single Row Condenser

The single row condenser is an air-cooled steam condenser located at the exhaust of a steam turbine. The exhaust steam is condensed inside only one single row of fin tubes made of oblong steel tube with cladded aluminum fins. Aluminum fins having a serpentine shape are brazed on the bare tube in a furnace.

4.7.4.11 Indirect Dry Cooling System

In an indirect dry cooling tower (IDCT) system, exhaust steam from the turbine is cooled by water in a condenser, which can be of surface type or direct contact jet type, and hot water is cooled by air in finned tube bundles utilizing natural draft tower or mechanical draft fans [118].

Indirect dry condensing system. An indirect dry condensing system connects a dry cooling tower with a steam condenser. The steam that flows from the low-pressure turbine is condensed in a surface condenser or a jet condenser. The heated water is pumped to vertically arranged heat exchangers that are located around the concrete cooling tower. The natural draft cooling tower generates the airflow across the heat exchangers. Indirect condensing system by natural draft cooling tower is shown in Figure 4.40.

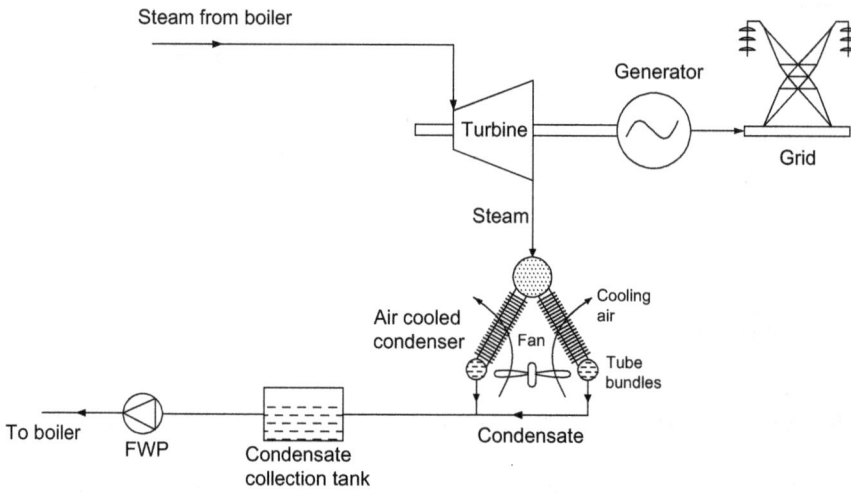

(a) Dry cooling system.

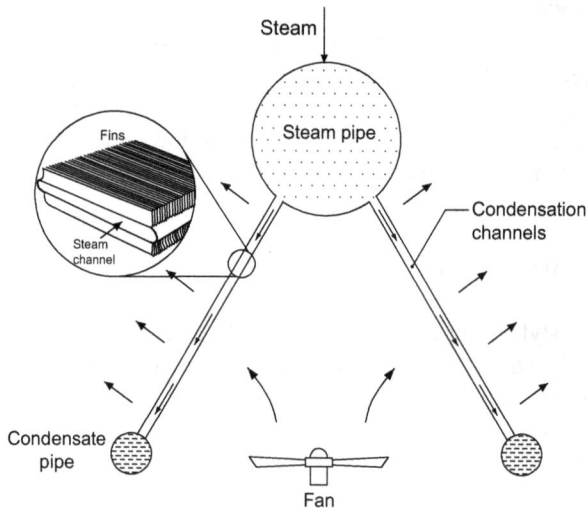

(b)   Details of air cooled heat exchanger (ACC).

FIGURE 4.40   (i) Direct dry cooling (a) 'A' - frame heat exchanger and (b) Details of 'A' frame heat exchanger unit.

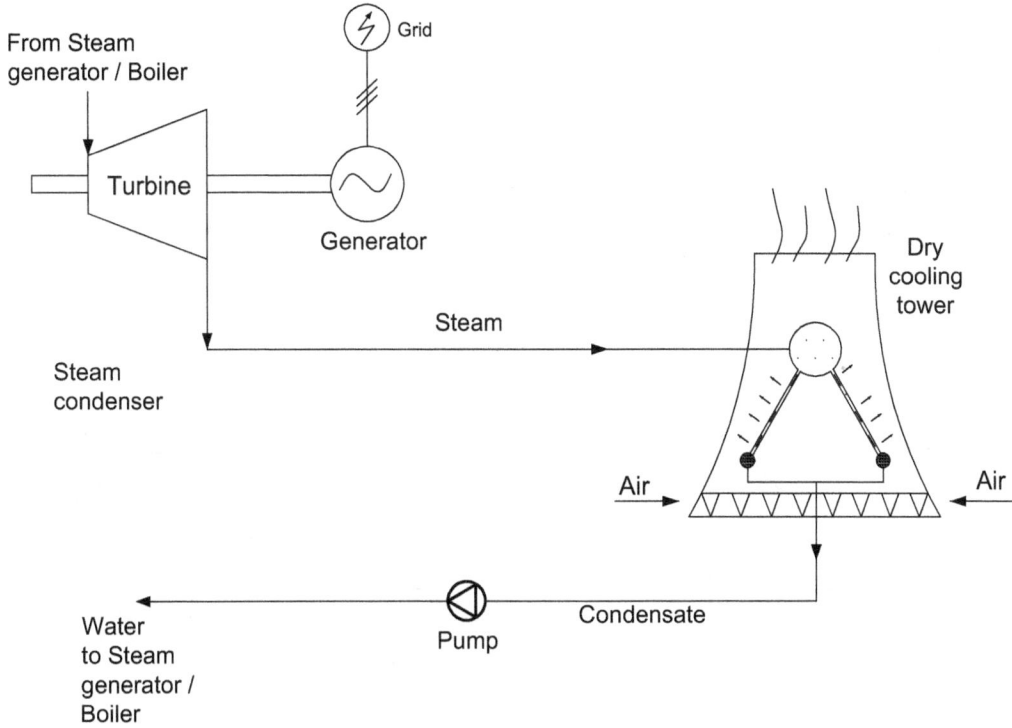

Direct dry cooling in a natural draft cooling tower.

FIGURE 4.40   (ii) Direct dry cooling ina natural draft cooling tower (Continued)

Indirect dry cooling by forced draft system. The mechanical indirect dry cooling tower is a particular type of IDCT where the cooling air flow is induced or forced by fans driven by electrical motors. Indirect dry cooling by natural draft cooling tower is shown in Figure 4.41a and in a 'A' unit is shown Figure 4.41(b).

4.7.4.12   Indirect Dry Cooling "Heller System"

Indirect dry cooling system known as the Heller System, uses a direct contact condenser instead of a surface condenser. The turbine exhaust steam is in direct contact with the cold water spray, and no condenser tubes are used. The resultant hot condensate and water mixture are pumped to an external air-cooled heat exchanger. The air-cooled heat exchanger can be either mechanical draft (induced or forced) or natural draft cooling tower type. The Heller system is usually associated with steam power plants; however, it can be used with other working fluids as well. Figure 4.42 illustrates the schematic arrangement of an air-cooled condensing equipment according to the Heller system. Figure 4.43 shows Heller system in which the hot water from the direct contact condenser is passed through air-cooled heat exchanger elements, which are installed around the periphery of the cooling tower and are cooled by a natural draft effect. The Heller system does not need any make-up cooling water and does not generate plume [119,120]. The cooling deltas (water to air fin tube heat exchangers) are arranged inside or outside the cooling tower and are divided into parallel sectors. In case of indirect dry cooling system, i.e., in Heller system, natural draft tower for cooling the hot water can be conveniently located away from the turbine hall.

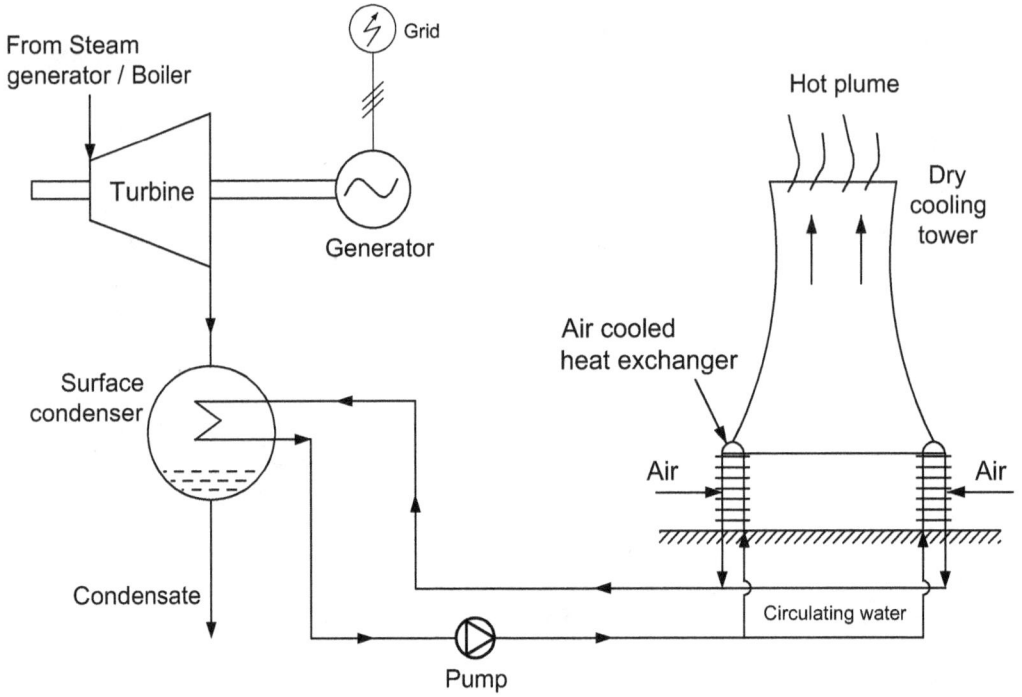

FIGURE 4.41a   Indirect dry cooling by natural draft cooling towe.

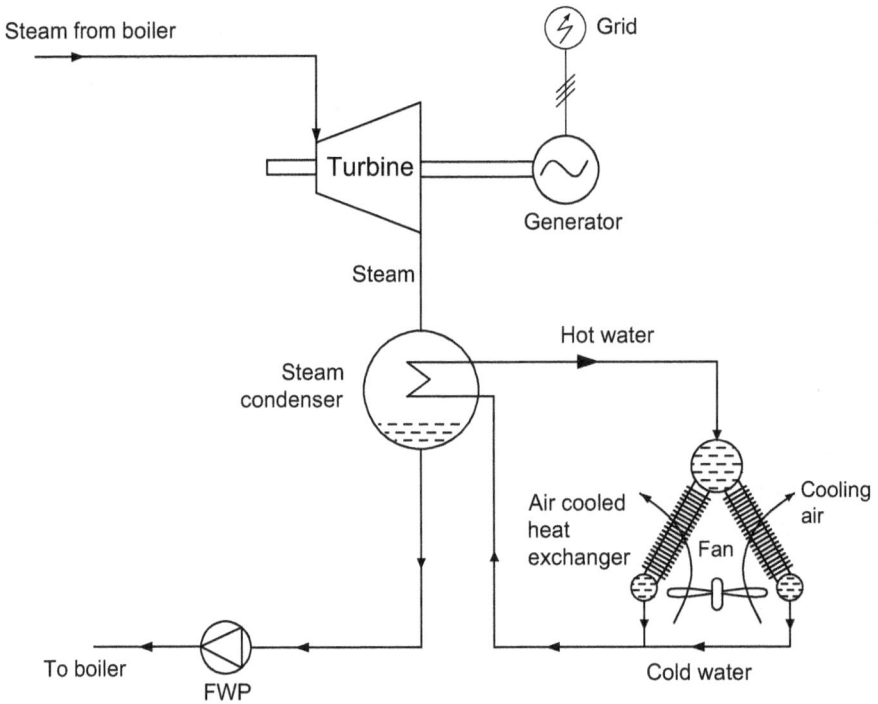

FIGURE 4.41b   Indirect dry cooling system by forced draft cooling system ('A' FRAME).

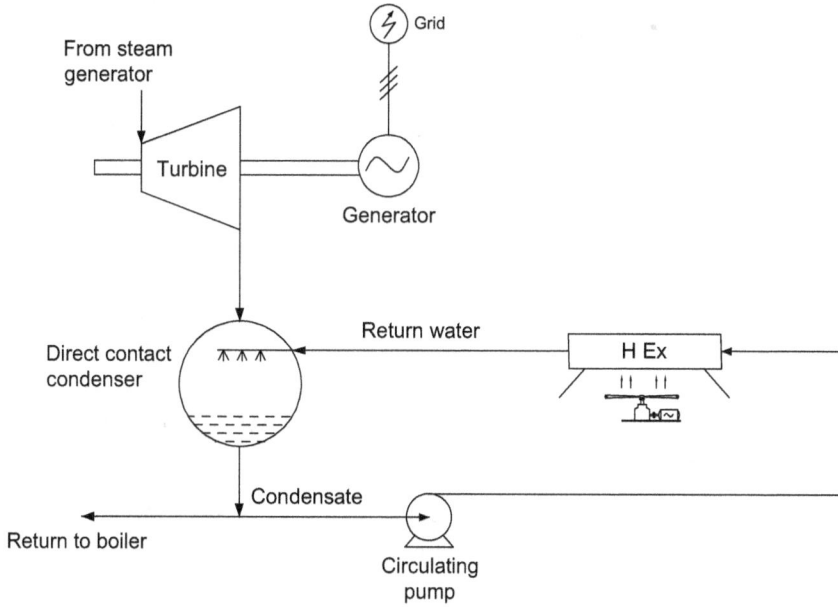

FIGURE 4.42 Concept of Hellar System (Indirect dry cooling).

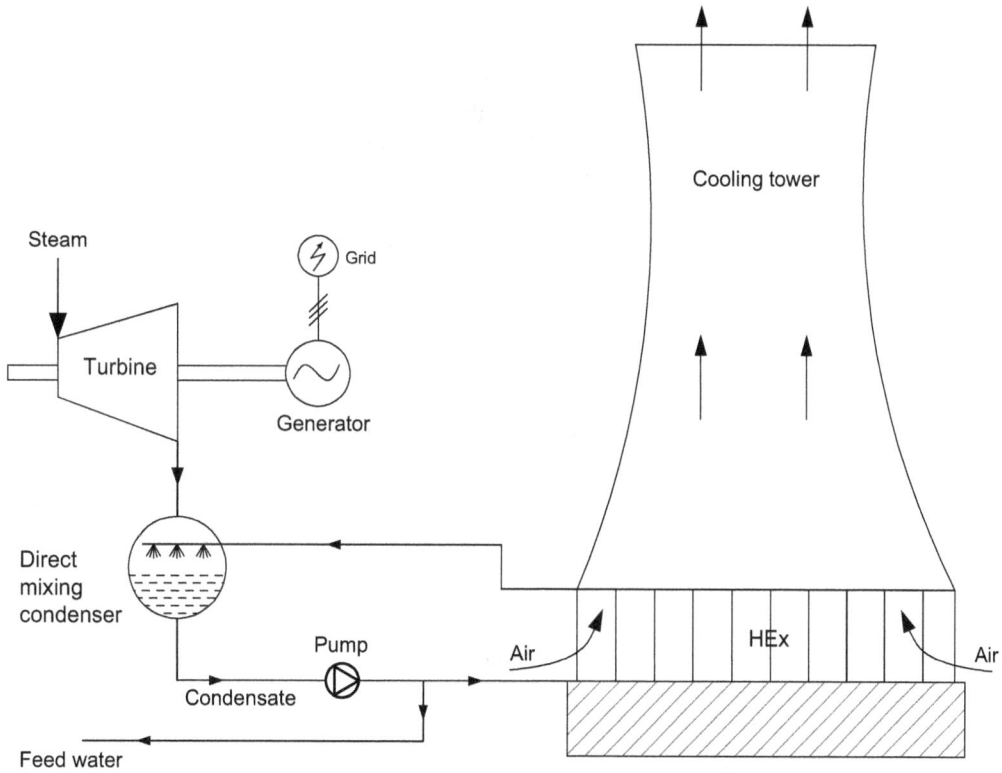

FIGURE 4.43 Heller system with a natural draft cooling tower (Indirect dry cooling).

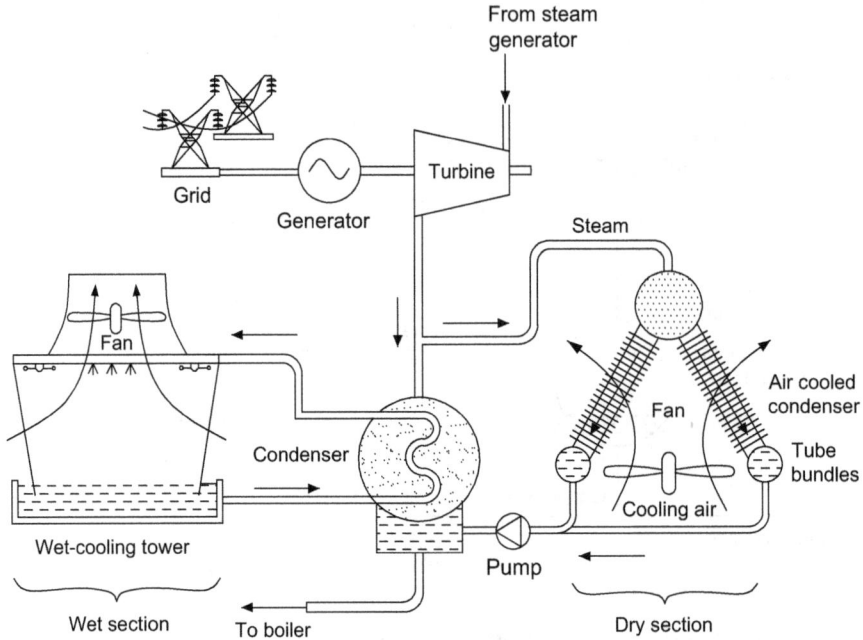

FIGURE 4.44   Hybrid Cooling (both dry and wet cooling) concept.

4.7.4.13   Hybrid Cooling Systems

Hybrid cooling systems have both dry and wet cooling elements that are used individually or together to achieve the best features of each; that is, the wet cooling performance on the hottest days of the year and the water conservation capability of dry cooling at other times. The most common configuration to date has been parallel, separate structures with direct dry cooling [120,121]. The drawback to hybrid cooling is that significant amounts of water are still required, particularly during the summer. Therefore, it is most suitable for sites that have significant water availability but not enough for all-wet cooling at all times of the year. Figure 4.44 shows a hybrid cooling concept.

4.7.5   SELECTION CONSIDERATIONS

Selecting the proper water-cooling equipment for a specific application requires consideration of cooling duty, economics, required services, environmental conditions, maintenance requirements, and aesthetics. Because a wide variety of water-cooling equipment may meet the required cooling duty, factors such as footprint, height, volume of airflow, fan and pump energy consumption, materials of construction, water quality, and availability influence final equipment selection. The optimum choice is generally made after an economic evaluation.

Sound

Sound has become an important consideration in the selection and siting of outdoor equipment such as cooling towers and other evaporative cooling devices.

4.7.6   TYPICAL COOLING TOWER INSTALLATIONS

Cooling towers are commonly of the induced-draft, crossflow variety, although counterflow, and forced-draft cooling towers are also represented. The cooling towers range in size from small to large capacity [110].

4.7.7 COMPONENTS OF A COOLING TOWER

The major parts of the tower include the tower framework, cold water basin, water distribution deck, fill, drift eliminator, louvers, etc. [109,110].

4.7.7.1 Basin and Cold Well

The basin is that portion of the cooling tower structure located under the tower that is used for collecting cooled water and which can be used as a location for adding make-up water. The cold well is a deepened portion of the basin that contains submerged water circulation pumps. The basin may be constructed of concrete, wood, metal, or fiberglass.

4.7.7.2 Water Distribution and Fan Deck

In a crossflow cooling tower, the hot water basin is used to distribute the warm return water flow uniformly over the tower fill. In a counterflow cooling tower, water sprays are used to distribute the warm water. The fan deck supports the motor and fan of the water spray system. The stack is the structure (typically a cylinder) that encloses the fan and directs warm, humid discharge air upward and out of the cooling tower.

Except where otherwise specified, all components of the cooling tower shall be fabricated of heavy-gauge steel, protected against corrosion by galvanizing. Since cooling towers are operated in open atmosphere and always has direct contact with the water they are subjected to corrosion and fouling. Proper material selection or proper water treatment is needed to keep the cooling tower in good working condition.

4.7.7.3 Fill

Fill is the internal part of a tower where air and water are mixed. The fill intercepts the downward fall of water. The water is mixed with the air contained in the fill material and water is evaporated and cooled. There are two types of fill: splash fill and film fill. The falling water hits the splash fill, splashes, and breaks up into smaller water droplets, resulting in an increased rate of evaporation. The splash fill is made of wooden slats or bars, plastic, or ceramic tile. Film fill is a compact plastic material, similar to a honeycomb, which causes water to flow over the fill material, creating a large wet surface that maximizes evaporation as air travels past the film surface.

Film fill provides more cooling capacity in a given space than splash fill. Splash fill can be partially or totally replaced with film fill to increase the capacity of an existing cooling tower. Because of the very close spacing, film fill is very susceptible to various types of deposition. Calcium carbonate scaling and fouling with suspended solids has occurred in some systems. Process contaminants, such as oil and grease, can be direct foulants and/or lead to heavy biological growth on the fill. Any type of deposition can severely reduce the cooling efficiency of the tower.

4.7.7.4 Louvers

Louvers are flat or corrugated members constructed of wood, plastic, cement board, or fiberglass, and installed across (horizontally) the open side of a tower. The main function of louvers is to prevent water from splashing out of the cooling tower through the openings where air enters the tower. Louvers are usually set at an angle to the direction of airflow.

4.7.7.5 Drift Eliminators

The drift eliminators efficiently remove water droplets from the air and return the recovered water to the cooling tower, thereby minimizing the loss of cooling tower water. They are located in areas that are situated after the fill and water sprays and just before the area where the air exits the cooling tower (see Figures 4.36–4.38).

4.7.8 Materials of Construction

Materials for cooling tower construction are usually selected to meet the expected water quality and atmospheric conditions [107,109, 110].

Wood

Wood has been used extensively for all static components except hardware. Redwood and fir predominate, usually with post-fabrication pressure treatment of waterborne preservative chemicals, typically chromated copper arsenate (CCA), or acid copper chromate (ACC). These microbicidal chemicals prevent the attack of wood-destructive organisms such as termites or fungi.

Metals
Galvanized iron is used for small and medium sized installations. Hot-dip galvanizing after fabrication is used for larger weldments. Hot-dip galvanizing and cadmium and zinc plating are used for hardware.
Brasses and bronzes are selected for special hardware, fittings, and tubing material.

Stainless steels principally SS 302, 304, and 316 are often used for sheet metal, drive shafts, and hardware in exceptionally corrosive atmospheres or to extend unit life. Stainless steel cold-water basins are increasingly popular.

Cast iron is a common choice for base castings, fan hubs, motor or gear reduction housings, and piping valve components.

Metals coated with polyurethane and PVC are used selectively for special components. Two-part epoxy compounds and epoxy-powdered coatings are also used for key components or entire cooling towers.

Plastics. Fiberglass reinforced plastic (FRP) materials are used for components such as structure, piping, fan cylinders, fan blades, casing, louvers, and structural connecting components. Polypropylene and acrylonitrile butadiene styrene (ABS) are specified for injection molded components, such as fill bars and flow orifices.

PVC is typically used as fill, eliminator, and louver materials. Poly-ethylene is now used for both hot- and cold-water basins. Reinforced plastic mortar is used in larger piping systems, coupled by neoprene 0-ring-gasketed bell and spigot joints.

Graphite Composites

Graphite composite drive shafts are available for use on cooling tower installations. These shafts offer a strong, corrosion-resistant alternative to stainless steel shafts and are often less expensive, more forgiving of misalignment, and transmit less vibration.

Concrete, Masonry, and Tiles

Concrete is typically specified for cold-water basins of field-erected cooling towers and is used in piping, casing and structural systems of the largest towers, primarily in the power and process industries. Special tiles and masonry are used when aesthetic considerations are important.

4.7.9 Cooling Tower Water System Concerns

Cooling towers are dynamic systems because of the nature of their operation and the environment they function within. Tower systems sit outside, open to the atmosphere, which makes them susceptible to dirt and debris carried by the wind. Their structure is also popular for birds and bugs to live in or around. Scaling will occur predominantly in the heat exchangers and in the fill-section of the tower structure, but may also occur in the piping or on the tower distribution deck. Below is a brief discussion on the four primary cooling system treatment concerns encountered in most open-recirculating cooling systems [103].

Corrosion—the process is enhanced by elevated dissolved mineral content in the water and the presence of oxygen, both of which are typical of most cooling tower systems. There are different types of corrosion encountered in cooling tower systems, including galvanic, pitting, microbiologically influenced corrosion and deposition, erosion-corrosion, among others.

Scaling—factors that contribute to scaling tendencies include water quality, pH, and temperature.

Fouling—common foulants include organic matter, process oils, and silt, i.e., fine dirt particles that blow into the tower system, or enter in the make-up water supply. Fouling in the tower fill can plug film fill reducing the evaporative surface area, leading to lower thermal efficiency of the system.

Microbiological activity—microbiological activity is microorganisms that live and grow in the cooling tower and cooling system. Cooling towers present an ideal environment for biological activity due to the warm, moist environment.

4.7.9.1 Water Loss

The water is lost in the cooling tower through evaporation, blowdown, and drift loss. Evaporation loss is directly related to heat load. Blowdown is used to control the concentration of the dissolved minerals in the circulating water. If the make-up water contains higher mineral content, then the blowdown volume will be higher. Generally, the cycles of concentration are maintained between 3 to 7.

The drift is water droplets carried out by the exiting air. Cooling towers also use a number of chemicals such as biocide to control biofouling, scale inhibitor to control scaling, and corrosion inhibitor to control corrosion.

Drift eliminators are a vital component of maintaining the efficiency of a cooling tower and minimizing the loss of water. The structure of a drift eliminator provides multiple deflections of the air stream, thus forcing water droplets, which are carried away with the airflow to deposit on the surface of eliminator and drip down to the collection basin.

Cooling towers also use a number of chemicals such as biocide to control biofouling, scale inhibitor to control scaling, and corrosion inhibitor to control corrosion. The blowdown may also contain these chemicals.

Deposition Control

There are many contaminants in cooling water that contribute to deposit problems. Three major types of deposition are discussed here: scaling, general fouling, and biological fouling [102].

Blowdown Control

Increasing blowdown to limit cycles of concentration is an effective way to reduce the scaling potential of circulating water. However, high rates of blowdown are not always tolerable and, depending on water quality, cannot always provide complete scale control. In many localities, supplies of fresh water are limited and costly.

4.7.9.2 Common Cooling Tower Water Issues—Deposition

Cooling towers are operated in open atmosphere and always have direct contact with the water and hence they are subjected to corrosion and fouling [109, 122].

Scale. Scale and scale-like deposits include calcium carbonate, calcium phosphate, magnesium silicate, silica, and other mineral compounds. They build up on heat exchanger tubes, reducing heat transfer.

Corrosion causes pitting and leaks in cooling systems and can lead to the replacement of pipes, pumps, heat exchanger tubes, and even entire cooling towers. Iron oxide, especially, contributes to fouling and deposition, which interfere with heat transfer.

Organic Fouling

Mud, sand, silt, clay, biological matter, and even oil can enter the system through its make-up supply or from the air. These suspended materials can accumulate and settle in the system, blocking flow and reducing efficiency

Microbiological Fouling

Biofilms severely restrict heat transfer. Slime masses bind inorganic and organic foulants and plug systems. Algae and fungi cause extensive plugging and fouling of heat exchanger tubes, water lines, tower spray nozzles, distribution pans, screens, and fill.

Foam

Cascading water, the continuous recycling of contaminants, and a high concentration of foam stabilizers can cause foam to overflow the tower sump, blow off the towers or even cause an air-lock in the water pumps. Worst of all, foam concentrates deposit-forming materials, increasing the chance of fouling in the system.

4.7.9.3   The pH Balance of Cooling Tower Water

Maintaining balanced pH levels for cooling tower water is critical for preventing scale and corrosion. A pH between 6.5 and 7.5 is generally considered the ideal range for reducing scale formation, though some nonacid treatments for scale prevention can increase the cooling tower pH range up to 8.5. The pH also depends on the cycles of concentration (CoC). Operating at higher CoC allows the tower water to have a higher pH, even up to 10.

One means of protecting against corrosion in towers made of stainless steel, copper, or steel is increasing the pH to 8.5 or above. Raising the CoC allows the carbonate concentration in the water to increase, boosting the alkalinity. Though higher alkalinity levels can cause scale formation, they prevent corrosion in certain types of metals and also inhibit bacterial growth.

Maintaining the ideal pH balance in tower water is a continual process that requires constant vigilance and control. Following all these steps will help ensure the proper pH balance of CT water [123–124]:

1. Determine water quality
   Start by measuring the water's pH, alkalinity, hardness and conductivity. These values will establish a baseline for treatment.
2. Establish target cycles of concentration
   Cycles of concentration = make-up volume ÷ blowdown volume
3. Monitor CoC and water performance
4. Automate the processes

Cut down on the over-application of chemicals to cooling tower water by installing an automated chemical dispenser that regulates the water chemistry automatically. The following devices will maximize efficiency while enabling real-time water monitoring:

Flow meters: these meters can measure the flow of blowdown and make-up lines to reduce water waste.

Conductivity controllers: conductivity controllers regulate blowdown. If the tower water exceeds the conductivity setpoint, the controller will release water from the tower.

Drift eliminators: drift eliminators are plastic devices that prevent vapor droplets from escaping the tower with the exhaust air. They prevent costly water loss in your system.

pH monitors: continual pH monitoring will give you the information you need to determine whether the water treatment methods you are using are working as they should.

5. Protect Equipment

Corrosion can occur even in well-controlled cooling tower environments. Use corrosion coupons to measure corrosion rates. Doing so will verify the effectiveness of the water treatment. Secondly, do not neglect regular inspections and repairs of your tower and all monitoring and chemical control equipment.

When the pH deviates from the prescribed range, several undesirable effects may occur:

i. White rust: if the pH rises above 8.3 and the water contains a high concentration of carbonate ions, cooling towers made of galvanized steel can develop white rust.
ii. Iron corrosion: with pH values between 7.5 and 8, iron and iron alloys in the cooling tower can experience corrosion.
iii. Corrosion from the atmospheric pollutants

4.7.10 Ways of Cooling Water Loss

Cooling towers dissipate heat from recirculating water to the ambient air. Heat is rejected to the environment from cooling towers through the process of evaporation. Therefore, by design, cooling towers use significant amounts of make-up water. The thermal efficiency and longevity of the cooling tower and equipment depend on the proper management of recirculated water. Water leaves or lost from a cooling tower system in one of four ways [125]:

1. Evaporation: the primary function of the tower and the method that transfers heat from the cooling tower system to the environment.
2. Drift: a small quantity of water may be carried from the tower as mist or small droplets. Drift loss is small compared to evaporation and blowdown and is controlled with baffles and drift eliminators.
3. Blowdown: when water evaporates from the tower, dissolved solids (such as calcium, magnesium, chloride, and silica) remain in the recirculating water. As more water evaporates, the concentration of dissolved solids increases. If the concentration gets too high, the solids can cause scale to form within the system. The dissolved solids can also lead to corrosion problems. The concentration of dissolved solids is controlled by removing a portion of the highly concentrated water and replacing it with fresh make-up water. Carefully monitoring and controlling the quantity of blowdown provides the most significant opportunity to conserve water in cooling tower operations.
4. Basin leaks or overflows: properly operated towers should not have leaks or overflows. Check float control equipment to ensure the basin level is being maintained properly, and check system valves to make sure there are no unaccounted for losses.

The sum of water that is lost from the tower must be replaced by make-up water:

$$\text{Make-up water} = \text{evaporation} + \text{blowdown} + \text{drift}$$

The concept of water loss is further discussed in Section 4.7.11

4.7.10.1 Cycles of Concentration (CoC)

One of the common terms used in describing the water use efficiency of cooling tower water systems is CoC. CoC represents the relationship between the make-up water quantity and blowdown quantity. CoC is a measure of the total amount of minerals that is concentrated in the cooling tower water relative to the amount of minerals in the make-up water or to the volume of each type of water. Because dissolved solids enter the system in the make-up water and exit the system in the blowdown water, the cycles of concentration are also approximately equal to the ratio of volume of make-up to blowdown water. From a water efficiency standpoint, maximize cycles of concentration. This will minimize blowdown water quantity and reduce make-up water demand. However, this can only be done within the constraints of make-up water and cooling tower water chemistry. The higher the CoC, the greater the water use efficiency. Most cooling tower systems operate with a CoC of 3 to 10, where 3 represents acceptable efficiency and 10 represents very good efficiency. It has been found that the range of 5 to 7 CoC represents the most cost-effective situation.

Dissolved solids increase as cycles of concentration increase, which can cause scale and corrosion problems unless carefully controlled. The concentration is usually reported as concentration cycles and refers to the number of times the compounds in the make-up water are concentrated in the blowdown water. For example, if the concentration in the make-up water were 125 ppm and the concentration of the blowdown were 563, the concentration cycles would be 563/125 or five cycles. Figure 4.14b shows the concept of calculation of CoC for a drum boiler which is similar to the CoC of cooling water in a cooling tower.

4.7.10.2 Low Cycle of Concentration

A low cycle of concentration often means greater cost due to greater water consumption and chemical usage. Common strategies for increasing cycles of concentration include removal or neutralization of various contaminants, such as hardness, iron, silica, and microbials, as well as controlling pH and alkalinity to limit scaling.

4.7.11 Relation Between, Make-Up, Cycles of Concentration (CoC) and Blowdown, After Refs. [106–109]

One of the common terms used in describing the water use efficiency of cooling tower water systems is cycles of concentration (CoC). The higher the CoC, the greater the water use efficiency. Most cooling tower systems operate with a CoC of 3 to 10, where 3 represents acceptable efficiency and 10 represents very good efficiency. It has been found that the range of 5 to 7 CoC represents the most cost-effective situation. The graphical representation of CoC is shown in Figure 4.45,

Calculating CoC by volume

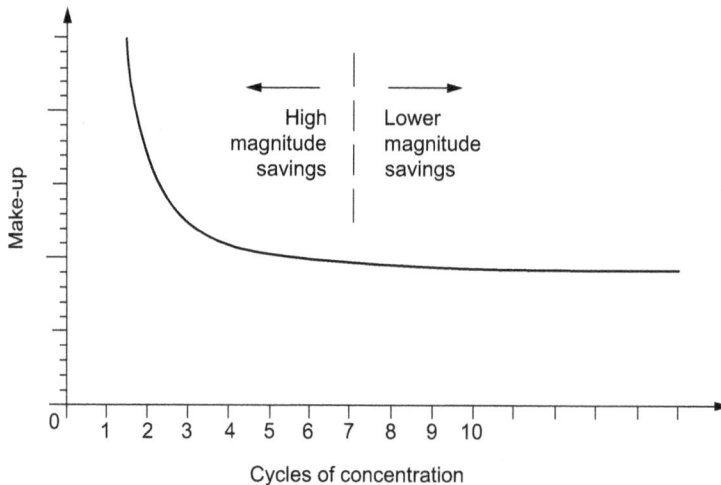

FIGURE 4.45 CoC curve.

If both make-up and blowdown water volumes are known, CoC, C by volume can be calculated. The term C is defined as

$$C = \frac{M}{B}$$

where
M is makeup water, kg/hr (gpm),
B is blowdown losses, kg/hr (gpm),

cycles of concentration, C.

a. The term cycles of concentration C or concentration ratio refers to the ratio of impurities or the total dissolved solids (TDS) in the circulating water to the TDS in the make-up water.

$$C = \frac{\text{TDS of circulating water}}{\text{TDS of make up water}} = \frac{\text{Conductivity of circulating water}}{\text{Conductivity of makeup water}}$$

b. Determining CoC by water analyses

To determine CoC, one must know the mineral content of both make-up and blowdown water. Conductivity is commonly measured in micromhos (µmhos).

c. Cycles of concentration, C is the ratio of the dissolved solids concentration in the circulating water compared to the dissolved solids concentration in the make-up water. The blow down rate, B can be calculated as follows:

$$\text{Blow down} = \frac{\text{Evaporation loss}}{\text{Cycles of concentration} - 1}$$

The relationship between blowdown, B, evaporation, E, and CoC, C is represented with this equation

$$B = \frac{E}{C-1}$$

where E is evaporation, kg/hr (gpm)

Water Balance, $M = E + B + D$

d. Controlling CoC

A simple rule: to increase CoC, decrease blowdown; to decrease CoC, increase blowdown.

Make-up—water supply needed to replace all losses due to evaporation, leaks, or discharge in cooling systems.

e. Blowdown

In all cooling equipment, operating in evaporative mode, the cooling is accomplished by evaporating a small portion of the recirculating water as it flows through the equipment. When this water evaporates, the impurities originally present in the water remain. Unless a small amount of water is drained from the system, known as blowdown, the concentration of dissolved solids will increase rapidly and lead to scale formation or corrosion or both. Also, since water is being lost from the system through evaporation and blowdown, this water needs to be replenished. The total amount of replenishment, known as make-up, is defined as

Make-up = evaporation loss + blowdown
In addition to the impurities present in the make-up water, any airborne impurities or biological matter are carried into the equipment and drawn into the recirculating water.
    System calculation[107]

    Water Balance

Without leakages taken into account, an elementary water balance equation is given by

$$\text{Make-up water} = \text{evaporation loss} + \text{blowdown} + \text{drift}$$

Figure 4.46 illustrates the concept of water loss due to blowdown, drift and evaporation and adding of make-up water.

1. Make-up = blowdown + evaporation

Make-up water quantity = evaporation loss + blowdown loss

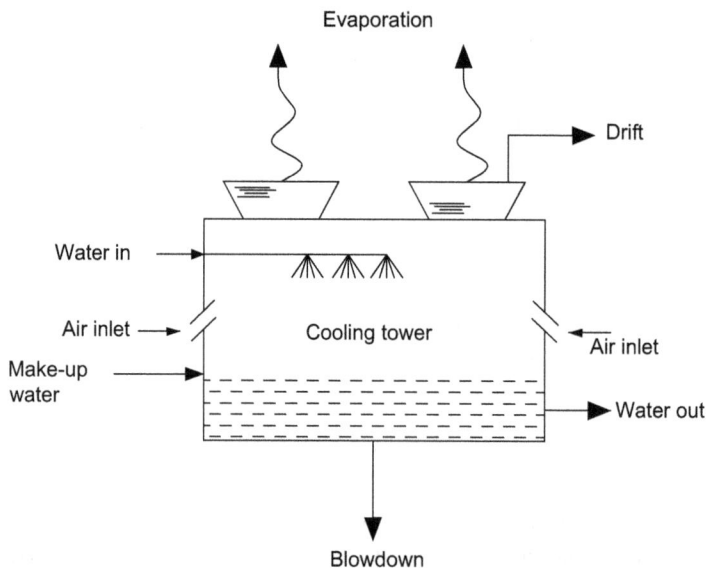

FIGURE 4.46 The concept of water loss due to blowdown, drift and evaporation and adding of make-up water.

2. The second principal relationship defines cycles of concentration, C in terms of make-up flow and blowdown flow:

$$C = \text{Make-up} \div \text{blowdown}$$

This equation can be rearranged to either of the following to solve for the make-up rate or blow-down rate [107]:

3. Blowdown = make-up ÷ cycles of concentration
4. Make-up = cycles of concentration × blowdown
5. Cycles of concentration × blowdown = blowdown + evaporation
6. Blowdown = evaporation ÷ (cycles of concentration—1)
7. Evaporation = blowdown × (cycles of concentration—1)
8. Summary
    i. All loses taken into account

$$\text{Make-up} = \text{evaporation} + \text{blowdown} + \text{drift} + \text{leaks/overflow}$$

    ii. Cooling tower make-up water with negligible drift and leaks:

$$\text{Make-up} = \text{evaporation} + \text{blowdown}$$

    iii. Cycles of concentration based on water use

$$C = \text{make-up water} \div \text{blowdown water}$$

    iv. Cooling tower cycles of concentration based on conductivity:

Cycles of concentration = conductivity (TDS) of blowdown water ÷ conductivity (TDS) of make-up water

Since TDS enter the system in the make-up water and exit the system in the blowdown water, the cycles of concentration are also approximately equal to the ratio of volume of make-up water to blowdown water.

To monitor microbiological activity a combination of testing, mathematical predictors, and online monitoring tools can be used. If an increase in microbiological activity is detected, adjustments in the microbiological treatment should be made to minimize potential fouling problems.

4.7.12 COOLING WATER PROBLEMS AND SOLUTIONS

Depending on the quality of available fresh water supply, waterside problems develop in cooling-water systems from

i.   sludge, deposition
ii. scaling
iii. corrosion
iv. biological growth

4.7.12.1 Corrosion and Scaling Tendencies after Ref. [110]

Corrosion

Red rust on steel components and white rust on galvanized surfaces may affect the longevity of system components.

Scale formation—scale not only reduces heat transfer and system efficiency, but also may lead to under-deposit corrosion. If scale is not controlled, it may continue building on critical components such as the fill and severely impact thermal performance.

Biological fouling—slime and algae formations may reduce heat transfer, promote corrosion, and harbor pathogens such as Legionella.

Types of Scale

Typical scales that occur in cooling-water systems are as follows [110]:

1. Calcium carbonate scale.
2. Calcium sulfate scale.
3. Calcium and magnesium silicate scale.
4. Calcium phosphate scale.

The most common type of scaling is formed by carbonates and bicarbonates of calcium and magnesium, as well as iron salts in water. Calcium dominates in fresh water while magnesium dominates in seawater.

4.7.12.2 Corrosion and Scale Control

To control corrosion and scale, maintain the water chemistry of the recirculating water within the parameters specified.

4.7.12.3 Scale Control

Scale can be controlled or eliminated by application of one or more proven techniques [110]:

1. Water softening—water softener, dealkalizer, ion exchange to remove scale forming minerals from make-up water.
2. Adjusting pH to lower values—scale-forming potential is minimized in acidic environment, i.e., lower pH.
3. Controlling cycles of concentration—limit the concentration of scale-forming minerals by controlling cycles of concentration. This is achieved by intermittent or continuous blow-down process, where a part of water is purposely drained off to prevent mineral build up.
4. Chemical dosage—apply scale inhibitors and conditioners in circulating water.
5. Physical water treatment methods—filtration, magnetic and de-scaling devices, etc.

4.7.12.4 Controlling Corrosion

Controlling corrosion involves controlling certain water properties, including pH, oxygen, temperature, water velocity, suspended solids, and dissolved solids.

4.7.12.5 Microbiological Fouling

Due to the warm conditions in cooling towers, biological fouling can potentially cause health risks or jeopardizing compliance with state mandates. Treatment of make-up or other water can help to remove or neutralize biological contaminants such as bacteria, fungi, and algae. Water treatment should meet the following requirements:

i. The chemicals must be compatible with the materials of construction used in the cooling system.

   ii. Chemicals should be fed into the recirculated water to avoid localized high concentrations, which may cause corrosion.

4.7.12.6 Controlling Microbiological Fouling

Controlling microbiological fouling depends upon effective control of all these parameters [110]:

water quality
water velocity
biocide treatment
biofilm control

To monitor microbiological activity a combination of testing, mathematical predictors, and online monitoring tools can be used. If an increase in microbiological activity is detected, adjustments in the microbiological treatment should be made to minimize potential fouling problems.

4.7.12.7 Chemistries for Microbiological Control

Chemicals used to control microbial populations can be grouped in three general classes based on their mode of action [110]:

   i. Oxidizing biocides—these break down organic substances and kill microorganisms, including bacteria, in cooling waters.
   ii. Nonoxidizing biocides—these control microbiological activity by interfering with cell structure and function.
   iii. Biodispersants—these break up and disperse deposits, such as slimes and biofilm.

4.7.13 Basics of Water Treatment

Cooling tower water treatment systems are used to remove contaminants present in feed water, circulation water, and/or blowdown water in order to prevent damage to cooling tower components. Water treatment is employed to prevent or minimize the following [126]:

   i. deposition and scaling;
   ii. corrosion of the system by oxygen or low pH;
   iii. attack of cooling tower wood by algae, bacteria, or fungi; and
   iv. fouling of heat exchanger surfaces by suspended solids and biological organisms.

A successful cooling-water treatment must address initial water conditioning, analysis of scale forming, fouling and corrosion control. The cooling water treatment program can be divided into four steps [127]:

   i. audit and assessment;
   ii. cooling water pretreatment;
   iii. chemical applications;
   iv. monitoring.

Before effective treatment can be provided, a thorough assessment of conditions of the cooling-water system must be made. Experienced field engineers backed by laboratory resources can audit your system, perform accurate tests to measure water quality and troubleshoot problems.

4.7.13.1 Details of Cooling Tower Water Treatment Program

Cooling tower water treatment systems are used to protect cooling tower components from damage due to contaminants present in feed water, circulation water, and/or make-up water. These contaminants may include chlorides, hardness, iron, biological materials, silica, sulfates, total dissolved solids (TDS), and/or total suspended solids (TSS). Untreated cooling tower feedwater can cause scaling, corrosion, biological growth, and fouling of cooling tower equipment, which can result in plant downtime, reduced productivity, and excessive maintenance or equipment replacement costs over time. A basic cooling tower water treatment system generally includes the following [126]:

   i.   makeup water intake
   ii.   clarification
   iii. filtration and/or ultrafiltration
   iv.   softening
   v.   chemical addition
   vi.   side-stream filtration
   vii. post-treatment
   xiii. automated monitoring and control.

Specific treatment processes vary depending on the requirements of the cooling tower and quality/chemistry of the feed and circulation water, but a typical cooling tower water treatment system will usually include the following steps:

Side-stream filtration
If the cooling tower water is going to be recirculated throughout the system, a side-stream filtration unit will be helpful in removing any problematic contaminants that have entered through drift contamination, leak, etc.

Blowdown treatment
Depending on how much water the cooling plant needs to circulate for proper cooling capacity, plants will choose to recycle and recover the water through some type of post-treatment in the form of reverse osmosis or ion exchange, especially in places where water might be scarce. This allows liquid and solid waste to be concentrated and removed while treated water can be returned to the tower and reused.

4.7.13.2   Critical Water Chemistry Parameters

Critical water chemistry parameters that require review and control include pH, alkalinity, conductivity, hardness, microbial growth, biocide, and corrosion inhibitor levels. Controlling these parameters allows water to be recycled through the system longer, thereby increasing cycles of concentration. Controlling blowdown using an automatic system provides a better opportunity to maximize cycles of concentration, as the TDS concentration can be kept at a more constant set point.

4.7.13.3   Cooling Tower Water Treatment System Control Parameters

A cooling tower water treatment system might be made up of the technologies necessary to regulate the level of [128]

   i.   alkalinity: will dictate potential of calcium carbonate scale
   ii.   chlorides: can be corrosive to metals; different levels will be tolerated based on materials of cooling tower and equipment
   iii. hardness: contributes to scale in the cooling tower and on heat exchangers
   iv.   iron: when combined with phosphate, iron can foul equipment
   v.   organic matter: promotes microorganism growth, which can lead to fouling, corrosion, and other system issues

vi. silica: known for causing hard scale deposits
vii. sulfates: like chlorides, can be extremely corrosive to metals
viii. total dissolved solids (TDS): contribute to scaling, foaming, and/or corrosion
ix. totals suspended solids (TSS): undissolved contaminants that can cause scaling, biofilms, and/or corrosion

4.7.13.4 Chemical Treatment Program Requirements

Chemical treatment programs must meet the following requirements [127]:

i. The chemicals must be compatible with the unit materials of construction as well as other materials used in the system (pipe, heat exchanger, etc.).
ii. Chemical scale and corrosion inhibitors, particularly acid (if used), should be introduced into the circulating water through automatic feeders.
iii. When chlorine is added to the system, free residual chlorine should not exceed 1 ppm, except as noted in start-up and shutdown section. Exceeding this limit may accelerate corrosion.

4.7.13.5 Chemical Applications

To control scale, corrosion, microbiological fouling, and foam, water quality must be maintained at all times, and the right microbicides must be applied in just the right doses. That takes a delicate balance of chemistries and application expertise. Typical use of chemicals include the below [126]:

4. corrosion inhibitors (e.g., bicarbonates) to neutralize acidity and protect metal components
ii. algaecides and biocide (e.g., bromine) to reduce the growth of microbes and biofilms
iii. scale inhibitors (e.g., phosphoric acid) to prevent contaminants from forming scale deposits

4.7.13.6 Multiple Chemical Treatments

Recirculating cooling waters are often treated by adding chemicals, which are selected to control one or more of the problems of biological growth, scale formation, corrosion, and fouling. The following types of chemicals are available [129]:

i. biocides, oxidizing biocides, corrosion inhibitors which form a protective film over metal areas,
ii. acids or other scale inhibitors which prevent mineral precipitation,
iii. conditioners which decrease the density of any scale particles which form, allowing the particles to be more easily carried off by the flowing water,
iv. dispersants which increase foulants' electrical charges, causing them to repel each other, and
v. wetting agents which reduce the water's surface tension so that particles are less likely to adhere to surfaces.

Maintaining correct water quality involves controlling the rates of blowdown and make-up water flow and involves adding chemicals in correct amounts at proper times. This, in turn, requires ensuring the compatibility of the chemicals, and requires monitoring and controlling pH and conductivity.

Chemical treatment carries with it the risks and responsibilities of storing and handling hazardous materials. In addition, it is undesirable to discharge toxic chemicals to aquatic ecosystems or to wastewater treatment plants that rely on bacterial activity.

Molybdate treatments provide effective corrosion protection and an environmentally acceptable alternative to chromate inhibitors.

Nitrite is another widely accepted nonchromate closed cooling-water inhibitor.

4.7.14 MAKE-UP WATER

Make-up water to the cooling tower is required to replace the water that evaporates to the atmosphere and water lost to blow down to control cycle of concentration. Supercritical units do not use boiler blowdown. But condensate polishing systems are required to treat the condensate pumped from the condenser hotwell and remove dissolved and suspended solids to improve feedwater quality.

4.7.15 MAKEUP WATER TREATMENT

Water treatment is employed to prevent or minimize:

i.   scaling or heat exchanger surfaces
ii. corrosion of the system by oxygen or low pH
iii. attack of cooling tower wood by algae, bacteria, or fungi, and
iv. deposition and fouling of heat exchanger surfaces by suspended solids, silts, fine particles and biological organisms.

4.7.16 COOLING TOWER OPERATION AND MAINTENANCE

4.7.16.1 Water Efficiency

To maintain water efficiency in operations and maintenance, follow these measures [125]:

i.   Calculate and understand "cycles of concentration." Check the ratio of conductivity of blow-down and make-up water. Many systems operate at two to four cycles of concentration, while six cycles or more may be possible. Increasing cycles from three to six reduces cooling tower make-up water by 20% and cooling tower blowdown by 50%.
ii. The actual number of cycles of concentration the cooling tower system can handle depends on the make-up water quality and cooling tower water treatment regimen. Typical treatment programs include corrosion and scaling inhibitors along with biological fouling inhibitors.
iii. Install a conductivity controller to automatically control blowdown.
iv. Install flow meters on make-up and blowdown lines. Check the ratio of make-up flow to blowdown flow.
v.   Read conductivity and flow meters regularly to quickly identify problems. Keep a log of make-up and blowdown quantities, conductivity, and cycles of concentration. Monitor trends to spot deterioration in performance.
vi. Consider using acid treatment such as sulfuric, hydrochloric, or ascorbic acid where appropriate.

4.7.16.2 Replacement Options

The following replacement options help to maintain water efficiency across facilities [125]:

i.   Get expert advice to help determine if a cooling tower replacement is appropriate.
ii. For specifics, consult with experts in the field.

4.7.16.3 Retrofit Options

The following retrofit options help maintain water efficiency across facilities [125]:

i.   Consider installing a side-stream filtration system. These systems filter silt and suspended solids and return the filtered water to the recirculating water.

ii. Install a make-up water or side-stream softening system when hardness (calcium and magnesium) is the limiting factor on cycles of concentration.

iii. Install covers on open distribution decks on top of the tower. Reducing the amount of sunlight on tower surfaces can significantly reduce biological growth such as algae.

iv. Consider alternative water treatment options, such as ozonation or ionization and chemical use. Be careful to consider the life cycle cost impact of such systems.

v. Install automated chemical feed systems on large cooling tower systems (more than 100 tons).

4.7.17 COOLING TOWER PERFORMANCE PARAMETERS

Cooling tower performance parameters include height of top of tower above sill, height at throat, height at top, dry bulb temperature, design wet bulb temperature, cold water outlet temperature, hot cooling water inlet temperature, PVC fill size, basin depth, etc.

4.7.18 MEASURES FOR EFFICIENCY IMPROVEMENT OF COOLING TOWERS

Maintenance practices/measures for improving efficiency of cooling towers [68]:

i. Clearances around cooling towers need to be adequate to ensure uninterrupted air intake or exhaust.

ii. Use right type of nozzles that do not clog and spray in a more uniform water pattern.

iii. Increase contact surface and contact time between air and water may be with the use of PVC film type fills by replacing splash bars.

iv. Clean water spray nozzles regularly.

v. Optimize the blowdown flow rate, taking into account the cycles of concentration (CoC) limit.

vi. Keep the cooling-water temperature to a minimum level by (a) segregating high heat loads like furnaces, air compressors, DG sets and (b) isolating cooling towers from sensitive applications like A/C plants, condensers of captive power plant etc.

vii. Monitor liquid to gas ratio and cooling-water flow rates and amend these depending on the design values and seasonal variations.

ix. Increase CoC improvement for water savings. The use of water treatment chemicals, pretreatment such as softening and pH adjustment, and other techniques can affect the acceptable range of cycles of concentration.

x. Check cooling water pumps regularly to maximize their efficiency.

4.7.19 COMMON COOLING TOWER PROBLEMS

Here are three common cooling tower problems [130]:

i. Blowdown.

ii. A low cycle of concentration.

iii. Noise and flume emissions from wet cooling towers and noise in dry cooling system.

iv. The risk of disease from pathogenic microorganisms for the wet cooling tower.

4.7.19.1 Drift and Plume Abatement

Drift consists of droplets of water entrained in the air leaving the top of the cooling tower, or blown from the sides by crosswinds. Louvres and drift eliminators are commonly used to minimize water loss. Drift eliminators are commonly installed in the top part of the tower to capture the entrained droplets.

4.7.20   Cooling Tower Timber Deterioration

Wood continues to be widely used for the construction of cooling towers. Wood deterioration can shorten the life of a cooling tower from an anticipated 10 years to 20–25 years or less. Cooling tower operation becomes inefficient and repair and replacement costs are excessive. Timber is a natural material and hence it is subjected to decay. The types of decay can be divided into the following categories [109, 131, 132]:

- i. Internal decay.
- ii. Biological attack. Organisms attack cooling tower wood and use cellulose as their source of carbon for growth and development. Consequently degrading cellulose quality and quantity.
- iii. Chemical attack. Usually caused by oxidizing agents and alkaline materials.
- Chromated copper arsenate (CCA) treatment of timber provides high immunity to chemical and biological decay and improves resistance to weather.

4.7.20.1   Pressure Treatment

Lumber used in cooling tower construction must be treated with a reliable preservative compound to prevent decay. The industry's current preferred treating chemical is Chromated Copper Arsenate (CCA), a waterborne formulation, which prevents biological attack by fungi or other microbes. Other preservatives are also available.

4.7.20.2   Control of Wood Deterioration

Preventive Maintenance. Preventive maintenance is the only effective method of protecting cooling towers from deterioration.

Periodic spraying with an antifungal is an effective preventive maintenance step if performed on a regular basis.

4.7.20.3 Sterilization

When a tower has suffered a serious infection that the normal corrective programs are not likely to control, consideration can be given to sterilization. In this process, wood temperature is elevated to 150°F for a period of 2 hr. Longer periods must be avoided to limit the loss of wood strength that occurs.

4.7.21   Drift Eliminator

For example, Brentwood's AccuPac® counterflow cellular drift eliminators as shown in Figure 4.47 are constructed of an alternating series of corrugated and wave PVC sheets, assembled to form closed cells. The closed cell structure yields the greatest surface area for droplet capture in a given volume. Brentwood's latest generation of cellular drift eliminators are specifically engineered for counterflow applications to maximize drift removal efficiency and minimize pressure drop [133].

4.7.22   Tower Location

Local heat sources upwind of the cooling tower can elevate the wet-bulb temperature of the air entering the tower. Interference occurs when a portion of the saturated air upwind of the tower

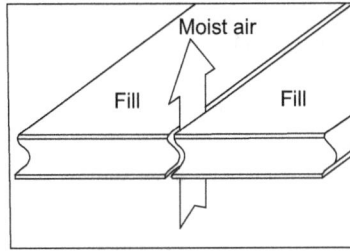

Nesting design.

FIGURE 4.47 Brentwood's AccuPac® counterflow cellular drift eliminators.

contaminates the ambient air of a downwind tower. Drift and condensed water can cause corrosion problems with downwind equipment.

Cooling Tower Water Saving Strategies

4.7.23 Water Loss in Cooling Towers

By design, cooling towers use significant quantities of water. Water leaves a cooling tower system in several ways [134, 135]:

i. evaporation;
ii. blowdown or bleed-off;
iii. drift;
iv. make-up water; and
v. leaks or overflows.

The above losses are illustrated in Figure 4.46. Water scarcity, water reuse, and environmental concerns present challenges for cooling-water systems.

Make-up water is commonly treated before use, and chemicals are added to inhibit scaling, corrosion, and fouling.

Blowdown can contain chemicals used in the cooling towers to control biofouling, scaling, and corrosion. Consequently, blowdown requires treatment before it can be reused or discharged (where allowed).

Drift is the small quantity of water that can be carried from the cooling tower as mist or water droplets.

Leaks or overflows should not occur in a properly operated cooling tower. Most plumbing and building codes require an overflow alarm be installed so that an alarm is activated when water is flowing into the overflow drain.

The amount of water needed by the cooling tower is dictated by the amount of water that is lost through evaporation, blowdown, drift, and leaks.

Fresh water is becoming a scarce resource even in formerly water-rich regions of the world due to development, population growth, and climate change.

4.7.24 Monitoring of Cooling Tower Performance

A continuous monitoring and preventive program is imperative for maximum cooling tower performance. The cooling tower provides the interface of the cooling water with the environment. Degradation in its performance affects the condenser and turbine from attaining optimal performance. Measuring thermal performance of the tower provides data about the effectiveness of the tower heat transfer. Measuring the chemical concentration and making adjustments to them provide the means to attain quality tower performance [11]:

i.   The first line of defense is to develop and maintain a quality water chemistry/chemical program.
ii. The next step is to monitor the performance of the tower to ensure that the program is working effectively.

4.7.24.1   Chemical Monitoring and Control

A successful chemical treatment program is crucial for cooling towers with high efficiency fills as they are very susceptible to fouling, which can lead to the plugging of the fills and reduce tower performance. With high efficiency fills, silt and mud alone will not cause fouling, but biological growth alone can lead to substantial fouling. The combination of silt, mud, and biological growth can lead to catastrophic fouling of the fills.

4.7.24.2   Microbiological Monitoring and Control

To monitor microbiological activity a combination of testing, mathematical predictors, and online monitoring tools can be used. If an increase in microbiological activity is detected, adjustments in the microbiological treatment should be made to minimize potential fouling problems.

4.7.24.3   Deposition Monitoring and Control

Deposition presents the most serious barrier to the transfer of heat through a surface and can be divided into two forms—fouling and scaling. Fouling, due to suspended solids in the water, is the accumulation of water suspended materials on tower fills or heat exchanger surfaces. Scale is a dense coating of inorganic materials and results from the precipitation of soluble minerals from supersaturated water. The formation of scale most often occurs when the water becomes over saturated or when water temperature increases. For some scale species, such as calcium compounds, solubility decreases with increasing temperature. As such, these deposits usually occur first at the outlet end of the condenser where the temperature is the highest. The rate of formation of the scale will depend on temperature, alkalinity or acidity, and the amount of scale-forming material in the water [11].

i.   The most common type of scale found in cooling-water systems is calcium carbonate. It is important to note that as temperature increases, the solubility of calcium carbonate decreases.
ii. Another common type of scale is calcium sulfate, which results from increased levels of sulfates in the cooling water.
iii. Other types of scale found in cooling-water systems include silica and magnesium compounds.

Deposition monitoring can be carried out by measuring the Scaling Indices and/or visual inspections.

4.7.24.4   Scaling Indices

A formation prediction can be calculated with the use of indices such as the Langelier Saturation Index (LSI). This index is described as follows:

$$LSI = pH_s - pH$$

where
pH is the measured water pH,
pH_s is the pH at saturation in calcite or calcium carbonate.

If the LSI calculation produces a positive value, the potential to form scale exists; and if a negative value is produced, no scale potential exists. At no time should an LSI value be greater than 2.2. If the values reach 2.2, changes in the treatment program need to be made immediately to decrease the potential for deposition. It is recommended that an LSI calculation be performed once a day, preferably in the afternoon when temperatures are warmest and conditions are most favorable for deposition.

4.7.24.5 Tower Film Fill Inspection and Cleaning

Inspection a cooling tower enables direct visual feedback about the cleanliness of the tower fill. If a steady increase in film fouling is detected, the current treatment program should be modified to prevent further fouling. Should the weight gain exceed 33% of the initial clean fill weight action should be taken to clean fouled fill [11, 108].

1. Online cleaning can be initiated. One method entails addition of hydrogen peroxide in combination with a surfactant and polymer.
2. Offline cleaning can be accomplished by removing sections of fill and manually cleaning.

4.7.24.6 Performance Monitoring Parameters

The recommended parameter for tracking cooling tower thermal performance is the cooling tower capability. The capability is defined as the ratio of the actual water flow rate to the predicted water flow rate at test conditions. The predicted water flow rate is the interpolated flow rate determined from the tower manufacturer's performance curves at the tested heat load, inlet wet-bulb temperature, and fan motor power. Cooling tower capability is determined using measurements of the following parameters [11, 136]:

i. circulating water flow rate
ii. fan motor power (for mechanical draft cooling towers)
iii. hot water temperature
iv. cold water temperature
v. inlet wet-bulb temperature and dry-bulb temperature (for natural draft cooling towers)
vi. barometric pressure

4.7.25 Inspection and Maintenance

Maintenance practices include [137]

i. Carry out frequent visual inspections and routine maintenance services during operation in subfreezing weather.
ii. Ensure all controls for capacity and freeze protection are set properly and functioning normally.
iii. Prevent excessively high water levels and possible overflow of the cold water basin due to over pumping, clogged strainers, or make-up valve malfunction.
iv. Some unit icing can be expected in very cold weather. Usually this will not affect the operation of the unit. Resolve any icing conditions that may damage the unit or the supports, impair the system performance or create a safety hazard.
v. If white rust forms on galvanized steel surfaces after the pH is returned to normal service levels, it may be necessary to repeat the passivation process.

4.7.26 Automation

Automation systems are available providing a broad range of capacities to control single or multiple parameters in the cooling system such as conductivity and blowdown control, pH control, and real-time chemical monitoring, and dosing. Blowdown controllers are available from several different commercial suppliers and offer a range of control points from simple conductivity/blowdown control, to timed or meter relay chemical dosing [107]. Many of them incorporate water meter inputs and alarm relays if threshold measurements are exceeded [138].

4.7.27 WATER SYSTEM MODELING

Software platforms provide the ability to model the system's scaling tendencies, corrosion characteristics, and view optimal chemical application. This can be a powerful tool to better understand if the system is operating at the maximum cycles of concentration possible, to see problem points in the system (low flow velocity, high-temperature heat exchangers, for example), and to see impacts of varying water characteristics [107].

4.7.28 ENVIRONMENTAL PROTECTION

Some of the environmental protection measures are hereunder:

i.   Chemicals are necessary for reducing the scale formation and fouling in the cooling tower. Adopt principle of green chemistry in water treatment, scaling and fouling control.
ii.   The cooling towers impact on the environment can also be reduced by using saltwater in the cooling process instead of freshwater.
iii.   Using advanced techniques is also good for reducing the impact of cooling towers on the environment.
iv.   Using an advanced filtration system keeps the process-water clean, which reduces the need for water treatment chemicals that can have a negative impact on the environment
v.   To prevent clogging of the cooling tower packing, it is crucial to keep the cooling tower's process water clean at all times. This is achieved by installing the right water filtration system, which will counter the issues caused by the collection of suspended solids, including corrosion and the growth of micro-bacteria.
vi.   The drift contains the same chemicals and micro-organisms contained in the cooling tower. These chemicals and micro-organisms carried in the droplets have a detrimental impact on the environment and become carriers of deadly pathogens such as Legionella pneumophila.
vii.   Key to preventing the release of hazardous droplets into the atmosphere is to install efficient drift eliminators inside the cooling tower,
viii. It is important to choose green options when cleaning a cooling tower to reduce the environmental impact. Blowdown has raised concerns in the past from environmental groups because of the potential release of these chemicals into the surrounding environment.
ix.   Eliminate cross-contamination between process fluid and cooling water- cross-contamination occurs when leaks from process equipment attached to a cooling tower introduce chemicals to the system that would otherwise not be present.

4.7.29 PRINCIPLES OF GREEN CHEMISTRY

Green chemicals such as corrosion inhibitor, oxygen scavengers, and neutralizers for cooling-water treatment products have been manufactured as per twelve tenets of green chemistry and offer "value proposition" by being environmental friendly [139].

4.7.30 ZERO LIQUID DISCHARGE (ZLD)

ZLD systems aim for complete recovery of cooling water, where no wastewater should leave the site and all waste should be converted to dry form for ultimate disposal. Though ZLD is ideal, not all power plants can achieve these systems, but waste minimization technologies are nonetheless desirable.

4.7.31 EFFLUENT PATHWAYS AND POTENTIAL TREATMENT PROGRAMS

Main types of effluents from a wet cooling tower include the following:

i. Drift—can contain heavy metals and organic compounds, and biological pathogens.
ii. Evaporation—can contain gaseous contaminates.
iii. Blowdown—can contain heavy metals and chemicals.
iv. Effluent from water treatment processes associated with the cooling circuit.
v. Sludge generated from cooling system effluent treatment plant- Contains inorganic and organic sediments generated by day to day operations. The waste must be analyzed for toxicity and hazards and disposed of as per environmental regulations.
vi. Miscellaneous items like chemicals spills, leaking of treatment chemicals from piping and treatment vessels, hazardous wastes from plant operation.

REFERENCES

1. Teir, S., and Kulla, A. Feedwater and Steam System Components, Helsinki University of Technology Department of Mechanical Engineering, Energy Engineering and Environmental Protection Publications, Steam Boiler Technology eBook, Espoo 2002.
2. Steam Its Generation and Use, Edited by J.B. Kitto and S.C. Stultz, The Babcock & Wilcox McDermott Company, Forty-first edition, First printing, 2005.
3. www.thermaxglobal.com/what-are-boilers-how-do-they-enhance-production-processes/
4. Chapter 2 Steam Generators, https://staff.emu.edu.tr/devrimaydin/Documents/MENG446/Chapter%202%20-%20Steam%20generators.pdf
5. Boiler Selection Consideration, Cleaver-Brooks, pp. 1–25; http://cleaverbrooks.com/reference-center/boiler-basics/Boiler%20Selection.pdf
6. www.Forbesmarshall.Com/Knowledge/Steampedia/Steam-Distribution/Carryover-And-Its-Causes
7. www.thermodyneboilers.com/fluidized-bed-combustion/
8. https://global.kawasaki.com/en/energy/solutions/energy_plants/waste_heat_boiler.html
9. www.doosanheavy.com/en/intro/randd/boiler_rnd/
10. www.wartsila.com/energy/learning-center/technical-comparisons/combined-cycle-plant-for-power-generation-introduction
11. Partnership to Advance Clean Energy—Deployment (PACE-D), PACE-D Technical Assistance Program, Best Practices Manual for Indian Supercritical Plants, February 2014, USAID.
12. Assessment of Requirement of Bag filter vis a vis Electrostatic Precipitator in Thermal Power Plants, Programme Objective Series Probes/105/2007, Central Pollution Control Board (Ministry of Environment & Forests, Govt. of India), Delhi, India, pp. 1–127.
13. Khan, S., and Khan, S. Boiler and its tangential fuel firing system, *Int. J. Automation and Control Eng. (IJACE)*, *3*, Issue 3 (August 2014)www.seipub.org/ijace doi: 10.14355/ijace.2014.0303.02 71
14. IMIA Working Group Paper 95(16) Supercritical Boilers 49th Annual IMIA Conference—Doha, Qatar, IMIA WG paper 95(16) Supercritical Boiler, pp. 1–39.
15. Study Material (Boiler) (Northern Region), National Power Training Institute, PGDC in TPPE—10th Batch (2005-06), pp. 1–37.
16. www.sciencedirect.com/topics/engineering/superheater
17. Parthiban, K. K. Boiler Consultant, Airheater for Industrial Boilers, Venus Energy Audit System, Mumbai, India,
18. www.nationalboard.org/PrintPage.aspx?pageID=164&ID=196
19. Chapter 41, Economizers and Air Heaters, Steam Its Generation and use, Edited by J.B. Kitto and S.C. Stultz, The Babcock & Wilcox McDermott company, Forty-first edition, First printing, 2005.
20. www3.epa.gov/ttn/chief/ap42/ch01/final/c01s01.pdf

21. www.watertechnologies.com/handbook/chapter-11-preboiler-and-boiler-corrosion-control, Chapter 11- Preboiler & Industrial Boiler Corrosion Control. VEOLIA Water Technologies.
22. www.watertechnologies.com/handbook/chapter-12-boiler-deposits-occurence-and-control VEOLIA Water Technologies. Chapter 12 Boiler Deposits: Occurrence and Control
23. www.lenntech.com/applications/process/boiler/corrosion.htm
24. www.rasmech.com/blog/how-to-fight-boiler-corrosion/
25. Boiler Feedwater Treatment (Part I): Why Water Treatment is Necessary—Technical Library—Sedifilt String-Wound Filter Cartridges
26. www.wilhelmsen.com/ships-service/water-treatment-solutions/corrosion-prevention-in-steam-boiler-systems/
27. http://cleanboiler.org/learn-about/boiler-efficiency-improvement/water-treatment-steam-boilers/
28. Cotton, I. J., and Hollander, O. Boiler Systems—Chemical Treatment, 2003, Springer, www.awt.org/pub/?id=015A80BB-9FCE-2EE8-B925-3EEBB1E13DF1
29. www.watertechonline.com/process-water/article/14293182/boiler-corrosion-mechanical-and-chemical-methods-for-control
30. www.forbesmarshall.com/fm_micro/news_room.aspx?Id=Boilers&nid=179
31. Ganapathy, V. Industrial Boilers and Heat Recovery Steam Generators Design, Applications, and Calculations, Marcel Dekker, New York, 2003.
32. Cuddihy, J. A., Jr., Simoneaux, W. J., Falgout, R. N. and Rauh, J. S. Boiler water treatment and related costs of boiler operation: An evaluation of boilers in the Louisiana sugar industry, *J. Amer. Soc. Sugar Cane Technologists*, 25, pp. 17–30 (2005).
33. https://rakhoh.com/en/a-guide-to-control-total-dissolved-solids-in-the-steam-boiler/
34. IAPWS, Technical Guidance Document: Corrosion Product Sampling and Analysis for Fossil and Combined Cycle Plants (2014). www.iapws.org.
35. http://ecoursesonline.iasri.res.in/mod/page/view.php?id=961
36. www.purolite.com/index/core-technologies/application/condensate-polishing
37. Condensate polishing technology, ANDERSON—WPT, Ancaster, ON L9G 4V5, Canada. Condensate Polishing with SeparIX External Regeneration.
38. www.veoliawatertech.com/vwst-southeastasia/ressources/files/1/44613-Condensate-Polishing_Veolia_Septem.pdf
39. www.degremont-technologies.com/~degremon/Condensate-Polisher-Systems-451
40. www.bousteadsalcon.com/Condensate%20Polishing.pdf
41. http://scitechconnect.elsevier.com/boiler-chemistry-feed-water/
42. www.emerson.com/documents/automation/flyer-steam-water-analysis-system-rosemount-en-69496.pdf
43. https://instrumentationtools.com/swas-analyzers/
44. www.hukseflux.com/uploads/inline/hukseflux_heat_flux_sensors_for_boiler_monitoring_v2304.pdf
45. www.eecpowerindia.com/codelibrary/ckeditor/ckfinder/userfiles/files/1_Boiler%20Tube%20failures.pdf
46. www.watertechnologies.com/handbook/chapter-14-boiler-system-failures. VEOLIA Water Technologies.
47. www.chicagotube.com/products-2/boiler-tubing/description-of-boiler-tube-grades-recommended-uses-per-the-asme-boiler-pressure-code-section-1-sa178-sa213-grades/
48. De Witt-Dickk, D., McIntye, S., and Hofilen, J. Ashland Specialty Chemical Company, Boonton, NJ, Presented at the 1st Annual International Water Conference, October 22–26 2002, Pittsburgh, PA.
49. www.flawtech.com/product/boiler-tube-damage-kit/
50. Koripelli, R. S. How to Alleviate Waterside Issues in Boilers and HRSGs, Feb 01, 2019, POWER; www.powermag.com/how-to-alleviate-waterside-issues-in-boilers-and-hrsgs/
51. www.lenntech.com/boiler/caustic-corrosion.htm#ixzz85Evli8o5
52. https://en.wikipedia.org/wiki/Caustic_embrittlement
53. www.corrosionpedia.com/definition/240/caustic-embrittlement
54. www.chardonlabs.com/resources/caustic-gouging-what-is-it-and-how-can-i-prevent-it/
55. Hydrocarbon Processing, January 1989 57 V Ganapathy, ABCO Industries, Abilene, Texas, pp. 57–59.
56. Boiler Caustic Embrittlement, Understanding the hazard, FM Global, 2003, Factory Mutua; Insurance Company, USA.

57. www.nationalboard.org/index.aspx?pageID=164&ID=196
58. www.evolvetcr.com/download/Boiler_tube_investigation.pdf; Boiler Tube Failure Investigation, TCR Advanced Engineering Company, A service partner company of TCR Engineering P. Ltd., 2012-16, NaviMubai, India.
59. www.epri.com/research/products/TR-102433-V1
60. Boiler System Failures, Analysis & Diagnostics Manual, Examples of issues in specific boiler system components, Metallurgy, Volume 2.0, Boiler Systems, ChemTreat 2014, ChemTreat, Inc., Glen Allen, Virginia, pp. 1–42.
61. www.babcock.com/home/about/resources/learning-center/finding-the-root-cause-of-boiler-tube-failure
62. www.tcradvanced.com/boiler-tube-failure.html
63. www.brighthubengineering.com/power-plants/74125-how-to-deal-with-tube-failures-in-boilers/
64. www.twi-global.com/technical-knowledge/faqs/what-is-high-temperature-hydrogen-attack-htha-hot-hydrogen-attack
65. www.osti.gov/servlets/purl/882855
66. Singh, P. M., and Pawel, S. J. Final Technical Report Stress-Assisted Corrosion in Boiler Tubes (DE-FC36-02ID14243), Oak Ridge National Laboratory, Oak Ridge, TN.
67. https://mmengineering.com/boiler-tube-failure/handbook/
68. http://npti.in/Download/Thermal/BoP/15%20V%20K%20Garg.pdf
69. Parthiban, K. K. Erosion Failures, Boiler Consultant, Case studies on boiler tube erosion, Venus Energy Audit System.
70. Boilers, Chapter 9, NAVEDTRA 14265A, 9-1-9-118; http://navybmr.com/study%20material/14265a/14265A_ch9.pdf
71. Bhattacharya, D. K. Non-Destructive Test Techniques For Boiler Systems, National Workshop on Boiler Corrosion, 11–12 April, 1995, NML, Jamshedpur India, pp. N-1–N-12; http://eprints.nmlindia.org/3562/1/N1-N12.PDF
72. Chaplin, R. A. Steam Turbine Steam System, Thermal Power Plants—Vol. III, Encyclopedia of Life Support Systems (EOLSS).
73a. https://deaerator.com/deaerator-purpose-types-functions/ 2022 Kansa City Deaerator Company
73b. The Two Main Types of Boiler Deaerators—Kansas City Deaerator
74. https://www.membranechemicals.com/water-treatment/deaeration-systems/
75. www.watertechnologies.com/handbook/chapter-10-boiler-feedwater-deaeration VEOLIA Water Technologies.
76. www.powermag.com/improve-plant-heat-rate-with-feedwater-heater-control/
77. Standards for Closed Feedwater Heaters, 9th Edition, 2015, Heat Exchange Institute (HEI), Cleveland, Ohio
78. TOSHIBA Steam Turbine November, 2007, TOSHIBA Plant Systems & Services Corporation, pp. 1–18.
78.1 www.toshiba.com/taes/products/steam-turbine-generator
79. www.thermopedia.com/content/756/ McNaught, J. M., Feedwater heaters, DOI: 10.1615/AtoZ.f.feedwater_heaters
80. Hommel, S. Part 2 of Series: Determining the Health of Feedwater Heaters, Published at January 2, 2019; www.fossilconsulting.com/2019/01/02/determining-health-of-feedwater-heaters/; Part 3 of Series: Level Control and Feed water Heater Problems, Published by Hommel at January 15, 2019, Scott Hommel, Author at Fossil Consulting Services, Inc.
81. https://inis.iaea.org/collection/NCLCollectionStore/_Public/47/056/47056962.pdf; Fulger, M., Mihalache, M., Velciuand, L., Vetelaru, I. Corrosion mechanisms of tubes from the Candu high pressure feed water heaters, pp.125–131. Lettersdistorted.
82. Birring, A. NDT for Non-Nuclear Power Generation: Pressure Vessels, Piping, and Turbines in Encyclopedia of Materials: Science and Technology, 2001
83. Standards for Steam Surface Condensers, 12th Edition, 2017.
84. www.watco-group.co/surface-condenser-in-thermal-power-plant/
85. Surface condenser–Operation, Maintenance and Installation Manual, GRAHAM CORPORATION, Batavia, New York., pp. 1–20.
86. https://savree.com/en/encyclopedia/steam-turbine-condenser

87. www.powermag.com/taming-condenser-tube-leaks-part-ii/

88. Bernstein, M. D. and Soler, A. I., The tubesheet analysis method in the new HEI condenser standards, *Trans. ASME J. Eng. Power, 100*, 363–368 (1978).

89. Singh, K. P. and Soler, A. I. (1984). Rectangular Tubesheets—Application to Power Plant Condensers. In: Mechanical Design of Heat Exchangers. Springer, Berlin, Heidelberg.

90. Structural Design Concepts for Increased Reliability and Safety in Power Plant Condensing Systems FP-507, Volume 2 Research Project 372-1 Final Report, September, Principal Investigator A. I. Soler Research Associate C. Shahravan Prepared for Electric Power Research Institute, Palo Alto, California.

91. Steam Generator Reference Book, Revision 1 Volume 1, EPRI Perspective, Project RP2858; RP4004 Project Managers: J. P. N. Paine. Based on work sponsored by The Steam Generator Owners Groups I and II, The Steam Generator Reliability Project and Electric Power Research Institute, J. Peter N. Paine, Editor Ulla E. Gustafsson, Associate Editor December 1994.

92. Diehl, T. H., Harris, M. A., Murphy, J. C., Hutson, S. S., and Ladd D. E., Methods for Estimating Water Consumption for Thermoelectric Power Plants in the United States, Scientific Investigations Report 2013–5188, U.S. Department of the Interior U.S. Geological Survey, USGS National Water Census and National Streamflow Information Program, pp. 1–78.

93. Steam Plant Surface Condenser Leakage Study, EPRI IMP-481 (Research Project 624-1), Final Report Volume I, March 1977, Prepared by BECHTEL CORPORATION, San Francisco, California, Electric Power Research Institute, Palo Alto, California; www.osti.gov/servlets/purl/7302881

94. www.powermag.com/condenser-tube-failure-mechanisms/

95. www.process-cooling.com/articles/88004-comparing-leak-detection-methods-for-power-generation-condensers

96. McMaster, R. C., ed., *Nondestructive Testing Handbook, Vol. 1, Leak Testing*, 2nd edn., American Society for Nondestructive Testing and American Society for Metals, Columbus, OH.

97. Anderson, G. L., Leak testing, in *Metal Handbook, Vol. 17, Nondestructive Evaluation and Quality Control*, 9th edn., American Society for Metals, Metals Park, OH, 1989, pp. 57–70.

98. www.applus.com/global/en/what-we-do/sub-service-sheet/mass-spectrometer-helium-leak-test

99. Carpenter, A. M. Water conservation in coal-fired power plants, CCC/275, February 2017, IEA Clean Coal Centre, Upper Richmond Road London SW15 2SH United Kingdom, pp. 1–72.

100. The Cooling Water Handbook A basic guide to understanding industrial cooling water systems and their treatment, 2017 Buckman Laboratories International, Inc., pp. 1–19. buckman.com.

101. www.watertechnologies.com/handbook/chapter-30-once-through-cooling VEOLIA Water Technologies.

102. www.watertechnologies.com/handbook/chapter-31-open-recirculating-cooling-systems VEOLIA Water Technologies.

103. Chapter 32—Closed Recirculating Cooling Systems, www.watertechnologies.com/handbook/chapter-32-closed-recirculating-cooling-systems VEOLIA Water Technologies.

104. Tsou, J. L., Maulbetsch, J., and Shi, J. EPR Power Plant Cooling System Overview for Researchers and Technology Developers, 3002001915, May 2013, Electric Power Research Institute, Inc., Palo Alto, California, pp. 1-1–4-4.

105. The Water-Energy Nexus: Challenges and Opportunities, U.S. Department of Energy, June 2014, pp. 1–240.

106. Water Requirements for Existing and Emerging Thermoelectric Plant Technologies, DOE/NETL-402/080108, August 2008, Revised April 2009, National Energy Technology Laboratory (NETL), pp. 1–17.

107. Cooling Towers: Understanding Key Components of Cooling Towers and How to Improve Water Efficiency, EERE Information Center, DOE/PNNL-SA-75820, February 2011, pp. 1–9. www.eere.energy.gov/informationcenter

108. Cooling Tower Fundamentals, John C Hensley (Editor), 2nd Edition, SPX Cooling Technologies, Inc., Kansas, pp. 1–116.

109. 39.2 2008 ASHRAE Handbook-HVAC Systems and Equipment, CHAPTER 39, COOLING TOWERS, pp. 39.1–39.20

110. Guyer, J. P. An Introduction to Cooling Tower Water Treatment, Course No: C05-019, Continuing Education and Development, Inc. 9 Greyridge Farm Court Stony Point, NY.

111. Common, Operation & maintenance manual, 2011 Baltimore Aircoil Company, Jessup, MD 20794, pp. 1–36.
112. Baltimore Aircoil International nv, Industriepark Zone A, B-2220 Heist-op-den-Berg, Belgium, pp. 1–154.
113. https://amarillogearservice.com/different-types-cooling-towers
114. https://spxcooling.com/library/induced-draft-vs-forced-draft-cooling-towers/
115. https://spgdrycooling.com/news/dry-cooling/
116. Case study - report on minimisation of water requirement in coal based thermal power stations, central electricity authority, Government of India, 2012, pp. 1–48.
117. www.chemicalonline.com/doc/single-row-condenser-0001
118. Ashwood, A. and Bharathan, D. Hybrid Cooling Systems for Low-Temperature Geothermal Power Production, Technical Report, NREL/TP-5500-48765, March 2011 U.S. Department of Energy Office of Scientific and Technical Information, Ridge, TN pp. 1–62.
119. www.bgrcorp.com/gct-indirect.php
120. Jaszay, T. Industrial review-Aus Der Industrie, The Air-Cooled Condensing Equipment "System Heller," A Comprehensive Survey, Polytechnical University, Budapest, Department of Energetic, pp. 389–402.
121. Balogh, A. and Szabo, Z. The advanced HELLER system, Technical Features & Characteristics, A paper for EPRI Conference on "Advanced Cooling Strategies/Technologies," June 2005, Sacramento (CA), GEA and EGI, pp. 1–53; www.nrc.gov/docs/ML0900/ML090090843.pdf
122. www.chardonlabs.com/resources/how-to-optimize-cooling-tower-water-ph-balance/
123. The Cooling Water Handbook - A Basic Guide to Understanding Industrial Cooling Water Systems and Their Treatment, 2017, pp. 1–19.
124. Cooling Water Problems and Solutions Course No: M05-009 Credit: 5 PDH A. Bhatia, Continuing Education and Development, Inc., Woodcliff Lake, NJ
125. www.energy.gov/eere/femp/best-management-practice-10-cooling-tower-management
126. An Introduction to Industrial Water Treatment Systems, a Publication of SAMCO Technologies; www.gmc.biz.vn/blogs/san-pham/cooling-water-treatment
127. The Cooling Water Handbook - A Basic Guide to Understanding Industrial Cooling Water Systems and Their Treatment, 2017, pp. 1–19.
128. www.gmc.biz.vn/blogs/san-pham/cooling-water-treatment
129. Ozonation of Cooling Tower Water: A Case Study by Stephen Osgood Water Conservation Unit, East Bay Municipal Utility District, June 1991, California Department of Water Resources, Water Conservation Office, pp.1–33.
130. KLM Technology Group Practical Engineering Guidelines for Processing Plant Solutions www.klmtechgroup.com Kolmetz Handbook of Process Equipment Design, Cooling tower selection, sizing and troubleshooting (engineering design guidelines) Rev: 02 May 2014, pp.1–72.
131. www.watertechnologies.com/handbook/chapter-29-cooling-tower-wood-maintenance
132. www.paharpur.com/wp-content/uploads/2017/04/Paharpur_Timber-Brochure.pdf
133. www.brentwoodindustries.com/cooling-tower/
134. Aull, R. Wet Cooling Towers: A Review of Current Technology, Products Under Development and Research Needs, ASME-IMECE Congress/NSF-EPRI Workshop on Advanced Power Plant Cooling: Reducing Water Consumption
135. Water Efficiency Management Guide Mechanical System EPA 832-F-17-016c, November 2017, U.S. Environmental Protection Agency (EPA), pp. 1–17.
136. Connecting Operating and Maintenance Instructions-RCT Cooling Towers, Baltimore Aircoil Int. nv, Industriepark—Zone A, B-2220 Heist-op-den-Berg, Belgium, pp. 1–16; www.baltimoreaircoil.eu/sites/BAC/files/BAC_MaintenanceRC_MRCv03EN.pdf
137. WWW.BALTIMOREAIRCOIL.COM
138. Tsou J. L., Maulbetsch, J., and Shi, J. Power Plant Cooling System Overview for Researchers and Technology Developers, 3002001915, May 2013, Electric Power Research Institute, Inc., Palo Alto, California, pp. 1-1-4-4.
139. www.thermaxglobal.com/thermax-chemicals/thermax-green-chemicals/ Thermax Global 2018.

BIBILOGRAPHY

Aherne, V. J. Editor, Water Conservation In Cooling Towers, Best Practice Guidelines, The Australian Institute of Refrigeration Air Conditioning and Heating Inc., AIRAH 2009, pp. 1–20. www.airah.org.au

Assessment of requirement of Bag filter vis a vis Electrostatic Precipitator in Thermal Power Plants, Programme Objective Series Probes/105/2007, Central Pollution Control Board (Ministry of Environment & Forests, Govt. of India), Delhi, India, pp. 1–127.

BEAMA, *Guide to Design of Feedwater Heating Plant*, 1968, The British Electrical and Allied Manufacturers' Association Ltd.,, London.

Handbook of Process Equipment Design, Boiler systems selection, sizing and troubleshooting (engineering design guidelines) Rev: 01 December 2011, PP. 1–63. KLM Technology Group Practical Engineering Guidelines for Processing Plant Solutions www.klmtechgroup.com Kolmetz

Hewitt, G. F, Shires, G. L., and Bott, T. R. (1994) *Process Heat Transfer*, CRC Press.

https://beeindia.gov.in/sites/default/files/Air%20Pre%20heater.pdf

www.globalspec.com/learnmore/manufacturing_process_equipment/heat_transfer_equipment/steam_generators_boilers

www.mechanicalbooster.com/2017/02/steam-condenser.html?expand_article=1

www.plymouth.com/downloads/selecting-feedwater-heater-tube-materials-for-greatest-efficiency-and-reliability/

www.plymouth.com/wp-content/uploads/2019/08/Selecting-FWH-Tube-API-PowerChem-2012.pdf

www.powerhx.com/kor/product/deaerator.php

www.powermag.com/how-to-alleviate-waterside-issues-in-boilers-and-hrsgs/

www.process-cooling.com/articles/88004-comparing-leak-detection-methods-for-power-generation-condensers

www.slideshare.net/SHIVAJICHOUDHURY/feedwater-heaters-in-thermal-power-plants

https://www.energy.gov/eere/amo/articles/minimize-boiler-blowdown

5 Heat Exchanger Installation, Operation, and Maintenance

Major types of heat exchangers in service are shell and tube, air cooled, tube fin, plate fin, plate exchanger, etc. Basic configuration of some types of heat exchangers are shown in Figure 5.1. A heat exchanger requires maintenance to keep it functioning efficiently. The major problems faced with heat exchangers in service are inadequate heat transfer and excessive pressure drop. Common reasons for inadequate heat transfer are that the plant is operating at enhanced capacity, gradual fouling buildup, variation in inlet temperatures of process fluids, air or gas binding resulting from improper piping installation or lack of suitable vents, operating conditions differing from design conditions, maldistribution of flow in the unit, fabrication defects like excessive clearances between the baffles and shell and/or tubes, poor orientation of baffle, wrong fitment of gaskets, excessive corrosion, blocked tubes/nozzles, leakage due to tube wall rupture and tube-to-tubesheet joint failures, lack of performance monitoring, improper maintenance procedures, lack of preventative maintenance schedules, etc. As a result, the existing heat exchanger equipment cannot achieve the desired heat duty. To ensure the designed performance level and to prevent leaks and failures, the heat exchanger installation, operation, maintenance, and periodical inspection must be carried out as per the manufacturer's operation and maintenance manual such as Ref. [1] or by experienced service providers or skilled personnel. Various issues involved procedures for installation and commissioning, safe operation, and maintenance of heat exchangers, repair, rehabilitation, plant life assessment, NDT methods for in-service inspection of heat exchangers and pressure vessels are discussed later.

5.1 STORAGE

If equipment arrives on site several months prior to the actual installation and start-up, it result in a potential for premature corrosion to occur unless the equipment is stored in a proper manner. Equipment should be stored so that it is not exposed to excessive construction debris and concrete dust, industrial contaminants, coastal contaminants, or high levels of humidity and moisture. Improper storage can lead to premature corrosion prior to start-up and can reduce the overall life of the equipment. Extra care should be taken to ensure that equipment located on the ground is above the maximum rain water level.

If the heat exchanger is not used immediately, take the precautions as follows:

5.1.1 Short-time storage

1. Keep the heat exchanger in a clean and dry room.
2. If its storage is outdoors or where humidity is high, seal the heat exchanger and wrap it with a vinyl sheet or the like.

DOI: 10.1201/9781003352068-5

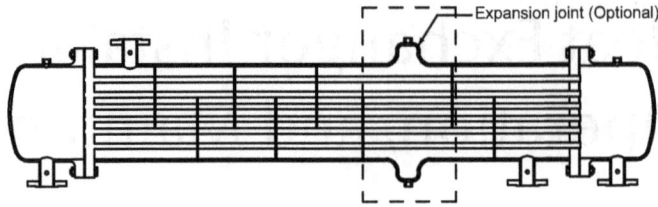

(a)   Shell and tube heat exchanger(STHE)

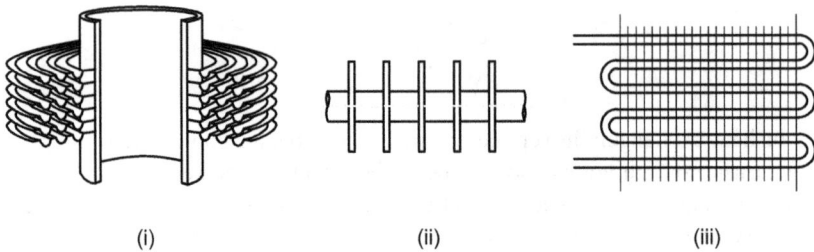

(i)   (ii)   (iii)

(b)   Finned tube heat exchanger

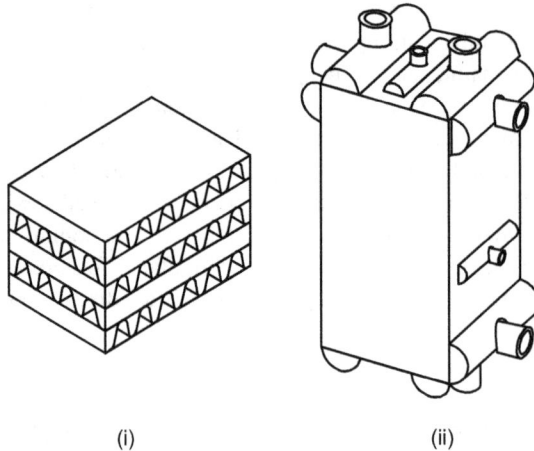

(i)   (ii)

(c)   Plate-fin heat exchanger(PFHE)

FIGURE 5.1   Basic configuration of some types of heat exchangers.

5.1.2   Long-time storage

The heat exchanger is normally shipped with its inside surfaces coated with rust-preventive oil. The rust-preventive coating normally stays active for a period of 2–3 months in a dry room. If the heat exchanger is to be stored longer, open its ports and apply the same type of rust-preventive oil where necessary every 2 or 3 months. If the storage is for more than a year, test the heat exchanger at its maximum service pressure and make sure it is in working order prior to operation.

(i)   Forced draft.   (ii)   Induced draft.

(d)   Air cooled heat exchanger(ACHE)

(e)   Micro channel heat exchanger   (f)   PHE

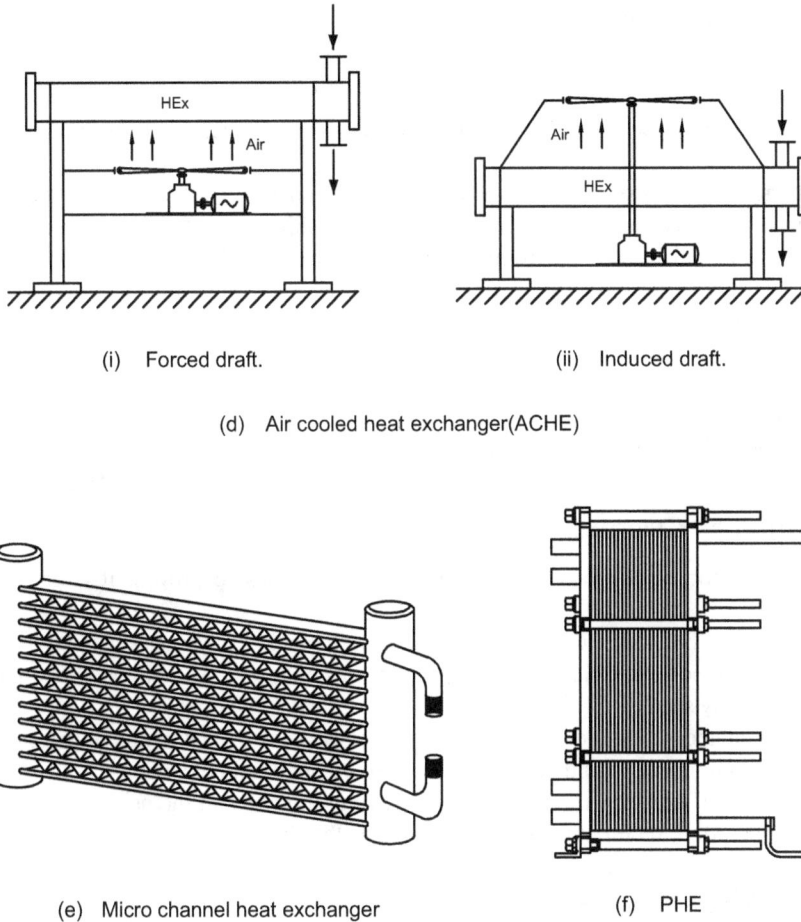

FIGURE 5.1   (Continued)

5.1.3   PHE storage

If the heat exchanger is to be placed in long-term storage, the following steps should be taken:

a. Store the heat exchanger in a closed room at 60°F–70°F.
b. The heat exchanger tightening bolts and plate pack shall be completely untightened and stress free.
c. The heat exchanger should be totally covered in black plastic to prevent light and dirt from adversely affecting the gaskets.
d. Avoid heat, ultraviolet, and welding light.
e. Apply a rust preventative.

5.2   INSTALLATION

The heat exchanger should be located in a clean, open area, with sufficient clearance at the head of the heat exchanger to remove the tube bundle. All heat exchanger openings should be inspected for foreign material. Protective plugs and covers should not be removed until just prior to installation. The entire system should be cleaned before starting operation. The foundation of the heat exchanger should be adequate so that the unit will not settle and cause piping strains.

5.2.1 LOCATIONAL FACTORS

Before installation consider the following:

a. Type of environment which has a bearing on the corrosion of heat exchangers

Potentially corrosive outdoor environments include areas adjacent to the seacoast, industrial sites, heavily populated urban areas, some rural locations, or combinations of any of these environments. Local environments called microenvironments must also be considered. For working in corrosive environments, such as sea coasts, industrial zones, and highly polluted areas, microchannel coils may require additional protection.

b. Wind directions
c. Vibration from near-by heavy equipment
d. Space for maintenance activities, space in terms of length for pulling of tube bundle

5.2.2 INSTALLATION GUIDELINES

1. When transporting the heat exchanger, take precautions to prevent it from tumbling or falling.
2. Fit the anchor bolts securely, taking into account the heat exchanger's weight.
3. When lifting the heat exchanger, use the two hooks on top of the main body or attach rope to the main body securely.
4. Ensure that the ropes are attached properly to prevent the heat exchanger from falling. Roping and hoisting should be carried out only by qualified personnel.
5. Install the heat exchanger on a level floor. Taking into account the heat exchanger's weight during operation, secure it using the specified anchor bolts.
6. Ensure that sufficient space is provided for removal of the drain plug.
7. Avoid applying excessive force to the nozzle when making connections. If subjected to excessive force, it may break.
8. Ensure that there is sufficient space around the heat exchanger for maintenance. It is particularly important to provide space for removal of the tube bundle for inspection, cleaning, and to carry out repairs.
9. Connect all nozzles, making sure that the inlet and outlet connections are made correctly. Ensure that all pipe joints are completely leakproof.
10. *U-tube heat exchangers*: for U-tube heat exchangers, with removable bundles, clear at the stationary head end to permit withdrawal of the tube bundle, or at the opposite end to permit removal of the shell.
11. *Foundations:* foundations must be adequate so that exchangers will not settle and impose excessive strains on the exchanger.
12. *Foundation bolts*: foundation bolts should be loosened at one end of the unit to allow free expansion of shells. Slotted holes in supports are provided for this purpose when saddles are provided.
13. *Fittings and piping*: it may be desirable to provide valves and bypasses in the piping systems to permit inspection and repairs.
14. *Vents:* proper venting is a start-up necessity. Improper venting usually occurs on start-up and is recognized by poor heat transfer and a high-pressure drop. Vent valves should be provided so units can be purged to prevent vapor or gas binding. Special consideration must be given to the discharge of hazardous or toxic fluids.
15. *Drains:* drains may discharge to atmosphere, if permissible, or into a vessel at lower pressure.

16. *Pulsation and vibration*: in all installations, care should be taken to eliminate or minimize the transmission of fluid pulsations and mechanical vibrations to the heat exchangers.

17. *Safety relief devices*: the ASME Code [3a, 3b] defines the requirements for safety relief devices. When specified by the purchaser, the manufacturer will provide the necessary connections for the safety relief devices. The purchaser will provide and install the required relief devices.

5.3 OPERATION

1. While starting, ensure that all operating conditions conform to the heat exchanger's specifications.
2. Do not use seawater with a heat exchanger that is designed for use with freshwater or vice versa.
3. On both sides, the fluid stream flow rate must be maintained within the specified limits.
4. The heat exchanger operating pressures must not exceed the specified values.
5. *Design and operating conditions*: equipment must not be operated at conditions that exceed the design parameters.
6. *Operating procedures*: before placing any exchanger in operation, reference should be made to the exchanger drawings, specification sheet(s), and name plate(s) for any special instructions. Local safety and health regulations must be considered. Improper start-up or shutdown sequences may cause leaking of tube-to-tubesheet joint and/or bolted flanged joints.
7. *Start-up operation*: shell and tube exchanger fluids must be introduced in such a manner as to minimize differential expansion between the shell and tubes. The exchanger may be placed into service by first filling the colder medium, followed by gradual introduction of the hotter medium. During start-up, all vent valves should be opened and left open until all passages have been purged of air and completely filled with fluid.
8. *Shutdown operation*: the unit must be shut down in a manner to minimize differential thermal expansion between shell and tubes. Exchangers may be shut down by first gradually stopping the flow of the hotter medium, then the flow of the colder medium through the unit. After shutting down the system, all units should be drained completely, especially when there is a likelihood of freezing or prone for corrosion. To guard against water hammer, condensate should be drained from steam heaters and similar apparatus during start-up or shutdown. To reduce water retention after drainage, the tubeside of the exchanger should be blown out with air.

5.4 MAINTENANCE

Follow the recommended procedure for operation. Quick start-up and shutdown is a major cause of heat exchanger damage. At regular intervals and as frequently as experience indicates, an examination should be made of the interior and the exterior of the heat exchanger. Exchangers subject to fouling or scaling should be cleaned periodically. Increase in pressure drop and/or reduction in performance usually indicates cleaning is necessary. When sacrificial anodes are provided, they should be inspected at regular intervals to determine whether they should be cleaned or replaced [1]. Other than the instructions on maintenance, a log of all plant operational events, start-ups, shutdowns, and malfunctions that affect the heat exchanger performance should be kept. Adopt a scheduled maintenance program as recommended by supplier of the heat exchanger. For example, as follows:

1. *Daily check*: check for oil or water or process fluid leaks if any in the sealed areas of the heat exchanger.
2. *Periodical maintenance*: ensure that the corrosion/fouling control is in place. Overhaul the heat exchanger and clean parts as per schedule. At the same time, check for visible cracks, if any, of heat exchanger components, integrity of liners, insulation, etc.

5.5 DETERIORATION IN SERVICE AND FAILURE MODES

A relatively large margin for reliability and safety is included in the design of pressure vessels and heat exchangers. However, lack of understanding of all service conditions in design, poor quality assurance during manufacture, and changes in service conditions can erode this margin. In general, conditions diminishing the safety margin can arise from inadequacies during design and manufacture, or from operational conditions such as corrosion and fouling, These are described in greater detail in the following, but with the major emphasis on service-induced causes and failures, inspection and testing by NDT methods, and repair and reclamation and/or refurbishment of ACHEs and STHEs.

5.6 DAMAGE MECHANISMS WITHIN PROCESS EQUIPMENT, AFTER API RP 571[4]

Damage mechanisms (also referred to as degradation mechanisms) is a general term referring to any cause of problems or failures within process equipment. Damage mechanisms are detailed and covered in API RP 571, *Damage Mechanisms Affecting Fixed Equipment in the Refining Industry* [4], is a recommended practice developed and published by the American Petroleum Institute (API) that provides an in-depth look at over 60 different damage mechanisms that can occur to process equipment in refineries. This recommended practice discusses damage mechanisms applicable to oil refineries; however, much of the information herein can also be applied to petrochemical and other industrial applications, as the user deems appropriate. It is up to the user to determine the applicability and appropriateness of the information contained herein as it applies to their facility. This recommended practice includes a general description of the damage mechanism, susceptible materials of construction, critical factors, inspection method selection guidelines, control factors, prevention and mitigation, related mechanism, etc. A list of damage mechanisms as per Ref. [4] is shown in Table 5.1. Some of the most common damage mechanisms in the refining and chemical processing industries are as follows [5]:

i.   Corrosion under insulation (CUI), which occurs when moisture builds up on the surface of insulated equipment.

ii. Sulfidation corrosion, a type of corrosion that occurs at temperatures above 500 °F (260 °C) due to sulfur compounds in crude.

iii. High-temperature hydrogen attack (HTHA), a mechanism that can affect equipment that is exposed to hydrogen at elevated temperatures (at least 400 °F or 204°C).

iv. Wet H_2S damage, which can occur when atomic hydrogen from wet H_2S corrosion reactions enter and weaken the steel.

v. CO_2 corrosion, which is a form of degradation that occurs when dissolved CO_2 in condensate forms carbonic acid, which corrodes steels.

vi. Hydrogen attack or hydrogen embrittlement, which happens when atomic hydrogen infuses into certain higher strength steels and causes them to become brittle.

vii. Brittle fracture, which is the sudden, very rapid fracture under stress where the material exhibits little or no evidence of ductility or plastic degradation before the fracture occurs.

The list of types of damage mechanisms covered by API RP 571, *Damage Mechanisms Affecting Fixed Equipment in the Refining Industry, 2020 is shown in* Table 5.1.

Some of the topics are discussed in Chapter 3 and Chapter 4 and a few topics are discussed below. For details on the general description of the damage mechanism, susceptible materials of construction, critical factors, inspection method selection guidelines, control factors, prevention and mitigation, related mechanism, etc., refer to API RP 571.

TABLE 5.1
List of Damage Mechanisms Covered by API RP 571–2020

1. 885 °F (475 °C) Embrittlement	36. High-temperature Hydrogen Attack
2. Amine Corrosion	37. Hydrochloric Acid Corrosion
3. Amine Stress Corrosion Cracking	38. Hydrofluoric Acid Corrosion
4. Ammonia Stress Corrosion Cracking	39. Hydrofluoric Acid Stress Corrosion Cracking of
5. Ammonium Bisulfide Corrosion (Alkaline Sour Water)	Nickel Alloys
6. Ammonium Chloride and Amine Hydrochloride	40. Hydrogen Embrittlement
Corrosion	41. Hydrogen Stress Cracking in Hydrofluoric Acid
7. Aqueous Organic Acid Corrosion	42. Liquid Metal Embrittlement
8. Atmospheric Corrosion	43. Mechanical Fatigue (Including Vibration-induced
9. Boiler Water and Steam Condensate Corrosion	Fatigue)
10. Brine Corrosion	44. Metal Dusting
11. Brittle Fracture	45. Microbiologically Influenced Corrosion
12. Carbonate Stress Corrosion Cracking	46. Naphthenic Acid Corrosion
13. Carburization	47. Nitriding
14. Caustic Corrosion	48. Oxidation
15. Caustic Stress Corrosion Cracking	49. Oxygenated Process Water Corrosion
16. Cavitation	50. Phenol (Carbolic Acid) Corrosion
17. Chloride Stress Corrosion Cracking	51. Phosphoric Acid Corrosion
18. CO_2 Corrosion	52. Polythionic Acid Stress Corrosion Cracking
19. Concentration Cell Corrosion	53. Refractory Degradation
20. Cooling Water Corrosion	54. Stress Relaxation Cracking (Reheat Cracking)
21. Corrosion Fatigue	55. Short-term Overheating-Stress Rupture (Including
22. Corrosion Under Insulation	Steam Blanketing)
23. Creep and Stress Rupture	56. Sigma Phase Embrittlement
24. Dealloying	57. Soil Corrosion
25. Decarburization	58. Sour Water Corrosion (Acidic)
26. Dissimilar Metal Weld Cracking	59. Spheroidization (Softening)
27. Erosion/Erosion-Corrosion	60. Strain Aging
28. Ethanol Stress Corrosion Cracking	61. Sulfidation
29. Flue Gas Dew-Point Corrosion	62. Sulfuric Acid Corrosion
30. Fuel Ash Corrosion	63. Temper Embrittlement
31. Galvanic Corrosion	64. Thermal Fatigue
32. Gaseous Oxygen-enhanced Ignition and Combustion	65. Thermal Shock
33. Graphitic Corrosion of Cast Irons	66. Titanium Hydriding
34. Graphitization	67. Wet H2S Damage (Blistering/HIC/SOHIC/SSC)
35. High-temperature H2/H2S Corrosion	

5.6.1 885 °F (475 °C) EMBRITTLEMENT

885°F (475°C) embrittlement is a loss of ductility and fracture toughness due to a metallurgical change that can occur in stainless steels containing a ferrite phase as the result of exposure in the temperature range 600°F to 1000°F (315°C to 540°C). The embrittlement can lead to cracking failure. Embrittlement can also result from heat treatment if the material is held within or cooled slowly through the embrittlement range.

Affected Materials

(a) 400 series SS (e.g., 405, 409, 410, 410S, 430, and 446).
(b) Duplex stainless steel alloys, such as 2205, 2304, and 2507.

(c) Austenitic stainless steel (300 series) weld metals, which normally contain up to about 10% ferrite phase to prevent hot cracking during welding.

Prevention/Mitigation

(a) The best way to prevent 885°F (475°C) embrittlement is to avoid exposing the susceptible material to the embrittling range or to use a nonsusceptible material.
(b) 885°F (475°C) embrittlement is reversible by heat treatment followed by rapid cooling.

5.6.2 Ammonia Stress Corrosion Cracking

Copper-zinc alloys, carbon steel, alloy steel, and high-strength, quenched and tempered steel under applied or residual stress and especially when cold formed and/or welded without subsequent thermal stress relief are subject to failure by stress corrosion cracking (SCC) in air-contaminated dry ammonia.

5.6.3 Brine Corrosion

Widespread or localized pitting occurring on the surface of equipment exposed to aqueous solutions of dissolved salts, typically a chloride or other halide salt. In chloride brine corrosion of stainless steels, pits are initiated where chlorides break down the passive layer. Carbon steel, alloy steel, or stainless steel, with carbon steel being the most common material exposed to brine solutions. Aluminum alloys are also susceptible to brine corrosion. Ni-Cr-Mo alloys have improved resistance to brine corrosion. The chloride pitting and crevice corrosion resistance of stainless steels and higher alloy Fe-Ni-Cr alloys is quantified by the pitting resistance equivalent number (PRE_N) given by

$$PRE_N = \%Cr + 3.3 \times (\%Mo + 0.5 \times \%W) + 16 \times \%N$$

Generally, a PRE_N of 40 or higher is considered necessary to resist corrosion in aerated seawater in ambient conditions. In stainless steels, an increase in molybdenum content improves resistance to chloride pitting and crevice corrosion in brine solutions, as can be seen by the increase in PRE_N. Nitrogen additions are also very potent in increasing resistance.

5.6.4 Carburization

Carbon is absorbed into a material at elevated temperature while in contact with a carbonaceous material or carburizing environment. Carburized steel is brittle and may spall or crack. Carburization can reduce (or eliminate) the remaining sound metal wall thickness and may also reduce the corrosion resistance of stainless steel.

5.6.5 Carbon Dioxide (CO_2) Corrosion

Carbon dioxide (CO_2) is found in oil and gas fields in varying concentrations. Dry CO_2, be it in gas phase or a supercritical fluid is not corrosive to metals and alloys. However, in the presence of water-containing produced fluids, severe corrosion of the infrastructure may result due to the formation of carbonic acid. Corrosion of materials in contact with CO_2-containing fluid is dependent on various factors such as (i) concentration of CO_2 (and other components like H_2S); (ii) water chemistry; (iii) operating conditions; and (iv) material type [6].

5.6.6 Caustic Corrosion/Gouging

Caustic corrosion occurs when caustic is concentrated and dissolves the protective magnetite (Fe_3O_4) layer, causing a loss of base metal and eventual failure. The following conditions appear to be necessary for this type of cracking to occur:

1. The metal must be stressed.
2. The boiler-water must contain caustic.
3. At least a trace of silica must be present in the boiler-water.
4. Some mechanisms, such as a slight leak, must be present to allow the boiler water to concentrate on the stressed metal.

Stem blanketing is a condition that occurs when a steam layer forms between the boiler water and the tube wall. Under this condition, insufficient water reaches the tube surface for efficient heat transfer. The water that reaches the overheated boiler wall is rapidly vaporized, leaving behind a concentrated caustic solution, which is corrosive. Boiler feedwater systems using demineralized or evaporated make up or pure condensate may be protected from caustic attack through coordinated phosphate/pH control [7].

5.6.7 CREEP

Creep is defined as slow deformation of material with time, with the deformation taking place at elevated temperatures without increase in stress. It is usually associated with the tertiary stage of creep, and brings about the onset of creep failure. It can, however, initiate at the relatively early stages of creep, and develop gradually throughout creep life. Creep damage is manifested by the formation and growth of creep voids or cavities within the microstructure of the material. Creep curve is shown in Figure 5.2. For metals, the creep curve consists of three regions:

1. The first is the primary or transient creep region, where the material experiences an increase in creep resistance or strain hardening.
2. The second stage is secondary or steady-state creep.
3. Lastly is tertiary creep, where the creep occurs at an accelerated rate as the material approaches ultimate failure.

Creep Test

ASTM E139-11(2018)
Standard Test Methods for Conducting Creep, Creep-Rupture, and Stress-Rupture Tests of Metallic Materials

Detection of creep—metallographic replication
Conventional NDT methods such as magnetic particle inspection or ultrasonic inspection techniques, are not able to detect creep damage prior to the formation of a creep crack. The most commonly

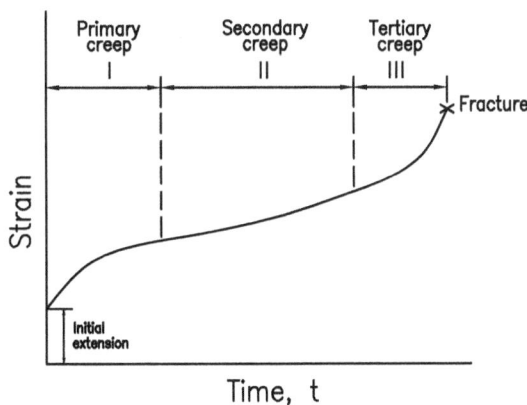

FIGURE 5.2 Creep curve.

(a)

(b)

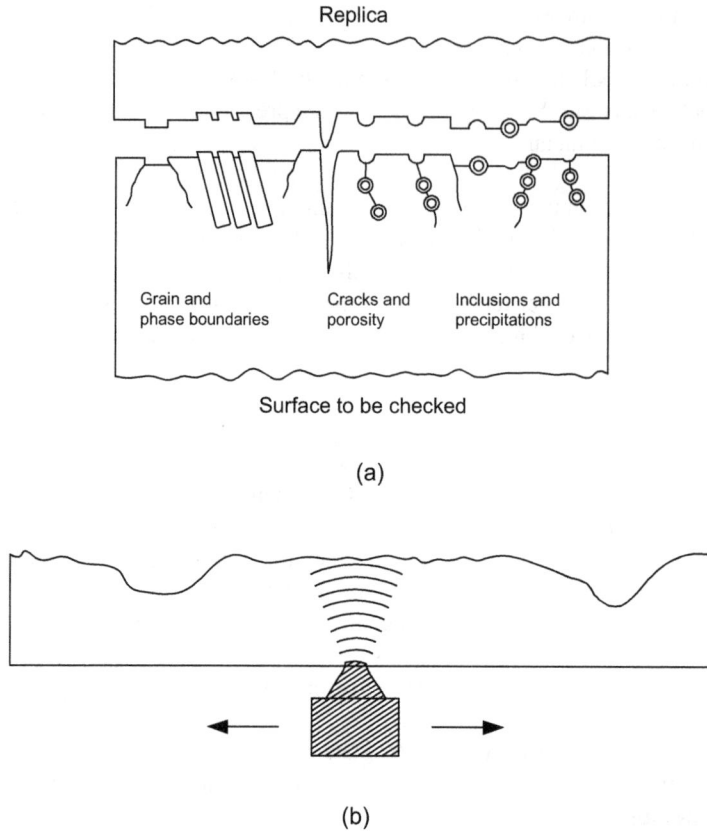

FIGURE 5.3   Principle of metallographic replication technique.

accepted technique for detecting creep damage is metallographic replication. Principle of metallographic replication technique is shown in Figure 5.3 and ASTM Standard for this technique is ASTM E1351-01:

ASTM E1351-01(2020) - Standard Practice for Production and Evaluation of Field Metallographic Replicas.

5.6.8   DECARBURIZATION

A condition where a steel loses strength due to the removal of carbon and carbides leaving only an iron matrix. Decarburization occurs during exposure to high temperatures, heat treatment, exposure to fires, or high-temperature service in a gaseous environment. Loss in room-temperature tensile strength and creep strength may potentially occur.

Affected Materials
Carbon steels and low-alloy steels.

5.6.9   DEW-POINT CORROSION

The combustion of most fossil fuels, natural gas being one exception, produces flue gases that contain sulfur dioxide, sulfur trioxide, and water vapor. At some temperature, these gases condense to form sulfurous and sulfuric acids. The exact dew point depends on the concentration of these gaseous species, but it is around 300°F. Thus surfaces cooler than this temperature are likely locations for dew point corrosion. Any point along the flue-gas path, from combustion in the furnace to the top of the chimney, is a possible site.

5.6.9.1 Equipment Affected by Dew-Point Corrosion

All fired process heaters and boilers that burn fuels containing sulfur have the potential for sulfuric acid dew point corrosion in the economizer sections and in the stacks. Heat recovery steam generators (HRSG's) that have 300 Series SS as feedwater heater materials may suffer chloride-induced stress corrosion cracking from the gas-side when the temperature of the inlet water is below the dew point of hydrochloric acid [8].

5.6.9.2 Methods to overcome dew-point corrosion

There are several ways in which flue gas dew-point corrosion can be avoided [9].

i. More resistant materials can be used in the construction of flues, which can prevent corrosion.
ii. Limiting the number of contaminants in fuels—since most fuels contain sulfur compounds and some are contaminated with chlorides.
iii. To prevent corrosion is to maintain the surface metal temperatures of exposed equipment above the dew point.
iv. It is possible to protect cooler surfaces by applying a coating that is resistant to the acidic condensate and will withstand the temperatures to which it is exposed.

5.6.10 Fuel Ash Corrosion

Fuel ash corrosion of components in process furnaces, utility boilers, and other equipment where accelerated high-temperature wastage of materials occurs when metal contaminants in the fuel form deposits and then melt on the metal surfaces inside fired heaters, boilers, and gas turbines. Corrosion typically occurs with fuel oil or coal that is contaminated with a combination of sulfur, sodium, potassium, and/or vanadium. Sodium plus vanadium in the fuel in excess of 100 ppm can be expected to form fuel ashes corrosive to metals by fuel oil ashes [10].
Fuel ash corrosion control.

A variety of methods have been used to control fuel ash corrosion of metallic members in process furnaces, utility boilers, and other combustion equipment. Among these are the following.

i. fuel oil additives;
ii. excess air control;
iii. refractory coatings;
iv. choice of alloys.

Use of excess air levels of 5% or less has been shown to reduce fuel ash corrosion in furnaces, most likely by stabilizing the vanadium as a refractory suboxide. High chromium alloys offer the best fuel ash corrosion resistance.

Affected Units or Equipment

(a) in any fired heater, boiler, or gas turbine utilizing fuels with the aforementioned contaminants.
(b) fired heaters burning vanadium- and sodium-contaminated fuel oils or residue.

5.6.11 Hydrogen Damage

Hydrogen damage is caused by the reaction of iron carbides on boiler steel tube, located in high heat flux zones, with hydrogen produced as a result of corrosion reactions, particularly those taking place in low pH water. Hydrogen damage is caused by three conditions existing simultaneously [11]:

flow disruption with high boiler deposits,
acidic contamination of the boiler water, and
high heat flux.
A tube with suspected hydrogen damage can be examined in the laboratory to confirm hydrogen damage by the use of a specific etch on the tube cross section.

Hydrogen damage due to wet H2S is discussed in Chapter 3.

5.6.12 Oxidation

Oxygen, present as a component of air (approximately 21%), reacts with carbon steel and other alloys at high temperature, converting the metal to oxide scale and thereby reducing the metal wall thickness.

Affected Materials

(a) All iron-based materials including carbon steel and low-alloy steels, both cast and wrought, are affected.
(b) All 300 series SS, 400 series SS, and nickel-based alloys also oxidize to varying degrees, depending on composition and temperature.

5.6.13 Refractory Degradation

Both thermally insulating and erosion-resistant refractories are susceptible to various forms of mechanical damage (cracking, spalling, and erosion) as well as corrosion due to oxidation, sulfidation, and other high-temperature mechanisms. High skin temperatures on the base metal being protected may result from refractory damage.

5.6.14 Stress Relaxation Cracking (Reheat Cracking)

Cracking of a metal due to stress relaxation via grain boundary strain in the creep temperature range during PWHT or in service at elevated temperatures. The temperature above which it occurs depends on the type of alloy. It is most often observed in heavy wall sections. At various times and in various situations, stress relaxation cracking (SRC) has also been called reheat cracking, stress relief cracking, creep embrittlement, low creep ductility cracking, stress-induced cracking, and stress-assisted grain boundary oxidation cracking. Stress relief cracking is a common cause of weld failures in many creep-resistant precipitation strengthened alloys [12, 12.1].

Affected Materials

(a) Cr-Mo steels, especially 2¼Cr-1Mo steel with vanadium added, 1Cr-½Mo, and 1¼Cr-½Mo.
(b) Types 304H, 316H, 321, and 347 SS.
(c) Nickel-based alloys, particularly Alloy 800H, 800HT, Alloy 617, etc.

5.6.15 TEMPER EMBRITTLEMENT

Temper embrittlement is the reduction in fracture toughness due to a metallurgical change that can occur in some low-alloy steels as a result of long-term exposure in the temperature range of about 345°C to 575°C (650°F to 1070°F). This change causes an upward shift in the ductile-to-brittle transition temperature as measured by Charpy impact test. Although the loss of toughness is not evident at operating temperature, equipment that is temper embrittled may be susceptible to brittle fracture during start-up and shutdown.

5.6.16 TITANIUM HYDRIDING

Hydriding of titanium is a metallurgical phenomenon in which hydrogen diffuses into the titanium and reacts to form an embrittling hydride phase. This can result in a complete loss of ductility and fracture toughness, potentially enabling a brittle fracture, with no noticeable sign of corrosion or loss in thickness.

5.6.17 WATER HAMMER

Water hammer is the cause of many problems such as ruptured pipes and damage to valves, brazed plate heat exchange and other equipment. Four conditions have been identified as causes of violent reactions from water hammer: (1) hydraulic shock; (2) thermal shock; (3) flow shock; and (4) differential shock [13].

Water hammer can occur in any steam or condensate line. Its effects can be even more pronounced in heterogeneous or condensate bi-phase systems. Water hammer occurs when the installation pipelines carry incompressible fluids, such as water, ethylene glycol, etc., and the fluid flow suddenly changes its velocity. Abruptly stopping the fluid flow produces a substantial pressure rise. High-intensity pressure waves will travel back and forth in the pipes between the point of closure and a point of relief, such as a larger diameter header, at extremely high speed. As it moves, the shock wave alternately expands and contracts the pipes. Types of water hammer, causes of water hammer and prevent of steam hammering are discussed in Chapter 4.

5.7 PRESSURE VESSEL

The term pressure vessel covers a wide variety of vessels including separation vessels, columns, storage vessels, reactors, and heat exchangers. Because of the risk associated with the accidental failure of pressure vessels, many countries have come up with regulations to govern pressure vessel designing and production. Pressure vessels are produced and used in a wide variety of geometrical shapes, capacities, and sizes for use in a large number of applications. Figure 5.4 shows a schematic of pressure vessel with some of the main constructional features and terminology.

Pressure vessels should be visually examined on a regular basis throughout their lives for impact marks, scrapes, corrosion, erosion, wear, cracks; anything that changes internal and/or external surfaces of vessels. Each finding should be evaluated for its effect on fatigue life of the vessel.

5.7.1 CONDITION MONITORING LOCATIONS

CMLs are designated locations on pressure vessels or piping systems where periodic external examinations are conducted to assess condition [14–18]. Examples of different types of CMLs include locations for thickness measurement, locations for stress corrosion cracking examinations, and locations for high-temperature hydrogen attack examinations. Whereas random inspection would provide information about current condition, repeated inspection at CMLs enables data

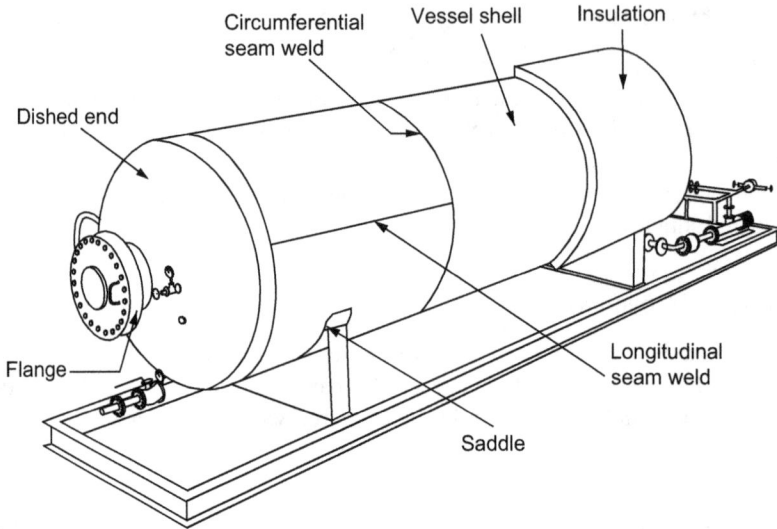

FIGURE 5.4 A schematic of pressure vessel with some of the main construction features.

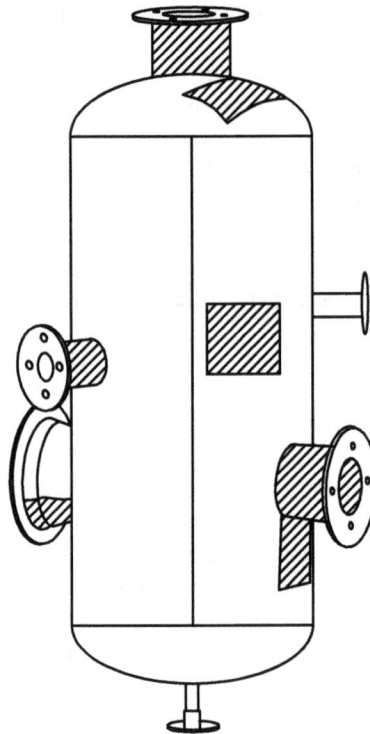

FIGURE 5.5 A schematic of a pressure vessel with some designated condition monitoring locations.

to be collected on change in condition, and ideally the rate of that change as well. This information allows more accurate risk-based inspection (RBI), optimized scheduling and safer more cost-effective operation. Figure 5.5 shows a schematic of a pressure vessel with some designated condition monitoring locations [17].

Each pressure vessel shall be monitored by conducting a representative number of examinations at CMLs to satisfy the requirements for an internal and/or on-stream inspection. For example, the thickness for all major components (shells, heads, cone sections) and a representative sample of vessel nozzles should be measured and recorded. Corrosion rates, the remaining life, and next inspection intervals should be calculated to determine the limiting component. CMLs with the highest corrosion rates and least remaining life shall be part of those included in the next planned examinations.

5.7.2 National Board Inspection Code (NBIC)

The *National Board Inspection Code* (NBIC)-2021 was first published in 1946 as a guide for chief inspectors. The NBIC provides standards for the installation, inspection, and repair and/or alteration of boilers, pressure vessels, and pressure relief devices. The NBIC is organized into four parts to coincide with specific post-construction activities involving pressure-retaining items [19]. Another guide on PV inspection is, "*API 510, Pressure Vessel Inspection Code: In-service Inspection, Rating, Repair, and Alteration, Tenth Edition, May 2014, with Addendum May 2017.*"

5.7.3 Pressure Vessel Failure

Pressure vessels, heat exchangers, and pressure pipings used in refineries, chemical processing plants, water treatment systems of boilers, etc., failure due to design errors, specifying or using improper materials, inadequacies in fabrication and welding, inadequacies in quality assurance program, improper heat treatment, damage during shipment, storage and erection and commissioning, excess operating pressure and temperature, etc. [20].

5.7.3.1 Causes of PV Failures

Pressure vessel failures can be due to the following [21]:

1. Material failure—improper selection of material; defects in material.
2. Design inadequacies—incorrect design data; inaccurate or incorrect design methods; inadequate shop testing.
3. Lacunae in fabrication—poor quality control; improper or insufficient fabrication procedures including welding; heat treatment or forming methods.
4. Service conditions different from design conditions
5. Change of service routine occasioned by user or conditions during operation is different from design conditions.

5.7.4 Failures Types in Pressure Vessels

Types of failures in a pressure vessel include the below [20]:

1. Elastic deformation—elastic instability or elastic buckling, vessel geometry, and stiffness as well as properties of materials are protection against buckling.
2. Fatigue.
2. Brittle fracture—can occur at low or intermediate temperatures.
3. Excessive plastic deformation.
4. Stress rupture—creep deformation as a result of fatigue or cyclic loading, i.e., progressive fracture.
5. Plastic instability.
6. High strain—low cycle fatigue is strain-governed and occurs mainly in lower-strength high-ductile materials.

7. Stress corrosion such as chloride stress corrosion cracking in stainless steels and caustic stress corrosion cracking in carbon steels.
8. Corrosion fatigue—occurs when corrosive and fatigue effects occur simultaneously. Corrosion can reduce fatigue life by pitting the surface and propagating cracks.
9. Other forms of corrosion which affect equipment, include pitting, line corrosion, general corrosion, grooving, galvanic corrosion, fatigue corrosion, low-/high-temperature corrosion, hydrogen embrittlement, etc.

Out of the many failure types mentioned above, fatigue failure and brittle fracture including fracture mechanics are discussed below.

5.7.5 FATIGUE

Fatigue cracking is one of the primary damage mechanisms of structural components. It results from cyclic stresses that are below the ultimate tensile stress, or even the yield stress of the material. The fatigue life of a component can be expressed as the number of loading cycles required to initiate a fatigue crack and to propagate the crack to critical size. The process of fatigue failure is characterized by three distinct steps:

i.   crack initiation,
ii. crack propagation,
iii. failure.

5.7.5.1 Factors Affecting Fatigue Life

In order for fatigue cracks to initiate, three basic factors are necessary:

1. The loading pattern must contain minimum and maximum peak values with large enough variation or fluctuation.
2. The peak stress levels must be of sufficiently high value.
3. The material must experience a sufficiently large number of cycles of the applied stress.

In addition to these three basic factors, there are a host of other variables, such as stress concentration, corrosion, temperature, overload, metallurgical structure, and residual stresses, which can affect the propensity for fatigue. Surface roughness is important because it is directly related to the level and number of stress concentrations on the surface. The higher the stress concentration the more likely a crack is to nucleate. Smooth surfaces increase the time to nucleation. Notches, scratches, and other stress risers decrease fatigue life. Surface residual stress will also have a significant effect on fatigue life. Compressive residual stresses from machining, shot blasting, and cold working will oppose a tensile load and thus lower the amplitude of cyclic loading.

5.7.5.2 Loading That Could Initiate a Fatigue Crack

Figure 5.6 shows several types of loading that could initiate a fatigue crack. The upper left figure shows sinusoidal loading going from a tensile stress to a compressive stress. Tensile stress is considered positive, and compressive stress is negative. In variable-amplitude loading, only those cycles exceeding some peak threshold will contribute to fatigue cracking [22].

The stress induced in a pressure vessel shell as pressure goes from atmospheric to operating pressure and back to ambient pressure is one stress cycle. Between increasing and decreasing pressure may be many pressure fluctuations (see Figure 5.7). Stresses associated with pressure fluctuations and number of cycles must be determined. As complexity of loadings on the vessel increases, difficulty of counting stress cycles increases and becomes increasingly onerous. Rainflow counting algorithm,

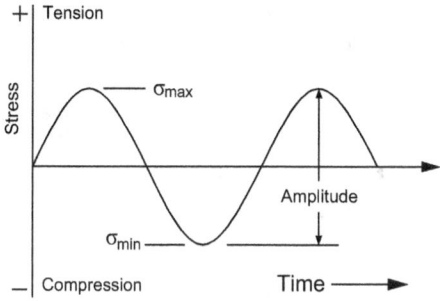

(a) Tension / Compression loading.

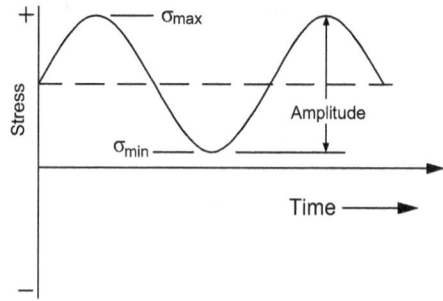

(b) Tension / Tension loading.

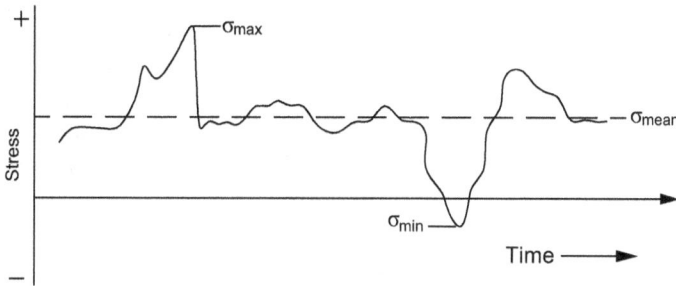

(c) Spectrum loading.

FIGURE 5.6 Types of loading that could initiate a fatigue crack.

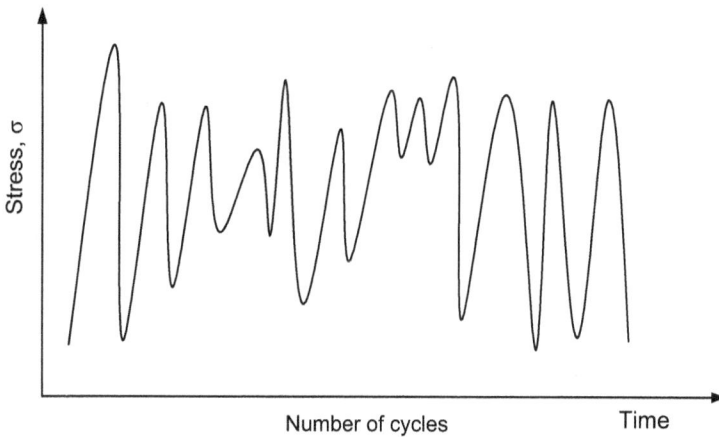

FIGURE 5.7 Stress fluctutation between increasing and decreasing stress.

or other methods complying with ASTM E1049, *Standard Practices for Cycle Counting in Fatigue Analysis*, is used to count stress cycles and combine partial cycles into complete cycles [22].

5.7.5.3 Miner's Rule

There are multiple stress cycles with different stress magnitudes and associated number of cycles. Fatigue damage from each stress cycle is cumulative, and the effect of all stress cycles must be determined. Miner's rule is used to evaluate effects of all stress cycles on vessel fatigue life. Miner's rule is given by

$$\frac{n_1}{N_1} + \frac{n_2}{N_2} + \frac{n_3}{N_3} + \dots \leq 1.0$$

where n_1 is the number of specified cycles at stress level 1, n_2 is the number of specified cycles at stress level 2, and n_3 is the number of specified cycles at stress level 3, and so on. N_1 is the number of permitted cycles at stress level 1, N_2 is the number of permitted cycles at stress level 2, and N_3 is the number of permitted cycles at stress level 3, and so on [22]. If the sum of the ratios is less than or equal to 1.0, vessel design is acceptable for that point in the vessel. This process must be repeated at other points in the vessel where there are high stresses. Vessel design is acceptable when all points in the vessel satisfy Miner's rule. As the vessel approaches the predicted end of life, visual examination should be supplemented by other NDE methods in search for cracks in highly stressed areas of the vessel.

5.7.5.4 S-N Curve

Fatigue properties of materials are often described using the fatigue limit or the S-N curve (fatigue curve, Wöhler curve). The S-N curve describes the relation between cyclic stress amplitude and number of cycles to failure. S-N curves are derived from fatigue tests. Tests are performed by applying a cyclic stress with constant amplitude (CA) on specimens until failure of the specimen. Test samples are highly polished round bars as identical to each other as manufacturing can make them. A test bar is rotated with load applied so a fiber at the surface of the bar is in tension and then in compression as the bar rotates such that there is a full reversal of stress [22–27] and a typical ideal S-N curve is shown in Figure 5.8.

The following terms are defined for the S-N curve:

a. Fatigue limit. Fatigue limit (also sometimes called the endurance limit) is the stress level, below which fatigue failure does not occur. This limit exists only for some ferrous (iron-base) and titanium alloys, for which the S–N curve becomes horizontal at higher N values. Other structural metals, such as aluminum and copper, do not have a distinct limit and will eventually fail even from small stress amplitudes. Studies of metal fatigue have indicated that the fatigue limit is sensitive to the microstructure, for example, the grain size. For a given material the fatigue limit depends on surface finish, size, type of loading, temperature, corrosive and other aggressive environments, mean stresses, residual stresses, and stress concentrations. Typical values of the limit for steels are 1/2 the ultimate tensile strength, to a maximum of 290.

b. Endurance strength

Endurance strength or fatigue strength refers to the highest stress value that a material can be subjected to for a specified number of cycles without resulting in failure. Fatigue strength can be improved by better surface finish, better component geometry, and compressive residual surface stress, etc. The endurance strength is determined in a fatigue test.

Fatigue strength is somewhat correlated with tensile strength. A "rule of thumb" often used is that fatigue strength equals one-third the tensile strength. But this approximation is not valid for many higher tempers of highly alloyed metals. The stronger tempers of some metals have lower fatigue

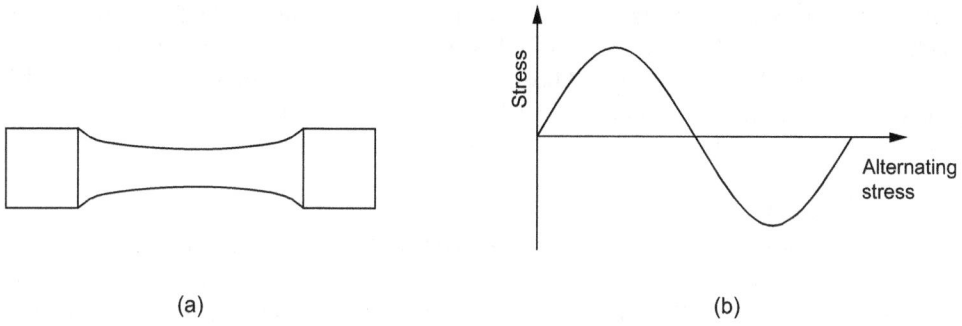

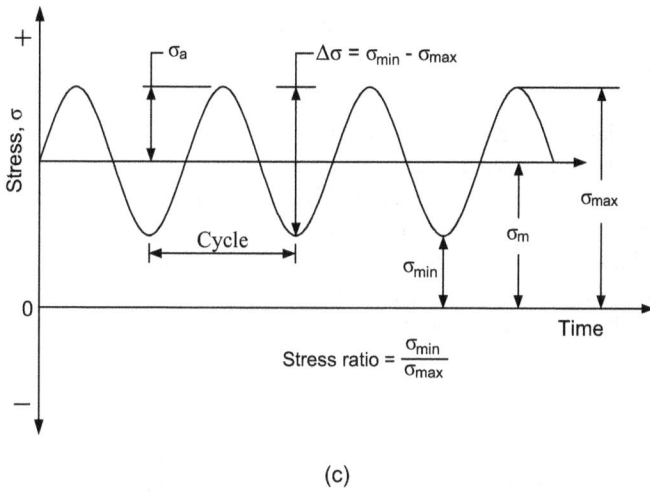

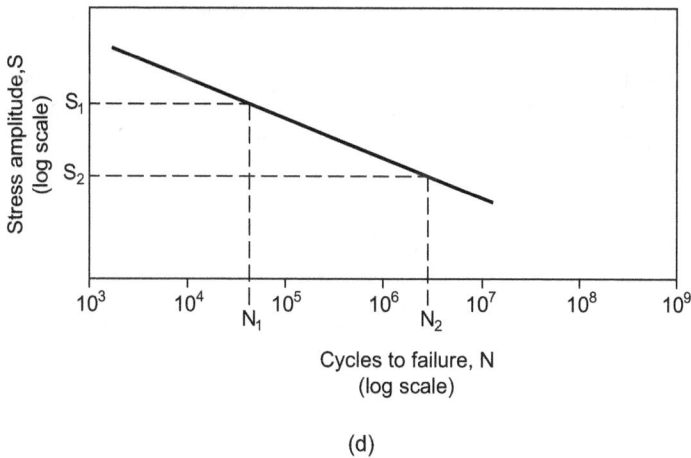

FIGURE 5.8 S-N Curve – (a) A test bar, (b) a stress cycle, (d) a full reversal of stress, and (d) a typical ideal S-N curve.

Source: https://www.nde-ed.org/Physics/Materials/Mechanical/Fatigue.xhtml

strength than their weaker tempers [26]. Typical values for steels are one half the ultimate tensile strength, to a maximum of 290 MPa (42 ksi). For iron, aluminum, and copper alloys, is typically 0.4 times the ultimate tensile strength. Maximum typical values for irons are 170 MPa (24 ksi), aluminums 130 MPa (19 ksi), and coppers 97 MPa (14 ksi). Note that these values are for smooth "un-notched" test specimens [27].

5.7.5.5 Fatigue Life

Fatigue life, N_f, is the number of stress cycles of a specified character that a specimen sustains before failure of a specified nature occurs. It is the number of cycles to cause failure at a specified stress level, as taken from the S–N plot. Fatigue life of a material is determined by testing many identical samples to failure.

5.7.5.6 Design Considerations for Better Fatigue Life

Design considerations for better fatigue life of engineering components include the following:

  i.   Select a material with high fatigue strength.
  ii. The holes, notches, and other stress raisers should be avoided.
  iii. The parts should be protected from corrosive atmosphere. A smooth finish of outer surface of the component increases the fatigue life.
  iv. The residual compressive stresses over the parts surface increases its fatigue strength.
  v.   Avoid corners, large radii are used in place of small radii for transitions from one geometric shape and make as gradual as possible.

5.7.6 FRACTURE MECHANICS

Fracture mechanics is the field of engineering mechanics concerned with the study of the propagation of cracks in materials. Fracture mechanics is a methodology that is used to predict and diagnose failure of a part with an existing flaw. The presence of a crack in a part magnifies the stress in the vicinity of the crack and may result in failure prior to that predicted using traditional strength-of-material methods. Fracture mechanics quantifies the relationship between material properties, stress level, crack length, and crack propagating mechanisms as shown in Figure 5.9 and it is a useful method of determining stress and flaw size, fracture toughness, fatigue crack growth, and stress-corrosion crack growth behavior. For basics of fracture mechanics, refer to Refs. [28–36].

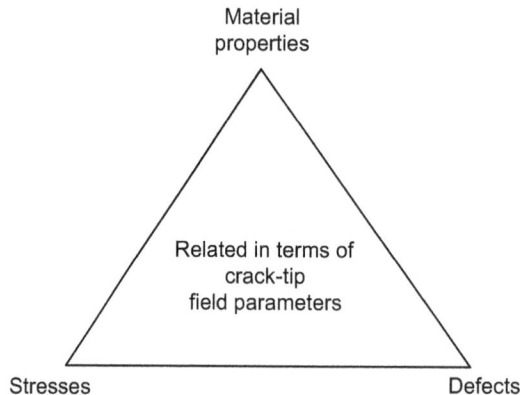

FIGURE 5.9 Fracture mechanics -The relationship between material properties, stress level and crack length.

Source: Adapted from William Leon, Ref.(28).

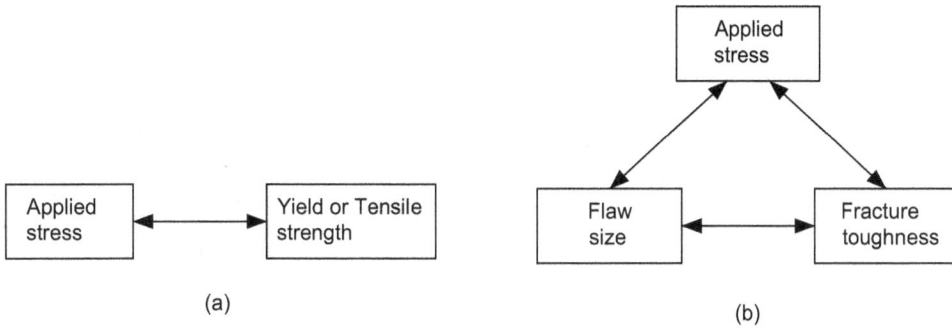

FIGURE 5.10   Comparison of the fracture mechanics approach to the design with the traditional stress-strain approach (a) Stress-strain approach and (b) Fracture mechanics approach.

Source: T.L. Anderson, Ref.(32).

The traditional approach to the design and analysis of a part is to use strength-of-material concepts. In this case, the stresses due to applied loading are calculated. Failure is determined to occur once the applied stress exceeds the material's strength (either yield strength or ultimate strength, depending on the criteria for failure). In fracture mechanics, a stress intensity factor is calculated as a function of applied stress, crack size, and part geometry. Failure occurs once the stress intensity factor exceeds the material's fracture toughness. Fracture mechanics analyzes flaws to determine which are safe and which are liable to propagate as cracks and cause failure of the flawed structure. There are two types of fracture mechanics:

i.   Linear-elastic fracture mechanics—the basic theory of fracture that deals with sharp cracks in elastic bodies.
ii. Elastic-plastic fracture mechanics—the theory of ductile fracture, usually characterized by stable crack growth (ductile metals).

Comparison of the fracture mechanics approach to the design with the traditional stress-strain approach is shown in Figure 5.10 [29].

5.7.6.1   Modes of Loading

There are three primary modes that define the orientation of a crack relative to the loading. A crack can be loaded in one mode exclusively, or it can be loaded in some combination of modes. Figure 5.11 shows the three primary modes of cracking.

i.   Mode I is called the opening mode and involves a tensile stress pulling the crack faces apart.
ii. Mode II is the sliding mode and involves a shear stress sliding the crack faces in the direction parallel to the primary crack dimension.
iii. Mode III is the tearing mode and involves a shear stress sliding the crack faces in the direction perpendicular to the primary crack dimension.

Engineering analysis almost exclusively considers Mode I because it is the worst-case situation and is also the most common. Cracks typically grow in Mode I, but in the case that the crack does not start in Mode I it will turn itself to become Mode I.

5.7.6.2   Fracture Toughness

Fracture toughness is defined as the stress-intensity factor at a critical point where crack propagation becomes rapid. It is given the symbol K_{Ic} and is measured in units of megapascals times the square

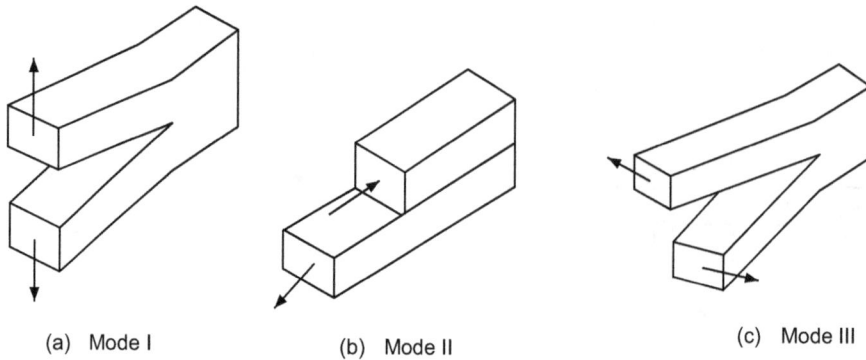

(a) Mode I   (b) Mode II   (c) Mode III

FIGURE 5.11   Fracture mechanics - The three primary modes of cracking.

root of the distance measured in meter. A material can resist applied stress intensity up to a certain critical value above which the crack will grow in an unstable manner and failure will occur. This critical stress intensity is the fracture toughness of the material. The fracture toughness of a material is dependent on many factors including environmental temperature, environmental composition (i.e., air, fresh water, salt water, etc.), loading rate, material thickness, material processing, and crack orientation to grain direction. It is important to keep these factors in mind when selecting a fracture toughness value to assume during design and analysis.

5.7.6.3   Fracture Toughness vs. Thickness

Fracture toughness decreases as material thickness increases until the part is thick enough to be in a plane-strain condition. Above this plane-strain thickness, the fracture toughness is a constant. A parameter called the critical stress-intensity factor (K_C) is used to represent the fracture toughness of most materials. A Roman numeral subscript indicates the mode of fracture and the three modes of fracture are Mode I, Mode II, and Mode III.

5.7.6.4   Role of Material Thickness

As the thickness of the specimen increases, fracture toughness decreases as shown in Figure 5.12 and the plain strain assumptions become more accurate. Specimens having different thicknesses produce different values for K_C. The value of K_C decreases with thickness until the thickness exceeds some critical dimension. At this point the value of K_I becomes relatively constant and this value, K_{IC}, is a true material property, which is called the plane-strain fracture toughness. K_{IC} is usually measured by the process specified in ASTM Standard E399. The relationship between stress intensity, K_I, and fracture toughness, K_{IC}, is similar to the relationship between stress and tensile stress [34].

The stress intensity K_I, represents the level of "stress" at the tip of the crack. The fracture toughness, K_{IC}, is the highest value of stress intensity that a material under very specific (plane-strain) conditions can withstand without fracture. As the stress intensity factor reaches the K_{IC} value, unstable fracture occurs. As with a material's other mechanical properties, K_{IC} is commonly reported in reference books and other sources.

5.7.6.5   Principal Fracture Path Directions

The fracture toughness of a material commonly varies with grain direction. Therefore, it is customary to specify specimen and crack orientations by an ordered pair of grain direction symbols. The first letter designates the grain direction normal to the crack plane. The second letter designates the grain direction parallel to the fracture plane. For flat sections of various products, e.g., plate, extrusions, forgings, etc., in which the three grain directions are designated (L) longitudinal; (T) transverse; and

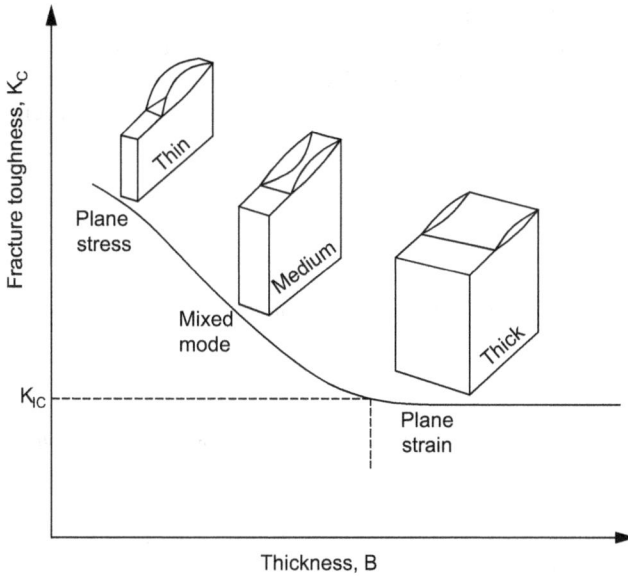

FIGURE 5.12 Relationship between K_C and specimen thickness.

Source: https://www.nde-ed.org/Physics/Materials/Mechanical/Fatigue.xhtml

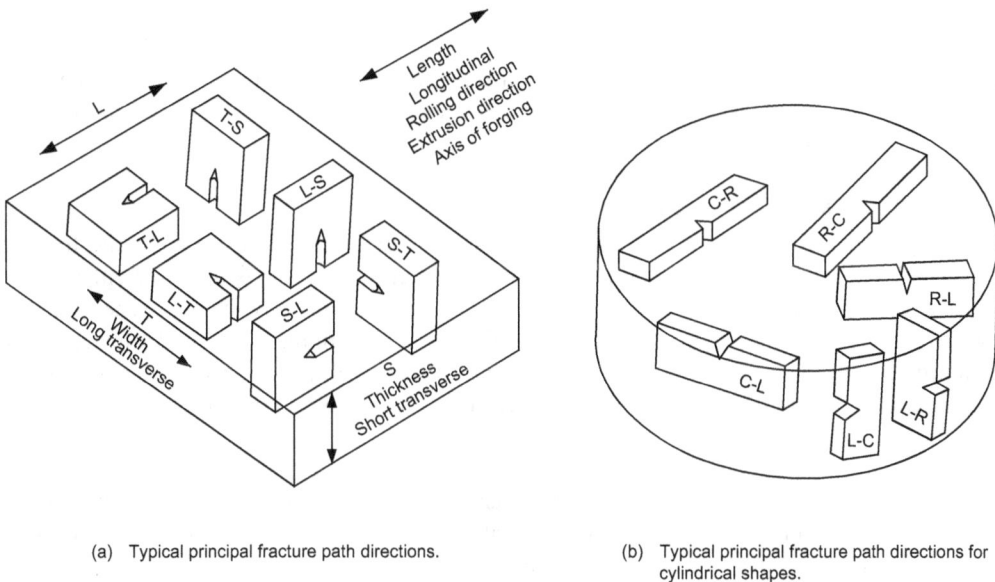

(a) Typical principal fracture path directions.

(b) Typical principal fracture path directions for cylindrical shapes.

FIGURE 5.13 Principal fracture path directions.

Source: Adopted from Richard C. Rice et al. [36]

(S) short transverse, the six principal fracture path directions are L-T, L-S, T-L, T-S, S-L, and S-T [36]. Figure 5.13 shows principal fracture path directions adopted from Ref. [36].

5.7.6.6 Flaw Evaluation by Fracture Mechanics

Some of the concepts utilized in making fracture mechanics analysis are briefly given here. Firstly it is assumed that a crack or a crack-like defect exists in the structure. The essence of the method then is to relate the stress field developed in the vicinity of the crack tip to the applied nominal stress on the structure, the material properties and the size of the defect necessary to cause failure. The elastic stress in the near vicinity of a crack tip is called stress intensity factor and is denoted by K_I. The critical value of K_I at which crack instability occurs is denoted by K_{IC}. The relationship between K_I and the applied stress is given by

$$K_I = C\sigma\sqrt{\pi a}$$

where
C is a constant dependent upon the crack geometry and the component geometry,
a is crack size,
σ is nominal service stress.

Values of C are available in various textbooks and handbooks. Fracture occurs when

$$K_I > K_{IC.}$$

5.7.6.7 The Brittle Fracture

Three basic factors contribute to a brittle-cleavage type of fracture. They are

1. a triaxial state of stress,
2. a low temperature, and
3. a high strain rate or rapid rate of loading.

All three of these factors do not have to be present at the same time to produce brittle fracture. A triaxial state of stress, such as exists at a notch, and low temperature are responsible for most service failures of the brittle type.

5.7.6.8 Brittle and Ductile Fracture

Fracture preceded by a significant amount of plastic deformation is known as ductile fracture, otherwise it is brittle fracture. Brittle fracture occurs when plastic flow is inhibited either by the effective locking of atomic dislocations by precipitates or elements or by the pre-existence or formation of cracks and imperfections acting as local stress raisers in the material. All materials can be embrittled if the temperature is lowered sufficiently. Glass, sealing wax, germanium, silicon and other materials though ductile at temperatures close to their melting point are brittle at ordinary temperatures.

Toughness is related to the **area under the stress-strain curve.** The stress-strain curve measures toughness under a gradually increasing load. Tensile toughness is measured in units of joule per cubic meter ($J\cdot m^{-3}$) in the SI system. The material must be both **strong** and **ductile** to be tough. Figure 5.14 shows a typical stress-strain curve of ductile and brittle materials [35]. For example, brittle materials (like ceramics) that are strong but with limited ductility are not tough; conversely, very ductile materials with low strengths are also not tough. A material should withstand both high stresses and high strains to be tough.

5.7.6.9 Notched-Bar Impact Tests

Various types of notched-bar impact tests are used to determine the tendency of a material to behave in a brittle manner. This type of test will detect differences between materials, which are not observable in a tension test. Two classes of specimens have been standardized for notched-impact testing, viz., Charpy bar specimens test and the Izod impact test.

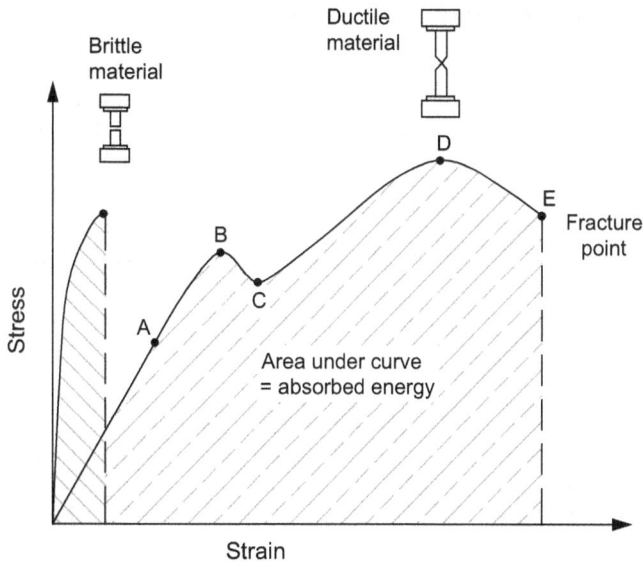

Ductile vs Brittle material stress-strain curve.

FIGURE 5.14 A typical stress-strain curve of ductile and brittle materials.

Source: Adapted from https://www.nuclear-power.com/nuclear-engineering/materials-science/material-properties/toughness/

5.7.6.10 Standard for Fitness-For-Service, FFS-1, 2021

Fitness-for-service (FFS) assessments are quantitative engineering evaluations that are performed to demonstrate the structural integrity of an in-service component that may contain a flaw or damage, or that may be operating under a specific condition that might cause a failure. This Standard provides guidance for conducting FFS assessments using methodologies specifically prepared for pressurized equipment. FFS is used to determine if equipment, such as boiler, HRSG, pressure vessels, chemical reactors, storage tanks, piping systems, and pipeline is fit to continue operation for some desired future period.

Fitness-for-service assessment is a multidisciplinary engineering approach that is used to determine if equipment such as boiler, HRSG, pressure vessels, chemical reactors, storage tanks, piping systems, and pipeline is fit to continue operation for some desired future period.

5.7.6.11 Approach For Crack-Like Flaws

The FFS Assessment procedures in the "Fitness-for-service" standard cover both the present integrity of the component given a current state of damage and the projected remaining life. The FFS Assessment procedures in this Standard cover both the present integrity of the component given a current state of damage and the projected remaining life. The FFS Assessment procedures in this Standard can be used to evaluate flaws commonly encountered in pressure vessels, piping, and tankage.

5.7.6.12 Fitness-for-Service ASME API 579-1/ASME FFS-1-2021

The ASME API 579-1/ASME FFS-1-2021 standard provides the means to carry out fitness-for-service by covering a broad list of assessment methods based on the condition of the equipment tested.

5.8   ASSET INTEGRITY MANAGEMENT (AIM) PROGRAM

Asset integrity is the ability of an asset to perform its required function effectively and efficiently whilst safeguarding life and the environment. AIM applies to the entirety of an asset's operation, from its design phase to its decommissioning and replacement [37]. The primary industry sector served by AIM is upstream oil and gas. However, the AIM process can be implemented for any asset of high value, including oil refineries, electrical power generation plants and chemical processing plants [37]. Eleven primary elements that make up an asset integrity management program (AIM). These elements cover all integrity aspects of the industry and the assets themselves. The elements are as follows [38]:

1. Risk assessment.
2. Engineering.
3. Construction and fabrication quality assurance.
4. Integrity operating windows.
5. In-service inspection.
6. Corrosion management.
7. Management of change.
8. Reliability centered maintenance.
9. Failure investigation and lessons learned.
10. Asset data management.
11. Assessments and audits.

Well run AIM strategies ensure that the people, systems, processes and resources that enable an asset to deliver its function are in place over the life cycle of the asset, while simultaneously maintaining health and safety and environmental legislation [39].

5.8.1   SCOPE OF AIM

Asset integrity work can relate to all phases of the asset lifecycle, including development, implementation, operation, training, and verification. It can include the following [40]:

i.   Technical integrity analysis.
ii.   Development and optimization of maintenance, inspection, and testing (MIT) plans.
iii.   Condition assessment.
iv.   Lifetime extension of aging assets.
v.   Barrier management.
vi.   Safety risk management.

5.8.2   ASSET PERFORMANCE MANAGEMENT (APM)

Asset performance management (APM) is a framework comprised of the methods used to ensure optimal performance throughout an asset's full lifecycle. It includes capabilities of data capture, integration, visualization, and analytics. These are integrated to improve reliability and availability of assets. Condition monitoring, predictive forecasting, operational forecasting and reliability-centered maintenance (RCM), risk-based inspection (RBI), quantitative risk analysis (QRA), etc., are also included in APM [41].

5.8.3   IMPORTANCE OF ASSET INTEGRITY MANAGEMENT (AIM)

Designing and maintaining equipment that is fit for its purpose and functions when needed is of paramount importance to process industries. AIM primarily involves

(1) inspections, tests, preventive maintenance, predictive maintenance, and repair activities;
(2) quality assurance processes.

5.9 PRESSURE VESSEL INSPECTION

Inspections are a crucial part of the maintenance process for pressure vessels. Pressure vessel inspections can refer to an inspection of the vessel's condition externally, internally, or both. In these inspections, inspectors may [42]

i.   prepare a checklist for external and internal inspection;
ii. collect visual data regarding the condition of the vessel, including the condition of insulation, welded joints, nozzle, structural connections, etc.;
iii. collect thickness data to know the material loss;
iv. conduct a stress analysis to determine whether the vessel is fit for further use;
v.   inspect the vessel's pressure relief valves to make sure that they function properly;
vi. conduct a hydrostatic pressure test.

5.9.1 PURPOSE OF PRESSURE VESSEL INSPECTION PROGRAM

The main purpose of implementing a pressure vessel inspection program is to ensure that each pressure vessel is safely operated and maintained. Some of the benefits that result from regularly scheduled pressure vessel inspections are listed below [42]:

i.   improvement of facility, personnel, and public safety;
ii. prevention of damage to the environment;
iii. improvement of reliability;
iv. reduction of operation and maintenance costs;
v.   minimization of unscheduled outages;
vi. minimization of liability.

5.9.2 REMOTE VISUAL INSPECTION (RVI)

As the quality of cameras and robotics continues to improve, inspectors are using RVI to collect visual data remotely instead of in person. An inspector might use RVI in their workflow as follows:

i.   An inspector sends a drone into a boiler and collects all of the visual data they need to evaluate its current condition.
ii. Once the visual data is collected, the inspector carefully reviews it, looking at all of the video footage to identify potential problem areas.

5.9.3 USE OF DRONES IN VISUAL INSPECTIONS

Inspectors have experimented with different confined space tools, like dropping cameras into confined spaces on ropes, or attached to robotic crawlers. But more and more inspectors have been turning to drone technology as a preferred RVI tool because it offers a high degree of control and a high degree of quality. Drones improve safety by removing the need for an inspector to enter a confined, potentially dangerous space in order to collect visual data.

5.9.4 INSPECTION PROCEDURE

A thorough inspection of a pressure vessel should include the following items [42]:

1. External examination of the pressure vessel and associated equipment including verification of the welded connections to determine the proper joint efficiency to employ during the stress analysis.
2. An ultrasonic thickness examination of the pressure vessel wall and dished heads and documentation for permanent record keeping.
3. An internal examination of the pressure vessel, if required.
4. Ultrasonic measurement techniques to determine the shell and dished head wall thicknesses for each pressure vessel.
5. A stress analysis based on actual wall thickness data acquired during the ultrasonic thickness survey, and the proper joint efficiencies, based on the type of construction used during fabrication of the pressure vessel.
6. A thorough inspection of the pressure relief valves and other safety devices to ensure the vessel is operating within its specified pressure range and is being adequately protected.
7. A hydrostatic pressure test to 1.5 times the maximum allowable working pressure should be performed if any repairs or alterations have been made to the pressure vessel.

The PV inspection method given below is based on Ref. [42]. For details of causes of PV failures, refer to *Sumio Yukawa, Guidelines For Pressure Vessel Safety Assessment, NIST Special Publication 780* [43].

5.9.5 External Inspection

The external inspection provides information regarding the overall condition of the pressure vessel. The following items should be reviewed [42]:

a. Insulation or other coverings.
b. Evidence of leakage.
c. Structural attachments. The pressure vessel mountings should be checked for adequate allowance for expansion and contraction.
d. Vessel connections. Manholes, reinforcing plates, nozzles, or other connections should be examined for cracks, deformations, or other defects. Bolts and nuts should be checked for corrosion or defects. Weep holes in reinforcing plates should remain open to provide visual evidence of leakage as well as to prevent pressure buildup between the vessel and the reinforcing plate. Accessible flange faces should be examined for distortion and to determine the condition of gasket seating surfaces.
e. Miscellaneous conditions. The surfaces of the vessel should be checked for erosion or dents.
f. Surface inspection. The surfaces of shells and heads should be examined for possible cracks, blisters, bulges, and other evidence of deterioration, giving particular attention to the skirt and to the support attachment and knuckle regions of the heads.
g. Welded joints. Welded joints and the adjacent heat affected zones should be examined for cracks or other defects. Magnetic particle and liquid penetrant examination is a useful means of examining suspect areas.

5.9.6 Thickness Survey

A thickness survey of the pressure vessels wall and dished heads should be perormed and documented by a certified inspector using ultrasonic testing equipment.

5.9.7 Stress Analysis

The thickness measurements for the shell and dished heads of the pressure vessels, which were obtained using ultrasonic techniques and the joint efficiencies based on the original fabrication of the pressure vessel should be used to perform a stress analysis to calculate the factor of safety. It is recommended that any pressure vessel with a factor of safety lower than 2 be replaced.

5.9.8 Potential Flaws in Pressure Vessel Nozzle Welds

A pressure vessel may consist of a large number of welded nozzles and other openings. For structural integrity, they must have a strength profile that matches or exceeds that of the pressure vessel itself. However, there are challenges that nozzle welds can cause, such as

i. cracking due to thermal fatigue, external load, or lack of penetration;
ii. stress corrosion as well as discontinuities.

Ultrasonic testing is an ideal method for ensuring that the weld is safe enough to hold the nozzle in place without having to compromise on the service life of the pressure vessel and jeopardizing production safety. Specifically, phased array ultrasonic testing (PAUT) technology allows technicians to utilize multiple transducers and analyze the reflected ultrasonic signals to check for any irregularities in the inspection area.

5.9.9 Internal Inspection

All parts of the vessel, viz., vessel connections, vessel closures, and vessel internals should be inspected for corrosion, erosion, hydrogen blistering, deformation, cracking, and laminations.

5.9.10 Inspection of Pressure Gages

When required, the accuracy of pressure gages should be verified by comparing the readings with a standard test gage or a dead weight tester.

5.9.11 Inspection of Safety Devices

The most important appurtenances on any pressurized system are the safety devices provided for overpressure protection of that system. These are devices such as safety valves, safety relief valves, pilot valves, and rupture disks or other nonreclosing devices that are called on to operate and reduce an overpressure condition. Periodic inspection and maintenance of these important safety devices is critical to ensure their continued functioning and to provide assurance that they will be available when called on to operate.

5.9.12 Pressure Testing

The only acceptable medium for pressure testing of reclamation pressure vessels is hydrostatic testing of the pressure vessel. The pressure test should not exceed 1.5 times the maximum allowable working pressure adjusted for temperature. When the original test pressure includes consideration of corrosion allowance, the test pressure may be further adjusted based on the remaining corrosion allowance.

5.9.13 Inspection of Piping Systems

Piping systems are designed for a variety of service conditions. Particular attention should be given to piping systems that are subject to corrosion, erosion, and fatigue and those that operate at high temperatures. Piping should be inspected to ensure there is the following:

    i. Provision for expansion.

    ii. Provision for adequate support.

    iii. No evidence of leakage.

    iv. Proper alignment of connections. The purpose is to determine if any changes of position have placed undue strain on the piping or other connections.

    v. Proper rating for the service conditions

    vi. No evidence of corrosion, erosion, or cracking or other detrimental conditions.

5.9.14 DETERIORATION OF PIPELINES

    a. Corrosion. Pipelines suffer from a number of corrosion forms: they range from pitting corrosion, uniform corrosion, galvanic corrosion, crevice corrosion, and microbiologically influenced corrosion to name a few.

    b. Cracks. Cracks may result from design and operating conditions that cause continual flexing.

    c. Erosion. Erosion may occur as a result of the abrasive action of a liquid or vapor.

    d. Leakage. A leak should be thoroughly investigated and corrective action initiated.

    e. Improper support. Visual inspection should include a check for evidence of improper support.

5.9.15 REPORTING

The following items may be included in the examination report and archived for future reference [44]: (1) Owner, location, type, and serial number of equipment/component examined. (2) Tube details. (3) Tube numbering system. (4) Extent of examination, for example, complete or partial coverage, which tubes, and to what length. (5) Qualification of personnel performing the examination. (6) NDT instruments details. (7) The tubing inspection system reference standard. (8) Brief outline of all inspection techniques used during the examination.

5.9.16 RECORD KEEPING

Post a copy of the inspector's approval or certification reports near or on the unfired pressure vessel. Also, maintain a permanent record for each pressure vessel.

5.10 PERIODICAL INSPECTION OF HEAT EXCHANGER UNIT

At regular intervals and as frequently as experience indicates, an examination should be made of the interior and exterior conditions of the unit. Negligence in keeping all tubes clean may result in complete or structural damage to other components. Sacrificial anodes, when provided, should be inspected to determine whether they should be cleaned or replaced.

5.10.1 INDICATIONS OF FOULING

Fouling of heat exchanger tubes can drastically reduce heat exchangers' efficiency, due to impaired heat transfer through the tube walls as well as reduced flow rates through the tubes.

    The unit should first be checked for air or vapor binding to confirm that this is not the cause for the reduction in performance. Since the difficulty of cleaning increases rapidly as the scale thickness or deposit increases, the intervals between cleanings should be optimum.

    Exchangers subject to fouling or scaling should be cleaned periodically. A light sludge or scale coating on the tube greatly reduces its efficiency. A marked increase in pressure drop and/or reduction in thermal performance usually indicate that cleaning is necessary.

Many different cleaning methods are often employed to remove fouling, depending on the operating environment of the specific heat exchanger and the typical mechanisms of fouling in such environments. For the cleaning process to be effective, it should be controlled and monitored properly, therefore ensuring the desired degree of cleanliness has been attained.

5.11 DETERIORATION OF HEAT EXCHANGER PERFORMANCE

5.11.1 AIR-COOLED HEAT EXCHANGERS

Many older ACHEs in service perform poorly compared to their original specified design due to several factors such as age, poor design, hot air recirculation, higher process fluid flow rates, deterioration of fin tubes, loss of thermal contact between fins and tubes, fouling both on the tubeside and airside, missing air seals, poor air distribution and/or airflow bypassing through bundle, etc. [45]. A step-by-step economical method by which an end user can systematically improve the heat transfer performance of the existing equipment is explained by Giammaruti [46], and the method is briefly mentioned here.

In the following sections, the analysis of the performance deterioration of air-cooled heat exchanger (ACHE), shell and tube heat exchanger, and compact heat exchanger and remedial measures to improve the performance are briefly dealt with.

5.11.1.1 Determine the Original Design Performance Data of the ACHE

Have an overview of the basic thermal and mechanical design data, materials of construction, and control information and compare the same against the current process parameters.

5.11.1.2 NDE of ACHE Tubing [44]

(1) Visual inspection; (2) near field testing array; (3) magnetic flux leakage testing; (4) Magnetic flux leakage array testing; (5) internal rotary inspection system; (6) tube end caliper; (7) acoustic pulse reflectometry, etc.

5.11.1.3 Inspection Plan for the ACHE Unit

Inspect tube bundle, tensioning device, integrity of frame header, tube support, fins, wiggle strips, header, condition of air seals, leaks at the plugs or cover plate gaskets, corrosion of fin tubes, etc. Based on the condition of the parts, measures should be taken as necessary to bring the ACHE back to its as-built condition. Table 5.2 shows a partial inspection plan for an ACHE.

Tube leaks in ACHE units are difficult to identify and repair due to the limited access to the tube sheet through the plug sheet. Typically when leaks in the tubes are identified they are plugged by using tapered hammer-in plugs, or in higher pressure ACHE units, the plugs have to be welded in place. In both cases, the damage frequently occurs to the tube sheet, plug sheet, and/or adjacent tubes.

5.11.1.4 API Standards for ACHE

a. API 661

API STD 661, Petroleum, Petrochemical, and Natural Gas Industries—Air-cooled Heat Exchangers 7th Edition, July 2013.

b. API STD 661-2013

Petroleum, Petrochemical, and Natural Gas Industries—Air-cooled Heat Exchangers.

TABLE 5.2
ACHE Inspection Plan (Sample)

No.	Activities
1.	Document review such as
	i. NDT written procedures
	ii. Calibration status of inspection and measuring instruments
2.	Supporting activities
	i. Scaffolding installation
	ii. heat exchanger core cleaning
3.	Tube internal inspection activities
	i. Borescope inspection
	ii. Vacuum test to identify punctured tube
	ii. Eddy current test to identify tube deterioration in service
4.	Tube plug installation activities
	i. Preparation and cleaning
	ii. Puncturing the tube to be plugged
	iii. Tube plug installation
5.	External inspection
	i. Stairways and walkways (if any)
	ii. Structures and supports
	iii. Integrity of pipe supports
	iii. Header, nozzle & plug connection
	iv. Coating and insulation (if any)
	v. Grounding of connection
6.	Pressure test—hydrostatic test
7.	Testing & Commissioning
	i. Leak Testing
	ii. Performance test

This standard gives requirements and recommendations for the design, materials, fabrication, inspection, testing, and preparation for shipment of air-cooled heat exchangers for use in the petroleum, petrochemical, and natural gas industries.

This standard is applicable to air-cooled heat exchangers with horizontal bundles, but the basic concepts can also be applied to other configurations.

5.11.1.5   Determine the Current ACHE Performance and Set Baseline

Measure the airflow and associated temperature rise across the bundle and hence calculate the amount of heat rejected to the air. The calculated airside heat transfer duty should then be compared to the design duty to determine how much more cooling is needed.

5.11.1.6   Install Upgrades

The following are the typical steps that can be taken in order to improve the tube bundle performance [46]:

Fans

1. In most applications, one of the quickest ways to obtain more duty out of the ACHE is to increase the airflow rate either by increasing the fan blade pitch or by increasing the fan

speed so long as the fan does not stall and enough motor horsepower is available to meet the increased airflow rate.
2. Reset tip clearances.
3. Install inlet bells.
4. Install fan seal disk.
5. *Install high-efficiency fans*: older ACHEs will usually have straight chord aluminum fan blades whose efficiency will be around 35%–55%. Today's more modern fans made of fiberglass or aluminum are more aerodynamic with a tapered chord and increasing pitch or twist from the blade tip to the fan hub with significantly greater efficiency of about 75%–85%. It is possible to obtain about 25%–40% more airflow with high-efficiency fans at the same or slightly more motor horsepower.

5.11.1.7 Tube Bundle

The heart of the ACHE is the tube bundle, and to improve its performance, the following measures are suggested: (1) clean the finned tube bundle—both airside and tubeside; (2) if a small number of fin tubes have been degraded due to overheating or excessive corrosion, retube the bundle with equivalent or high-performance fin tubes. Integral fin tubes typically have the maximum heat transfer ability over a sustained period of time. Overall, about 10%–50% increase in ACHE duty can be achieved depending on the level of fin tube deterioration. If a major part of the tube bundle is deteriorated, replace the bundle by a high-performance unit. Test the unit to confirm the increase in ACHE thermal duty achieved.

5.12 SHELL AND TUBE HEAT EXCHANGER MAINTENANCE

5.12.1 CAUSES OF HEAT EXCHANGER TUBE FAILURE

Based on causes, there are four types of heat exchanger failures that can occur [47, 48]:

i. mechanical,
ii. chemically induced corrosion,
iii. combination of mechanical and chemically induced corrosion, and
iv. scale, mud, and algae fouling.

Mechanical—these failures can take different forms including metal erosion, steam or water hammer, flow-induced vibration, thermal fatigue, freeze-up, thermal fatigue and loss of cooling water.

Corrosion—these failures result from the complex chemical interaction between the materials of the heat exchanger and the fluids circulated through it. There are seven types of chemically induced corrosion failures: general corrosion, pitting corrosion, stress corrosion, dezincification, galvanic corrosion, crevice corrosion, and condensate grooving.

Condensate Grooving—this problem occurs on the outside of steam-to-water heat exchanger tubes particularly in the U-bend area. It is recognized by an irregular groove, or channel cut in the tube as the condensate drains from the tubing in the form of rivulets. A corrosion cell usually develops in the wetted area because of the electrical potential difference between the dry and wet areas.

Scale, Mud, and Algae Fouling—Various marine growths deposit a film or coating on the surfaces of heat transfer tubes. The film acts as an insulator, restricting heat flow and protecting corrodents. As a result of this insulating effect, tube wall temperatures go up and corrosion increases.

5.12.1.1 Major Causes of Heat Exchanger Tube Failure

Causes of heat exchanger tube failure include (1) corrosion; (2) erosion including U-bend erosion and inlet tube-end erosion; (3) steam or water hammer; (4) thermal fatigue; (5) flow-induced vibration damage; and (6) pitting.

5.12.2 QUALITY AUDITING OF EXISTING HEAT EXCHANGER

First step in conducting performance auditing is to collect information about the present thermal performance of the heat exchanger and compare it with design conditions, Gulley [49]. The performance auditing of existing heat exchanger normally covers the following: (1) assess the present operating conditions and compare it with the design conditions; review the operations data sheet; (2) observe actual performance test data; (3) observe for terminal temperatures and flow rates of shellside and tubeside fluids; (4) examine the periodic maintenance procedure; review the methods of cleaning and results of heat exchanger performance after cleaning; (5) examine the *in situ* water treatment programs for adequacy of corrosion control (inadequate monitoring can result in calcium deposit on hot heat exchanger tube; Langelier index is a common water quality chemistry analysis that can be used to reduce the likelihood of scaling due to calcium deposits); (6) observe for the existing biofouling control mechanism and its adequacy; (7) verify that the firm had evaluated the potential for water hammer in susceptible heat exchangers and undertaken appropriate measure to address it; (8) verify that the number of plugged tubes is within permitted limit based on thermal duty; (9) examine, if available, the eddy current test and visual inspection records to determine the structural integrity of the heat exchanger (also review the records of online monitoring of structural integrity of vessels or heat exchangers by acoustic emission if in place); (10) verify that adequate controls on flow and operational limits are in place to prevent heat exchanger degradation due to excessive flow-induced vibration during operation; (11) observe tube-to-tubesheet joints for leaks, and shell and channel for cracks; and (12) perform weep-hole inspection for any telltale signs of leaks as discussed later.

5.12.3 LEAK DETECTION: WEEP-HOLE INSPECTION

Pressure vessels that involve multilayer shells or a single shell with a liner, and that are designed to Section VIII of the ASME Boiler and Pressure Vessel Code, generally are provided with weep holes. The purpose of the weep holes is to allow the detection of any leak that develops in the inner liner of the vessel [50].

5.12.4 REMOVAL OF TUBE BUNDLE, TRANSPORTATION, AND CLEANING

The heat exchanger maintenance consists in cleaning the bundle of inner tubes and the outer cleaning of the bundle. The maintenance procedure is carried out in four phases:

1. tube bundle extraction;
2. tube bundle transport;
3. tube bundle cleaning;
4. tube bundle insertion.

Large number of models of tube bundle pulling kits and devices, tube bundle handling devices, and/or band saw for the dismantling and refurbishment of heat exchanger tubesheets are available from heat exchanger service providers and/or tube-to-tubesheet rolling/welding equipment manufactures like Maus, Italia, Powermaster, India, Teco Tube Expanders, the Netherlands, etc. Two types of tube bundle pulling devices are shown in Figure 5.15a through c, transportation unit is shown in Figure 5.16, and heat exchanger tube bundle gripper is shown in Figure 5.17.

FIGURE 5.15 (a) Aerial tube budle extractor—45 ton capacity and (b) Self-propelled bundle extractor. (c) Truck mounted bundle extractor. (Courtesy of Peinemann Equipment, Hoogvliet, the Netherlands.)

5.12.5 Preparing Tube Bundles for Tube ID Inspection

Tube cleanliness can have a significant impact on the performance and probability of detection of some NDE methods. The IRIS technique generally provides the most accurate data however it requires rigorous cleaning. Electromagnetic technique accuracy decreases exponentially with tube fouling. Most electromagnetic inspection methods require hydroblast, grit blast, or chemical cleaning where no more than 15% of the tube ID is obscured by deposits or other debris for most electromagnetic techniques.

5.12.6 Cleaning Tube Bundles

The heat transfer surfaces of heat exchangers should be kept reasonably clean to assure satisfactory performance. Convenient means for cleaning should be made available. Heat exchangers may be cleaned either online or offline by chemical or mechanical methods. Various cleaning methods

FIGURE 5.16 On-site self-propelled tube bundle transporter. (Courtesy of Maus Italia F. Agostino &C.s.a.s., BagnoloCremasco (Cr), Italy.)

FIGURE 5.17 Bundle gripper. (Courtesy of Peinemann Equipment, Hoogvliet, the Netherlands.)

FIGURE 5.18 Automatic high-pressure external cleaning robot and multiple nozzle carrier robot for the automatic internal cleaning of tube bundles. (Courtesy of Maus Italia F. Agostino &C.s.a.s., BagnoloCremasco (Cr), Italy.) (c) Outside tube bundle cleaners. (Courtesy of Peinemann Equipment, Hoogvliet, the Netherlands.)

had been dealt with in Chapter 2. Additionally, automatic high-pressure external cleaning robot and multiple nozzle carrier robot for the automatic internal cleaning of tube bundles are shown in Figure 5.18, and pneumatic shafting tube cleaners for tube bundle heat exchangers of Maus, Italia, make are shown in Figure 5.19.

5.12.6.1 Tube Bundle Cleaning Methods

Hydroblasting is perhaps the most widely used method. The cleaning is accomplished by directing water into each tube separately with a cleaning lance that usually has a proprietary tip on the end of the lance for multiple-directional flow to increase the efficiency and effectiveness of the cleaning process. Hydroblasting can be classified into at least four categories, depending upon pressure [44]:

1. Low-pressure hydroblasting (less than 10k psi).
2. Medium-pressure hydroblasting (up to 20k psi).
3. High-pressure hydroblasting (up to 30k psi).
4. Ultrahigh-pressure hydroblasting (up to 40k psi).

5.12.7 PRIMARY TUBE FAILURE MECHANISMS OF STHE AND INSPECTION METHODS

Nondestructive testing is used extensively for the condition monitoring and remaining life estimates of heat exchanger tubing. API RP-571 outlines the principles, applications, capabilities, and limitations of commercially available NDE methods for shell and tube and air-cooled heat exchangers. Primary tube failure mechanisms of STHE and inspection methods are given in Table 5.3.

5.12.8 LOCATING TUBE LEAKS

Shellside leak testing by hydrostatic method is the preferred method to locate perforated or split tubes and leaking tube-to-tubesheet joint. In most cases, the entire front face of each tubesheet will be accessible for inspection. The point where water escapes indicates a defective tube or tube-to-tubesheet joint.

5.12.9 GASKET REPLACEMENT

Gaskets and gasket surfaces should be thoroughly cleaned and should be free of scratches and other defects. Gaskets should be properly positioned before attempting to tighten bolts. It is

---

TABLE 5.3
Primary Tube Failure Mechanisms of STHE and Inspection Methods

Primary tubing failures/damage mechanisms	Tubing inspection techniques
1. Uniform internal or external wall loss	a. visual inspection
2. ID or OD pitting	b. remote visual inspection
3. ID grooving	c. multifrequency eddy current testing
4. galvanic corrosion	d. segmented eddy current array
5. erosion	e. rotational eddy current
6. erosion-corrosion	f. full saturation eddy current testing
7. impingement attack external wall loss	g. partial saturation eddy current
8. tube end erosion	h. remote field testing
9. cracking (thermal, fatigue, stress corrosion cracking, etc.)	i. remote field testing array
10. OD baffle wear (baffle fretting wear)	j. nearfield testing
11. dents, gouges	k. near field array
12. flow-induced vibration damage: tube to tube collision damage, tube-to-shell collision damage, fretting wear	l. internal rotating probe ultrasonic inspection system
13. metallurgical changes, dealloying, sigma phase deposition in duplex stainless steel,	m. magnetic flux leakage testing/magnetic flux leakage array
14. tube-to-tubesheet joint weld defects like cracking, lack of penetration, nonfusion, porosity, etc.	n. acoustic pulse reflectometry
15. through wall holes	o. measurement of tube end by caliper
16. tube blockage	
17. tube warpage	

Source: API RP-586, NDT Methods for Equipment Damage Mechanisms [44].

recommended that when a heat exchanger is dismantled for any cause, it be reassembled with new gaskets.

5.12.10 STHE Repair

Various types of heat exchanger maintenance can be done on the tube bundles including, but not limited to the following [51]:

i.   Shell and shellside tube bundle cleaning;
ii. Water box or channel cleaning;
iii. Tubesheet repair;
iv. Tube sleeving;
v.   Retubing of heat exchanger bundles;
iii. Tubesheet, water box, or channel coatings;
iv. Cleaning inside the tubes with various methods: drilling out hard deposits, shooting projectiles, rotary cleaning with brushes;
v.   Pressure testing the tubes for leakage pneumatically or hydro-statically;
vi, Pressure testing the heat exchanger shellside pneumatically or hydro-statically;
vii. Tubesheet coating;
viii. Tube plugging;
ix. Inspection and reporting on overall conditions of the heat exchanger.

5.12.11 Preparing Tube Bundles for Tube Internal Inspection

Tube cleanliness can have a significant impact on the performance and probability of detection of some NDE methods. The IRIS technique generally provides the most accurate data however it requires rigorous cleaning. Electromagnetic technique accuracy decreases exponentially with tube fouling.

Tube cleanliness can have a significant impact on the performance and probability of detection of defects in some NDE methods. The IRIS technique generally provides the most accurate data however it requires rigorous cleaning. Electromagnetic technique accuracy decreases exponentially with tube fouling. There are numerous methods of cleaning heat exchanger tubulars in preparation for inspection, but they can be roughly grouped into four categories: hydro-blasting, abrasive blasting, chemical cleaning, and other methods (e.g., tube scrapers/pigs, tube brushes) as briefed below:

5.12.12 Tube Bundle Cleaning Methods

High-pressure hydro-jetting system ranging from 100 bar to 2800 bar pressure consisting high-pressure reciprocating plunger pump is used for cleaning. There are many advantages of using high-pressure hydro-jetting equipment over chemical cleaning, brush cleaning, and several other conventional methods.

External cleaning. To clean bundles externally, use jet guns and outside bundle cleaners. This cleaning equipment cleans the outside of bundles effectively. A nozzle head moves both horizontally and vertically, and forwards and backwards in order to clean the bundle with water under high pressure. The equipment is remote-controlled from the safe environment of a cabin.

Chemical cleaning. Bundles can be cleaned in a container or *in situ* with a combination of chemicals to clean the contamination of a specific bundle.

Ultrasonic technology. The equipment will be submerged in a container in a liquid composed for this purpose. In the container, transducers create sound waves, after which cavitation makes the contamination come loose from the equipment.

5.12.13 ASME PCC-2-2018, Repair of Pressure Equipment and Piping

This Standard provides methods for repair of equipment, piping, pipelines, and associated ancillary equipment within the scope of ASME Pressure Technology Codes and Standards after they have been placed in service. A few of the repair methods are discussed below [52]:

   a.   Tube-to-tubesheet joints

(i) replacement of a damaged tube; *(ii)* plugging a tube; *(iii)* removing the tube and performing weld buildup of the tube hole and then remachining the tube hole to the original tube hole diameter; *(iv)* seal welding or strength welding of the joint, etc.

   b.   Tube pullout testing

Tube pullout testing is performed to determine the amount of torque required to achieve the maximum tube-to-tubesheet joint strength in condenser retubing project. Pullout testing is recommended if retubing project is utilizing a different tube material than previously installed, or if there are questions regarding the joint strength.

   c.   Tubeside repair by plugging

Repair of tubes may be accomplished by plugging the tube with a welded or mechanical attachment.

   (i)   All tubes that are plugged should be pierced to provide for venting and draining as shown in Figure 5.20. Large temperature differential between tubeside and shellside may require the tube to be cut into two.
   (ii) Tapered plugs that are installed where tubes are not pierced can present a serious safety hazard.

   d.   Tubeside repair by sleeving tubes

The methods of sleeve installation include the following: forcing a ball through the sleeve, welding the ends of the sleeve, roller expansion, explosive bonding or hydraulic expansion of the sleeve. A mock-up may be advisable to ensure weld or expansion quality. A map should be developed to record the number and location of tubes that have been sleeved.

   e.   Tubeside repair by ferrule installation

Ferrules may be held in place by a flanged end with a tight fit to the tube inside diameter, by expanding the ends of the ferrule into the tube at the tubesheet, or by welding. Uniform contact with the tube may be achieved by roller or hydraulic expansion of the ferrule. A mock-up may be advisable to ensure weld or expansion quality. Figure 5.21 shows a few types of ferrule configurations.

   f.   Tubeside repair by pulling tubes.
   g.   Tubeside repair by tube replacement.
   h.   Tubesheet repair—weld repair of cracks and face repair by overlay and machining.
   i.   Tube-to-tubesheet joint repair.

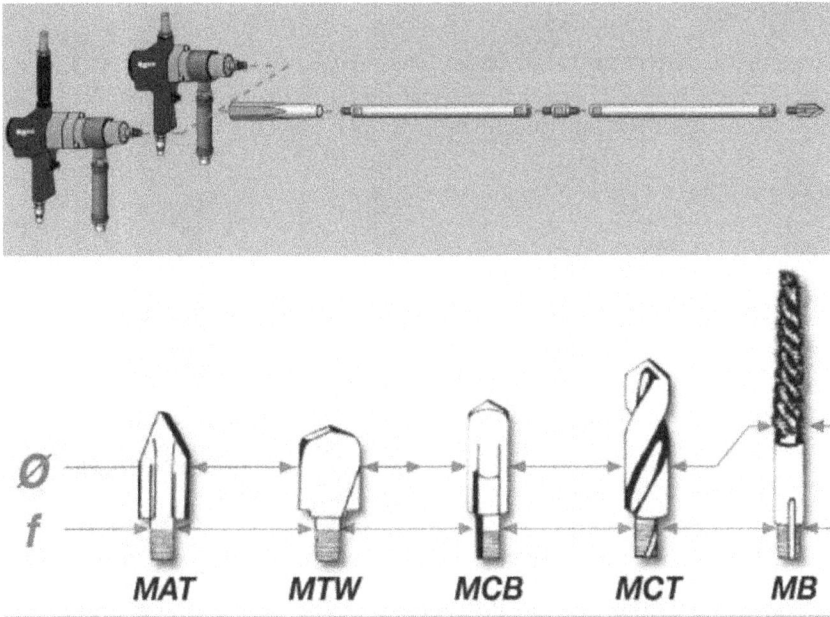

FIGURE 5.19 Pneumatic shafting tube cleaners for tube bundle heat exchangers. (Courtesy of Maus Italia F. Agostino &C.s.a.s., BagnoloCremasco (Cr), Italy.)

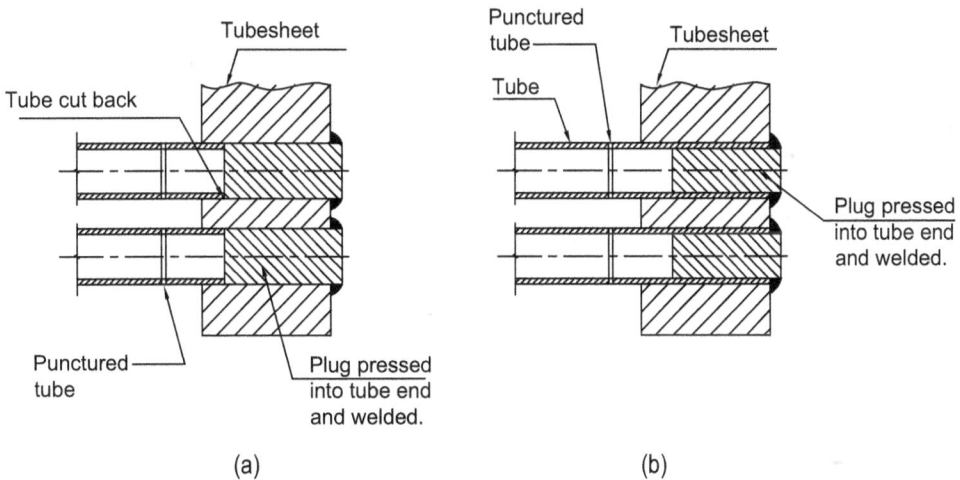

FIGURE 5.20 Plugged and punctured tube.

Re-expanding of tube-to-tubesheet joints with leaks of a small flow rate, otherwise known as weeping tubes, where no apparent damage has occurred to the seating surfaces, may be re-expanded to obtain an acceptable mechanical fit of tube to tubesheet, Weld buildup repair of tube holes and machining, seal welding or strength welding of tubes, etc.

j. Repair of pressure containment components.
k. Examination of pressure containment components.

Examination of the shell, channel, and other pressure containment components, excluding the tubes, may be performed by any of the methods VT, RT, UT, and MT.

5.12.14 Tubes Repair

Failure of high-pressure (HP) feedwater heater tubes is one of the major causes of forced outages in fossil-fired power plants. The problem areas in the feedwater heaters include [8,9] (1) normal and abnormal operating conditions (maximum); (2) tube plugging due to leaks; (3) drain cooler zone-level control; (4) steam impingement desuperheat zone, condensing zone; (5) tube vibration; (6) inlet end erosion; (7) loss of impingement plates, etc.

5.12.15 Heat Exchanger Tube Failures

Tube-to-tubesheet joint failure and flow-induced vibration are leading causes of leaks. These issues are best dealt with through careful thermal design, performing flow-induced vibration analysis using state-of-the-art software, mechanical design, selection of the proper materials of construction, appropriate construction of the equipment, including careful inspection during fabrication and to carry out necessary heat treatment process of components at appropriate stages of construction, etc.

5.12.16 NDT Methods for Heat Exchanger and Boiler Tube Inspection

Common tubing NDE methods for straight shell and tube heat exchanger tubes are as follows [53]:

- Multifrequency eddy current testing (MFECT ECT)
- Segmented eddy current array (ECA)
- Remote field eddy current (RFT)
- Partial saturation eddy current (PSET)
- Full saturation eddy current (FSET)
- Magnetic flux leakage testing (MFL)
- Magnetic flux leakage array testing (MFA)
- Near field testing (NFT)
- Near field testing array (NFA)
- Internal rotary inspection system (IRIS)
- Acoustic pulse reflectometry (APR)
- Tube end calliper

5.12.17 Tube Inspection by ECT, RFET, NFT, ECA, MFL, IRIS

Eddy Current Testing Methods [54]
ECT is a simple, high-speed, high sensitive, versatile, and reliable NDT technique and is popularly used in many engineering industries.

Remote Field Eddy Current Testing
RFT is primarily used to inspect ferromagnetic tubing since conventional ECT have difficulty in inspecting the full thickness of the tube wall due to the strong skin effect in ferromagnetic materials.

Near Field Testing
Near field testing (NFT) technology is a rapid and cost-effective solution intended specifically for fin-fan carbon-steel tubing inspection.

Eddy Current Array
Eddy current array (ECA) along with bobbin probe technology is the solution to perform high speed inspection in a single pass.

Magnetic Flux Leakage (MFL)
Magnetic flux leakage (MFL) uses powerful magnets to magnetize the conductive material under test where a magnetic field is created around defects like corrosion or material loss.

IRIS—Internal Rotary Inspection System
Internal rotary inspection system (IRIS) is an ultrasonic method for testing of pipes and tubes. The ultrasonic beam allows detection of metal loss from the inside and outside of the tube wall. A three-dimensional picture of the defect is obtained, thus the defect profile and its depth is provided. Interpretation of results is easier than in the other techniques.

With high inspection speed and low cleaning requirements, these cost-effective inspection technologies are able to inspect ferromagnetic, nonferromagnetic, and fin-fan tubes.

5.12.18 ACOUSTIC PULSE REFLECTOMETRY (APR)

Acoustic pulse reflectometry (APR) has emerged recently as a very fast and effective method for inspecting heat exchanger tubes, demonstrating high sensitivity to variations in cross section. This makes it extremely suitable for assessing the internal cleanliness of such tubes. The basic concept behind APR is to inject a wideband acoustic pulse into the medium inside a tube—air, in this case. This pulse acts as a form of "virtual probe." As long as the pulse encounters no discontinuities, it continues to propagate down the tube. Whenever a discontinuity is encountered, such as a blockage, expansion (due to wall loss, for example) or hole—a reflection is created. The reflected waves propagate back down the tube where they are recorded and stored on disk. The inspection process involves the following [55]:

i. Probe injects acoustic pulse down the tube.
ii. Reflected echoes are recorded and analyzed.
iii. A set of patented algorithms identifies and reports exact location, type, and size of inner diameter side defects.

The inspection is fast (10 seconds per tube, up to 2000 tubes inspection in one shift of 12 hours), noninvasive and provides inbuilt computerized signal analysis tools to quickly identify tube faults, such as pitting wall loss, erosions, hole leakage, blockage, bulging in heat exchangers, condensers, boilers, chillers, reactor tubes, etc.

5.12.18.1 Use of APR to Detect Fouling
When applying traditional tube inspection methods such as eddy current or ultrasound, the tubes must necessarily be traversed by a probe. In such cases, the disturbance caused by fouling can range from a minor annoyance in light cases, to the point where it precludes any possibility of inspection, in the heavier fouling cases. When applying APR, however, randomly distributed fouling appears simply as a multitude of blockages of different sizes.

5.12.19 HEAT EXCHANGER TUBE PLUGGING

Individual tubes that have been found to be leaking can be plugged by pop-a-plug device or replaced with solid rods in place of tubes. Sometimes gradual patterns of tube leaks develop, which can suggest plugging the tubes for entire rows or regions within the heat exchanger. As a practical guideline, plugged tube count can be about 10% of the tube count. Plug and seal materials are available for shell and tube exchangers such as *feedwater heaters, condensers, air coolers, steam boilers, process heat exchangers, chillers, etc.* Sizes suitable for inner diameter range from 10 to 51 mm (0.40–2.0 in.).

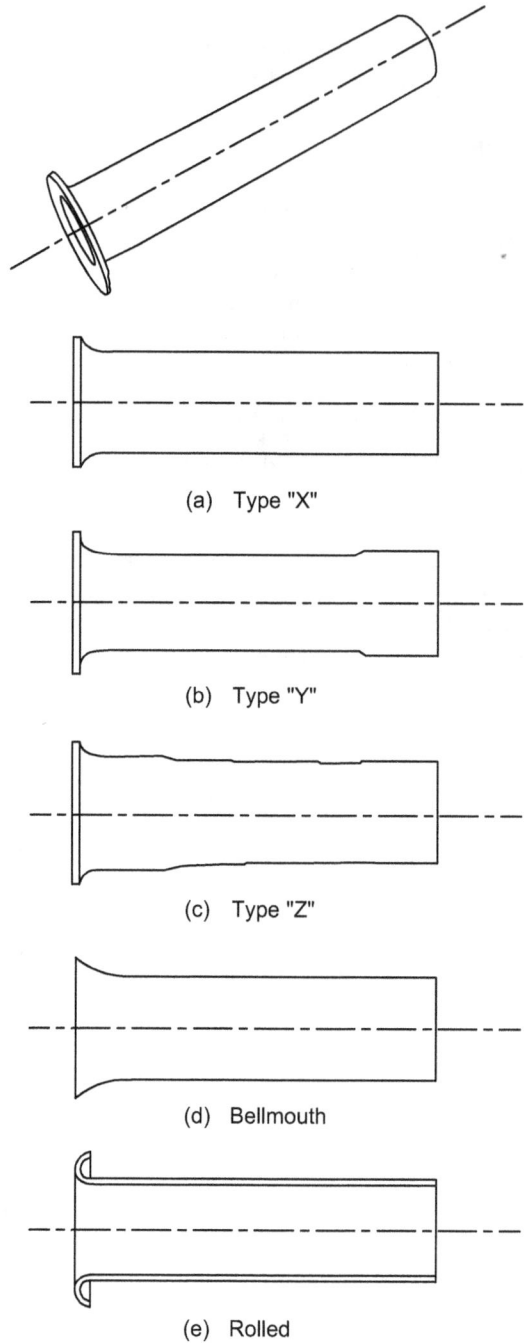

(a)   Type "X"

(b)   Type "Y"

(c)   Type "Z"

(d)   Bellmouth

(e)   Rolled

FIGURE 5.21   Few types of ferrules configurations.

Figures 5.22 through 5.26 show various tube plug models readily available in the market. These plugs are installed by either pulling the taper pin of the plug or by the application of torque.

Figure 5.20 shows tube welded with a plug. Another option is the replacement of part of or all of an existing heat exchanger's tubes. In some instances, it will be best to design and procure a new heat exchanger for the current and projected plant requirement.

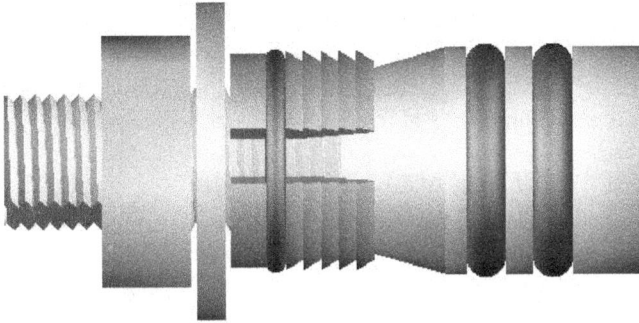

a.

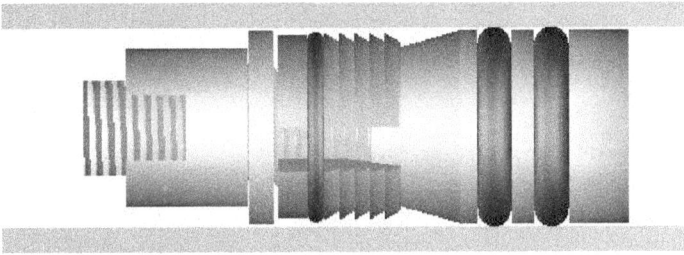

b.

FIGURE 5.22   Mechanical seal plug. (a) Schematic and (b) plug installed in a tube. (Courtesy of Powerfect, Brick, NJ.)

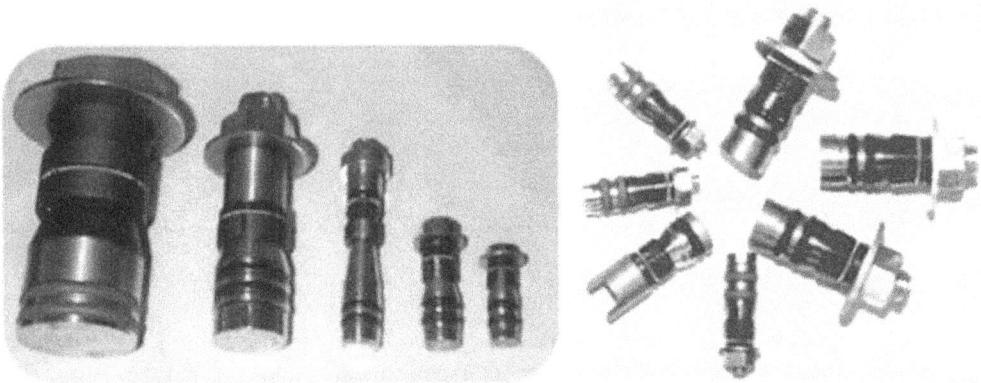

FIGURE 5.23   (a–b) Mechanical seal plugs. (Courtesy of Powerfect, Brick, NJ.)

5.12.19.1 TEMA Guidelines for Tube Plugging
RGP-RCB-2 PLUGGING TUBES IN TUBE BUNDLES [56]

In U-tube heat exchangers, and other exchangers of special design, it may not be possible or feasible to remove and replace defective tubes. Under certain conditions as indicated therein, the manufacturer may plug either a maximum of 1% of the tubes or two tubes without prior agreement.

5.12.19.2 Explosive Welded Plugs in a Feedwater Heater

Mechanically fitted or fusion welded plugs have been used in the past for effecting repairs to leaking heat exchanger tubes and/or joints. Welding a plug into a feedwater heater on-site is an extremely difficult job. Also in recent years, the use of high-pressure/temperature exchangers has revealed shortcomings in those traditional methods of repair. The tube bore or tubesheet hole requires little preparation but it is important to remove surface oxides and scale from the weld area, no machining of the hole is required. The plug containing an explosive charge which is inserted into the prepared tube/tube plate hole, and the charge initiated to produce an explosive weld between the plug and the tube or tube plate [57].

5.12.19.3 Tube Inserts

Tube inserts are used in shell and tube heat exchangers to provide protection to the tubes. The inserts are placed at the ends of heat exchanger tubes to

- protect them from inlet-end erosion;
- protect against corrosion;
- protect against impingement attack;
- transfer extremely high heat past the tube sheet to the exchanger tubes.

5.12.19.4 Sleeves

The power generation industry has long used sleeving of tubular heat exchangers to extend the life of steam surface condensers and closed feedwater heaters. Installing thin sleeves in tubes in which the tube walls have become so thin that failure is imminent is a maintenance technique useful for extending heat exchanger life. One should consider sleeving when [58]

i.   it would be very costly to replace aged exchangers because of the installation position or the shutdown time required to remove the existing one and install a replacement exchanger or tube bundle;

ii. replacement tubing delivery is so long that unscheduled shutdowns for plugging failed tubes become so frequent as to cause serious loss of production; and

iii. some decrease in the effective heat transfer surface is acceptable.

Ferrules and sleeves are typically 6" to 18" long sections of cut to length from annealed tube stock, flared and chamfered at opposite ends. The installed ferrule/sleeve serves as a wear plate to decrease the damage done to the parent tube due to inlet erosion. The ferrule or sleeve can be replaced after it has outlived its usefulness.

Hydraulic expansion of tube end ferrules and sleeves is the premier method for installation, expanding the complete length of the ferrule to intimate contact with the parent tube. This method is achieved by introducing water at a high internal pressure within a sleeve, between two seals, to expand the sleeve to a point of contact between the sleeve and the parent tube.

Full-length tube liners. Full-length tube liners are designed to restore tubes that are suffering from isolated or full-length corrosion (pitting, cracking). Leaking or damaged tubes requiring plugging are returned to service.

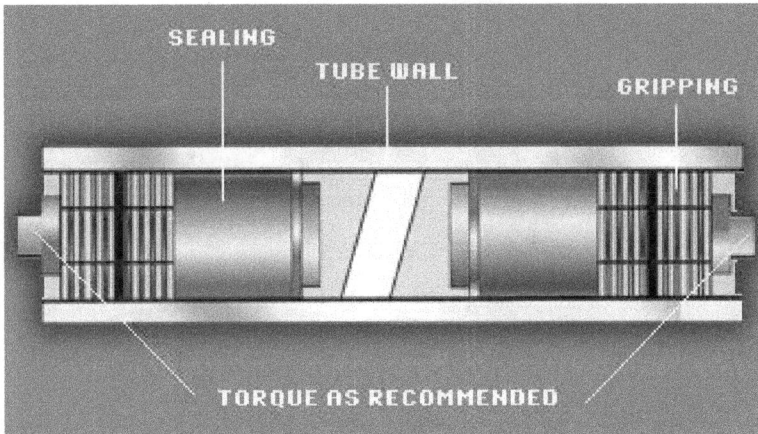

b.

c.

d.

FIGURE 5.24 Plugging of tube holes. (a) Tube plugs, (b) tube plugs installed inside a tube (schematic), (c) insertion of tube plug into a tube, and (d) mechanism of installation of tube plug. (Courtesy of Conco Systems, Inc., Verona, PA.)

5.12.19.5 Pop-A-Plug® Heat Exchanger Tube Plugging System

Pop-A-Plug® Heat Exchanger Tube Plugging System (refer to Figure 3.24) offers significant advantages over hammer-in or welded plug applications for the following reasons:

i.   Average installation time: 2 minutes per plug.
ii. No expert welders required.
iii. Consistent and repeatable installation.
iv. Permanent sealing solution with removability option.

5.12.19.6 Tube Plug Sleeving

Most heat exchangers are designed with a certain amount of extra tube capacity to allow a percentage of tubes to be plugged during the life of the exchanger. Two primary causes of tube failure, which often result in tubes being plugged or replaced, are erosion and corrosion. In fact, heat exchangers are often completely retubed to repair tube failures that occur only at the tube ends. The HydroSwage® Strain Control Sleeving System 43101, offers a method of installing sleeves (ferrules) as shown in Figure 5.27 to restore lost tube material or as sacrificial barriers to the damaging elements that cause the tube loss [59].

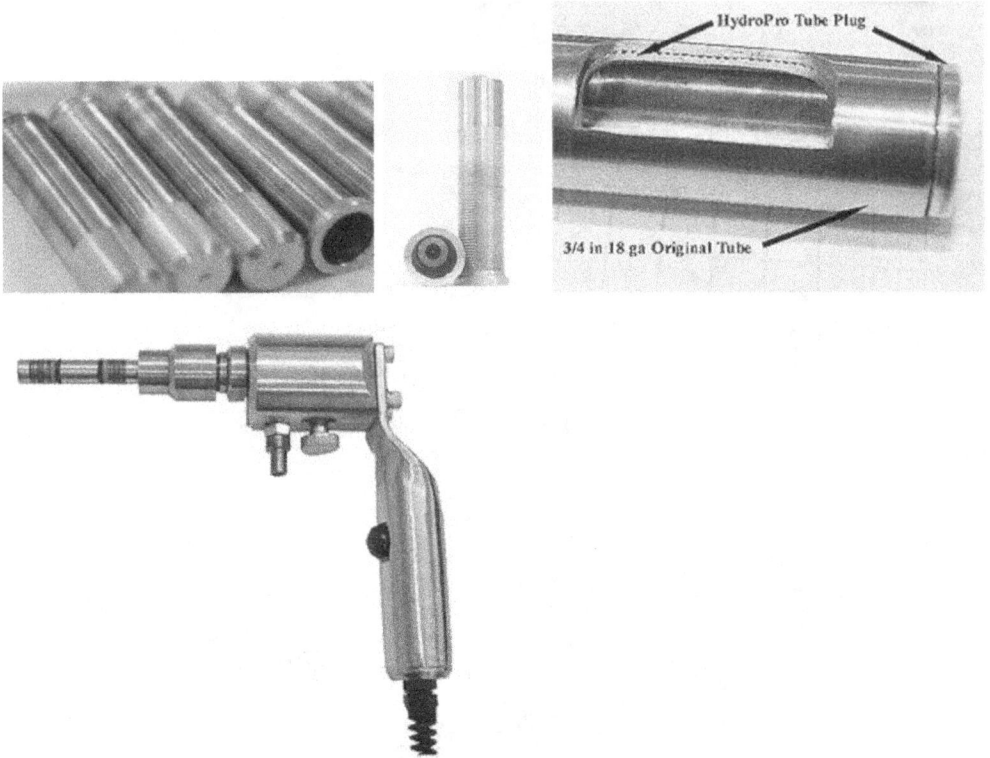

FIGURE 5.25 Tube plugs. (a) Unexpanded tube plugs, (b) cutaway showing a HydroPro tube plug in place in a ¾ in. 18 gauge original tube, and (c) HydroPro Tube Plug Installation Mandrel Assembly and Mandrel Holder (Gun). (Courtesy of HydroPro, Inc., San Jose, CA.)

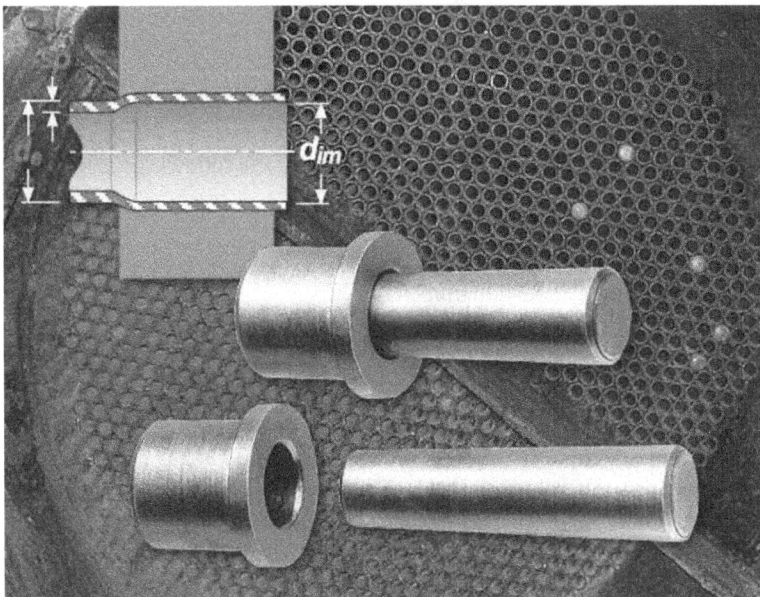

FIGURE 5.26 Tube plugs showing ring and a tapered pin. (Courtesy of Maus Italia F. Agostino &C.s.a.s., BagnoloCremasco (Cr), Italy.)

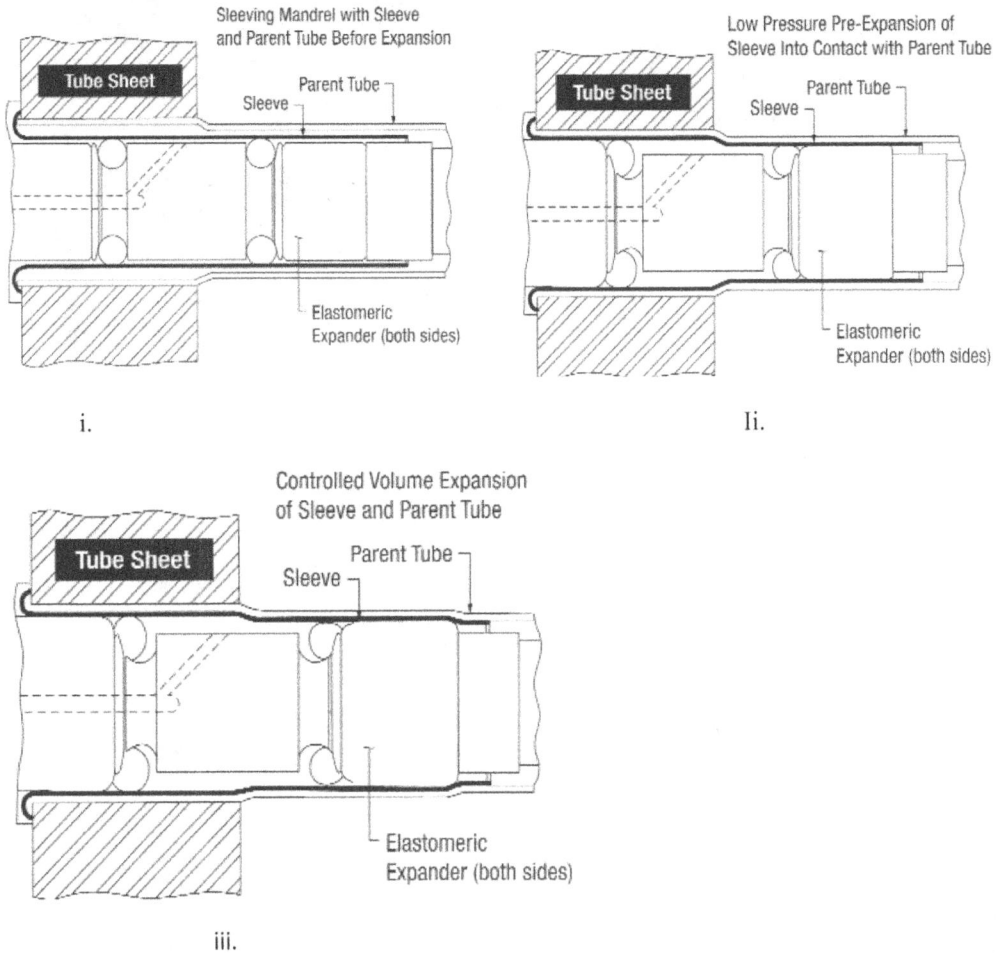

FIGURE 5.27 (i-iii) HydroSwage Strain Control SleevingSystem. (Courtesy of Haskel International, Inc., Burbank, CA. All rights reserved.)

5.12.19.7 Hydrostatic Testing Equipment

Hydrostatic testing equipment makes possible the testing of individual tubes inside any shell and tube heat exchanger to a pressure within the tube design ASME code limits and provides a positive means of determining the integrity of each tube. Weak tubes will fail during the test, rather than when the unit is placed back in service, thus avoiding expensive forced outages. A typical individual hydrostatic tube testing equipment kit and test plugs are shown in Figure 5.28.

5.12.20 Brazed Aluminum Plate-Fin Heat Exchanger

Problems with brazed aluminum plate-fin heat exchangers are rare. However, if problems develop, advice should be sought from the manufacturer.

5.12.20.1 Leak Detection

External leaks will be evident by the appearance of localized freeze spots or vapor cloud on the outer casing of the insulation. Or the smell or sound of the escaping fluid may also be discernible. When a leak is suspected, it should be investigated fully and immediately, and the manufacturer's repair procedure be put in action as soon as practicable.

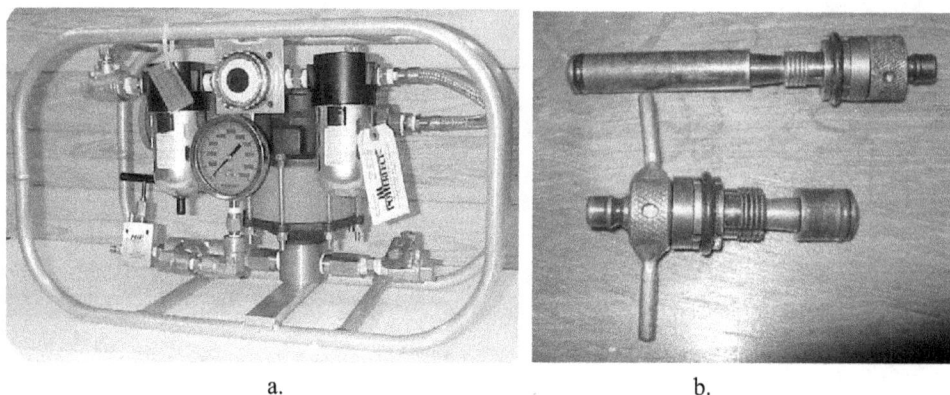

a.   b.

FIGURE 5.28 Individual tube hydrostatic testing equipment kit. (a) Equipment and (b) hydrostatic test plugs. (Courtesy of Powerfect, Brick, NJ.)

5.12.20.2 Repair of Leaks

Detected external leaks such as cracked pipe welds may be repaired by rewelding using an approved weld procedure and by a qualified aluminum alloy welder. Should the need arise to locate and repair a leak associated with the brazed aluminum plate-fin heat exchanger, then the manufacturer is to be consulted [60].

5.13 NDT METHODS TO INSPECT AND ASSESS THE CONDITION OF HEAT EXCHANGER AND PRESSURE VESSEL COMPONENTS

5.13.1 NONDESTRUCTIVE EXAMINATION

Nondestructive examination (NDE) provides the mechanism for quality control of base metals and weldments. There are few predominant methods of nondestructive examination and their variations that pertain to the pressure vessels in service: visual examination (VT), radiography (RT), ultrasonics (UT), eddy current test (ECT), magnetic particle inspection (MT), and dye penetrant inspection (PT). Radiography and ultrasonics are used for volumetric examination, while magnetic particle and dye penetrant inspection techniques provide for surface examination. Some of the common NDT methods mentioned above are shown in Figure 5.29(a–c). Few NDT methods are briefly discussed below based on Ref. [61].

5.13.2 REMOTE VISUAL INSPECTION (RVI)

RVI is a cost-effective inspection technique used to capture real-time views and images from inside voids, such as tubes, pipes, rotating machinery, engines, heat exchangers, tray towers, pressure vessels, refractory-lined vessels, and enclosed structures. RVI can be a complementary technique to other NDT disciplines and is frequently used as the primary or initial inspection screening method to find localized corrosion and erosion.

5.13.3 VIDEO BORESCOPES

The most portable to the most capable, all designed for ease-of-use while delivering video images of unsurpassed quality.

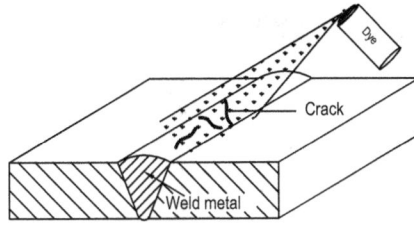

(a) PT Principle.

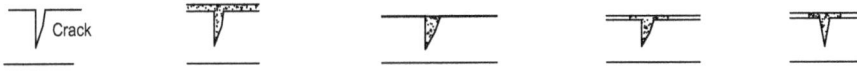

| (1) cleaned surface | (2) Penetrant covered | (3) Excess penetrant removed | (4) Developer applied | (5) Interpretation |

(b) PT - Interpretation.

(i) PT

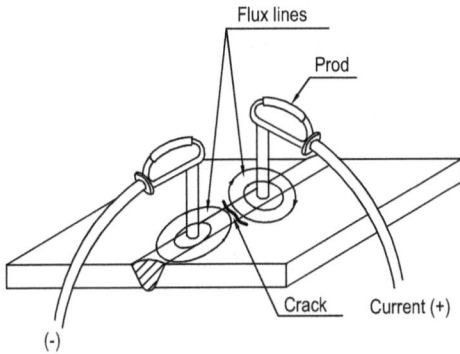 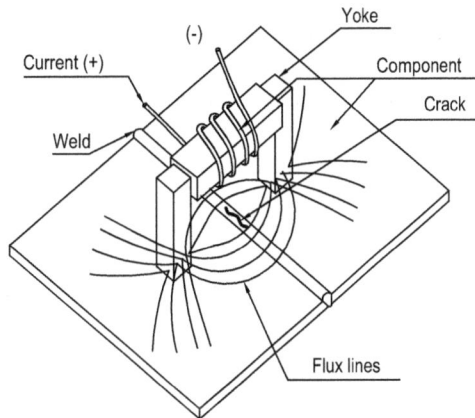

(a) (b)

(ii) MT

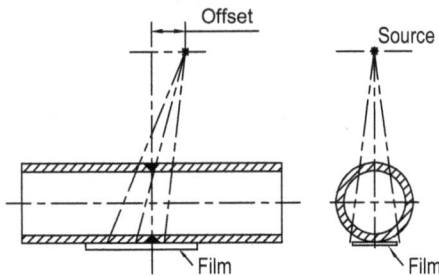

(iii) RT

FIGURE 5.29 (i-vii) NDT methods for heat exchanger examination.

Note: VT-visual testing, PT-dye penetrant testing, MT-magnetic particle testing, RT-radiography testing, UT-ultrasonic testing, AE or AET-acoustic emission testing, ET-eddy current testing , and LT-leak testing.

(iv) ECT

(v) UT

(vi) AE

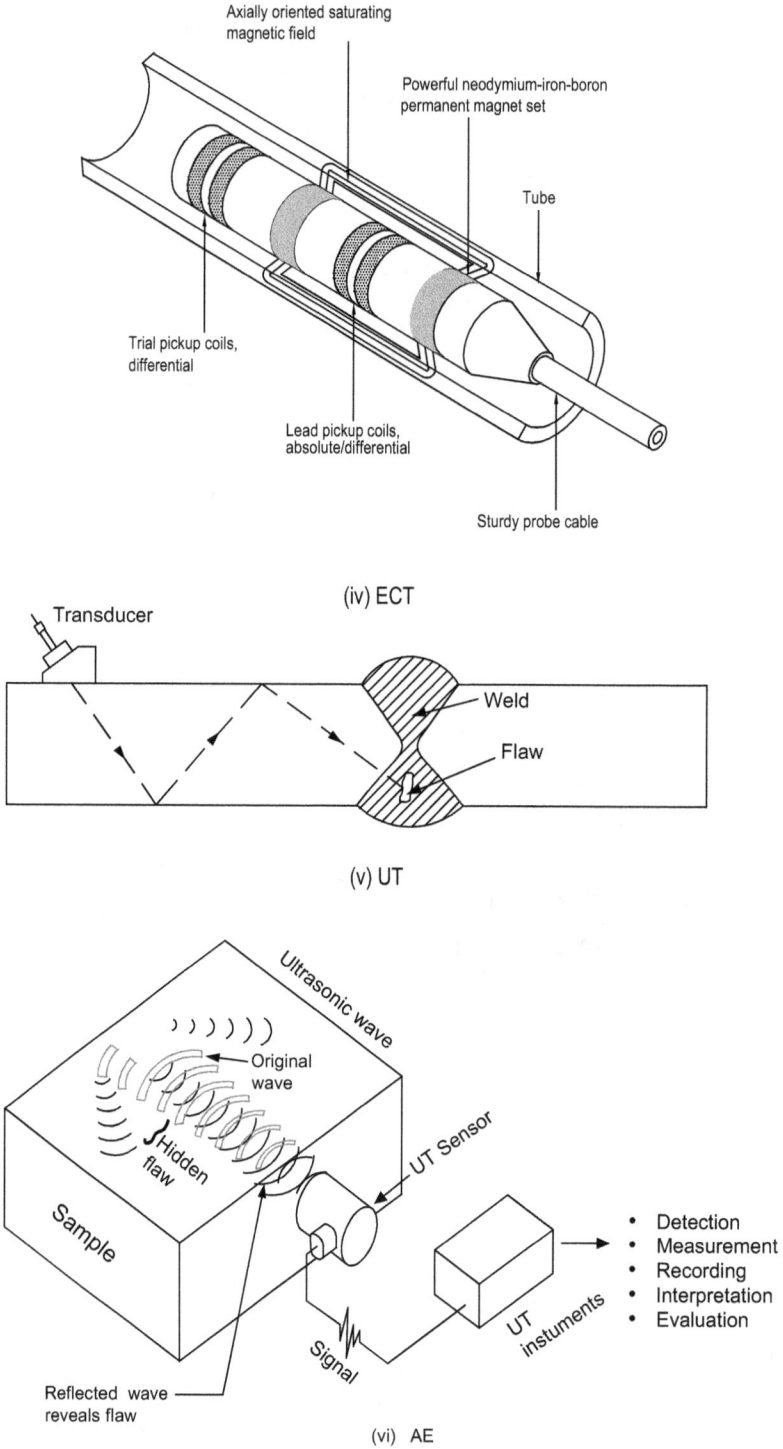

FIGURE 5.29 (iv-vii) (Continued)

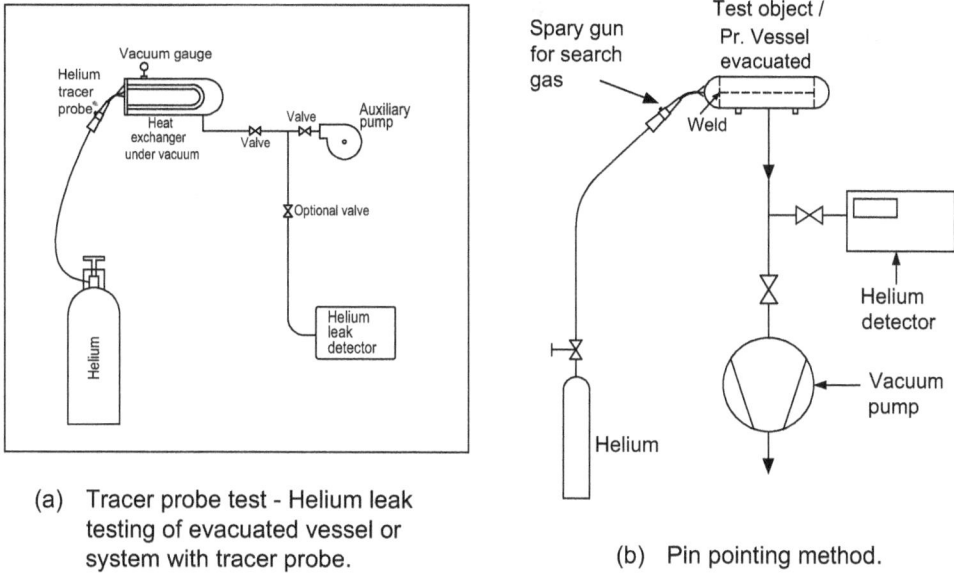

(a) Tracer probe test - Helium leak
testing of evacuated vessel or
system with tracer probe.

(b) Pin pointing method.

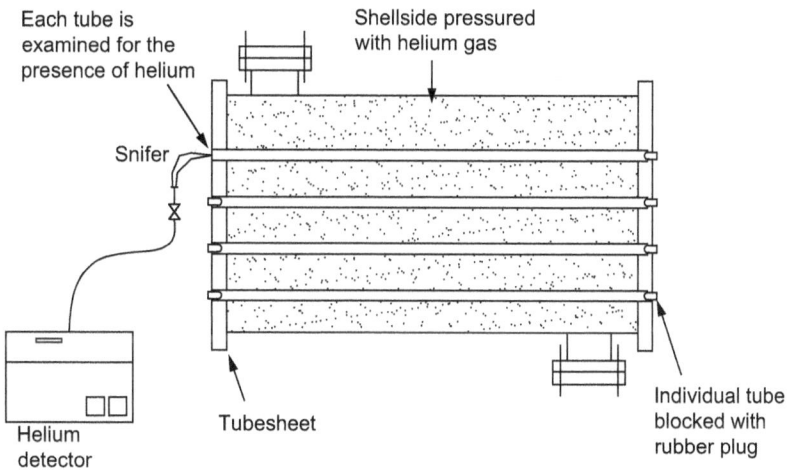

(c) Leak test of heat exchanger.

(vii) LT

FIGURE 5.29   (vii) (Continued)

5.13.4 ULTRASONIC TESTING (UT)

Ultrasound technique has been used in nondestructive testing is used to find a variety of defects or
nonconformities within almost any kind of solid material: (i) Thickness gages for corrosion detection; (ii) conventional flaw detectors; and (iii) phased array flaw detectors.

Ultrasonic phased array inspection continues to evolve rapidly and can be applied to many
corrosion and erosion applications for upstream, midstream, and downstream plant assets.
Corrosion Flaw Detector

Ultrasonic phased array instruments dedicated to locating and sizing pits can significantly improve inspection productivity.

5.13.5 Radiographic Testing (RT)

Various RT techniques are conventional film radiography, digital technologies, including computed radiography and direct radiography, portable or stationary X-ray sources, 3D computed tomography, and analytical X-ray. RT technique utilizes a variety of digital detector arrays to ensure optimal image quality and throughput for each application.

5.13.6 Tube Inspection

Major causes of heat exchanger tube failures are mentioned under Section 5.12. Possible tube defects in service are shown in Figure 5.30.

5.13.6.1 Ultrasonic Internal Rotary Inspection System

Ultrasonic internal rotary inspection system (IRIS) is based on the principle of measuring thickness using ultrasonic waves. The IRIS method is mostly used for inspection of carbon steel tubes. The method is very accurate for thickness measurement as well as for detecting internal and external pits. IRIS can, however, miss pinholes, and this method is not recommended for the detection of cracks [62]. Figure 5.31 shows IRIS inspection method.

5.13.6.2 Eddy Current Testing

Both the conventional eddy current testing (ECT) and remote field ECT are used for inspection. Conventional ECT is used for inspection of nonferromagnetic tubing such as stainless steel, copper–nickel alloys, titanium, etc. Remote field ECT is applicable for inspection of ferromagnetic tubing such as carbon steel and nickel. A specialized version of conventional ECT is full saturation ECT. This technique is applicable for thin ferromagnetic tubes such as Seacure in condenser tubes and partially ferromagnetic tubing materials such as Monel and Alloy 2205 [62].

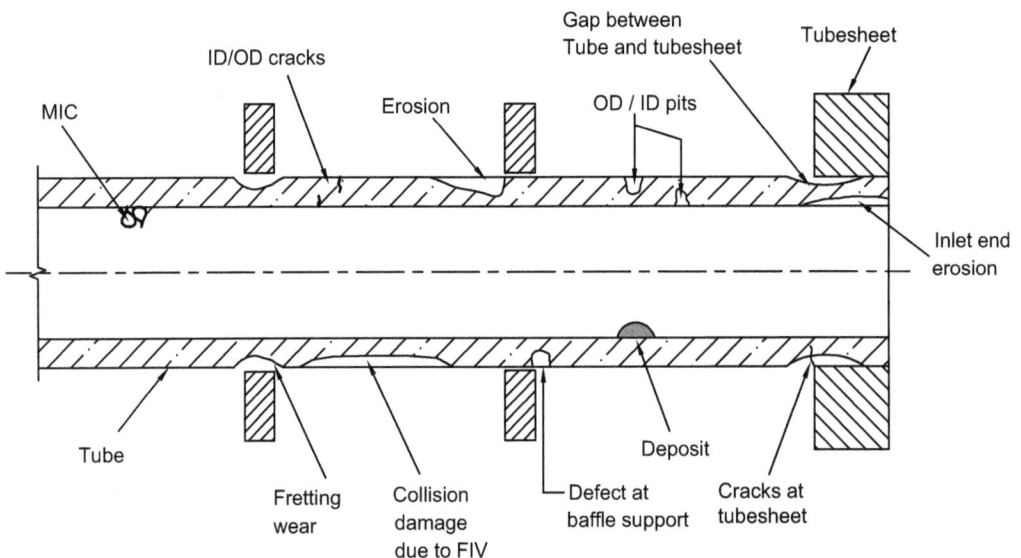

FIGURE 5.30   Typical heat exchanger tube defects (in service). (Courtesy of Anmol Birring, Houston, TX.)

Vessels in petrochemical plants: the two major applications of ECT in vessel inspection are crack detection and clad thickness measurement. The technique is used as an alternative to penetrant testing (PT) as it can detect tight cracks, and the technique is significantly faster than PT. Clad measurement is performed on vessels with carbon steel shell and stainless steel clad. NDE Associates, Inc., uses the Hocking Phasec 2200 and the Nortec 1000 for such inspection.

Piping systems: stainless steel is susceptible to stress corrosion cracking. ECT is a highly effective technique for the detection of SCC cracking on the OD surface of SS piping. The technique detects tight cracks that can be missed by dye PT. ECT will also detect cracks that are just below the surface but within the eddy current skin depth.

Bolt hole inspection: bolt holes are to be inspected for cracks. Cracks in bolt holes can be caused by fatigue and corrosion. Inspection is done with a bolt hole scanner interfaced to an ECT machine.

Bellows: ECT is applied on bellows for crack detection. Cracks in bellows are caused by stress corrosion. Both stainless steel and Inconel bellows are inspected for the detection of ID and OD surface cracks.

5.13.6.3   In-Service Examination of Heat Exchangers for Detection of Leaks

In-service examination of leaks, using either pressurized air or vacuum technique or bubbler, is shown schematically in Figure 5.32. Additionally, individual tube can be tested by the following two methods:

1. Individual tube leak testing. by Helium mass (a) Sniffing method and (b) vacuum method is shown in Figure 5.33.
2. Individual through the tube testing method as shown in Figure 5.34

Fractured tubes at the interior section can be plugged using tube stabilizers (Figure 5.35), marketed by M/s Expando Seal Tools.

5.13.6.4   Remote Field Eddy Current Testing

The remote field eddy current testing (RFECT) is generally used for the detection of wall loss in carbon steel tubes. Pits can be detected using differential coils; however, the reliability for pit detection is limited. Pinholes can be missed by RFECT. Pit detection sensitivity can be improved by

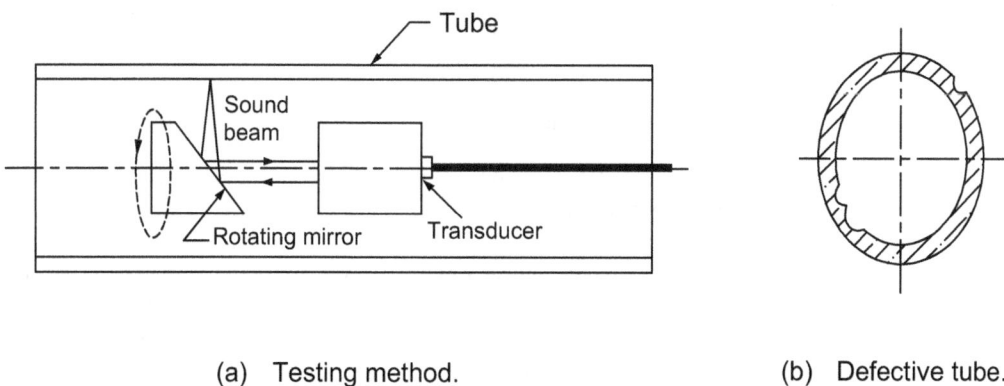

(a) Testing method.   (b) Defective tube.

FIGURE 5.31   IRIS inspection method.

(a)

(b)

(c) Soap bubble test.

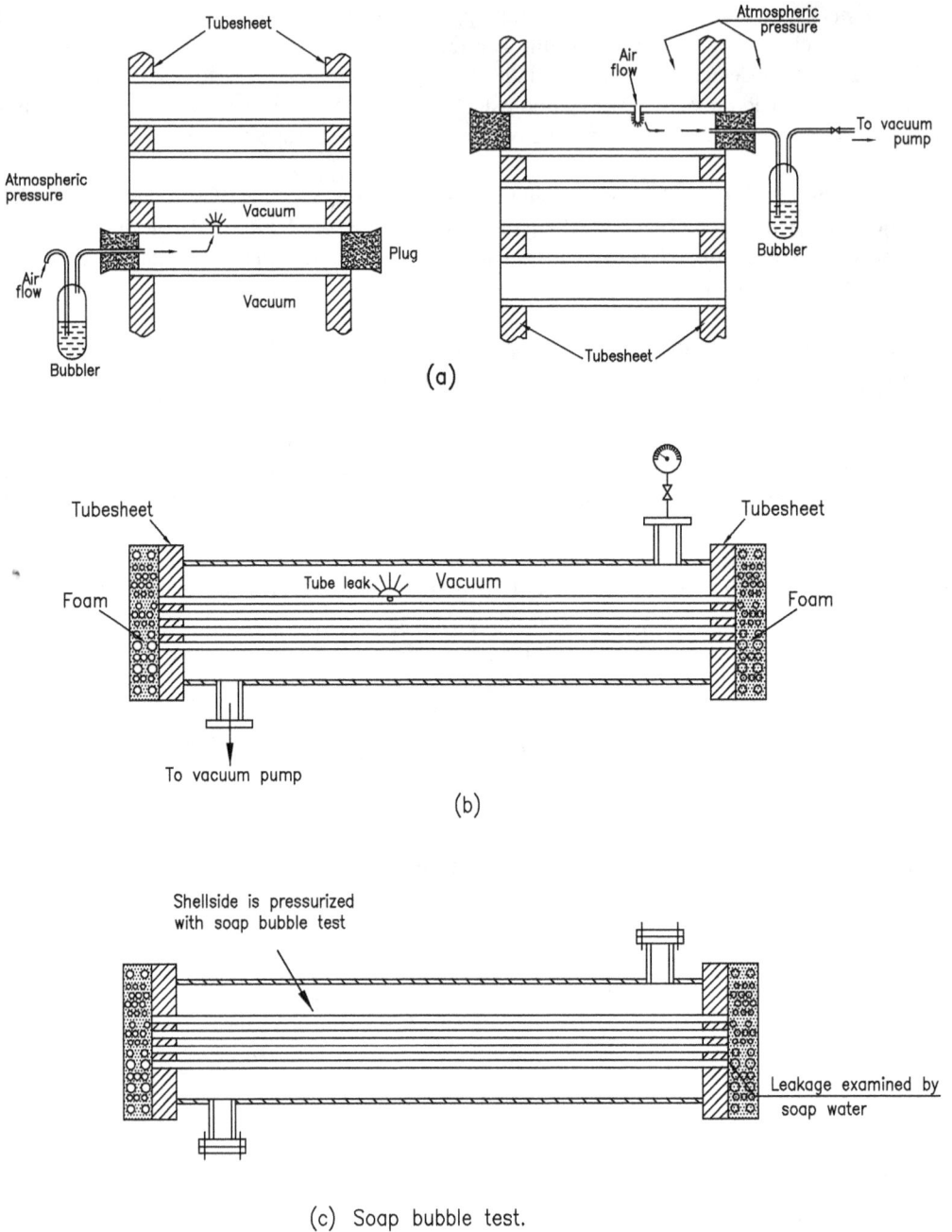

FIGURE 5.32 In-service examination of leaks: (a) by vacuum technique using bubbler, (b) by vacuum technique using soap foam, (c) pressure testing using soap foam.

reducing the pull speed. The inspection speed with RFECT is significantly lower than conventional ECT. While conventional ECT can easily be performed at a speed up to 6 ft/s, RFECT is limited to about 10 in./s. Higher pull speeds will miss small defects and increase background noise. Flaw sizing with RFECT is done using the voltage–plane curves. These curves are used to size tube wall loss. The curves relate flaw depth, flaw length, and the flaw circumference to the phase of the remote field signal in the absolute channel [62].

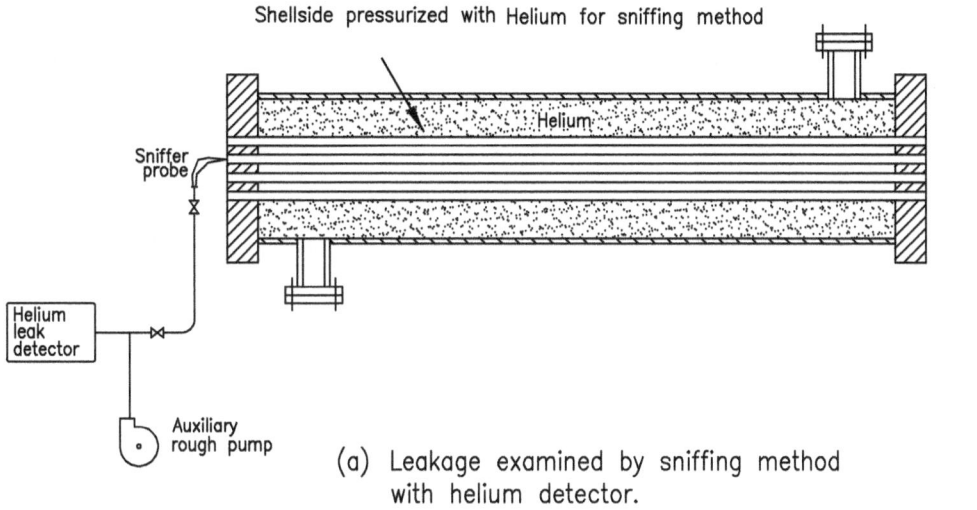

(a) Leakage examined by sniffing method with helium detector.

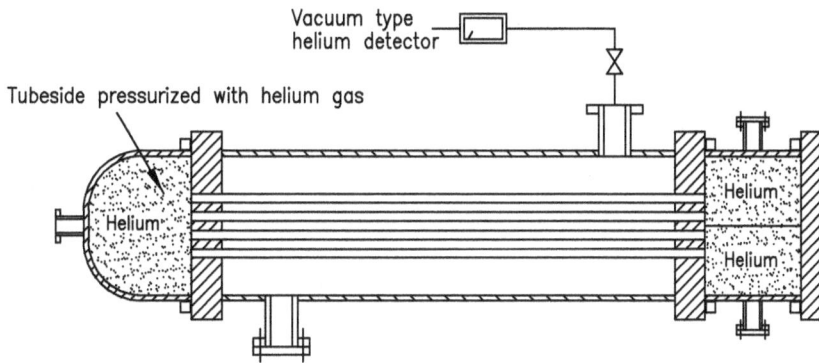

(b) Leakage examined by vacuum type helium detector. To examine tubeside is to be filled one side at a time.

FIGURE 5.33 Helium mass Leak testing (a) Sniffing method and (b) Vacuum method.

5.14 RESIDUAL LIFE ASSESSMENT OF HEAT EXCHANGERS BY NDT TECHNIQUES

The presence of flaws in critical components will affect its structural integrity and increase the likelihood of failure. In-service NDT can detect and quantify the damage and provide input for remedial actions. Also the residual life can be assessed. It has been shown that the sound attenuation in steel or any other construction material can be used to quantify the level of degradation of the material's mechanical properties. Knowing this, the remaining life of an affected plant can be estimated. There are a number of inspection methods available. Some of the methods based on ultrasonics are [63] as follows:

1. Ultrasonic echo attenuation method.
2. Amplitude-based backscatter method.
3. Creep waves (i.e., time-of-flight measurement).

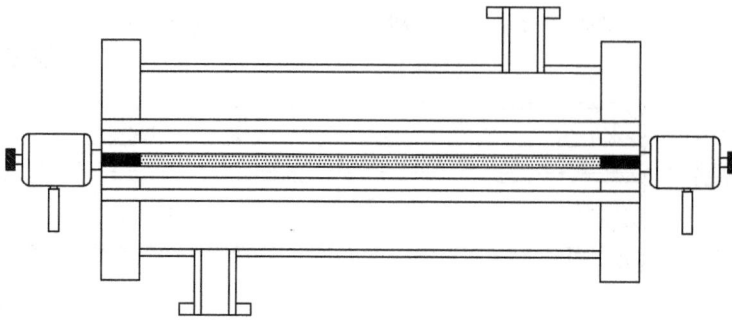

(a)

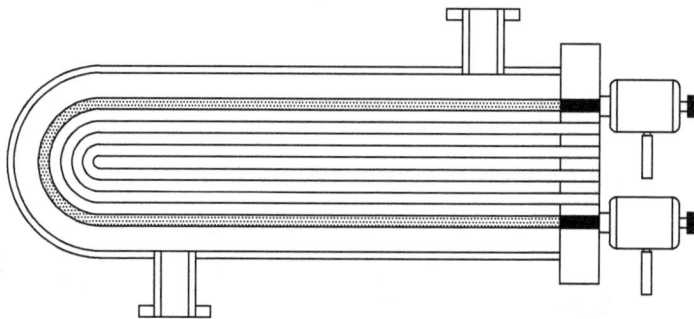

(b)

FIGURE 5.34 Individual through the tube testing. (Courtesy of Maus Italia F. Agostino &C.s.a.s., BagnoloCremasco (Cr), Italy.)

4. Pitch-catch mode shear wave velocity.
5. Ultrasonic method based on backscatter and velocity ratio measurement.
6. Advanced ultrasonic backscatter techniques (AUBT).
7. Method based on thickness mapping, backscatter, and velocity ratio (TOFD).
8. Photon-induced positron annihilation (PIPA) and distributed source positron annihilation (DSPA).
9. *In situ* metallographic replicas.

Other methods are based on radiography, replication technique, etc. Basic principles of some of the methods are mentioned later.

5.14.1 CREEP WAVES

This technique uses ultrasonic waves that travel along the surface of a component. As it propagates, it converts to a mode that travels into the component at an angle to the surface. The latter wave will convert back to a surface wave if it hits a surface parallel to the surface on which it originated.

5.14.2   HYDROGEN ATTACK DETECTION BY ULTRASONIC METHOD BASED ON BACKSCATTER AND VELOCITY RATIO MEASUREMENT

Both the base metal and weld HAZ should be inspected for hydrogen attack. Base metal attack is detected by using a combination of ultrasonic backscatter and velocity measurements [60]. Hydrogen attack increases the ultrasonic backscatter and reduces the ultrasonic velocity in the material. The backscatter and velocity ratio measurements are applied to detect hydrogen attack. Application of ultrasonic backscatter for the detection of high-temperature hydrogen attack is shown in Figure 5.36. The ultrasonic backscatter technique was developed by Birring [62].

5.14.3   PULSED EDDY CURRENTS

This is a technique for detecting corrosion and erosion and measuring average remaining wall thickness.

5.14.4   FLASH RADIOGRAPHY

Originally developed to image rapidly moving dynamic object, flash radiography has found application in the detection of corrosion on pipe outside diameter under insulation.

5.14.5   LOW-FREQUENCY ELECTROMAGNETIC TEST

Low-frequency electromagnetic test (LFET) is a technique by which changes in the wall thickness of carbon steel storage tank floor plates can be detected. The technique involves passing a low-frequency magnetic field into the plate and using scanner-mounted pickup coils to detect the induced AC magnetic field in the plate martial. In the presence of wall thickness loss and pitting, a measurable distortion is induced in the field that can be detected.

5.14.6   PHOTON-INDUCED POSITRON ANNIHILATION AND DISTRIBUTED SOURCE POSITRON ANNIHILATION

PIPA and DSPA technologies detect fatigue, embrittlement, and other forms of structural damage in materials at the atomic level. These techniques can determine the remaining life of various materials precisely.

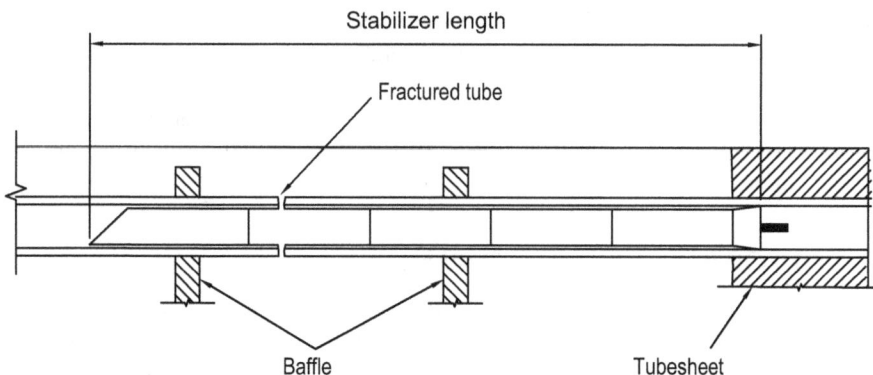

FIGURE 5.35   Plugging fractured tube using tube stabilizer.

Source: M/s Expando Seal Tools.

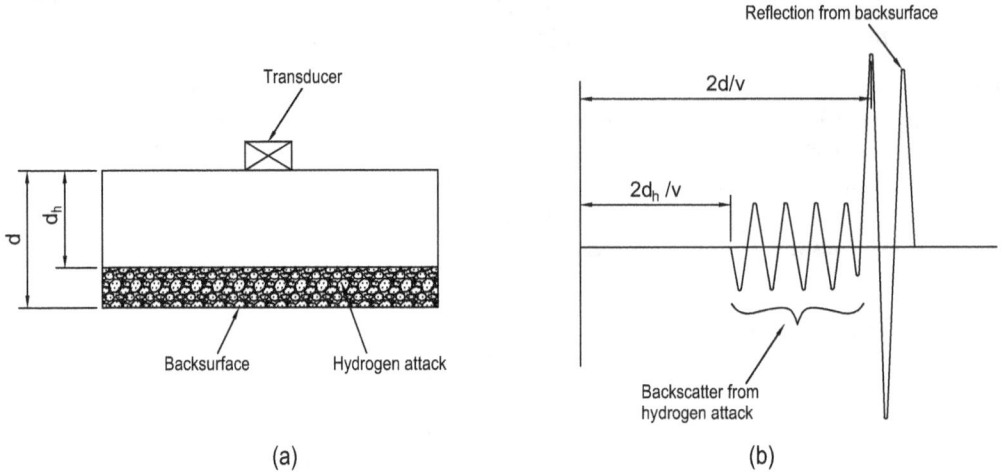

FIGURE 5.36   Application of ultrasonic backscatter for the detection of high-temperature hydrogen attack.

Source: Adopted and modified from Birring, A. S., Selection of NDT techniques for inspection of heat exchangers, Presented at the *ASNT International Conference on Petroleum Industry Inspection Conference,* **Houston, TX, June 1999.**

5.14.7   REPLICATION TECHNIQUES

When a component is subject to localized overheating, as a result of either a fire or an upset operating condition, severe damage can result. Or it may also involve metallurgical changes to the component that, in the worst case, will result in failure. Generally, the most effective method of evaluating the effects of these conditions is *in situ* metallurgical examination (known as replication technique) in conjunction with hardness testing. In the replication technique, a small area on the surface of the subject vessel is ground sufficiently, and a replica of the ground spot is taken and the microstructure is studied with a reference standard [50]. Replication principle is shown in Figure 5.2.

5.14.8   CREEP DETERMINATIONS BY NONDESTRUCTIVE TESTING METHOD

Conditions of creep occur in many pressure vessels, heat exchangers, boilers, etc. Because of a number of catastrophic failures, detection of creep by nondestructive examination techniques has become increasingly important. In recent years, creep has been measured by metallurgical techniques referred as replication technique. These methods generally are costly in that they require removal of insulation and cooling of the equipment to room temperature to see the location where creep is suspected. A more effective nondestructive examination technique that recently has been developed involves direct creep measurements with capacitive strain gauges attached to the surface of the pressure vessel or heat exchanger [50].

5.14.9   NDT FOR PLANT LIFE ASSESSMENT

NDT for plant life assessment deals with application of NDT techniques to detect discontinuities in an industrial manufacturing process that can affect the mechanical strength of a product and may cause its premature failure. Plant life assessment in many cases means the remaining life assessment of a structure, component or product. Assessment of structural integrity requires three inputs [64]:

i. material properties (e.g., yield strength, fracture toughness, etc.),
ii. flaw characteristics (type, location, size, shape, orientation), and
iii. stresses.

5.15 ON-SITE METALLURGY SERVICES

On-site metallurgy services meet needs for in-service inspection and asset integrity management. Site-based metallurgy techniques include the following [65]:

i. *In situ* metallography—a range of methods used to identify the microstructure of objects that are too big for laboratory testing or are unable to have samples removed.
ii. Replication—a method for taking a polymer mold of a surface *in situ*. The mold is taken to the laboratory for further analysis using optical microscopy or alicona.
iii. Portable hardness testing—assessing how hard the metal is using equotip or ultrasonic hardness testing techniques.
iv. Ferrite scope—specialist testing for the percentage of ferrite in a structure or equipment to ensure the balance of the microstructure of ferrite and austenite is correct.

5.16 PROFESSIONAL SERVICE PROVIDERS FOR HEAT EXCHANGERS

A number of professional service providers are available in the open market. Such service providers include among others, Westingouse Electric Company, PA, Powerfect, Brick, NJ, American Power Services Inc, Erlanger, KY, and NDE Associates, Inc., Webster, TX. They can provide professional assistance in the areas of performance assessment, residual life assessment, condition monitoring of condenser, heat exchanger and feedwater heater, and their repair. Typical services offered by them include tube plugging and sleeving, leak testing, tube removal, on-site troubleshooting, visual inspection and nondestructive testing such as eddy current tube testing/remote field tube testing, ultrasonic testing, magnetic particle testing, metallurgical analysis, hydrostatic testing, etc.

REFERENCES

1. www.jfdcoil.com/pdf/JFD%20Installation%20Manual%20Online.pdf
2. Sanitary shell and tube heat exchanger specification, *Installation Instructions, Operation Instructions and Maintenance Procedures*, Allegheny Bradford Corporation, Bradford, PA.
3a. ASME Boiler and Pressure Vessel Code, *Section VIII, Division 1—Pressure Vessels*, American Society of Mechanical Engineers, New York, 2021.
3b. ASME Boiler and Pressure Vessel Code, *Section VIII, Division 2, Pressure Vessels—Alternative Rules*, American Society of Mechanical Engineers, New York, 2021.
4. API RP 571, *Damage Mechanisms Affecting Fixed Equipment in the Refining Industry, 2020 i*
5. https://inspectioneering.com/tag/damage+mechanisms
6. www.twi-global.com/what-we-do/research-and-technology/technologies/materials-and-corrosion-management/corrosion-testing/co2-corrosion
7. www.lenntech.com/boiler/caustic-corrosion.htm#ixzz7pzQTAL8Y
8. www.corrosionclinic.com/types_of_corrosion/What-is-dew-point-corrosion.htm
9. https://inspectioneering.com/tag/flue+gas+dew+point+corrosion
10. Bonar, J. A. *Fuel Ash Corrosion,* Esso Research and Engineering *Go.*, Florham Park, NJ.
11. Noble, C. *Hydrogen Damage Test, How to Spot Hydrogen Damage.*
12.1 Rezumat, *Stress Relief Cracking of Ferritic Steels: Part One*, Date Published: Aug 2021
12.2 www.totalmateria.com/page.aspx?ID=CheckArticle&site=kts&LN=RO&NM=571
13. www.enggcyclopedia.com/2012/05/steam-hammering/

14. www.coursehero.com/file/p2rulfru/56-Condition-Monitoring-Locations-CMLs-561-General-CMLs-are-designated-areas-on/

15. https://inspectioneering.com/tag/condition+monitoring+locations

16. www.gate.energy/the-brainery/choosing-amp-using-condition-monitoring-locations-cmls

17. https://pinnaclereliability.com/learn/topics/condition-monitoring-location-cml-optimization/

18. www.aiche.org/conferences/southwest-process-technology-conference/2012/proceeding/paper/optimizing-location-condition-monitoring-locations-cmls

19. National Board Inspection Code, ANSI/NB-23, 2015.

20. www.sherwoodengineering.com.au/design-criteria-failure-modes-pressure-vessels/

21. MOSS, D. R. *Pressure Vessel Design Manual*, Illustrated procedures for solving major pressure vessel design problems, Third Edition, Gulf Professional Publishing is an imprint of Elsevier, Burlington, MA (2004).

22. Brown, F. Fall 2010 National Board *BULLETIN*.

23. Harvey, J. F. *Theory and Design of Modern Pressure Vessels*. New York: Van Norstrand Reinhold Company, 1974.

24a. https://material-properties.org/what-is-fatigue-life-s-n-curve-woehler-curve-definition/

24b. https://endurica.com/wohler-curves-or-fracture-mechanics/

25. https://community.sw.siemens.com/s/article/what-is-a-sn-curve

26. https://en.wikipedia.org/wiki/Fatigue_limit#

27. www.copper.org/applications/industrial/DesignGuide/performance/factfatigue03.html

28. William, L. *Basics of Fracture Mechanics as Applied to Structural Integrity of RPVs, 2067-2, Joint ICTP/IAEA Workshop on Irradiation-induced Embrittlement of Pressure Vessel Steels*, 23–27 November 200, ATI Consulting, Pinhurst, NC.

29. http://ksm.fsv.cvut.cz/~sejnom/download/pm10_tisk.pdf

30. *Non-Destructive Testing for Plant Life Assessment* International Atomic Energy Agency, Vienna, Austria, 2005, pp. 1–61.

31. Roylance, D. *Introduction to Fracture Mechanics* Department of Materials Science and Engineering Massachusetts Institute of Technology Cambridge, MA 02139 June 14, 2001, pp. 1–17.

32. Anderson, T. L. *FRACTURE MECHANICS, Fundamentals and Applications*, Third Edition, 2005, CRC Press, Taylor & Francis Group, Boca Raton, FL.

33. www.totalmateria.com/page.aspx?ID=CheckArticle&site=kts&NM=291. Date Published: Oct 2010

34. www.nde-ed.org/Physics/Materials/Mechanical/FractureToughness.xhtml

35. Stress-strain-curves-Ductile-vs-Brittle-Material.png (733×666) (nuclear-power.com)

36. Rice, R. C., Jackson, J. L., Bakuckas, J., Thompson, S. *Metallic Materials Properties Development and Standardization*. Tech. Rep. Federal Aviation Administration, Office of Aviation Research, January 2003.

37. www.arcweb.com/blog/what-asset-integrity-management-aim

38. https://inspectioneering.com/journal/2020-10-29/9411/11-primary-elements-to-ensure-asset-integrity-in-the-lifecycle-of-oil-and-gas-fa

39. www.twi-global.com/technical-knowledge/faqs/asset-integrity-management

40. www.dnv.com/services/asset-integrity-management

41. www.dnv.com/software/FAQ/What-is-APM.html

42. McStraw, B. *Inspection of Unfired Pressure Vessels Facilities- Instructions, Standards and Techniques*, Volume 2-9, August 2001, Minor Revision 1/2016, United States Department of the Interior Bureau of Reclamation, Hydroelectric Research and Technical Services Group, Denver, Colorado, pp. 1–29.

43. Yukawa, S. *Guidelines For Pressure Vessel Safety Assessment*, NIST Special Publication 780, 1990, U.S. Department of Commerce, pp. 1–67.

44. API RP-586, NDT Methods for Equipment Damage Mechanism

45. Ellmer, M., *A Practical Guide for Identifying and Solving Air-Cooled Heat Exchanger Performance Problems in the Field*, Reprinted from *Hydrocarb. Eng.*, pp. 1–6, April 2008.

46. Giammaruti, R., *Performance Improvement to Existing Air-Cooled Heat Exchangers*, Paper No. TP04-13, Dry cooling, Presented at the *Cooling Technology Institute Annual Conference*, Houston, TX, February 2–11, 2004, pp. 1–4.

47. www.fluiddynamics.com.au/articles-case-studies/six-causes-of-heat-exchanger-tube-failure

48. Schwartz, M. P. *Chief Product Engineer, ITT Bell & Gossett a unit of Fluid Handling Div.*, International Telephone & Telegraph Corp, Skokie IL

49. Gulley, D., Troubleshooting shell and tube heat exchangers, *Hydrocarb. Process., 75*(9), 91–98 (1996).

50. JobShop.com, Nondestructive testing techniques on plant reliability, Technical Information, 1996. www.jobshop.com

51. www.projectiletube.com/heat-exchanger-tube-cleaning/

52. ASME PCC-2-2018, Repair of pressure equipment and piping, *ASME,* 2018, pp. 1–270

53. BaherElsheikh, Heat Exchangers, Nov. 2020

54. www.ndts.co.in/tube-inspection-by-ect-reft-nft-eca-mfl-iris/

55. Using acoustic pulse reflectometry for quality control of heat exchanger cleaning, N. Amir1 and D. Bobrow1, *Proceedings of International Conference on Heat Exchanger Fouling and Cleaning—* 2011 (Peer-reviewed) June 05—10, 2011, Crete Island, Greece Editors: M.R. Malayeri, A.P. Watkinson and H. Müller-Steinhagen Published online www.heatexchanger-fouling.com

56. TEMA, *Standards of Tubular Exchanger Manufacturers Association*, 10th edn., Tubular Exchanger Manufacturers Association, Tarrytown, NY, 2019.

57. Hattingh, S., Peter, J., Paine, N. and Gustafsson, U. E., *Steam Generator Reference Book*, Revision 1, Volume 1, EPRI PERSPECTIVE, PROJECT RP2858; RP4004, Based on work sponsored by The Steam Generator Owners Groups I and II, The Steam Generator Reliability Project and Electric Power Research Institute, December 1994.

58. www.chemengonline.com/full-length-sleeving-for-process-heat-exchanger-tubes/

59. https://favfittings.com/tube-inserts-pipe-ferrules/

60. ALPEMA Standard, *The Brazed Aluminium Plate-Fin Heat Exchanger Manufacturer's Association*, 3rd edn., Didcot, Oxon, U.K., 2010, www.alpema.org

61. Corrosion & Erosion Inspection Solutions For Detection, *Sizing & Monitoring*, 2010, General Electric Company, pp. 1–15.

62a. Birring, A. S., *Selection of NDT Techniques for Inspection of Heat Exchangers*, Presented at the *ASNT International Conference on Petroleum Industry Inspection Conference*, Houston, TX, June 1999.

62b. Birring, A. S. and Kawano, K., Ultrasonic detection of hydrogen attack in steels, *Corrosion*, March, 1989.

63. Kot, R., Hydrogen attack, detection, assessment and evaluation, *10th Asia-Pacific Conference on Non-Destructive Testing*, September 2001, Brisbane, Queensland, Australia.

64. Non-destructive testing for plant life assessment, International Atomic Energy Agency, Vienna, Austria, 2005, pp. 1–61.

65. www.intertek.com/non-destructive-testing/materials-testing/on-site-metallurgy/

BIBILOGRAPHY

1. Electric Power Research Institute, *Feedwater Heater Survey*, Palo Alto, CA, August 1991, GS 7417.

2. Electric Power Research Institute, *High-Reliability Feedwater Heater Study*, Palo Alto, CA, June 1988, CS-5856.

3. www.flyability.com/visual-inspection

4. Metallurgical Consultants, Pressure Vessel Failure, Altamonte Springs, FL, 2009, pp. 1–2, www.materialsengineer.com/DA-Pressure-Vessel.htm

5. www.pumpsandsystems.com/why-steam-hammer-happens-how-eliminate-it

6. www.retubeco.com/condenser-solutions/tube-pullout-testing/

7. Reclamation Safety and Health Standards, United States Department of Interior, Bureau of Reclamation, 1995.

Index

For Product Safety Concerns and Information please contact our EU
representative GPSR@taylorandfrancis.com
Taylor & Francis Verlag GmbH, Kaufingerstraße 24, 80331 München, Germany

www.ingramcontent.com/pod-product-compliance
Lightning Source LLC
Chambersburg PA
CBHW080129220326
41598CB00032B/5000